MOMENT OR TORQUE

1 lb-ft = 1.36 N-m

POWER

$$
\begin{aligned}
1\ \text{W} &= 1.0\ \text{J/s} = 860.42\ \text{cal/hr} \\
1\ \text{hp} &= 550.0\ \text{ft-lb/s} = 745.7\ \text{W} \\
1\ \text{kW} &= 1 \times 10^3\ \text{W} \\
1\ \text{Btu/hr} &= 778.16\ \text{ft-lb/hr} = 0.2931\ \text{W}
\end{aligned}
$$

ENERGY

$$
\begin{aligned}
1\ \text{J} &= 1.0\ \text{N-m} = 1 \times 10^7\ \text{ergs} \\
1\ \text{erg} &= 1\ \text{dyne-cm} \\
1\ \text{cal} &= 4.1855\ \text{J} \\
1\ \text{ft-lb} &= 1.356\ \text{J} \\
1\ \text{Btu} &= 1055.06\ \text{J}
\end{aligned}
$$

VISCOSITY

$1\ \text{lb}_m/\text{ft-s} = 1.488\ \text{N-s/m}^2$

SPECIFIC HEAT

$1\ \text{Btu/lb}_m\text{-}°\text{F} = 4.184\ \text{kJ/kg-}°\text{C}$

GAS CONSTANT

$1\ \text{ft-lb/lb}_m\text{-}°\text{R} = 5.382\ \text{J/kg-K}$

THERMAL CONDUCTIVITY

$1\ \text{Btu/hr-ft-}°\text{F} = 1.731\ \text{W/m-}°\text{C}$

HEAT TRANSFER COEFFICIENT

$1\ \text{Btu/hr-ft}^2\text{-}°\text{F} = 5.6786\ \text{W/m}^2\text{-}°\text{C}$

BULK MODULUS

$1 \times 10^6\ \text{psi} = 6.895 \times 10^9\ \text{Pa}$

PHYSICAL CONSTANTS

Standard Acceleration of Gravity
$g = 9.80665\ \text{m/s}^2 = 32.1742\ \text{ft/s}^2$

Speed of Light
$c = 2.998 \times 10^8\ \text{m/s}$

Planck's Constant
$h_p = 6.626 \times 10^{-34}\ \text{J-s}$

Stefan-Boltzmann Constant
$\sigma = 5.673 \times 10^{-8}\ \text{W/m}^2\text{-K}^4$
$= 0.1712 \times 10^{-8}\ \text{Btu/h-ft}^2\text{-}°\text{R}^4$

Universal Gas Constant
$R = 8.3143\ \text{J/gmole-K}$
$= 1.9859\ \text{Btu/lbmole-}°\text{R}$

Theory and Design for Mechanical Measurements

WILEY

John Wiley & Sons, Inc.

Theory and Design for Mechanical Measurements

Fourth Edition

Richard S. Figliola
Clemson University

Donald E. Beasley
Clemson University

WILEY

John Wiley & Sons, Inc.

ACQUISITIONS EDITOR	Joe Hayton
PRODUCTION EDITOR	Janine Rosado
SENIOR MEDIA EDITOR	Tom Kulesa
SENIOR MARKETING MANAGER	Frank Lyman
SENIOR DESIGNER	Kevin Murphy
COVER PHOTO	Art courtesy of Suzanne Figliola

This book was set in 10/12 Times Roman by Thomson Digital and printed and bound by Hamilton Printing Company. The cover was printed by Phoenix Color Corp.

This book is printed on acid free paper. ∞

To order books or for customer service please, call 1-800-CALL WILEY (225-5945).

Library of Congress Cataloging in Publication Data:

Figliola R. S.
    Theory and design for mechanical measurements / Richard S. Figliola, Donald E. Beasley.– 4th ed.
        p. cm.
    Includes bibliographical references and index.
    ISBN 978-0-471-44593-7 (cloth : alk. paper)

    1. Mensuration–Textbooks. I. Beasley, Donald E. II. Title.

T50.F54 2006
681′.2–dc22   2005047492

Printed in the United States of America

10 9 8 7 6 5

Preface

We are pleased to offer this fourth edition of our text. The text provides a well-founded background in the theory of engineering measurements. Integrated throughout are the necessary elements for the design of measurement systems and measurement test plans, with an emphasis on the role of statistics and uncertainty analyses in design. The measurements field is very broad, but through careful selection of the topical coverage we establish the physical principles and practical techniques for most engineering applications. Our aim is not to offer a manual for instrument construction and assembly, but rather to develop concepts and decision bases for the design and assessment of measurement systems and on interpreting the results obtained from such systems. The text is appropriate for undergraduate and graduate level study in engineering, but is also suitably advanced and oriented to serve as a reference source for professional practitioners. The pedagogical approach invites independent study.

The organization of the text develops from our view that certain aspects of measurements can be generalized, such as test plan design, signal reconstruction, or dynamic response. Topics such as response, statistics, and uncertainty analysis do require a basic development of principles but are then best illustrated by integrating these topics throughout the text material. Other aspects are better treated in the context of the measurement of a specific physical quantity, such as strain or temperature.

This edition builds from our experiences and feedback from our readers. We have tried to maintain the text to a manageable size (and cost), so the material coverage is not exhaustive. For those using this text as part of a developed course, we anticipate that instructors will augment the material with their own expertise in certain areas. But we have expanded our coverage of mechatronics concepts, substantially revised the uncertainty coverage for consistency with test standards, and have rewritten many sections, either to update, to improve pedagogy, or to condense. We have added an additional Appendix on Laplace transforms. We have expanded the Glossary.

With this edition, we have added access to some new software based on National Instruments Labview® for exploring some of the text concepts, while retaining our previous efforts using Matlab®. The software will be accessable to purchasers of the text over the Wiley website; please note that registration and a passcode will be required. The Labview programs are available as executables so they can be run directly. With a current edition of Labview or Matlab, programs can be modified or adapted for specific applications. We intend to have an interactive area on this website so that users can exchange ideas and supplement materials and Labview or Matlab programs. We intend to periodically update and add material to the website.

Readers familiar with prior editions of this text will notice some changes in terminology in regards to uncertainty analysis. Our focus has not changed. The changes bring the text into harmony with the terminology now used in the international metrology (ISO GUM) standard and the United States test (ASME PTC 19.1) standard. We have replaced our use of the terms precision and bias errors with the terms random and systematic errors. We also discuss Type A and B errors and reconcile these differences between the Standards. Both of us have been involved with developing the Uncertainty Test Standards in some capacity: Dr. Figliola serves on the ASME PTC 19.1 Committee that maintains and revises the ASME/ANSI Test Uncertainty Standard; Dr. Beasley was a principal reviewer of PTC 19.1-2005. We do point out that there is a difference between documenting a Standard, which is a handbook for a veteran practitioner, and teaching and learning a new general concept, which is the purpose of a textbook. Where we deviate from the methodology presentation of the Standards, we do so for pedagogical reasons.

Chapters 1 through 5 provide an introduction to measurement theory. Chapter 1, Basic Concepts of Measurement Methods, stresses the importance of designing a test plan, as well as presenting the concepts of measuring systems, calibration, and standards. We have added new material, such as using significant digits. Chapter 2, Static and Dynamic Characteristics of Signals, examines the basis by which information is transmitted throughout a measurement system. Both the static and dynamic components of signals are treated and characterized through their time-invariant values and their time-dependent values of amplitude and frequency. In Chapter 3, Measurement System Behavior, the response of measurement systems to signals is explored and the concepts of sensitivity, time response, and frequency response are presented.

An awareness of the measurement system design process and calibration procedures motivates the introduction of the statistical nature of physical variables and for uncertainty analysis in Chapter 4, Probability and Statistics, and Chapter 5, Uncertainty Analysis. The importance of these two chapters cannot be overstressed, as practicing engineers are finding that their use goes beyond engineering measurements. Chapter 4 provides a sufficient introduction to probability and statistics to comprehend and to report the behavior of measured variables. An associated course laboratory in the measurement of engineering variables would provide an excellent context for this. Chapter 5 provides a complete treatment of uncertainty analysis. It establishes a methodology for performing uncertainty analyses based on the evolution of available information so typical of a real test process. A working knowledge is developed through carefully selected example problems. Further examples are integrated throughout the remainder of the text. However, the concepts associated with uncertainty analysis are greatly enhanced through application, such as through associated laboratory exercises.

Chapter 6, Analog Electrical Devices and Measurements, presents a few basic analog electrical measurements and circuits. Many of these form important components of most practical measurement systems. This section has been updated with discussion of some older methods eliminated or greatly reduced. We have supplemented the discussion with more use of op-amp circuits. Chapter 7, Sampling and Digital Devices, covers sampling concepts, practical data acquisition and data transmission. In addition to developing the criteria for discrete sampling and the role of the Fourier transform in signal reconstruction, we present some of the basic components required for analog-to-digital or digital-to-digital communication. We present computer-based data acquisition systems and methods for serial and parallel communication in detail.

Chapters 8 through 12 describe the principles and practice for measuring important variables such as, temperature, pressure, flow, displacement, force, power, and strain. Our goals in writing these chapters were both to provide an understanding of the physical principles used in such instrumentation and to provide sufficient practical information to select components in assembling a measurement system. With each instrument discussed we have noted practical ranges of operation, required supporting equipment, and types of errors and magnitudes of uncertainty to be expected, as well as other practical information.

Chapter 12 has been expanded for coverage on basic ideas of mechatronics in modern measurements. We discuss the integration of sensors, actuators, controllers, and feedback schemes.

Many users report to us that they use different course structures, so many that it makes a preferred order of presentation difficult. To accommodate this, we have written the text in a manner that allows any instructor to customize the order of material presentation. While the material of Chapters 4 and 5 are integrated throughout, every other chapter stands on its own. The text is flexible and can be used in a variety of course structures at both the undergraduate and graduate levels. We use it in both of these forums, as well as professional development courses, and simply rearrange material and emphasis to suit.

We express our sincerest appreciation to the students, teachers, and engineers who have used our earlier editions. We are indebted to the many who have written us with their constructive comments and encouragement. We thank the reviewers of this edition and past editions. We also thank the Wiley editorial staff for their welcome assistance. We are so very grateful to our wives and daughters, Suzanne and Elizabeth, and Leigh and Sarah for their continued patience, understanding, and valued help.

Richard S. Figliola
Donald E. Beasley
Clemson, South Carolina

Contents

Chapter 1

Basic Concepts of Measurement Methods

1.1 INTRODUCTION

We make measurements everyday. For example, we routinely measure our body weight on a scale or read the temperature of an outdoor thermometer. We put little thought into the selection of instruments for these routine measurements. After all, the direct use of the data is clear to us, the type of instruments and techniques are familiar to us, and the outcome of these measurements is not important enough to merit much attention to features like improved accuracy or alternative methods, so long as the chosen methods meet our needs. But when the stakes become greater, the selection of measurement equipment and techniques, and the interpretation of the measured data can demand considerable attention. Just contemplate how you might verify that a new engine is built as designed and meets the power and emissions performance specifications required.

But first things first. The objective in any test is to answer a question. So we take measurements to establish the value or the tendency of some variable, the results of which will help answer our question. But the information acquired is based on the value or the tendency suggested by the measurement device. So just how does one establish the relationship between the real value of a variable and the value actually measured? How can a measurement or test plan be devised so that the measurement provides the unambiguous information we seek? How can a measurement system be used so that the engineer can easily interpret the measured data and be confident in their meaning? There are procedures that address these measurement questions.

At the onset, we want to stress that the subject of this text is real-life oriented. Specifying a measurement system and measurement procedures represents an open-ended design problem whose outcome will not have a unique solution. That means there may be several approaches to solving a measurement problem, and some will be better than others. This text emphasizes accepted procedures for analyzing a measurement problem to assist in the selection of equipment, methodology, and data analysis to meet the design objectives. Perhaps more than in any other technical field, the approach taken in measurement design and the outcome achieved will often depend on the attention and experience of the designer.

1.2   GENERAL MEASUREMENT SYSTEM

We begin with a model of the generic measurement system. A *measurement*[1] is an act of assigning a specific value to a physical variable. That physical variable is the *measured variable*. A measurement system is a tool used for quantifying the measured variable. As such, it is used to extend the abilities of the human senses that, while they can detect and recognize different degrees of roughness, length, sound, color, and smell, are limited and relative; they are not very adept at assigning specific values to sensed variables. A general template for a measurement system is illustrated in Figure 1.1. Basically such a system consists of part or all of four general stages: (1) Sensor-transducer stage; (2) Signal-conditioning stage; (3) Output stage; and (4) Feedback-control stage. These stages form the bridge between the input to the measurement system and the system output, a quantity that is used to infer the value of the physical variable measured. We discuss later how the relationship between the input information, as acquired by the sensor, and the system output is established by a calibration.

The *sensor* is a physical element that employs some natural phenomenon by which it senses the variable being measured. The *transducer* converts this sensed information into a detectable signal, which might be electrical, mechanical, optical, or otherwise. The goal is to convert the sensed information into a form that can be easily quantified. For example, the liquid contained within the bulb on the common bulb thermometer of Figure 1.2 exchanges energy with its surroundings until the two are in thermal equilibrium. At that point they are at the same temperature. This energy exchange is the input signal to this measurement system. The phenomenon of thermal expansion of the liquid results in its movement up and down the stem, which in this case is the output signal from which we determine temperature. The liquid in the bulb acts as the sensor. By forcing the expanding liquid into a narrow capillary, this measurement system transforms thermal information into a mechanical displacement. Hence, the bulb's internal capillary design acts as a transducer.

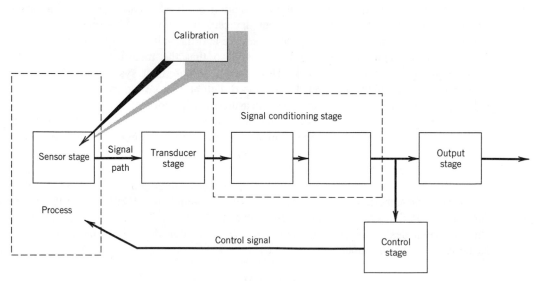

Figure 1.1 Components of a general measurement system.

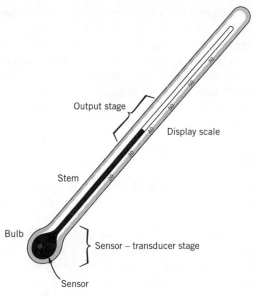

Figure 1.2 Components of bulb thermometer equivalent to sensor, transducer, and output stages.

It is worth noting that the term "transducer" is also often used in reference to a packaged device, which may contain a sensor, transducer, and even some signal conditioning elements. While such terminology is not true to our presentation, the context in which the term is used prevents ambiguity.

Sensor selection, placement, and installation are particularly important, because the input to the measurement system is the information sensed by the sensor. Accordingly, the interpretation of all information passed through and indicated by the system depends on that which is actually sensed by the sensor. For example, the interpretation of the output of a medical thermometer depends on where its sensor is placed.

Signal conditioning equipment takes the transducer signal and modifies it to a desired magnitude. This optional intermediate stage might be used to perform tasks such as increasing the magnitude of the signal through amplification, removing portions of the signal through some filtering technique, and/or providing mechanical or optical linkage between the transducer and the output stage. For example, the translational displacement of a mechanic's caliper (sensor) is often converted into a rotational displacement of a pointer. This stage can consist of one or more devices, which are often connected in series. For example, the diameter of the thermometer capillary relative to the bulb volume determines how far up the stem the liquid moves with increasing temperature. It "conditions" the signal by amplifying the liquid displacement.

The *output stage* indicates or records the value measured. This might be a simple readout display, a marked scale, or even a recording device, such as a computer disk drive. The readout scale of the bulb thermometer in Figure 1.2 serves as the output stage of that measurement system.

In those measurement systems involved in process control, a fourth stage, the *feedback-control stage*, contains a controller that interprets the measured signal and makes a decision regarding the control of the process. This decision results in a signal that changes the process parameter that affects the magnitude of the sensed variable. In simple controllers, this decision is based on the magnitude of the signal of the sensed variable, usually whether it exceeds some high or low set point, a value set by the system operator. For example, a simple measurement system with control stage is a household furnace thermostat. The operator fixes the set point for temperature on the thermostat display, and the furnace is activated as the local temperature at the thermostat, as determined by the sensor within the device, rises or falls about the set point. In a more sophisticated controller, a signal from a measurement system can be used as an input to an "expert system" controller that, through an artificial intelligence algorithm, determines the optimum set conditions for the process. *Mechatronics* deals with the interfacing of mechanical and electrical components with microprocessors, controllers, and measurements. We will discuss some features of mechatronic systems in this text with extended treatment given in Chapter 12.

1.3 EXPERIMENTAL TEST PLAN

An experimental test serves to answer a question. So the way it is designed and executed should be to answer that question and that question alone. This, of course, is not so easy to do. Let's consider an example.

Suppose you wanted to answer the question, "What is the fuel use of my new car?" through a test. What might be your test plan? In a test plan, you want to identify the variables that you will measure but you also need to look closely at other variables that will influence the result. Two important variables to measure would be distance and fuel volume consumption. Obviously, the accuracy of the odometer will affect the distance measurement and the way you fill your tank will affect your estimate of the fuel volume. But what other variables might influence your results? If your intended question is to estimate the average fuel usage to expect over the course of ownership, then the driving route you choose would play a big role in the results and is a variable. Obviously, only highway driving will impose a different trend on the results than only city driving so if you drive a mix you might want to randomize your route by using various types of driving conditions. If more than one driver uses the car, then the driver becomes a variable because each individual drives somewhat differently. Certainly weather and road conditions influence the results and you might want to consider this in your plan. So we see that the utility of the measured data is very much impacted by variables beyond the primary ones measured. In developing your test, the question you propose to answer will be a factor in developing your test plan and you should be careful in defining that question so as to meet your objective. Imagine how your test conduct would need to be different if you were interested instead in providing values used to advertise the expected average fuel use of a model of car. Also, you need to consider just how good an answer you need. Is 2 liters/100 km or 1 mi/gal close enough? If not, then the test might require much tighter controls. Lastly, as a concomitant check, you might compare your answer with information provided by the manufacturer or independent agency to make sure your answer seems reasonable. Interestingly, this one example contains all the same elements of any sophisticated test. If you can conceptualize the factors influencing this test and how you will plan around them then you are on track to handle almost any test. Before we move into the details of measurements, we focus here on some important concepts germane to all measurements and tests.

Experimental design involves itself with developing a measurement test plan. A test plan draws from the following three steps:[2]

1. *Parameter Design Plan.* Test objective and identification of process variables and parameters and a means for their control. Ask: "What question am I trying to

---

[2]These three strategies are similar to the bases for certain design methods [15] used in engineering systems design.

answer? What needs to be measured?" "What variables and parameters will affect my results?"

2. ***System and Tolerance Design Plan.*** The selection of a measurement technique, equipment, and test procedure based on some preconceived tolerance limits for error.[3] Ask: "In what ways can I do the measurement and how good do the results need to be to answer my question?"

3. ***Data Reduction Design Plan.*** Plan ahead on how to analyze, present, and use the anticipated data. Ask: "How will I interpret the resulting data? How will I use the data to answer my question? How good is my answer? Does my answer make sense?"

Going through all three steps in the test plan before any measurements are taken is a useful habit for a successful engineer. Often, Step 3 will force you to reconsider Steps 1 and 2! In this section, we focus on the concepts related to Step 1 but will discuss and stress all three throughout the text.

Variables

Once we define the question that we want the test to answer, the next step is to "Identify the relevant process parameters and variables." Variables are entities that influence the test. In addition to the targeted measured variable, there may be other variables pertinent to the measured process that will affect the outcome. All known process variables should be evaluated for any possible cause and effect relationships. If a change in one variable will not affect the value of some other variable, the two are considered independent of each other. A variable that can be changed independently of other variables is known as an *independent variable*. A variable that is affected by changes in one or more other variables is known as a *dependent variable*. Normally, the variable that we measure depends on the value of the variables that control the process. A variable may be continuous, in that its value is able to change in a continuous manner, such as stress under a changing load or temperature in a room, or it may be discrete in that it takes on discrete values or can be quantified in a discrete way, such as the value of the role of dice or a test run by a single operator.

The *control* of variables is important. A variable is controlled if it can be held at a constant value or at some prescribed condition during a measurement. Complete control of a variable would imply that it can be held to an exact prescribed value. Such complete control of a variable is not usually possible. We will use the adjective "controlled" to refer to a variable that can be held as prescribed, at least in a nominal sense. The cause and effect relationship between the independent variables and the dependent variable is found by controlling the values of the independent variables while measuring the dependent variable.

Variables that are not or cannot be controlled during measurement, but that affect the value of the variable measured are called *extraneous variables*. Their influence can confuse the clear relation between cause and effect in a measurement. Would not the driving style affect the fuel consumption of a car? Then unless controlled, this influence will affect the result. Extraneous variables can introduce differences in repeated

---

[3]The Tolerance Design Plan strategy used in this text draws on uncertainty (sensitivity) analyses. Sensitivity methods are common in design optimization.

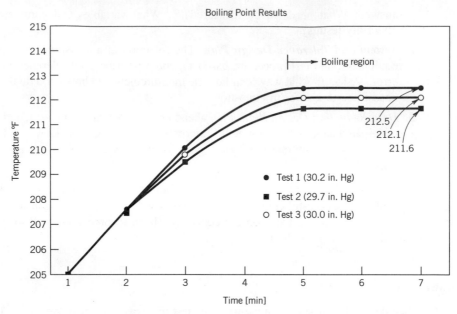

Figure 1.3 Results of a boiling point test for water.

measurements of the same measured variable taken under seemingly identical operating conditions. They can also impose a false trend onto the behavior of that variable. The effects due to extraneous variables can take the form of signals superimposed onto the measured signal with such forms as noise and drift.

Consider a thermodynamics experiment to establish the boiling point of water. The apparatus for measuring the boiling point might yield the results shown in Figure 1.3 for three test runs conducted on separate days. Notice the different outcome for each test.

Why should the data from three seemingly identical tests show such different results?

Suppose we determine that the measurement system accuracy accounts for only 0.1°F of the test data scatter. So another plausible contributing factor is the effect of an extraneous variable. Indeed, a close examination of the test data shows a measured variation in the barometric pressure, which would affect the boiling temperature. The pressure variation is consistent with the trend seen in the boiling point data. Because the local barometric pressure was not controlled (i.e., it was not held fixed between the tests), the pressure acted as an extraneous variable adding to the differences in outcomes between the test runs. *Control important variables or be prepared to solve a puzzle!*

Parameters

In this text, we define a *parameter* as a functional grouping of variables. For example, a Moment of Inertia, or a Reynolds number has its value determined from the values of a grouping of variables. A parameter that has an effect on the behavior of the measured variable is called a *control parameter*. Available methods for establishing control parameters based on known process variables include similarity and dimensional analysis techniques (e.g., [1–3]), and physical laws. A parameter is controlled if its value can be maintained during a set of measurements.

As an example, the flow rate, Q, developed by a fan depends on rotational speed, n, and the diameter, d, of the fan. A control parameter for this group of three variables, found by similarity methods, is the fan flow coefficient, $C_1 = Q/nd^3$. For a given fan, d is fixed (and therefore controlled), and if speed is somehow controlled, the fan flow rate associated with that speed can be measured and the flow coefficient can be determined. Parameters will be affected by extraneous variables.

Noise and Interference

Just how extraneous variables affect measured data can be delineated into noise and interference. *Noise* is a random variation of the value of the measured signal as a consequence of the variation of the extraneous variables. Noise increases data scatter. *Interference* imposes undesirable deterministic trends on the measured value. Any uncontrolled influence that causes the signal or test outcome to behave in a manner different from its true behavior is interference.

A common interference in electrical instruments comes from an ac power source and is seen as a sinusoidal wave superimposed onto the measured signal path. Hum and acoustic feedback in public address and audio systems are ready examples of interference effects that are superimposed onto a desirable signal. Sometimes the interference is obvious. But if the period of the interference is longer than the period over which the measurement is made, the false trend may go unnoticed. So we want to either control the source of interference or break up its trend.

Consider the effects of noise and interference on the signal, $y(t) = 2 + \sin 2\pi t$. As shown in Figure 1.4, noise adds to the scatter of the signal. Through statistical techniques and other means, we can sift through the noise to get at the desirable signal information. But interference imposes a trend onto the signal. The measurement plan should be devised to break up such trends so that they appear as random variations in the data set. Although this will increase the scatter in the measured values of a data set, noise can be handled by statistics. It is far more important to eliminate false trends in the data set.

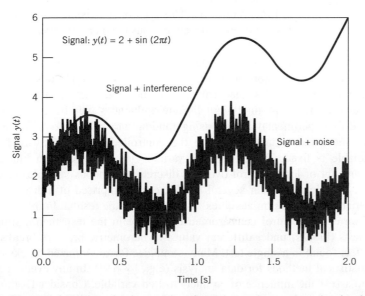

Figure 1.4 Effects of noise and interference superimposed on the signal $y(t) = 2 + \sin 2\pi t$.

With this discussion in mind, recall the boiling point example earlier. Barometric pressure caused interference in each individual test. The barometric pressure did not change over the conduct of any one test. But we could discern the effect only because we showed the results of several tests over which the value of this uncontrolled variable did change. This is a form of randomization. Randomization methods are available that can be easily incorporated into the measurement plan and will minimize or eliminate interference trends. Several methods are discussed in the paragraphs that follow.

Random Tests

Recall our car fuel use example in which the question is: "What fuel usage should I expect from this car?" Let y be the fuel use, which depends on x_a, fuel volume consumption, and x_b, distance traveled. We determine y by varying these two variables (that is, we drive the car). But the test result can be affected by discrete extraneous variables such as the route, driver, and weather and road conditions. For example, driving only on highways would impose a false (untypical) trend on our intended average fuel estimate, so we could drive on different types of roads to break up this trend. This approach introduces a random test strategy.

In general, consider the situation in which the dependent variable, y, is a function of several independent variables, x_a, x_b, ... However, the measurement of y can also be influenced by several extraneous variables, z_j, where $j = 1, 2, . . .$, such that $y = f(x_a, x_b, . . . ; z_j)$. To find the dependence of y on the independent variables they are varied in a controlled manner. Although the influence of the z_j variables on these tests cannot be eliminated, the possibility of their introducing a false trend on y can be minimized by a proper test strategy. A random test is one such strategy.

We define a *random test* by a measurement matrix that sets a random order to the change in the value of the independent variable applied. Trends normally introduced by the coupling of a relatively slow and uncontrolled variation in the extraneous variables with a sequential application in values of the independent variable applied will be broken up. A random test is a form of randomization. This type of plan is effective for the local control of extraneous variables that change in a continuous manner. Consider Examples 1.1 and 1.2.

Discrete extraneous variables are treated a little differently. The use of different instruments, different test operators, and different test operating conditions are examples of discrete extraneous variables that can affect the outcome of a measurement. Randomizing a test matrix to minimize discrete influences can be done efficiently through the use of experimental design using random blocks. A block consists of a data set of the measured variable in which the controlled variable is varied but the extraneous variable is fixed. The extraneous variable is varied between blocks. This enables some amount of local control over the discrete extraneous variable. In the fuel usage example, we might consider several blocks, each comprised of a different driver (extraneous variable) driving similar routes, and averaging the results. In the example of Figure 1.3, if we cannot control the barometric pressure in the test, then a strategy of using several tests (blocks) under different values of barometric pressure breaks up the interference effect found in a single test. Many strategies for randomized blocks exist, as do advanced statistical methods for data analysis (e.g., [4–6,9]). In any event, a random test is useful to assess the influence of an uncontrolled variable. Consider Examples 1.3 and 1.4.

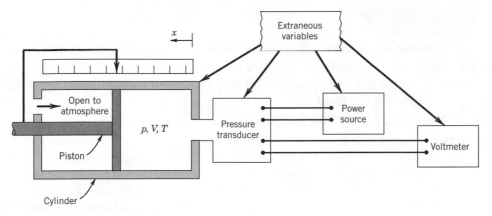

Figure 1.5 Pressure calibration system.

EXAMPLE 1.1

In the pressure calibration system shown in Figure 1.5, a sensor–transducer is exposed to a known pressure, p. The transducer, powered by an external supply, converts the sensed signal into a voltage that is measured by a voltmeter. The measurement approach is to control the applied pressure by the measured displacement of a piston that is used to compress a gas contained within the piston-cylinder chamber. The gas chosen closely obeys the ideal gas law. Hence, piston displacement, x, which sets the chamber volume, $\forall = (x \times \text{area})$, is easily related to chamber pressure. Identify the independent and dependent variables in the calibration and possible extraneous variables.

KNOWN Pressure calibration system of Figure 1.5

FIND Independent, dependent, and extraneous variables

SOLUTION The control parameter for this problem can be formed from the ideal gas law: $p\forall/T =$ constant, where T is the gas temperature. An independent variable in the calibration is the piston displacement that sets the volume. This variable can be controlled by locking the piston into position. From the ideal gas law, gas pressure will also be dependent on temperature, and therefore temperature is also an independent variable. However, T and $\forall$ are not in themselves independent according to the control parameter. Since volume is to be varied through variation of piston displacement, T and $\forall$ can be controlled provided a mechanism is incorporated into the scheme to maintain a constant gas temperature within the chamber. This will also maintain chamber area constant, a relevant factor in controlling the volume. In that way, the applied variations in $\forall$ will be the only effect on pressure, as desired. The dependent variable is the chamber gas pressure. The pressure sensor is exposed to the chamber gas pressure and, hence, this is the pressure it senses. Examples of likely extraneous variables would include noise effects due to the room temperature, z_1, and line voltage variations, z_2, which would affect the excitation voltage from the power supply and the performance of the voltmeter. Connecting wires between devices will act as an antenna and possibly will introduce interference, z_3, superimposed onto the electrical signal, an effect that can be reduced by proper electrical shielding. This list is not exhaustive but illustrative. Hence,

$$p = f(\forall, T; z_1, z_2, z_3), \text{ where } \forall = f_1(x, T).$$

COMMENT Even though we might try to keep the gas temperature constant, even slight variations in the gas temperature would affect the volume and pressure and, hence, will act as an additional extraneous variable!

EXAMPLE 1.2

Develop a test matrix that will minimize the interference effects of any extraneous variables in Example 1.1.

KNOWN $p = f(\forall, T; z_1, z_2, z_3)$, where $\forall = f_1(x; T)$. Control variable $\forall$ is changed. Dependent variable p is measured.

FIND   Randomize possible effects of extraneous variables

SOLUTION   Part of our test strategy is to vary volume, control gas temperature, and measure pressure. An important feature of all test plans is a strategy that minimizes the superposition of false trends onto the data set by the extraneous variables. Since z_1, z_2, and z_3 and any inability to hold the gas temperature constant are continuous extraneous variables, their influence on p can be randomized by a random test. This entails shuffling the order by which $\forall$ is applied. Say that we pick six values of volume: $\forall_1$, $\forall_2$, $\forall_3$, $\forall_4$, $\forall_5$, and $\forall_6$, where the subscripts correspond to an increasing sequential order of the respective values of volume. Any random order will do fine. One possibility, found by using the random function features of a hand-held calculator, is

$$\forall_2 \quad \forall_5 \quad \forall_1 \quad \forall_4 \quad \forall_6 \quad \forall_3$$

If we perform our measurements in a random order, interference trends will be broken up.

EXAMPLE 1.3

The manufacture of a particular composite material requires mixing a percentage by weight of binder with resin to produce a gel. The gel is used to impregnate a fiber to produce the composite material in a manual process called the lay-up. The strength, σ, of the finished material depends on the percent binder in the gel. However, the strength may also be lay-up operator dependent. Formulate a test matrix by which the strength to percent binder–gel ratio relationship under production conditions can be established.

KNOWN

$$\sigma = f(\text{binder; operator})$$

ASSUMPTION   Strength is affected only by binder and operator

FIND   Test matrix to randomize effects of operator

SOLUTION   The dependent variable, σ, is to be tested against the independent variable, percent binder–gel ratio. The operator is an extraneous variable in actual production. As a simple test, we could test the relationship between three binder–gel ratios, A, B, and C, and measure strength. We could also choose three typical operators (z_1, z_2, and z_3) to produce N separate composite test samples for each of the three binder–gel ratios. This gives the 3-block test pattern:

Block				
1	z_1:	A	B	C
2	z_2:	A	B	C
3	z_3:	A	B	C

In the analysis of the test, all of these data can be combined. The results of each block will include each operator's influence as a variation. We can assume that the order used within each block is unimportant. But if only the data from one operator are considered, the results may show a trend consistent with the lay-up technique of that operator. The test matrix above will randomize the influence of any one operator on the strength test results by introducing the influence of several operators.

EXAMPLE 1.4

Suppose following lay-up, the composite material of Example 1.3 is allowed to cure at a controlled but elevated temperature. We wish to develop a relationship between the binder–gel ratio and the cure temperature and strength. Develop a suitable test matrix.

KNOWN $\sigma = f$ (binder, temperature, operator)

ASSUMPTION   Strength is affected only by binder, temperature, and operator

FIND   Test matrix to randomize effect of operator

SOLUTION   We develop a simple matrix to test for the dependence of composite strength on the independent variables of binder–gel ratio and cure temperature. We could proceed as in Example 1.3 and set up three randomized blocks for ratio and three for temperature for a total of 18 separate tests. Suppose instead we choose three temperatures, T_1, T_2, and T_3, along with three binder–gel ratios, A, B, and C and three operators, z_1, z_2, and z_3, and set up a 3×3 test matrix representing a single randomized block. If we organize the block such that no operator runs the same test combination more than once, we randomize the influence of any one operator on a particular binder–gel ratio, temperature test.

	z_1	z_2	z_3
A	T_1	T_2	T_3
B	T_2	T_3	T_1
C	T_3	T_1	T_2

COMMENT   The suggested test matrix not only randomizes the extraneous variable, it has reduced the number of tests by one-half over the direct use of three blocks for ratio and for temperature. However, either approach is fine. The above matrix is referred to as a Latin square [4–6, 9].

If we wanted to include our ability to control the independent variables in the test data variations, we could duplicate the Latin-square test several times to build up a significant database.

Replication and Repetition

In general, the estimated value of a measured variable improves with the number of measurements. For example, a bearing manufacturer would obtain a better estimate of the mean diameter and the variation in the diameters of a batch of bearings by measuring many bearings rather than just a few. Repeated measurements made during any single test run or on a single batch are called *repetitions*. Repetition helps to quantify the variation in a measured variable as it occurs during any one test or batch while the operating conditions are held under nominal control. However, repetition will not permit an assessment of how exact the operating conditions can be set.

If the bearing manufacturer was interested in how closely bearing mean diameter was controlled in day-in and day-out operations with a particular machine or test operator, duplicate tests run on different days would be needed. An independent duplication of a set of measurements using similar operating conditions is referred to as a *replication*. Replication allows for quantifying the variation in a measured variable as it occurs between different tests, each having the same nominal values of operating conditions.

Finally, if the bearing manufacturer were interested in how closely bearing mean diameter was controlled when using different machines or different machine operators, duplicate tests using these different configurations holds the answer. Here, replication provides a means to randomize the interference effects of the different bearing machines or operators mentioned.

Replication allows an assessment of the control on setting the operating conditions, that is, the ability to reset the conditions to some desired value. Ultimately, replication estimates our control over the procedure used.

EXAMPLE 1.5

Consider a room furnace thermostat. Set to some temperature, we can make repeated measurements (repetition) of room temperature and come to a conclusion about the average value and the variation in room temperature at that particular thermostat setting. Repetition allows us to estimate the variation in this measured variable. This repetition permits an assessment of how well we can maintain (control) the operating condition.

Now suppose we change the set temperature to some arbitrary value but sometime later return it to the original setting and duplicate the measurements. The two sets of test data are replications of each other. We might find that the average temperature in the second test differs from the first. The different averages suggest something about our ability to set and control the temperature in the room. Replication permits the assessment of how well we can duplicate a set of conditions.

Concomitant Methods

Is my test working? What value of result should I expect? To help answer these, a good strategy is to incorporate *concomitant methods* in a measurement plan. The goal is to obtain two or more estimates for the result, each based on a different method, which can be compared as a check for agreement. This may affect the experimental design in that additional variables may need to be measured. Or the different method could be an analysis that estimates an expected value of the measurement. For example, suppose we want to establish the volume of a cylindrical rod of known material. We could simply measure the diameter and length of the rod to compute this. Alternatively, we could measure the weight of the rod and compute volume based on the specific weight of the material. The second method complements the first and provides an important check on the adequacy of the first estimate.

1.4 CALIBRATION

A *calibration* applies a known input value to a measurement system for the purpose of observing the system output value. It establishes the relationship between the input and output values. The known value used for the calibration is called the *standard*.

Static Calibration

The most common type of calibration is known as a static calibration. In this procedure, a known value is input to the system under calibration and the system output is recorded. The term "static" implies that the values of the variables involved remain constant; that is, they do not vary with time or space. In static calibrations, only the magnitudes of the known input and the measured output are important.

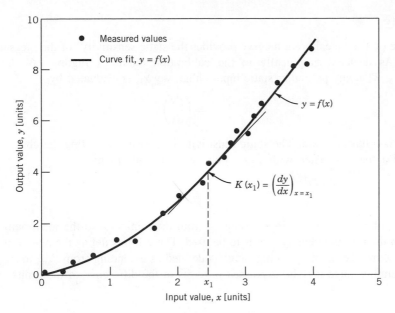

Figure 1.6 Representative static calibration curve.

By applying a range of known input values and by observing the system output values, a direct calibration curve can be developed for the measurement system. On such a curve the input, x, is plotted on the abscissa against the measured output, y, on the ordinate, such as indicated in Figure 1.6. In a calibration the input value is usually a controlled independent variable, while the measured output value is the dependent variable of the calibration.

The static calibration curve describes the static input–output relationship for a measurement system and forms the logic by which the indicated output can be interpreted during an actual measurement. For example, the calibration curve is the basis for fixing the output display scale on a measurement system, such as that of Figure 1.2. Alternatively, a calibration curve can be used as part of developing a functional relationship, an equation known as a correlation, between input and output. A correlation will have the form $y = f(x)$ and is determined by applying physical reasoning and curve fitting techniques to the calibration curve. The correlation can then be used in later measurements to ascertain the unknown input value based on the output value, the value indicated by the measurement system.

Dynamic Calibration

When the variables of interest are time (or space) dependent and such varying information is sought, we need dynamic information. In a broad sense, dynamic variables are time (or space) dependent in both their magnitude and frequency content. A *dynamic calibration* determines the relationship between an input of known dynamic behavior and the measurement system output. Usually, such calibrations involve applying either a sinusoidal signal or a step change as the known input signal. The dynamic nature of signals and measurement systems is explored fully in Chapter 3.

Static Sensitivity

The slope of a static calibration curve provides the static sensitivity[4] of the measurement system. As depicted graphically in the calibration curve of Figure 1.6, the static sensitivity, K, at any particular static input value, say x_1, is evaluated by

$$K = K(x_1) = \left(\frac{dy}{dx}\right)_{x=x_1} \tag{1.1}$$

where K is a function of x. The static sensitivity is a measure relating the change in the indicated output associated with a given change in a static input.

Range

A calibration applies known inputs ranging from the minimum to the maximum values for which the measurement system is to be used. These limits define the operating *range* of the system. The input operating range is defined as extending from x_{min} to x_{max}. The input operating range may be expressed in terms of the difference of the limits as

$$r_i = x_{max} - x_{min} \tag{1.2}$$

This is equivalent to specifying the output operating range from y_{min} to y_{max}. The output span or full-scale-operating range (FSO) is expressed as

$$r_o = y_{max} - y_{min} \tag{1.3}$$

It is important to avoid extrapolation beyond the range of known calibration during measurement since the behavior of the measurement system is uncharted in these regions. As such, the range of calibration should be carefully selected.

Resolution

The *resolution* represents the smallest increment in the measured value that can be discerned. In terms of a measurement system, it is quantified by the smallest scale increment or least count (least significant digit) of the output readout indicator.

Accuracy and Error

The exact value of a variable is called the *true value*. The value of the variables as indicated by a measurement system is called the *measured value*. The *accuracy* of a measurement refers to the closeness of agreement between the measured value and the true value. But the true value is never known exactly and various influences, called errors, have an effect on both of these values. So the concept of the accuracy of a measurement is a qualitative one.

An appropriate approach to stating this closeness of agreement is to identify the measurement errors and to quantify them by the value of their associated uncertainties, where an uncertainty is the estimated range of value of an error. We define an *error, e*, as the difference between the measured value and the true value, that is

$$e = \text{measured value} - \text{true value} \tag{1.4}$$

---

[4]Some texts refer to this as the static gain.

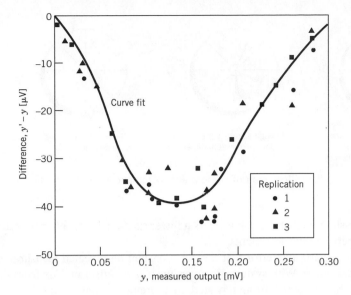

Figure 1.7 Calibration curve in the form of a deviation plot for a temperature sensor.

While the true value is not known, this definition serves as a reference definition. It may not be possible to exactly qualify an error but in fact they exist and they have a magnitude as given by equation (1.4). The concept is something we discuss next and then develop extensively in Chapter 5. Often an estimate for the value of error is based on a reference value used during the instrument's calibration as a surrogate for the true value. A relative error based on this reference value is estimated by

$$A = \frac{|e|}{\text{reference value}} \times 100 \tag{1.5}$$

A few vendors may still refer to this term as the "relative accuracy."

A special form of a calibration curve is the *deviation plot*, such as shown in Figure 1.7. Such a curve plots the error or deviation between a reference or expected value, y', and the measured value, y, versus the measured value. Deviation curves are extremely useful when the differences between the reference and the measured value are too small to suggest possible trends on direct calibration plots. As an example, a deviation plot of the calibration of a temperature-sensing thermocouple is given in Figure 1.7. The voltage output during calibration is compared with the expected values obtained from reference tables. The trend of the errors can be correlated, as shown by the curve fit. We see the maximum errors are two orders of magnitude smaller than the measured values but also see that the errors are smallest at either limit of the measuring range.

Random and Systematic Errors and Uncertainty

Random error is a measure of the random variation found during repeated measurements of a variable. An estimate of a measurement system random error does not require a calibration, per se. But note that a system that repeatedly indicates the same wrong value upon repeated application of a particular input would be considered to have small random error contributions regardless of its accuracy. The *repeatability* of a measurement system refers its ability to indicate the same value on repeated measurements for a specific value of input. The term "precision" is sometimes used as a measure of the repeatability of a

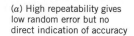

(a) High repeatability gives
low random error but no
direct indication of accuracy

(b) High accuracy means low
random and systematic errors

(b) Systematic and random errors
lead to poor accuracy

Figure 1.8 Throws of a dart: illustration of random and systematic errors and accuracy.

measurement or system; higher precision infers a lower random error, better repeatability, or less variation between measurements.

The portion of the absolute error that remains constant on repeated measurements is called the *systematic error*. With systematic error, there is an offset or *bias* from the true value that cannot be discerned from repeated measurements; lower bias infers lower systematic error in a measurement. Both random and systematic errors affect the measure of a system's accuracy.

The concepts of accuracy and of systematic and random errors in instruments and measurement systems can be illustrated by the throw of darts. Consider the dart board in Figure 1.8 where the goal will be to throw the darts into the bull's-eye. For this analogy, the bull's-eye can represent the true value and each throw can represent a measurement value. In Figure 1.8(a), the thrower displays good repeatability (i.e., small random error) in that each throw repeatedly hits the same spot on the board, but the thrower is not accurate in that the dart misses the bull's-eye each time. This thrower is precise, but we see that small random error alone is not a measure of accuracy. The error in each throw can be computed from the distance between the bull's-eye and each dart. The average value of the error yields the systematic error. This thrower has an offset to the left of the target. If the systematic error could be reduced, then this thrower's accuracy would improve. In Figure 1.8(b), the thrower displays a high accuracy, hitting the bull's-eye on each throw. Both scatter and offset are near zero. High accuracy must imply both small random and systematic errors as shown. In Figure 1.8(c), the thrower does not show good accuracy with errant throws scattered around the board. Each throw contains a different amount of error. While the systematic error is the average of the errors in the throws, random error is related to the varying amount of error in the throws. The random and systematic errors of the thrower can be computed using the statistical methods that are discussed in Chapter 4 and the methods of comparison discussed in Chapter 5. Both random and systematic errors quantify the error in any set of measurements and are used to estimate accuracy.

Suppose a measurement system were used to measure a variable whose value was kept constant and known exactly, as in a calibration. For example, ten independent measurements are made with the results, as shown in Figure 1.9. The variations in the measurements, the observed scatter in the data, would be related to the random error associated with the measurement of the variable. This scatter is mainly due to (1) the measurement system and the method of its use, and (2) any uncontrolled variations in the variable. However, the offset between the apparent average of the readings and the true value would provide a measure of the systematic error to be expected from this measurement system.

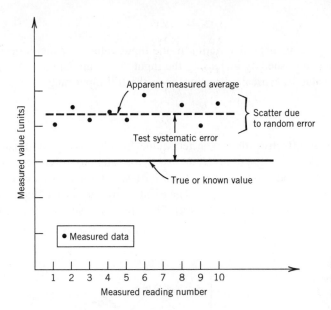

Figure 1.9 Effects of random and systematic errors on calibration readings.

In a measurement, the error cannot be known exactly since the true value is not known. But based on the results of a calibration, the operator might feel confident that the error is within certain bounds, a plus or minus range of the indicated reading. *Uncertainty* refers to the estimate of the effects of the errors on the result of a measurement. Uncertainty is brought about by all of the errors that are present in the measurement system, its calibration, and measurement technique. Individual errors are properties of the instruments and measurement system. Uncertainty is a property of the test result. In Figure 1.9, we see that we might estimate the random error, i.e. the random uncertainty, from the data scatter. The systematic uncertainty might be based on a comparison against a concomitant method. A method of estimating the uncertainty in the result of a test is treated in detail in Chapter 5.

The uncertainty assigned to an instrument or measurement system is the result of several interacting random and systematic errors inherent to the measurement system, the calibration procedure, and the standard used to provide the known value. These errors can be delineated and quantified as the uncertainties of elemental errors through the use of particular calibration procedures and data reduction techniques. An example is given for a typical pressure transducer in Table 1.1.

Table 1.1 Manufacturer's Specifications: Typical Pressure Transducer

Operation	
Input range	0–1000 cm H_2O
Excitation	±15 V dc
Output range	0–5 V
Performance	
Linearity error	$\pm0.5\%$ FSO
Hysteresis error	Less than $\pm0.15\%$ FSO
Sensitivity error	$\pm0.25\%$ of reading
Thermal sensitivity error	$\pm0.02\%$ /°C of reading
Thermal zero drift	$\pm0.02\%$ /°C FSO
Temperature range	0–50 °C

Sequential Test

A *sequential test* applies a sequential variation in the input value over the desired input range. This may be accomplished by increasing the input value (upscale direction) or by decreasing the input value (downscale direction) over the full input range.

Hysteresis

The sequential test is an effective diagnostic technique for identifying and quantifying hysteresis error in a measurement system. *Hysteresis error* refers to differences between an upscale sequential test and a downscale sequential test. The hysteresis error of the system is given by $e_h = (y)_{\text{upscale}} - (y)_{\text{downscale}}$. The effect of hysteresis in a sequential test calibration curve is illustrated in Figure 1.10(a). Hysteresis is usually specified for a

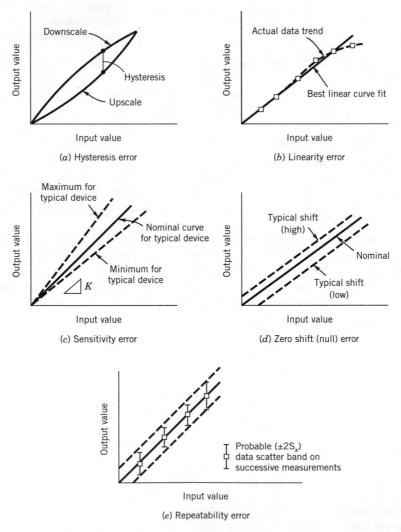

Figure 1.10 Examples of some common elements of instrument error. (*a*) Hysteresis error. (*b*) Linearity error. (*c*) Sensitivity error. (*d*) Zero shift (null) error. (*e*) Repeatability error.

measurement system in terms of the maximum hysteresis error as a percentage of full-scale output range (FSO):

$$\%e_{h_{max}} = \frac{e_{h_{max}}}{r_o} \times 100 \tag{1.6}$$

such as the value indicated in Table 1.1. Hysteresis occurs when the output of a measurement system is dependent on the previous value indicated by the system. Such dependencies can be brought about through some realistic system limitations such as friction or viscous damping in moving parts or residual charge in electrical components. Some hysteresis is normal for any system and affects the repeatability of the system.

Random Test

A *random test* applies a random order in the values of a known input over the intended calibration range. The random application of input tends to reduce the impact of interference. It breaks up hysteresis effects and observation errors. It ensures that each application of input value is independent of the previous. As such, it reduces calibration systematic error converting it to random error. Generally, such a random variation in input value will more closely simulate the actual measurement situation.

A random test provides an important diagnostic for the delineation of several measurement system performance characteristics based on a set of random calibration test data. In particular, linearity error, sensitivity error, zero error, and instrument repeatability error, as illustrated in Figure 1.10(b)–(e), can be quantified from a static random test calibration.

Linearity Error

Many instruments are designed to achieve a linear relationship between the applied static input and indicated output values. Such a linear static calibration curve would have the general form

$$y_L(x) = a_0 + a_1 x \tag{1.7}$$

where the curve fit $y_L(x)$ provides a predicted output value based on a linear relation between x and y. However, in real systems, truly linear behavior is only approximately achieved. As a result, measurement device specifications usually provide a statement as to the expected linearity of the static calibration curve for the device. The relationship between $y_L(x)$ and measured value $y(x)$ is a measure of the nonlinear behavior of a system:

$$e_L(x) = y(x) - y_L(x) \tag{1.8}$$

where $e_L(x)$ is the *linearity error* that arises in describing the actual system behavior by equation (1.7). Such behavior is illustrated in Figure 1.10(b) in which a linear curve has been fit through a calibration data set. For a measurement system that is essentially linear in behavior, the extent of possible nonlinearity in a measurement device is often specified in terms of the maximum expected linearity error as a percentage of full-scale output range:

$$\%e_{L_{max}} = \frac{e_{L_{max}}}{r_o} \times 100 \tag{1.9}$$

This is how the linearity error for the pressure transducer in Table 1.1 was estimated. Statistical methods of quantifying data scatter about a line or curve fit are discussed in Chapter 4.

Sensitivity and Zero Errors

The scatter in the data measured during a calibration affects the precision in predicting the slope of the calibration curve. As shown for the linear calibration curve in Figure 1.10(c) in which the zero intercept is fixed, the scatter in the data about the curve fit are random errors. The *sensitivity error*, e_K, is a statistical measure of the random error in the estimate of the slope of the calibration curve (we discuss the statistical estimate further in Chapter 4). The static sensitivity of a device is also temperature dependent, and this is often specified. In Table 1.1, the sensitivity error reflects calibration results at a constant reference ambient temperature, whereas the thermal sensitivity error was found by calibration at different temperatures.

If the zero intercept is not fixed but the sensitivity is constant, then a drift in the zero intercept introduces a vertical shift of the calibration curve, as shown in Figure 1.10(d). This shift is known as the *zero error*, e_z. Zero error can usually be reduced by periodically adjusting the output from the measurement system under a zero input condition. However, some random variation in the zero intercept is common, particularly with electronic and digital equipment subjected to temperature variations (e.g., thermal zero drift in Table 1.1).

Instrument Repeatability

The ability of a measurement system to indicate the same value upon repeated but independent application of the same input is known as the instrument *repeatability*. Specific claims of repeatability are based on multiple calibration tests (replication) performed within a given lab on the particular unit. Repeatability, as shown in Figure 1.10(e), is based on a statistical measure (developed in Chapter 4) called the standard deviation, S_x, a measure of the variation in the output for a given input. The value claimed is usually in terms of the maximum expected error as a percentage of full-scale output range:

$$\%e_{R_{\max}} = \frac{2S_x}{r_o} \times 100 \qquad (1.10)$$

The instrument repeatability reflects only the error found under controlled calibration conditions. It does not include the additional errors introduced during measurement due to variation in the measured variable or due to procedure.

Reproducibility

The term "reproducibility," when reported in instrument specifications, refers to the closeness of agreement in results obtained from duplicate tests carried out under changed conditions of measurement. Manufacturer claims of instrument reproducibility must be based on multiple tests (replication) performed in different labs on a single unit or model of instrument.

Instrument Precision

The term "instrument precision," when reported in instrument specifications, refers to the results of separate repeatability tests. The precision is another term for the random error of an instrument. Manufacturer claims of instrument precision must be based on multiple tests (replication) performed on different units of the same manufacture, either performed in the same lab (same–lab precision) or preferably, performed in different labs (between–lab precision).

Overall Instrument Error

An estimate of the *overall instrument error* is made by combining all known errors. When instrument errors are combined, the overall result is an uncertainty. This uncertainty is often misleadingly referred to as the instrument accuracy in some instrument specifications. An estimate is computed from the square root of the sum of the squares of all known errors. For M known errors, the overall instrument error, u_c, is estimated by

$$u_c = \left[e_1^2 + e_2^2 + \cdots + e_M^2 \right]^{1/2} \tag{1.11}$$

For example, for an instrument having known hysteresis, linearity, and sensitivity, the instrument error is estimated by

$$u_c = \left[e_h^2 + e_L^2 + e_K^2 \right]^{1/2} \tag{1.12}$$

1.5   STANDARDS

When a measurement system is calibrated, its indicated value is compared directly with a reference value. This reference value forms the basis of the comparison and is known as the *standard*. This standard may be based on the output from a piece of equipment, from an object having a well-defined physical attribute to be used as a comparison, or from a well-accepted technique known to produce a reliable value. Let us explore how certain standards come to be and how these standards are the foundation of all measurements.

A *dimension* defines a physical variable that is used to describe some aspect of a physical system. A *unit* defines a quantitative measure of a dimension. For example, mass, length, and time describe base dimensions with which we associate the units of kilogram, meter, and second. A *primary standard* defines the value of a unit. It provides the means to describe the unit with a unique number that can be understood throughout the world. The primary standard then assigns a unique value to a unit by definition! As such it must define the unit exactly. In 1960, the 11th General Conference on Weights and Measures, the international agency responsible for maintaining exact uniform standards of measurements, formally adopted the International System of Units (SI) as the international standard of units. The system has been adopted worldwide. Other unit systems are commonly used in the consumer market and so deserve mention. These other unit systems are not standards and are to be treated as conversions from SI. Examples of these include the inch-pound (I-P) unit system found in the United States and the gravitational mks (meter-kilogram-second) unit system common to most of the world.

Primary standards are necessary, because the value assigned to a unit is actually arbitrary. For example, over 4500 years ago the Egyptian cubit was used as a standard of length and was based on the length from outstretched fingertips to the elbow. It was later codified with a master of marble, a stick about 52 cm in length, on which scratches were etched to define subunits of length. So whether today's standard unit of length, the meter, is the length of a king's arm or the distance light travels in a fraction of a second depends only on how we want to define it. To avoid confusion, units are defined by international agreement through the use of primary standards. Once agreed upon, the primary standard forms the exact definition of the unit until it is changed by some later agreement. Important features sought in any standard should include: global availability, continued reliability, and stability with minimal sensitivity to external environmental sources. Next we examine some basic dimensions and the primary standards that form the definition of the units that describe them [10].

Base Dimensions and Their Units

Mass

The dimension of mass is defined by the kilogram. Originally, the unit of the kilogram was defined by the mass of one liter of water at room temperature. But today an equivalent yet more consistent definition defines the kilogram exactly as the mass of a particular platinum-iridium cylindrical bar that is maintained under very specific conditions at the International Bureau of Weights and Measures located in Sèvres, France. This particular bar (consisting of 90% platinum and 10% iridium by mass) forms the primary standard for the kilogram. It remains as the only basic unit still defined in terms of a material object.

In the United States, the inch-pound (I-P) unit system (also referred to as the U.S. customary units) remains widely used. In the I-P system, *mass* is defined by the pound-mass, lb_m, which is derived directly and exactly from the definition of the kilogram:

$$1 \ 1b_m = 0.4535924 \ kg \tag{1.13}$$

Equivalent standards for the kilogram and other standards units are maintained by the U.S. National Institute of Standards and Technology (NIST) in Gaithersburg, Maryland and other national labs around the globe. NIST claims that their mass standard is accurate to within 1 mg in 27,200 kg.

Time and Frequency

The dimension of time is defined by the unit of a second. One second is defined [7] as the time elapsed during 9,192,631,770 periods of the radiation emitted between two excitation levels of the fundamental state of cesium-133. Despite this seemingly unusual definition, this primary standard can be reliably reproduced at suitably equipped laboratories throughout the world to within 2 parts in 10 trillion.

The Bureau International de l'Heure (BIH) in Paris maintains the primary standard for clock time. Periodically, adjustments to clocks around the world are made relative to the BIH clock so as to keep time synchronous.

The standard for cyclical frequency is derived from the time standard. The standard unit is the hertz (1 Hz = 1 cycle/second). The cyclical frequency is related to the circular frequency (radians/second) by

$$1 \, Hz = \frac{2\pi \ rad}{1 \ s} \tag{1.14}$$

Both time and frequency standard signals are broadcast worldwide over designated radio stations for use in navigation and as a source of a standard for these dimensions.

Length

The meter is the standard unit for length. New primary standards are defined when the ability to determine the new standard becomes more accurate than the existing standard.

In 1982, a new primary standard was adopted to define the unit of a meter. One meter (m) is defined exactly as the length traveled by light in 1/299,792,458 of a second, a number derived from the velocity of light in a vacuum (defined as 299,792,458 m/s).

The inch-pound system unit of the inch and the related unit of the foot are derived exactly from the meter.

$$1 \text{ ft} = 0.3048 \text{ m}$$
$$1 \text{ in} = 0.0254 \text{ m}$$

(1.15)

Temperature

The kelvin, K, is the SI unit of thermodynamic temperature and is the fraction 1/273.16 of the thermodynamic temperature of the triple point of water. A temperature scale was devised by William Thomson, Lord Kelvin (1824–1907), and forms the basis for the absolute practical temperature scale in common use. This scale is based on polynomial interpolation between the equilibrium phase change points of a number of common pure substances from the triple point of equilibrium hydrogen (13.81 K) to the freezing point of pure gold (1337.58 K). Above 1337.58 K, the scale is based on Plank's law of radiant emissions. The details of the standard have been modified over the years but are governed by the International Temperature Scale—1990 [8].

The I-P unit system uses the absolute scale of Rankine (°R). This and the common scales of Celsius (°C), used in the metric system, and Fahrenheit (°F) are related to the Kelvin scale by the following:

$$(^\circ C) = (K) - 273.15$$
$$(^\circ F) = (^\circ R) - 459.67$$
$$(^\circ F) = 1.8 \times (^\circ C) + 32.0$$

(1.16)

Current

The SI unit for current is the ampere. One ampere (A) is defined as that constant current which, if maintained in two straight parallel conductors of infinite length and of negligible circular cross section and placed 1 m apart in vacuum, would produce a force equal to 2×10^{-7} newtons per meter of length between these conductors.

Measure of Substance

The unit of quantity of a substance is defined by the mole. One mole (mol) is the amount of substance of a system that contains as many elementary entities as there are atoms in 0.012 kilogram of carbon 12.

Luminous Intensity

The intensity of light is defined by the candela. One candela (cd) is the luminous intensity, in a given direction, of a source that emits monochromatic radiation of frequency 5.40×10^{14} hertz and that has a radiant intensity in that direction of 1/683 watt per steradian.

Derived Units

Other dimensions and their associated units are defined in terms of and derived from the base dimensions and units [10, 11].

Force

From Newton's law, force is proportional to mass times acceleration:

$$\text{force} = \frac{\text{mass} \times \text{acceleration}}{g_c}$$

where g_c is a proportionality constant.

Force is defined by a derived unit called the newton (N), which is derived from the base dimensions of mass, length, and times:

$$1\,\text{N} = 1\frac{\text{kg-m}}{\text{s}^2} \tag{1.17}$$

So for this system the value of g_c must be 1.0 kg-m/s^2-N. Note that the resulting expression for Newton's second law does not explicitly require the inclusion of g_c to make units match and so it can be ignored.

However, in I-P units, the units of force and mass are related through the definition: One pound-mass (lb$_\text{m}$) exerts a force of 1 pound (lb) in a standard gravitational field. With this definition,

$$1\,\text{lb} = \frac{(1\,\text{lb}_\text{m})(32.174\,\text{ft/s}^2)}{g_c} \tag{1.18}$$

and g_c must take on the value of 32.174 lb$_\text{m}$-ft/lb-s^2. In I-P units, the pound is a defined quantity and g_c must be derived through Newton's law. Similarly, in the gravitational mks system, which uses the kilogram-force (kg$_\text{f}$),

$$1\text{kg}_\text{f} = \frac{(1\text{kg})(9.80665\ \text{m/s}^2)}{g_c} \tag{1.19}$$

and the value for g_c takes on a value of exactly 9.80665 kg-m/s^2.

Many engineers have some difficulty with using g_c in the non-SI systems. Actually, whenever force and mass appear in the same expression, just remember to relate them using g_c through Newton's law:

$$g_c = \frac{mg}{F} = 1\frac{\text{kg-m/s}^2}{\text{N}} = 32.174\frac{\text{lb}_\text{m}\text{-ft/s}^2}{\text{lb}} = 9.80665\frac{\text{kg-m/s}^2}{\text{kg}_\text{f}} \tag{1.20}$$

Other Derived Dimensions and Units

Energy is defined as force times length and uses the unit of the joule (J) which is derived from base units as:

$$1\,\text{J} = 1\frac{\text{kg-m}^2}{\text{s}^2} = 1\text{N} - \text{m} \tag{1.21}$$

Power is defined as energy per unit time in terms of the unit of the watt (W), which is derived from base units as:

$$1\,\text{W} = 1\frac{\text{kg-m}^2}{\text{s}^3} = 1\frac{\text{J}}{\text{s}} \tag{1.22}$$

Stress and *pressure* are defined as force per unit area, where area is length squared, in terms of the pascal (Pa), which is derived from base units as

$$1 \text{ Pa} = 1 \frac{\text{kg}}{\text{m-s}^2} = 1 \text{ N/m}^2 \tag{1.23}$$

Electrical Dimensions

The units for the dimensions of electrical potential, and resistance, charge, and capacitance are based on the definitions of the absolute volt (V), and ohm (Ω), coulomb (C), and farad (F), respectively. Derived from the ampere, 1 ohm absolute is defined by 0.9995 times the resistance to current flow of a column of mercury that is 1.063 m in length and has a mass of 0.0144521 kg at 273.15 K. The volt is derived from the units for power and current, $1 \text{ V} = 1 \text{ N-m/s-A} = 1 \text{ W/A}$. The ohm is derived from the units for electrical potential and current, $1 \Omega = 1 \text{ kg-m}^2/\text{s}^3\text{-A}^2 = 1 \text{ V/A}$. The coulomb is derived from the units for current and time, $1 \text{ C} = 1 \text{ A-s}$. One volt is the difference of potential between two points of an electical conductor when a current of 1 ampere flowing between those points dissipates a power of 1 watt. The farad (F) is the standard unit for capacitance derived from the units for charge and electric potential, $1 \text{ F} = 1 \text{ C/V}$.

On a practical level, working standards for resistance and capacitance take the form of certified standard resistors and capacitors or resistance boxes and are used as standards for comparison in the calibration of resistance measuring devices. The practical potential standard makes use of a standard cell consisting of a saturated solution of cadmium sulfate. The potential difference of two conductors connected across such a solution is set at 1.0183 V at 293 K. The standard cell maintains constant electromotive force over very long periods of time, provided that it is not subjected to a current drain exceeding 100 μA for more than a few minutes. The standard cell is typically used as a standard for comparison for voltage measurement devices.

A chart for converting between units is included inside the text cover. Table 1.2 lists the basic standard and some derived units used in SI and the corresponding units used in the I-P and gravitational metric systems.

Hierarchy of Standards

The known value applied to a measurement system during its calibration becomes the standard on which the calibration is based. So how do we pick this standard, and how good is it? Obviously, actual primary standards are impractical as standards for normal calibration use. But they serve as a reference for exactness. It would not be reasonable to travel to France to calibrate an ordinary laboratory scale using the primary standard for mass (nor would it likely be permitted!). So for practical reasons, there exists a hierarchy of reference and secondary standards used to duplicate the primary standards. Just below the primary standard in terms of absolute accuracy are the national reference standards maintained by designated standards laboratories throughout the world. These provide a reasonable duplication of the primary standard but allow for worldwide access to an extremely accurate standard. Next to these, we develop transfer standards. These are used to calibrate individual laboratory standards that might be used at various facilities within a country. Laboratory standards serve to calibrate working standards. Working standards are used to calibrate everyday devices used in manufacturing and research facilities. In the United States, NIST maintains primary, reference, and secondary standards and recommends standard procedures for the calibration of measurement systems.

Table 1.2 Dimensions and Units*

Unit	Dimension	
	SI	IP
Primary		
Length	meter (m)	inch (in)
Mass	kilogram (kg)	pound-mass (lb$_m$)
Time	second (s)	second (s)
Temperature	kelvin (K)	rankine (°R)
Current	ampere (A)	ampere (A)
Substance	mole (mol)	mole (mol)
Light intensity	candela (cd)	candela (cd)
Derived		
Force	newton (N)	pound-force (lb)
Voltage	volt (V)	volt (V)
Resistance	ohm (Ω)	ohm (Ω)
Capacitance	farad (F)	farad (F)
Inductance	henry (H)	henry (H)
Stress, Pressure	pascal (Pa)	pound-force/inch^2 (psi)
Energy	joule (J)	British thermal unit (BTU)
Power	watt (W)	foot pound-force (ft-lb)

*SI dimensions and units are the international standards. IP units are presented for convenience.

Each subsequent level of the hierarchy is derived by calibration against the standard at the previous higher level. Table 1.3 lists an example of such a lineage for standards from a primary or reference standard maintained at a national standards lab down to a working standard used in a typical laboratory or production facility to calibrate everyday working instruments. If the facility does not maintain a local (laboratory or working) standard, then the instruments must be sent off and calibrated elsewhere. In such a case, a standards traceability certificate would be issued for the instrument.

As one moves down through the standards lineage, the degree of exactness by which a standard approximates the primary standard from which it is derived deteriorates. That is, increasing elements of error are introduced into the standard as one goes from one generation of standard to the next. As a common example, an institution might maintain its own working standard (for some application) that is used to calibrate the measurement devices found in the individual laboratories throughout the institution. Periodic calibration of the working standard might be against the institution's well-maintained local

Table 1.3 Hierarchy of Standards*

Primary Standard	Maintained as Absolute Unit Standard
Transfer Standard	Used to calibrate Local Standards
Local Standard	Used to calibrate Working Standards
Working Standard	Used to calibrate local instruments

*There may be additional intermediate standards between each hierarchy level.

Table 1.4   Example of a Temperature Standard Traceability

	Standard	
Level	Method	Error [°C]*
Primary	Fixed thermodynamic points	0
Transfer	Platinum resistance thermometer	±0.005
Working	Platinum resistance thermometer	±0.05
Local	Thermocouple	±0.5

* Typical instrument systematic and random errors.

standard. The local standard would be periodically sent off to be calibrated against the NIST (or appropriate national standards lab) transfer standard (and traceability certificate issued). NIST will periodically calibrate its own transfer standard against its reference or primary standard. This is illustrated for a temperature standard traceability hierarchy in Table 1.4. The uncertainty in the approximation of the known value increases as one moves down the hierarchy. It follows, then, that since the calibration determines the relationship between the input value and the output value, the accuracy of the calibration will depend in part on the accuracy of the standard. But if typical working standards contain some error, how is accuracy ever determined? At best, this closeness of agreement is quantified by the errors and the estimates of the uncertainties in the calibration. And the confidence in that estimate will depend on the quality of the standard and the calibration techniques used.

Test Standards and Codes

The term "standard" is also applied in other ways in engineering. *Test standards* refer to well-defined test procedures, technical terminology, methods to construct test specimens or test devices, and/or methods for data reduction. The goal of a test standard is to provide consistency in the conduct and reporting of a certain type of measurement between test facilities. Similarly, *test codes* refer to procedures for the manufacture, installation, calibration, performance specification, and safe operation of equipment.

Diverse examples of test standards and codes are illustrated in readily available documents (e.g., [12–15]) from professional societies, such as the American Society of Mechanical Engineers (ASME), the American Society of Testing and Materials (ASTM), and the International Standards Organization (ISO). For example, ASME Power Test Code 19.5 provides detailed designs and operation procedures for flow meters, while ASTM Test Standard F558-88 provides detailed procedures for evaluating vacuum cleaner cleaning effectiveness and controls the language for product performance claims. *Test standards and test codes are legal instruments that must be observed by engineers.*

1.6   PRESENTING DATA

Since we use several plotting formats throughout this text to present data, it is best to introduce these formats here. Data presentation conveys significant information about the relationship between variables. Software is readily available to assist in providing high-quality plots or plots can be generated manually using graph paper. Several forms of plotting formats are discussed next.

Rectangular Coordinate Format

In rectangular grid format, both the ordinate and the abscissa have uniformly sized divisions providing a linear scale. This is the most common format used for constructing plots and establishing the form of the relationship between the independent and dependent variable.

Semilog Coordinate Format

In a semilog format, one coordinate has a linear scale and one coordinate has a logarithmic scale. Plotting values on a logarithmic scale performs a logarithmic operation on those values, e.g, plotting $y = f(x)$ on a logarithmic x-axis is the same as plotting $y = \log f(x)$ on rectangular axes. Logarithmic scales are advantageous when one of the variables spans more than one order of magnitude. In particular, the semilog format may be convenient when the data approximately follow a relationship of the form $y = ae^x$ or $y = a10^x$ as a linear curve will result in each case. A natural logarithmic operation can also be conveyed on a logarithmic scale, as the relation $\ln y = 2.3 \log y$ is just a scaling operation.

Full-log Coordinate Format

The full-log or log-log format has logarithmic scales for both axes and is equivalent to plotting $\log y$ vs. $\log x$ on rectangular axes. Such a format is preferred when both variables contain data values that span more than one order of magnitude. With data that follow a trend of the form $y = ax^n$, a linear curve will be obtained in log-log format.

Significant Digits

Significant digits refer to the number of digits found before and after the decimal point in a reported number. While leading zeros are not significant, all trailing zeros are significant. For example, 42.0 has three significant digits. The number of significant digits needed depends on the problem and the close discretion of the engineer. But in the course of working a problem, there should be consistency in the number of significant digits used and reported. The number of digits reported reflects a measure of the accuracy in the value assigned by the engineer. So to determine the number of significant digits required, just ask yourself: "What value makes sense to this problem?" For example, to maintain a relative accuracy of within 1%, we need to report values to three significant figures. But because rounding errors tend to accumulate during calculations, we would perform all intermediate calculations to at least four significant figures, and then to round the final result down to three significant figures.

1.7   SUMMARY

During a measurement the input signal is not known but is inferred from the value of the output signal from the measurement system. We discussed the process of calibration as the means to relate a measured input value from the measurement system output value and the role of standards in that process. An important step in the design of a measurement system is the inclusion of a means for a reproducible calibration that closely simulates the type of signal to be input during actual measurements. A test is the process of "asking a question." The idea of a test plan was developed to answer that question. However, a measured output signal can be affected by many variables that will introduce variation and trends and confuse that answer. Careful test planning is required to reduce such effects. A number of test plan strategies were developed, including randomization. The popular term "accuracy" was explained in terms of the more useful concepts of random error, systematic error, and uncertainty and their affects on a measured value. We also explored the idea of test standards and engineering codes, legal documents that influence practically every manufactured product around us.

REFERENCES

1. Bridgeman, P. W., *Dimensional Analysis*, 2d ed., Yale University Press, New Haven, CN, 1931; paperback   Y-82, 1963.
2. Duncan, W. J., *Physical Similarity and Dimensional Analysis*, Arnold, London, 1953.
3. Massey, B. S., *Units, Dimensions and Physical Similarity*, Van Nostrand-Reinhold, New York, 1971.
4. Lipsen, C., and N. J. Sheth, *Statistical Design and Analysis of Engineering Experimentation*, McGraw-Hill, New York, 1973.
5. Peterson, R. G., *Design and Analysis of Experiments*, Marcel-Dekker, New York, 1985.
6. Mead, R., The Design of Experiments: Statistical Principles for Practical Application, Cambridge Press, New York, 1988.
7. NBS Technical News Bulletin 52:1, January 1968.
8. International Temperature Scale—1990, *Metrologia* 27:3, 1990.
9. Montgomery, D., Design and Analysis of Experiments, 5th ed., Wiley, New York, 2001.
10. B. Taylor, *Guide for the Use of the International System of Units*, NIST Special Publication 330, 2001.
11. B. Taylor, NIST Guide to SI Units, NIST Special Publication 811, 1995.
12. *ASME Power Test Codes*, American Society of Mechanical Engineers, New York.
13. *ASHRAE Handbook of Fundamentals*, American Society of Heating, Refrigeration and Air Conditioning Engineers, New York, 1982.
14. *American National Standard*, American National Standards Institute, New York.
15. *ASTM Standards*, American Society for Testing and Materials, Philadelphia, PA.
16. Peace, G. S., *Taguchi Methods*, Addison-Wesley, Reading, MA, 1993.

NOMENCLATURE

e	absolute error	u_c	instrument uncertainty
e_h	hysteresis error	x	independent variable; input value; measured variable
e_K	sensitivity error	y	dependent variable; output value
e_L	linearity error	y_L	linear polynomial
e_R	repeatability error	A	relative error; relative accuracy
e_z	zero error	K	static sensitivity
p	pressure [$ml^{-1} t^{-2}$]	S_x	standard deviation of x
r_i	input span	T	temperature [°]
r_o	output span	$\forall$	volume [l^3]

PROBLEMS

1.1 Discuss your understanding of the hierarchy of standards beginning with the primary standard. In general, what is meant by the term standard? Can you cite examples of standards in everyday use?

1.2   What is the purpose of a calibration? Suppose an instrument is labeled as "calibrated." What should this mean to you as an engineer?

1.3   Suppose you found a dial thermometer in a stockroom. Discuss several methods by which you might estimate random and systematic error in the thermometer? Its uncertainty?

1.4   Consider the example described in Figure 1.3. Discuss the effect of the extraneous variable, barometric pressure, in terms of noise and interference relative to any one test and relative to several tests. Discuss how the interference effect can be broken up into noise.

1.5   How would the resolution of the display scale of an instrument affect its uncertainty? Suppose the scale was somehow offset by one least count of resolution. How would this affect its uncertainty? Explain in terms of random and systematic error.

1.6   How would the hysteresis of an instrument affect its uncertainty? Explain in terms of random and systematic error.

1.7   Select three different types of measurement systems with which you have experience and identify which attributes of the system comprise the measurement system stages of Figure 1.1.

1.8   Identify the measurement system stages for the following systems (refer back to Figure 1.1 and use other resources, such as a library or Internet search, as needed to learn more about each system):

   a. Room thermostat

   b. Automobile speedometer

   c. Portable CD stereo system

   d. Antilock braking system (automobile)

   e. Audio speaker

1.9   What is the range of the calibration data of Table 1.5 below.

1.10   For the calibration data of Table 1.5, plot the results using rectangular and log-log scales. Discuss the apparent advantages of either presentation.

1.11   For the calibration data of Table 1.5, determine the static sensitivity of the system at (a) $X = 5$; (b) $X = 10$; (c) $X = 20$. For which input values is the system more sensitive? Explain what this might mean in terms of a measurement and in terms of measurement errors.

1.12   Consider the voltmeter calibration data in Table 1.6. Plot the data using a suitable scale. Specify the percent maximum hysteresis based on full-scale range. Input X is based on a standard known to be accurate to better than 0.05 mV.

1.13   Three clocks are compared to a time standard at three successive hours. The data are given in Table 1.7. Using Figure 1.9 as a guide, arrange these clocks in order of estimated accuracy. Discuss your choices.

1.14   Each of the following equations can be represented as a straight line on a x-y plot by choosing the appropriate axis scales. Plot them both in rectangular coordinate format and

Table 1.5   Calibration Data

X [cm]	Y [V]	X[cm]	Y [V]
0.5	0.4	10.0	15.8
1.0	1.0	20.0	36.4
2.0	2.3	50.0	110.1
5.0	6.9	100.0	253.2

Table 1.6   Voltmeter Calibration Data

Increasing Input [mV]		Decreasing Input [mV]	
X	Y	X	Y
0.0	0.1	5.0	5.0
1.0	1.1	4.0	4.2
2.0	2.1	3.0	3.2
3.0	3.0	2.0	2.2
4.0	4.1	1.0	1.2
5.0	5.0	0.0	0.2

Table 1.7   Clock Calibration Data

	Standard Time		
Clock	1:00:00	2:00:00	3:00:00
	Indicated Time		
A	1:02:23	2:02:24	3:02:25
B	1:00:05	2:00:05	3:00:05
C	1:00:01	1:59:58	3:00:01

then in an appropriate format to yield a straight line. Explain how the plot operation yields the straight line. Variable y has units of volts. Variable x has units of meters (use a range of $0.01 \leq x \leq 10.0$). Note: this is easily done using a spreadsheet program where you can compare the use of different axis scales.

a. $y = x^2$　　　　　**d.** $y = 10x^4$

b. $y = 1.1x$　　　　　**e.** $y = 10e^{-2x}$

c. $y = 2x^{0.5}$

1.15　Plot $y = 10e^{-5x}$ volts on in semilog format (use three cycles). Determine the slope of the equation at $x = 0$; $x = 2$; and $x = 20$.

1.16　Plot the following data on appropriate axes. Estimate the static sensitivity K at each X.

Y [V]	X [V]
2.9	0.5
3.5	1.0
4.7	2.0
9.0	5.0

1.17　The following data have the form $y = ax^b$. Plot the data in an appropriate format to estimate the coefficients a and b. Estimate the static sensitivity K at each value of X. How is K affected by X?

Y [cm]	X [m]
0.14	0.5
2.51	2.0
15.30	5.0
63.71	10.0

1.18 For the calibration data given, plot the calibration curve using suitable axes. Estimate the static sensitivity of the system at each X. Then plot K against X. Comment on the behavior of the static sensitivity with static input magnitude for this system.

Y [cm]	X [kPa]
4.76	0.05
4.52	0.1
3.03	0.5
1.84	1.0

1.19 A bulb thermometer hangs outside a window and is used to measure the outside temperature. Comment on some extraneous variables that might affect the difference between the actual outside temperature and the indicated temperature from the thermometer.

1.20 A synchronous electric motor test stand permits either the variation of input voltage or output shaft load and the subsequent measurement of motor efficiency, winding temperature, and input current. Comment on the independent, dependent, and extraneous variables for a motor test.

1.21 The transducer specified in Table 1.1 is chosen to measure a nominal pressure of 500 cm H_2O. The ambient temperature is expected to vary between 18 and 25°C during tests. Estimate the magnitude of each elemental error affecting the measured pressure.

1.22 A force measurement system (weight scale) has the following specifications:

Range:	0 to 1000 N
Linearity error:	0.10% FSO
Hysteresis error:	0.10% FSO
Sensitivity error:	0.15% FSO
Zero Drift:	0.20% FSO

Estimate the overall instrument uncertainty for this system based on available information.

1.23 An engineer ponders over a test plan thinking: "What strategy should I include to estimate any variation over time of the measured variable?" "What strategy should I include to estimate my control of the independent variable during the test?" What does the engineer mean?

1.24 If the outcome of a test were suspected to be dependent on the ability to control the test operating conditions, what strategy should be incorporated into the test plan to estimate this effect?

1.25 State the purpose of using randomization methods during a test. Develop an example to illustrate your point.

1.26 Provide an example of repetition and replication in a test plan from your own experience.

1.27 Develop a test plan that might be used to estimate the average temperature that could be maintained in a heated room as a function of the heater thermostat setting.

1.28 Develop a test plan that might be used to evaluate the fuel efficiency of a production model automobile. Explain your reasoning.

1.29 A race engine shop has just completed two engines of the same design. How might you determine which engine performs better: (i) on a test stand (engine dynamometer) or (ii) at the race track? Describe some measurements that you feel might be useful and how you might use that information. Discuss possible differences between the two tests and how these might influence the results (e.g., you can control room conditions on a test stand but not at a track).

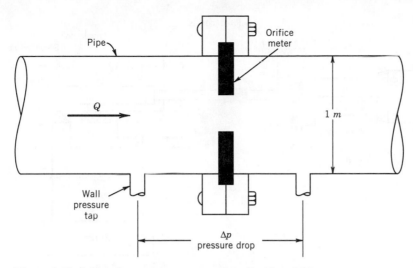

Figure 1.11 Orifice flow meter set-up used for Problem 1.33.

1.30 A large batch of carefully made machine shafts can be manufactured on one of four lathes by one of 12 quality machinists. Set up a test matrix to estimate the tolerances that can be held within a production batch. Explain your reasoning.

1.31 Suggest an approach(es) to estimate the linearity error and the hysteresis error of a measurement system.

1.32 Suggest a test matrix to evaluate the wear performance of four different brands of aftermarket passenger car tires of the same size, load, and speed ratings on a fleet of eight cars of the same make. If the cars were not of the same make, what would change?

1.33 The relation between the flow rate through a pipeline of area, A, and the pressure drop, Δp, across an orifice-type flow meter inserted in that line (Figure 1.11) is given by

$$Q = CA\sqrt{\frac{2\Delta p}{\rho}}$$

For a pipe diameter of 1 m and a flow range of 20°C water between 2 and 10 m³/min and $C = 0.75$, plot the expected form of the calibration curve for flow rate versus pressure drop over the flow range. Is the static sensitivity a constant? Incidentally, such an instrument test is described by ANSI/ASME Test Standard PTC 19.5.

1.34 For the orifice meter calibration in Problem 1.32: Would the term "linearity error" have a meaning for this system? Explain. Also, list the dependent and independent variables in the calibration.

1.35 A piston engine manufacturer uses four different subcontractors to plate the pistons for a make of engine. Plating thickness is important in quality control (performance and part life). Devise a test matrix to assess how well the manufacturer can control plating under its current system.

1.36 A simple thermocouple circuit is formed using two wires of different alloy: One end of the wires is twisted together to form the measuring junction, while the other ends are connected to a voltmeter and form the reference junction. A voltage is set up by the difference in temperature between the two junctions. For a given pair of alloy material and reference junction temperature, the temperature of the measuring junction is inferred

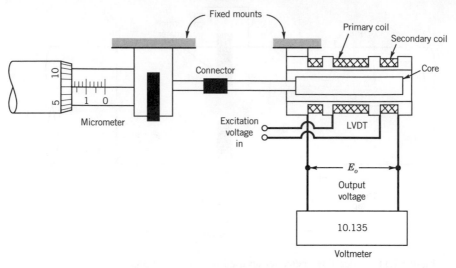

Figure 1.12 LVDT set-up used for Problem 1.37.

from the measured voltage difference. For a measurement, what variables need to be controlled? What are the dependent and independent variables?

1.37   A Linear Variable Displacement Transducer (LVDT) senses displacement and indicates a voltage output, which is linear to the input. Figure 1.12 shows an LVDT set-up used for static calibration. It uses a micrometer to apply the known displacement and a voltmeter for the output. A well-defined voltage powers the transducer. What are the independent and dependent variables in this calibration? Can you suggest any extraneous variables? What would be involved in a replication?

1.38   For the LVDT calibration of the previous problem, what would be involved in determining the repeatability of the instrument? The reproducibility? What effects are different in the two tests? Explain.

1.39   A manufacturer wants to quantify the expected average fuel mileage of a product line of automobile. They decide that they can put one or more cars on a chassis dynamometer and run the wheels at desired speeds and loads to assess this or they can use drivers and drive the cars over some selected course instead. (i) Discuss the merits of either approach considering the control of variables and identifying extraneous variables. (ii) Can you recognize that somewhat different tests might provide answers to different questions? For example, discuss the difference in meanings possible from the results of operating one car on the dynamometer and comparing it to one driver on a course. Cite other examples. (iii) Are these two test methods examples of concomitant methods?

1.40   You estimate your car's fuel use by comparing fuel volume used over a known distance. Your brother, who drives the same model car, disagrees with your claimed results based on his own experience. How might you justify the differences based on the concepts of control of variables, interference and noises effects, and test matrix used?

1.41   In discussing concomitant methods, we cited an example of computing the volume of a cylindrical rod based on its average dimensions versus known weight and material properties. While we should not expect too different of an answer with either technique, identify where noise and interference effects will affect the result of either method.

1.42   When a strain gauge is stretched under uniaxial tension, its resistance varies with the imposed strain. A resistance bridge circuit is used to convert the resistance change into a

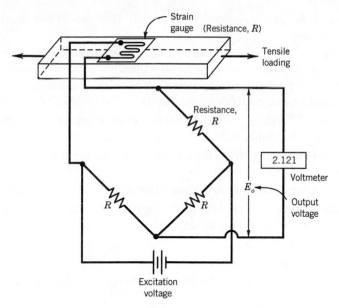

Figure 1.13 Strain gauge set-up used for Problem 1.42.

voltage. Suppose a known tensile load were applied to the system shown in Figure 1.13 and the output measured on a voltmeter. What are the independent and dependent variables in this calibration? Can you suggest any extraneous variables? What would be involved in a replication?

1.43 For the strain gauge calibration of the previous problem, what would be involved in determining the repeatability of the instrument? The reproducibility? A statement on instrument precision? What effects are different in the tests? Explain.

1.44 A major tennis manufacturer is undertaking a test program for shoes, tennis balls, and tennis strings. Develop a test plan for the situations described below. In each case provide details for how the test should be conducted, and describe expected difficulties in interpreting the results. In each case the tests are to be conducted during college tennis matches. Four teams of six players each are to test the products under match (as opposed to practice) conditions. A tennis team consists of six players, and a match consists of six singles matches and three doubles matches.

- Tennis shoes: Two different sole designs and materials are to be wear tested. The life of the soles are known to be strongly affected by court surface and playing style.
- Tennis strings: A new tennis string material is to be tested for durability. String failure occurs due to breakage or due to loss of tension. String life is a function of the racquet, the player's style, and string tension.
- Tennis balls: Two tennis balls are to be play tested for durability. The condition of a tennis ball can be described by the coefficient of restitution, and the total weight (since the cover material actually is lost during play). A player's style and the court surface are the primary determining factors on tennis ball wear.

1.45 The acceleration of a cart down a plane inclined at an angle α to horizontal can be determined by measuring the change in speed of the cart at two points, separated by a distance s, along the inclined plane. Suppose two photocells are fixed at the two points along the plane. Each photocell measures the time for the cart, which has a length, L, to pass it. Identify the important variables in this test. List any assumptions that you feel are intrinsic to such a test. Suggest a concomitant approach. How would you interpret the data to answer the question?

1.46 Is it more fuel efficient to drive a car in warm, humid weather with the air conditioning on and windows rolled up closed or air conditioning off with the windows rolled down for ventilation? Develop a test plan to address this question. Include a concomitant approach (experimental or analytical) that might assess the validity of your test results.

1.47 Explain the potential differences in the following evaluations of an instrument's accuracy. Figure 1.9 will be useful, and you may refer to ASTM E177, if needed.

   (i) The closeness of agreement between the true value and the average of a large set of measurements.

   (ii) The closeness of agreement between the true value and an individual measurement.

1.48 Suggest a reasonable number of significant digits for reporting the following common values and give some indication as to your reasoning:

   (i) Your body weight for a passport

   (ii) A car's fuel usage (use liters per 100 km)

   (iii) The weight of a bar of pure (at least 99.5%) gold (consider a 1 kg_f bar and a 100 oz bar)

   (iv) Distance traveled by a body in 1 second if moving at 1 m/s (use meters)

1.49 Research the following test codes (these are available in most libraries). Write a short (200-word) report that describes the intent and an overview of the code:

   (a) ASTM F 558-88 (Air Performance of Vacuum Cleaners)

   (b) ANSI Z21.86 (Gas Fired Space Space Heating Appliances)

   (c) ISO 10770-1:1998 (Test Methods for Hydraulic Control Valves)

   (d) ANSI/ASME PTC19.1-1998 (Measurement Uncertainty: Instruments and Apparatus)

   (e) ISO 7401:1988 (Road Vehicles: Lateral Response Test Methods)

   (f) Your local municipal building code or housing ordinance

   (g) Any other code assigned by your instructor

1.50 Show how the following functions can be transformed into a linear curve of the form $Y = a_1 X + a_o$ where a_1 and a_0 are constants. Let m, b, c be constants.

   (a) $y = bx^m$

   (b) $y = be^{mx}$

   (c) $y = b + c\sqrt[m]{x}$

Chapter 2

Static and Dynamic Characteristics of Signals

2.1 INTRODUCTION

A measurement system takes an *input* quantity and transforms it into an *output* quantity that can be observed or recorded, such as the movement of a pointer on a dial or the magnitude of a digital display. Our goal in this chapter is to understand the characteristics of both the input signals to a measurement system and the resulting output signals. The shape and form of a signal are often referred to as its *waveform*. The waveform contains information about the *magnitude* and *amplitude*, which indicate the size of the input quantity, and the *frequency*, which indicates the way the signal changes in time. An understanding of waveforms is required for the selection of measurement systems and the interpretation of measured signals.

2.2 INPUT/OUTPUT SIGNAL CONCEPTS

Two important tasks that engineers face in the measurement of physical variables are (1) selecting a measurement system and (2) interpreting the output from a measurement system. A simple example of selecting a measurement system might be the selection of a tire gauge for measuring the air pressure in a bicycle tire or in a car tire, as shown in Figure 2.1. The gauge for the car tire would be required to indicate pressures up to 275 kPa (40 lb/in^2), but the bicycle tire gauge would be required to indicate higher pressures, maybe up to 700 kPa (100 lb/in^2). This idea of the *range* of an instrument, its lower to upper measurement limits, is fundamental to all measurement systems and demonstrates that some basic understanding of the nature of the input signal, in this case the magnitude, is necessary in evaluating or selecting a measurement system for a particular application.

A much more difficult task is the evaluation of the output of a measurement system when the time or spatial behavior of the input is not known. The pressure in a tire does not change while we are trying to measure it, but what if we wanted to measure pressure in a cylinder in an automobile engine? Would the tire gauge or another gauge based on its operating principle work? We know that the pressure in the cylinder varies with time. If our task was to select a measurement system to determine this time-varying pressure, information about the pressure variations in the cylinder would be necessary. From thermodynamics and the speed range of the engine it may be possible to estimate the magnitude of pressures to be expected, and the rate with which they change. From that we can select an appropriate measurement system. But to do this, we need to develop a way to express this idea of the magnitude and rate of change of a variable.

Many measurement systems exhibit similar responses under a variety of conditions, which suggests that the performance and capabilities of measuring systems may be

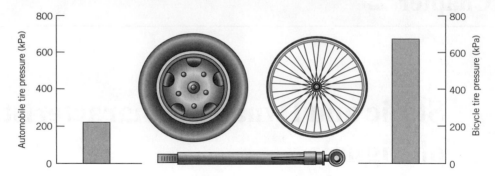

Figure 2.1 Measurement system selection based on input signal range.

described in a generalized way. To examine further the generalized behavior of measurement systems, we first examine the possible forms of the input and output signals. We will associate the term "signal" with the "transmission of information." A *signal* is the physical information about a measured variable being transmitted between a process and the measurement system, between the stages of a measurement system, or the output from a measurement system.

Classification of Waveforms

Signals may be classified as either analog, discrete time, or digital. *Analog* describes a signal that is continuous in time. Because physical variables tend to be continuous in nature, an analog signal provides a ready representation of their time-dependent behavior. In addition, the magnitude of the signal is continuous and thus can have any value within the operating range. An analog signal is shown in Figure 2.2(*a*); a similar continuous signal would result from a recording of the pointer rotation with time for the output display shown in Figure 2.2(*b*). Contrast this continuous signal with the signal shown in Figure 2.3(*a*). This format represents a *discrete time signal*, for which information about the magnitude of the signal is available only at discrete points in time. A discrete time signal usually results from the sampling of a continuous variable at finite time intervals.

Because information in the signal shown in Figure 2.3(*a*) is available only at discrete times, some assumption must be made about the behavior of the measured variable during

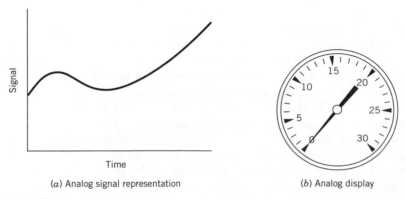

(*a*) Analog signal representation   (*b*) Analog display

Figure 2.2 Analog signal concepts.

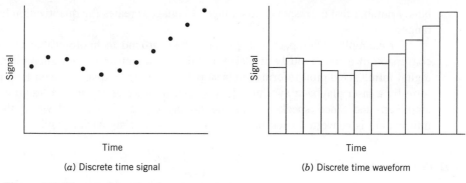

(a) Discrete time signal   (b) Discrete time waveform

Figure 2.3 Discrete time signal concepts.

the times when it is not available. One approach is to assume the signal is constant between samples. The waveform that results from this assumption is shown in Figure 2.3(*b*). Clearly, as the time between samples is reduced the difference between the discrete variable and the continuous signal it represents decreases.

Digital signals are particularly useful when data acquisition and processing are performed by using a digital computer. A digital signal has two important characteristics. First, a digital signal exists at discrete values in time, like a discrete time signal. Second, the magnitude of a digital signal is discrete, determined by a process known as quantization at each discrete point in time. Quantization assigns a single number to represent a range of magnitudes of a continuous signal.

Figure 2.4(*a*) shows digital and analog forms of the same signal where the magnitude of the digital signal can have only certain discrete values. Thus, a digital signal provides a quantized magnitude at discrete times. The waveform that would result from assuming that the signal is constant between sampled points in time is shown in Figure 2.4(*b*).

As an example of quantization, consider a digital watch that displays time in hours and minutes. For the entire duration of 1 min, a single numerical value is displayed until it is updated at the next discrete time step. As such, the continuous physical variable of time is quantized in its conversion to a digital display.

Sampling of an analog signal to produce a digital signal can be accomplished by using an analog-to-digital (A/D) converter, a solid-state device that converts an analog voltage signal to a binary number system representation. The limited resolution of the

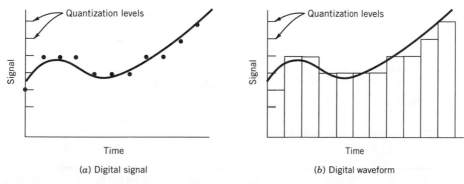

(a) Digital signal   (b) Digital waveform

Figure 2.4 Digital signal representation and waveform.

binary number that corresponds to a range of voltages creates the quantization levels and ranges.

For example, a compact disk player is built around technology that relies on the conversion of a continuously available signal, such as music from a microphone, into a digital form. The digital information is stored on a compact disk and later read in digital form by a laser playback system [1]. However, to serve as input to a traditional stereo amplifier, and since speakers and the human ear are analog devices, the digital information is converted back into a continuous voltage signal for playback.

Signal Waveforms

In addition to classifying signals as analog, discrete time, or digital, some description of the waveform associated with a signal is useful. Signals may be characterized as either static or dynamic. A *static signal* does not vary with time. The diameter of a shaft is an example. Many physical variables change slowly enough in time, compared to the process with which they interact, that for all practical purposes these signals may be considered static in time. For example, the voltage across the terminals of a battery is approximately constant over its useful life. Or consider measuring temperature by using an outdoor thermometer; since the outdoor temperature does not change significantly in a matter of minutes, this input signal might be considered static when compared to our time period of interest. A mathematical representation of a static signal is given by a constant, as indicated in Table 2.1. In contrast, often we are interested in how the measured variable changes with time. This leads us to consider time-varying signals further.

A *dynamic signal* is defined as a time-dependent signal. In general, dynamic signal waveforms, $y(t)$, may be classified as shown in Table 2.1. A *deterministic signal* varies in time in a predictable manner, such as a sine wave, a step function, or a ramp function, as shown in Figure 2.5. A signal is *steady periodic* if the variation of the magnitude of the signal repeats at regular intervals in time. Examples of steady periodic behaviors would

Table 2.1   Classification of Waveforms

I. Static	$y(t) = A_0$
II. Dynamic	
Periodic waveforms	
Simple periodic waveform	
	$y(t) = A_0 + C\sin(\omega t + \phi)$
Complex periodic waveform	
	$y(t) = A_0 + \sum_{n=1}^{\infty} C_n \sin(n\omega t + \phi_n)$
Aperiodic waveforms	
Step[a]	
	$y(t) = A_0 U(t)$
	$\quad = A_0 \quad \text{for } t > 0$
Ramp	
	$y(t) = Kt \quad \text{for } 0 < t < t_f$
Pulse[b]	
	$y(t) = A_0 U(t) - A_0 U(t - t_1)$
III. Nondeterminisitic waveform	
	$y(t) \approx A_0 + \sum_{n=1}^{\infty} C_n \sin(\omega_n t + \phi_n)$

[a]$U(t)$ represents the unit step function, which is zero for $t < 0$ and for $t \geq 0$.
[b]t_1 represents the pulse width.

Deterministic variables

Nondeterministic variable

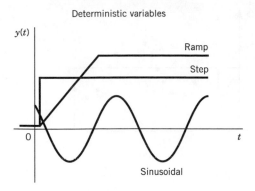

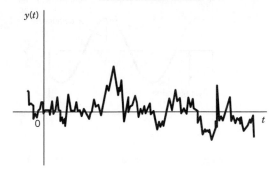

Figure 2.5 Examples of dynamic signals.

include the motion of an ideal pendulum, or the temperature variations in the cylinder of an internal combustion engine, under steady operating conditions. Periodic waveforms may be classified as simple or complex. A *simple periodic waveform* contains only one frequency. A *complex periodic waveform* contains multiple frequencies and is represented as a superposition of multiple simple periodic waveforms. *Aperiodic* is used to describe deterministic signals that do not repeat at regular intervals, such as a step function.

Also described in Figure 2.5 is a *nondeterministic signal* that has no discernible pattern of repetition. A nondeterministic signal cannot be prescribed before it occurs, although certain characteristics of the signal may be known in advance. As an example, consider the transmission of data files from one computer to another. Signal characteristics such as the rate of data transmission and the possible range of signal magnitude are known for any signal in this system. However, it would not be possible to predict future signal characteristics based on existing information in such a signal. Such a signal is properly characterized as nondeterministic. Nondeterministic signals are generally described by their statistical characteristics or a model signal that represents the statistics of the actual signal.

2.3   SIGNAL ANALYSIS

In this section, we consider concepts related to the characterization of signals. A measurement system produces a signal that may be analog, discrete time, or digital. An analog signal is continuous with time and has a magnitude that is analogous to the magnitude of the physical variable being measured. Consider the analog signal shown in Figure 2.6(a), which is continuous over the recorded time period from t_1 to t_2. The average or mean value[1] of this signal is found by

$$\bar{y} \equiv \frac{\int_{t_1}^{t_2} y(t)dt}{\int_{t_1}^{t_2} dt} \tag{2.1}$$

The mean value, as defined in equation (2.1), provides a measure of the static portion of a signal over the time $t_2 - t_1$. It is sometimes called the *dc component* or dc offset of the signal.

---

[1]Strictly speaking, for a continuous signal the mean value and the average value are the same. This is not true for discrete time signals.

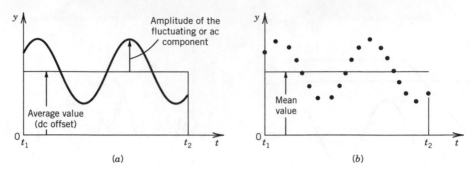

Figure 2.6 Analog and discrete representations of a dynamic signal.

The mean value does not provide any indication of the amount of variation in the dynamic portion of the signal. The characterization of the dynamic portion, or *ac component*, of the signal may be illustrated by considering the average power dissipated in an electrical resistor through which a fluctuating current flows. The power dissipated in a resistor due to the flow of a current is

$$P = I^2 R$$

where

P = power dissipated

I = current

R = resistance

If the current varies in time, the total electrical energy dissipated in the resistor over the time t_1 to t_2 would be

$$\int_{t_1}^{t_2} P dt = \int_{t_1}^{t_2} [I(t)]^2 R dt \tag{2.2}$$

The current $I(t)$ would, in general, include both a dc component and a changing ac component. Consider finding the magnitude of a constant effective current, I_e, that would produce the same total energy dissipation in the resistor as the time-varying current, $I(t)$, over the time period t_1 to t_2. Assuming that the resistance, R, is constant, this current would be determined by equating $(I_e)^2 R(t_2 - t_1)$ with equation (2.2) to yield

$$I_e = \sqrt{\frac{1}{t_2 - t_1} \int_{t_1}^{t_2} [I(t)]^2 dt} \tag{2.3}$$

This value of the current is called the *root-mean-square value* or *rms value* of the current; clearly it is the average value of the square of the current over the measured time period. Based on this reasoning, the rms value of any continuous analog variable $y(t)$ over the time, $t_2 - t_1$, is expressed as

$$y_{\text{rms}} = \sqrt{\frac{1}{t_2 - t_1} \int_{t_1}^{t_2} y^2 dt} \tag{2.4}$$

A time-dependent analog signal, $y(t)$, can be represented by a discrete set of N numbers over the time period from t_1 to t_2 through the conversion

$$y(t) \rightarrow \{y(r\delta t)\} \quad r = 1, \ldots, N$$

which uses the sampling convolution

$$\{y(r\delta t)\} = y(t)\delta(t - r\delta t) = \{y_i\} \quad i = 1, 2, \ldots, N$$

Here, $\delta(t - r\delta t)$ is the delayed unit impulse function, δt is the sample time increment between each number, and $N\delta t = t_2 - t_1$ gives the total sample period over which the measurement of $y(t)$ takes place. The effect of discrete sampling on the original analog signal is demonstrated in Figure 2.6(b) in which the analog signal has been replaced by $\{y(r\delta t)\}$, which represents N values of a discrete time signal for $y(t)$.

For either a discrete time signal or a digital signal, the mean value can be estimated by the discrete equivalent of equation (2.1) as

$$\bar{y} = \frac{1}{N}\sum_{i=1}^{N} y_i \tag{2.5}$$

where each y_i is a discrete number in the data set of $\{y(r\delta t)\}$. The mean approximates the static component of the signal over the time interval t_1 to t_2. The rms value can be estimated by the discrete equivalent of equation (2.4) as

$$y_{\text{rms}} = \sqrt{\frac{1}{N}\sum_{i=1}^{N} y_i^2} \tag{2.6}$$

The rms value takes on additional physical significance when either the signal contains no dc component or the dc component has been subtracted from the signal. The rms value of a signal having a zero mean is a statistical measure of the magnitude of the fluctuations in the signal, relative to the mean value.

Effects of Signal-Averaging Period

The choice of a time period, t_f, where $t_f = t_2 - t_1$, for signal analysis depends on the intended purpose of the analysis. For simple periodic functions, averaging the signal over a time exactly equal to the period of the function results in a true representation of the long-term average value of the signal. This is also true for the rms value. Averaging a simple periodic signal over a time period that is not exactly the period of the function can produce misleading results. However, as the averaging time period becomes long relative to the signal period, the resulting values will accurately represent the signal. Complex waveforms and nondeterministic signals have a range of frequencies present; as such, no single averaging period will produce an exact representation of the signal. In such cases, the signal should be analyzed over a period of time that is long compared to the longest period contained within the signal waveform. This can be tested by comparing the resulting mean and rms values determined over increasingly longer periods.

Consider the signal corresponding to the time-dependent fuel consumption of an automobile shown in Figure 2.7. Certainly, the mean value determined for the t_f used in Figure 2.7(a) will differ from that used in Figure 2.7(b). However, a mean value over the smaller time interval of Figure 2.7(b) might yield information on performance under very specific conditions relevant to that portion of the signal, such as during the climbing of a hill. There are no rules to assist in the selection of t_f beyond the development of a measurement plan based on a careful evaluation of the measurement objective and the intended use of the results.

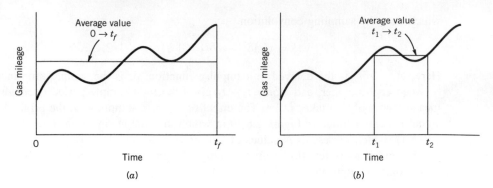

Figure 2.7 Effect of time period on mean value for a nondeterministic signal.

DC Offset

When the ac component of the signal is of primary interest, the dc component can be removed. This procedure often allows a change of scale in the display or plotting of a signal such that fluctuations that were a small percentage of the dc signal can be more clearly observed without the superposition of the large dc component. The enhancement of fluctuations through the subtraction of the average value is illustrated in Figure 2.8.

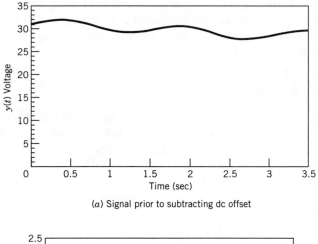

(a) Signal prior to subtracting dc offset

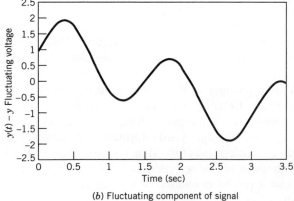

(b) Fluctuating component of signal

Figure 2.8 Effect of subtracting dc offset for a dynamic signal.

Figure 2.8(*a*) contains the entire complex waveform consisting of both static and dynamic parts. When the dc component is removed, the characteristics of the fluctuating component of the signal are readily observed, as shown in Figure 2.8(*b*). The statistics of the fluctuating component of the signal dynamic portion may contain important information for a particular application.

EXAMPLE 2.1

Suppose the current passing through a resistor can be described by

$$I(t) = 10 \sin t$$

where I represents the time-dependent current in amperes. Establish the mean and rms values of current over a time from 0 to t_f, with $t_f = 2\pi$ and then with $t_f = 2\pi$. How do the results relate to the power dissipated in the resistor?

KNOWN $I(t) = 10 \sin t$ A

FIND $\bar{I}$ and I_{rms} with $t_f = \pi$ and 2π

SOLUTION   The average value for a time from 0 to t_f is found from equation (2.1) as

$$\bar{I} = \frac{\int_0^{t_f} I(t)dt}{\int_0^{t_f} dt} = \frac{\int_0^{t_f} 10 \sin t dt}{t_f}$$

Evaluation of this integral yields

$$\bar{I} = \frac{1}{t_f}[-10 \cos t]_0^{t_f}$$

With $t_f = \pi$, the average value, $\bar{I}$, is $20/\pi$ A. For the value of $t_f = 2\pi$, the evaluation of the integral yields an average value of zero.

The rms value for the time period 0 to t_f is given by the application of equation (2.4), which yields

$$I_{\mathrm{rms}} = \sqrt{\frac{1}{t_f} \int_0^{t_f} I(t)^2 dt} = \sqrt{\frac{1}{t_f} \int_0^{t_f} (10 \sin t)^2 dt}$$

This integral is evaluated as

$$I_{\mathrm{rms}} = \sqrt{\frac{100}{t_f} \left(-\frac{1}{2} \cos t \sin t + \frac{t}{2} \right) \Big|_0^{t_f}}$$

For $t_f = \pi$, the rms value is $\sqrt{50}$ A. Evaluation of the integral for the rms value with $t_f = 2\pi$ also yields $\sqrt{50}$ A.

COMMENT   Although the average value over the period 2π is zero, the power dissipated in the resistor must be the same over both the positive and negative half-cycles of the sine function. Thus, the rms value of current is the same for the time period of π and 2π and is indicative of the power dissipated.

2.4   SIGNAL AMPLITUDE AND FREQUENCY

A key factor in measurement system behavior is the nature of the input signal to the system. A means is needed to classify waveforms for both the input signal and the resulting output signal relative to their magnitude and frequency. It would be very helpful if the behavior of measurement systems could be defined in terms of their response to a

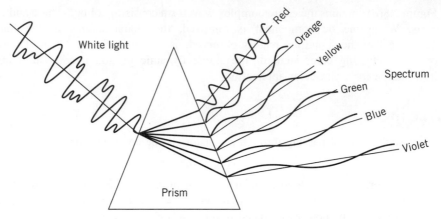

Figure 2.9 Separation of white light into its color spectrum. Color corresponds to a particular frequency or wavelength; light intensity corresponds to varying amplitudes.

limited number and type of input signals. This is, in fact, exactly the case: A very complex signal, even one that is nondeterministic in nature, can be approximated as an infinite series of sine and cosine functions, as suggested in Table 2.1. The method of expressing such a complex signal as a series of sines and cosines is called *Fourier analysis*.

Nature provides some experiences that support our contention that complex signals can be represented by the addition of a number of simpler periodic functions. For example, combining a number of different pure tones can generate rich musical tones. And an excellent physical analogy for Fourier analysis is provided by the separation of white light through a prism. Figure 2.9 illustrates the transformation of a complex waveform, represented by white light, into its simpler components, represented by the colors in the spectrum. In this example, the colors in the spectrum are represented as simple periodic functions that combine to form white light. Fourier analysis is roughly the mathematical equivalent of a prism and yields a representation of a complex signal in terms of simple periodic functions.

The representation of complex and nondeterministic waveforms by simple periodic functions allows measurement system response to be reasonably well defined by examining the output resulting from a few specific input waveforms, one of which is a simple periodic. As represented in Table 2.1, a simple periodic waveform has a single, well-defined amplitude and a single frequency. Before a generalized response of measurement systems can be determined, an understanding of the method of representing complex signals in terms of simpler functions is necessary.

Periodic Signals

The fundamental concepts of frequency and amplitude can be understood through the observation and analysis of periodic motions. Although sines and cosines are by definition geometric quantities related to the lengths of the sides of a right triangle, for our purposes sines and cosines are best thought of as mathematical functions that describe specific physical behaviors of systems. These behaviors are described by differential equations that have sines and cosines as their solutions. As an example, consider a mechanical vibration of a mass attached to a linear spring, as shown in Figure 2.10. For a linear spring, the spring force, F, and displacement, y, are related by $F = ky$, where k is

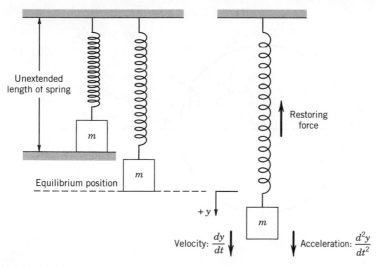

Figure 2.10 Spring-mass system.

the constant of proportionality, called the spring constant. Application of Newton's second law to this system yields a governing equation for the displacement, y, as

$$m\frac{d^2y}{dt^2} + ky = 0 \tag{2.7}$$

This linear, second-order differential equation with constant coefficients describes the motion of the idealized spring mass system when there is no external force applied. The general form of the solution to this equation is

$$y = A\cos\omega t + B\sin\omega t \tag{2.8}$$

where $\omega = \sqrt{k/m}$. Physically we know that if the mass is displaced from the equilibrium point and released, it will oscillate about the equilibrium point. The time required for the mass to finish one complete cycle of the motion is called the *period*, and is generally represented by the symbol T.

   Frequency is related to the period and is defined as the number of complete cycles of the motion per unit time. This frequency, f, is measured in cycles per second (Hz; 1 cycle/s = 1 Hz). The term ω is also a frequency, but instead of having units of cycles per second it has units of radians per second. This frequency, ω, is called the *circular frequency* since it relates directly to cycles on the unit circle, as illustrated in Figure 2.11. The relationship between ω, f, and the period, T, is

$$T = \frac{2\pi}{\omega} = \frac{1}{f} \tag{2.9}$$

In equation 2.8, the sine and cosine terms can be combined if a phase angle is introduced such that

$$y = C\cos(\omega t - \phi) \tag{2.10a}$$

or

$$y = C\sin(\omega t - \phi^*) \tag{2.10b}$$

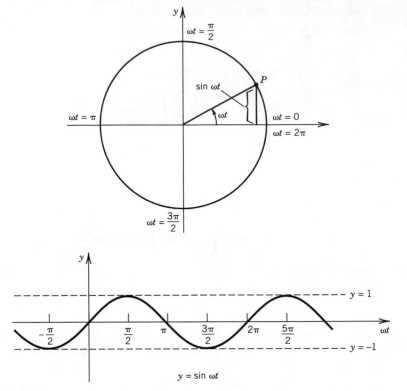

Figure 2.11 Relationship between cycles on the unit circle and circular frequency.

The values of C, ϕ, and ϕ^* are found from the following trigonometric identities:

$$A \cos \omega t + B \sin \omega t = \sqrt{A^2 + B^2} \cos(\omega t - \phi)$$

$$A \cos \omega t + B \sin \omega t = \sqrt{A^2 + B^2} \sin(\omega t + \phi^*) \qquad (2.11)$$

$$\phi = \tan^{-1} \frac{B}{A} \qquad \phi^* = \tan^{-1} \frac{A}{B} \qquad \phi^* = \frac{\pi}{2} - \phi$$

The size of the maximum and minimum displacements from the equilibrium position, or the value C, is the amplitude of the oscillation. The concepts of amplitude and frequency are essential for the description of time-dependent signals.

Frequency Analysis

Many signals that result from the measurement of dynamic variables are nondeterministic in nature and have a continuously varying rate of change. These signals, having *complex waveforms*, present difficulties in the selection of a measurement system and in the interpretation of an output signal. However, it is possible to separate a complex signal, or any signal for that matter, into a number of sine and cosine functions. *In other words, any complex signal can be thought of as made up of sines and cosines of differing periods and amplitudes, which are added together in an infinite trigonometric series.* This representation of a signal as a series of sines and cosines is called a *Fourier series*. Once a signal is broken down into a series of periodic functions, the importance of each frequency can be easily determined. This information about frequency content allows proper choice of a measurement system, and precise interpretation of output signals.

In theory, Fourier analysis allows essentially all mathematical functions of practical interest to be represented by an infinite series of sines and cosines.[2]

The following definitions relate to Fourier analysis:

1. A function $y(t)$ is a *periodic function* if there is some positive number T such that

$$y(t + T) = y(t)$$

The *period* of $y(t)$ is T. If both $y_1(t)$ and $y_2(t)$ have period T, then

$$ay_1(t) + by_2(t)$$

also has a period of T (a and b are constants).

2. A *trigonometric series* is given by

$$A_0 + A_1 \cos t + B_1 \sin t + A_2 \cos 2t + B_2 \sin 2t + \cdots$$

where A_n and B_n are the coefficients of the series.

EXAMPLE 2.2

As a physical (instead of mathematical) example of frequency content of a signal, consider stringed musical instruments, such as guitars and violins. When a string is caused to vibrate by plucking or bowing, the sound is a result of the motion of the string and the resonance of the instrument. (The concept of resonance will be explored in Chapter 3.) The musical pitch for such an instrument is the lowest frequency of the string vibrations. Our ability to recognize differences in musical instruments is primarily a result of the higher frequencies present in the sound, which are usually integer multiples of the fundamental frequency. These higher frequencies are called *harmonics*.

The motion of a vibrating string is best thought of as composed of several basic motions that together create a musical tone. Figure 2.12 illustrates the vibration modes associated with a string plucked at its center. The string vibrates with a fundamental frequency and odd-numbered harmonics, each having a specific phase relationship with the fundamental. The relative strength of each harmonic is graphically illustrated through its amplitude in Figure 2.12. Figure 2.13 shows the motion, which is caused by plucking a string one-fifth of the distance from a fixed end. The resulting frequencies are illustrated in Figure 2.14. Notice that the fifth harmonic is missing from the resulting sound.

Musical sound from a vibrating string is analogous to a measurement system input or output, which contains many frequency components. Fourier analysis and frequency spectra provide insightful and practical means of reducing such complex signals into a combination of simple waveforms. Next, we explore the frequency and amplitude analysis of complex signals.

Program Sound.vi uses your computer's microphone and sound board to sample ambient sounds and to decompose them into harmonics (try humming a tune).

---

[2]A periodic function may be represented as a Fourier series if the function is piecewise continuous over the limits of integration and the function has a left- and right-hand derivative at each point in the interval. The sum of the resulting series is equal to the function at each point in the interval except points where the function is discontinuous.

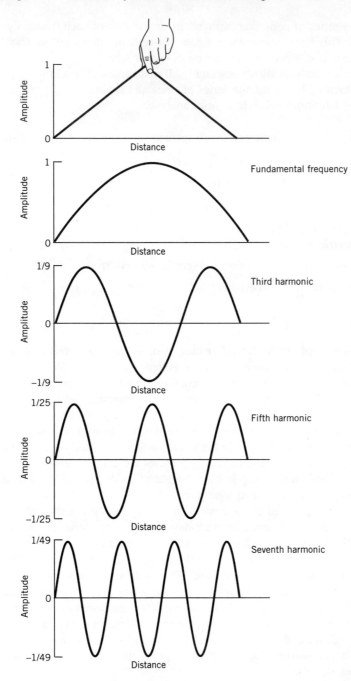

Figure 2.12 Modes of vibration for a string plucked at its center.

Fourier Series and Coefficients

A periodic function $y(t)$ with a period $T = 2\pi$ is to be represented by a trigonometric series, such that for any t,

$$y(t) = A_0 + \sum_{n=1}^{\infty} (A_n \cos nt + B_n \sin nt) \qquad (2.12)$$

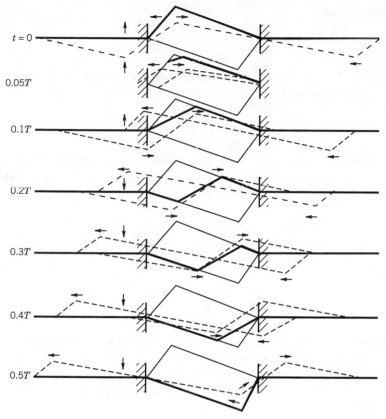

Figure 2.13 Motion of a string plucked one-fifth of the distance from a fixed end. [From N. H. Fletcher and T. D. Rossing, *The Physics of Musical Instruments*. Copyright © 1991 by Springer-Verlag, New York. Reprinted by permission.]

With $y(t)$ known, the coefficients A_n and B_n are to be determined. For A_0 to be determined, equation (2.12) is integrated from $-\pi$ to π:

$$\int_{-\pi}^{\pi} y(t)dt = A_0 \int_{-\pi}^{\pi} dt + \sum_{n=1}^{\infty} \left(A_n \int_{-\pi}^{\pi} \cos ntdt + B_n \int_{-\pi}^{\pi} \sin ntdt \right) \qquad (2.13)$$

Since

$$\int_{-\pi}^{\pi} \cos ntdt = 0 \qquad \int_{-\pi}^{\pi} \sin ntdt = 0$$

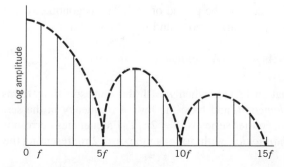

Figure 2.14 Frequency spectrum for a string plucked one-fifth of the distance from a fixed end. [From N. H. Fletcher and T. D. Rossing, *The Physics of Musical Instruments*. Copyright © 1991 by Springer-Verlag, New York. Used by permission.]

equation (2.13) yields

$$A_0 = \frac{1}{2\pi} \int_{-\pi}^{\pi} y(t)dt \tag{2.14}$$

The coefficient A_m may be determined by multiplying equation (2.12) by cos mt and integrating from $-\pi$ to π. The resulting expression for A_m is

$$A_m = \frac{1}{\pi} \int_{-\pi}^{\pi} y(t)\cos mtdt \tag{2.15}$$

Similarly, multiplying equation (2.12) by sin mt and integrating from $-\pi$ to π yields B_m. Thus, for a function $y(t)$ with a period 2π, the coefficients of the trigonometric series representing $y(t)$ are given by the Euler formulas:

$$A_0 = \frac{1}{2\pi} \int_{-\pi}^{\pi} y(t)$$

$$A_n = \frac{1}{\pi} \int_{-\pi}^{\pi} y(t)\cos ntdt \tag{2.16}$$

$$B_n = \frac{1}{\pi} \int_{-\pi}^{\pi} y(t)\sin ntdt$$

The trigonometric series corresponding to $y(t)$ is called the Fourier series for $y(t)$, and the coefficients A_n and B_n are called the Fourier coefficients of $y(t)$. Functions represented by a Fourier series normally do not have a period of 2π. However, the transition to an arbitrary period can be affected by a change of scale, yielding a new set of Euler formulas described below.

Fourier Coefficients for Functions Having Arbitrary Periods

The coefficients of a trigonometric series representing a function of frequency ω are given by the Euler formulas:

$$A_0 = \frac{1}{T} \int_{-T/2}^{T/2} y(t)dt$$

$$A_n = \frac{2}{T} \int_{-T/2}^{T/2} y(t)\cos n\omega tdt \tag{2.17}$$

$$B_n = \frac{2}{T} \int_{-T/2}^{T/2} y(t)\sin n\omega tdt$$

where $n = 1, 2, 3, \ldots$, and $T = 2\pi/\omega$ is the period of $y(t)$. The trigonometric series that results from these coefficients is a Fourier series and may be written as

$$y(t) = A_0 + \sum_{n=1}^{\infty} (A_n \cos n\omega t + B_n \sin n\omega t) \tag{2.18}$$

In the series for $y(t)$ in equation (2.18), when $n = 1$ the corresponding terms in the Fourier series are called *fundamental*, and have the lowest frequency in the series. The *fundamental frequency* for this Fourier series is $\omega = 2\pi/T$. Frequencies corresponding to $n = 2, 3, 4, \ldots$ are known as *harmonics*, with, for example, $n = 2$ representing the second harmonic.

A series of sines and cosines may be written as a series of either sines or cosines through the introduction of a phase angle, so that the Fourier series in equation (2.18),

$$y(t) = A_0 + \sum_{n=1}^{\infty} (A_n \cos n\omega t + B_n \sin n\omega t)$$

may be written as

$$y(t) = A_0 + \sum_{n=1}^{\infty} C_n \cos (n\omega t - \phi_n) \tag{2.19}$$

or

$$y(t) = A_0 + \sum_{n=1}^{\infty} C_n \cos (n\omega t + \phi_n^*) \tag{2.20}$$

where

$$C_n = \sqrt{A_n^2 + B_n^2}$$
$$\tan \phi_n = \frac{B_n}{A_n} \quad \tan \phi_n^* = \frac{A_n}{B_n} \tag{2.21}$$

Even and Odd Functions

A function $g(t)$ is even if it is symmetric about the origin, which may be stated, for all t,

$$g(-t) = g(t)$$

A function $h(t)$ is odd if, for all t,

$$h(-t) = -h(t)$$

For example, $\cos nt$ is even, while $\sin nt$ is odd. A particular function or waveform may be even, odd, or neither even nor odd.

Fourier Cosine Series

If $y(t)$ is even, its Fourier series will contain only cosine terms:

$$y(t) = \sum_{n=1}^{\infty} A_n \cos \frac{2\pi nt}{T} = \sum_{n=1}^{\infty} A_n \cos n\omega t \tag{2.22}$$

Fourier Sine Series

If $y(t)$ is odd, its Fourier series will contain only sine terms

$$y(t) = \sum_{n=1}^{\infty} B_n \sin \frac{2\pi nt}{T} = \sum_{n=1}^{\infty} B_n \sin n\omega t \tag{2.23}$$

Functions that are neither even nor odd result in Fourier series that contain both sine and cosine terms.

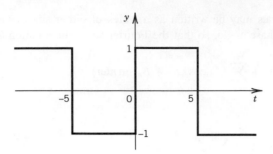

Figure 2.15 Function represented by a Fourier series in Example 2.3.2.

EXAMPLE 2.3

Determine the Fourier series that represents the function shown in Figure 2.15.

KNOWN $T = 10\,(-5\,\text{to}\,+5)$
$A_0 = 0$

FIND   The Fourier coefficients $A_1, A_2, \ldots$ and $B_1 B_2, \ldots$

SOLUTION   Since the function shown in Figure 2.15 is odd, the Fourier series will contain only sine terms [see equation (2.23)]:

$$y(t) = \sum_{n=1}^{\infty} B_n \sin \frac{2\pi n t}{T}$$

where

$$B_n = \frac{2}{10}\left[\int_{-5}^{0} (-1)\sin\left(\frac{2\pi n t}{10}\right)dt + \int_{0}^{5} (1)\sin\left(\frac{2\pi n t}{10}\right)dt\right]$$

$$B_n = \frac{2}{10}\left\{ \left[\frac{10}{2n\pi}\cos\left(\frac{2n\pi t}{10}\right)\right]_{-5}^{0} + \left[\frac{-10}{2n\pi}\cos\left(\frac{2n\pi t}{10}\right)\right]_{0}^{5}\right\}$$

$$B_n = \frac{2}{10}\left\{\frac{10}{2n\pi}[1 - \cos(-n\pi) - \cos(n\pi) + 1]\right\}$$

The reader can verify that all $A_n = 0$ for this function. For even values of n, B_n is identically zero and for odd values of n,

$$B_n = \frac{4}{n\pi}$$

The resulting Fourier series is then

$$y(t) = \frac{4}{\pi}\sin\frac{2\pi}{10}t + \frac{4}{3\pi}\sin\frac{6\pi}{10}t + \frac{4}{5\pi}\sin\frac{10\pi}{10}t + \cdots$$

Note that the fundamental frequency is $\omega = 2\pi/10$ rad/s and the subsequent terms are the odd-numbered harmonics of ω.

COMMENT   Consider the function given by

$$y(t) = 1 \quad 0 < t < 5$$

We may represent $y(t)$ by a Fourier series if we extend the function beyond the specified range (0–5) either as an even periodic extension or as an odd periodic extension. (Because we are interested only in the Fourier series representing the function over the range

$0 < t < 5$, we can impose any behavior outside of that domain that helps to generate the Fourier coefficients!) Let's choose an odd periodic extension of $y(t)$; the resulting function remains identical to the function shown in Figure 2.15.

EXAMPLE 2.4

Find the Fourier coefficients of the periodic function

$$y(t) = -5 \quad \text{when} \quad -\pi < t < 0$$
$$y(t) = +5 \quad \text{when} \quad 0 < t < \pi$$

and $y(t + 2\pi) = y(t)$. Plot the resulting first four partial sums for the Fourier series.

KNOWN   Function $y(t)$ over the range $-\pi$ to π

FIND   Coefficients A_n and B_n

SOLUTION   The function as stated is periodic, having a period of $T = 2\pi$ (i.e., $\omega = 1$ rad/s), and is identical in form to the odd function examined in Example 2.3. Since this function is also odd, the Fourier series contains only sine terms, and

$$B_n = \frac{1}{\pi} \int_{-\pi}^{\pi} y(t)\sin n\omega t\, dt = \frac{1}{\pi} \left[\int_{-\pi}^{0} (-5)\sin ntdt + \int_{0}^{\pi} (-5)\sin ntdt \right]$$

which yields upon integration

$$\frac{1}{\pi} \left\{ \left[(5)\frac{\cos nt}{n} \right]_{-\pi}^{0} - \left[(5)\frac{\cos nt}{n} \right]_{0}^{\pi} \right\}$$

Thus,

$$B_n = \frac{10}{n\pi}(1 - \cos n\pi)$$

which is zero for even n, and $20/n\pi$ for odd values of n. The Fourier series can then be written[3]

$$y(t) = \frac{20}{\pi} \left(\sin t + \frac{1}{3}\sin 3t + \frac{1}{5}\sin 5t + \cdots \right)$$

Figure 2.16 shows the first four partial sums of this function, as they compare to the function they represent. Note that at the points of discontinuity of the function, the Fourier series takes on the arithmetic mean of the two function values.

COMMENT   As the number of terms in the partial sum increases, the Fourier series approximation of the square wave function becomes even better. For each additional term in the series,

---

[3]If we assume that the sum of this series most accurately represents the function y at $t = \pi/2$, then

$$y\left(\frac{\pi}{2}\right) = 5 = \frac{20}{\pi}\left(1 - \frac{1}{3} + \frac{1}{5} - + \cdots\right)$$

or

$$\frac{\pi}{4} = 1 - \frac{1}{3} + \frac{1}{5} - \frac{1}{7} + - \cdots$$

This series approximation of π was first obtained by Gottfried Wilhelm Leibniz (1646–1716) in 1673 from geometrical reasoning.

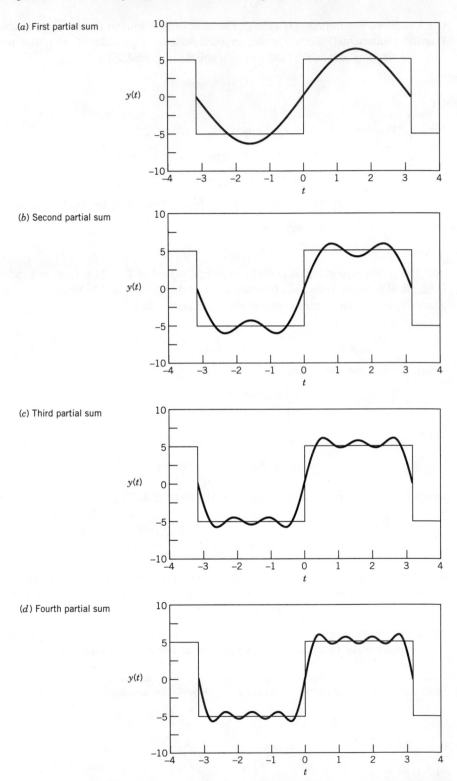

Figure 2.16 First four partial sums of the Fourier series $(20/\pi)(\sin t + 1/3 \sin 3t + 1/5 \sin 5t + \cdots)$ in comparison with the exact waveform.

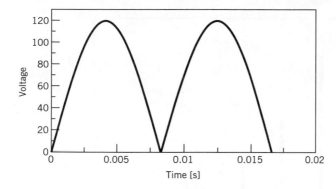

Figure 2.17 Rectified sine wave.

the number of "humps" in the approximate function in each half-cycle corresponds to the number of terms included in the partial sum.

The Matlab® program file *FourCoef* with the companion software illustrates the behavior of the partial sums for several waveforms. The LabView® program Waveform-Generation.vi creates signals from trignometric series.

EXAMPLE 2.5

As an example of interpreting the frequency content of a given signal, consider the output voltage from a rectifier. A rectifier functions to "flip" the negative half of an alternating current (ac) into the positive half plane, resulting in a signal that appears as shown in Figure 2.17. For the ac signal the voltage is given by

$$E(t) = 120 \sin 120\,\pi t$$

The period of the signal is 1/60 s, and the frequency is 60 Hz.

KNOWN The rectified signal can be expressed as

$$E(t) = |120 \sin 120\,\pi t|$$

FIND The frequency content of this signal as determined from a Fourier series analysis.

SOLUTION The frequency content of this signal can be determined by expanding the function in a Fourier series. The coefficients may be determined using the Euler formulas, keeping in mind that the rectified sine wave is an even function. The coefficient A_0 is determined from

$$A_0 = 2\left[\frac{1}{T}\int_0^{T/2} y(t)dt\right] = \frac{2}{1/60}\int_0^{1/120} 120 \sin 120\pi t dt$$

with the result that

$$A_0 = \frac{2 \times 60 \times 2}{\pi} = 76.4$$

The remaining coefficients in the Fourier series may be expressed as

$$A_n = \frac{4}{T}\int_0^{T/2} y(t)\cos\frac{2n\pi t}{T}dt$$

$$= \frac{4}{1/60}\int_0^{1/120} 120 \sin 120\pi t \cos n\pi t dt$$

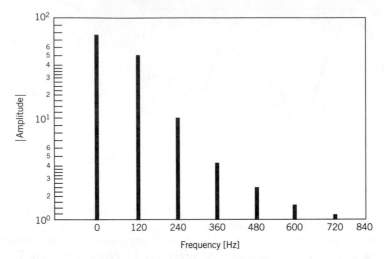

Figure 2.18 Frequency content of the function $y(t) = |120 \sin 120\pi t|$ displayed as an amplitude-frequency spectrum.

For values of n that are odd, the coefficient A_n is identically zero. For values of n that are even, the result is

$$A_n = \frac{120}{\pi}\left(\frac{-2}{n-1} + \frac{2}{n+1}\right) \tag{2.24}$$

The Fourier series for the function $|120 \sin 120\pi t|$ is

$$76.4 - 50.93 \cos 240\pi t - 10.10 \cos 480\pi t - 4.37 \cos 720\pi t \ldots$$

Figure 2.18 shows the amplitude versus frequency content of the rectified signal, based on a Fourier series expansion.

COMMENT The frequency content of a signal is determined by examining the amplitude of the various frequency components that are present in the signal. For a periodic mathematical function, expanding the function in a Fourier series and plotting the amplitudes of the contributing sine and cosine terms can illustrate these frequency contributions.

The LabView® program WaveformGeneration.vi follows Examples 2.4 and 2.5 and provides the spectra for a number of signals. Several Matlab® programs are provided also (e.g., *FourCoef, FunSpect, DataSpect*) to explore the concept of superposition of simple periodic signals to create a more complex signal. The software allows you to create your own functions and then explore their frequency and amplitude content.

2.5 FOURIER TRANSFORM AND THE FREQUENCY SPECTRUM

The previous discussion of Fourier analysis demonstrates that an arbitrary, but known, function can be expressed as a series of sines and cosines known as a Fourier series. The coefficients of the Fourier series specify the amplitudes of the sines and cosines, each having a specific frequency. Unfortunately, in most practical measurement applications the input signal may not be known in functional form. Therefore, although the theory of Fourier analysis demonstrates that any function can be expressed as a Fourier series, the analysis presented so far has not provided a specific technique for analyzing measured signals. Such a technique for the decomposition of a measured dynamic signal in terms of amplitude and frequency is now described.

Recall that the dynamic portion of a signal of arbitrary period can be described from equation (2.17).

$$A_n = \frac{2}{T} \int_{-T/2}^{T/2} y(t) \cos n\omega t dt$$

$$B_n = \frac{2}{T} \int_{-T/2}^{T/2} y(t) \sin n\omega t dt$$

(2.25)

where the amplitudes A_n and B_n correspond to the nth frequency of a Fourier series.

If we consider the period of the function to approach infinity, we can eliminate the constraint on Fourier analysis that the signal be a periodic waveform. In the limit as T approaches infinity the Fourier series becomes an integral. The spacing between frequency components becomes infinitesimal. This means that the coefficients A_n and B_n become continuous functions of frequency and can be expressed as $A(\omega)$ and $B(\omega)$ where

$$A(\omega) = \int_{-\infty}^{\infty} y(t) \cos \omega t dt$$

$$B(\omega) = \int_{-\infty}^{\infty} y(t) \sin \omega t dt$$

(2.26)

The Fourier coefficients $A(\omega)$ and $B(\omega)$ are known as the components of the Fourier transform of $y(t)$. To develop the Fourier transform, consider the complex number defined as

$$Y(\omega) \equiv A(\omega) - iB(\omega)$$

(2.27)

where $i = \sqrt{-1}$. Then from equations (2.26) it follows directly that

$$Y(\omega) \equiv \int_{-\infty}^{\infty} y(t)(\cos \omega t - i \sin \omega t) dt$$

(2.28)

Introducing the identity

$$e^{-i\theta} = \cos \theta - i \sin \theta$$

leads to

$$Y(\omega) \equiv \int_{-\infty}^{\infty} y(t) e^{i\omega t} dt$$

(2.29)

Recalling from equation (2.9) that the cyclical frequency, f, in hertz, is related to the circular frequency and its period by

$$f = \frac{\omega}{2\pi} = \frac{1}{T}$$

equation (2.29) is rewritten as

$$Y(f) \equiv \int_{-\infty}^{\infty} y(t) e^{-i2\pi ft} dt$$

(2.30)

Equation (2.29) or (2.30) provides the two-sided *Fourier transform* of $y(t)$. The significance of the Fourier transform, $Y(f)$, is that it describes the signal as a continuous function of frequency. If $y(t)$ is known or measured, then its Fourier transform will provide the amplitude-frequency properties of the signal which otherwise are not readily apparent in its time-based form. We can think of the Fourier transform as a decomposition of $y(t)$ into amplitude versus frequency information. This property is analogous to the optical properties displayed by the prism in Figure 2.9.

If $Y(f)$ is known or measured, we can recover the signal $y(t)$ from

$$y(t) = \frac{1}{2\pi} \int_{-\infty}^{\infty} Y(f)e^{i2\pi ft} df \qquad (2.31)$$

Equation 2.31 describes the *inverse Fourier transform* of $Y(f)$. It suggests that given the amplitude-frequency properties of a signal we can reconstruct the original signal $y(t)$.

The Fourier transform is a complex number having a magnitude and a phase,

$$Y(f) = |Y(f)|e^{i\phi(f)} = A(f) - iB(f) \qquad (2.32)$$

The magnitude of $Y(f)$, also called the modulus, is given by

$$|Y(f)| = \sqrt{\mathrm{Re}[Y(f)]^2 + \mathrm{Im}[Y(f)]^2} \qquad (2.33)$$

and the phase by

$$\phi(f) = \tan^{-1}\frac{\mathrm{Im}[Y(f)]}{\mathrm{Re}[Y(f)]} \qquad (2.34)$$

As noted earlier, the Fourier coefficients are related to cosine and sine terms. Then the amplitude of $y(t)$ can be expressed by its amplitude-frequency spectrum, or simply amplitude spectrum

$$C(f) = \sqrt{A(f)^2 + B(f)^2} \qquad (2.35)$$

and its phase shift by

$$\phi(f) = \tan^{-1}\frac{B(f)}{A(f)} \qquad (2.36)$$

Thus the introduction of the Fourier transform provides a method to decompose a measured signal $y(t)$ into its amplitude-frequency components. Later we will see how important this method is, particularly when digital sampling is used to measure and interpret an analog signal.

A variation of the amplitude spectrum is the power spectrum, which is given by magnitude $C(f)^2/2$. Further details concerning the properties of the Fourier transform and spectrum functions can be found in [2, 3]. An excellent historical account and discussion of the wide-ranging applications are found in [4].

Discrete Fourier Transform

As a practical matter, it is likely that if $y(t)$ is measured and recorded then it will be stored in the form of a discrete time series. A data-acquisition system with digital computer is the most common approach for this. Data are acquired over a finite period of time rather than the mathematically convenient infinite period of time. A discrete data set containing N values representing a time interval from 0 to t_f will accurately represent the signal provided that the measuring period has been properly chosen and is sufficiently long. We will deal with the details for such period selection in a discussion on sampling concepts in Chapter 7. The preceding analysis will now be extended to accommodate a discrete series.

Consider the time-dependent portion of the signal $y(t)$, which is measured N times at equally spaced time intervals δt. In this case, the continuous signal $y(t)$ will be replaced by the discrete time signal given by $y(r\delta t)$ for $r = 1, 2, \ldots, N$. In effect, the relationship between $y(t)$ and $\{y(r\delta t)\}$ is described by a set of impulses of an amplitude determined by the value of $y(t)$ at each time step $r\delta t$. This transformation from a continuous to discrete time signal is described by

$$\{y(r\delta t)\} = y(t)\delta(t - r\delta t) \quad r - 1, 2, \ldots, N \tag{2.37}$$

where $\delta(t - r\delta t)$ is the delayed unit impulse function and $\{y(r\delta t)\}$ refers to the discrete data set given by $y(r\delta t)$ for $r = 1, 2, \ldots, N$.

An approximation to the Fourier transform integral of equation (2.30) for use on a discrete data set is the discrete Fourier transform (DFT). The DFT is given by

$$Y(f_k) = \frac{2}{N} \sum_{r=1}^{N} y(r\delta t) e^{-i2\pi rk/N} \quad k = 1, 2, \ldots \frac{N}{2} \tag{2.38}$$

where the harmonics $f_k = k\delta f$ with $k = 1, 2, \ldots$, N/2 and with $\delta f = 1/N\delta t$. Here, δf is the frequency resolution of the DFT with each $Y(f_k)$ spaced in frequency increments of δf. In developing equation (2.38) from equation (2.30), t was replaced by $r\delta t$ and f replaced by $k/N\delta t$. The factor $2/N$ scales the transform when it is obtained from a data set of finite length.

The DFT as expressed by equation (2.38) yields $N/2$ discrete values of the Fourier transform of $\{y(r\delta t)\}$. This is the so-called one-sided or half-transform as it assumes that the data set is one-sided, extending from 0 to t_f, and it returns only positive valued frequencies.

Equation (2.38) performs the numerical integration required by the Fourier integral. Equations (2.35), (2.36), and (2.40) demonstrate that the application of the DFT on the discrete series of data, $y(r\delta t)$, permits the decomposition of the discrete data in terms of frequency and amplitude content. Hence, by using this method a measured discrete signal of unknown functional form can be reconstructed as a Fourier series through Fourier transform techniques.

Software for computing the Fourier transform of a discrete signal is included in the companion software. The time required to compute directly the DFT algorithm described in this section increases at a rate that is proportional to N^2. This makes it inefficient for use with data sets of large N. A fast algorithm for computing the DFT, known as the *fast Fourier transform* (FFT), was developed by Cooley and Tukey [5]. This method is widely available and is the basis for most Fourier analysis software packages. The FFT algorithm is discussed in most advanced texts on signal analysis (e.g., [6, 7]). The accuracy of discrete Fourier analysis depends on the frequency content[4] of $y(t)$ and on the Fourier transform frequency resolution. An extensive discussion of these interrelated parameters is given in Chapter 7.

EXAMPLE 2.6

Convert the continuous signal described by $y(t) = 10 \sin 2\pi t$ V into a discrete set of eight numbers using a time increment of 0.125 s.

KNOWN   The signal has the form $y(t) = C_1 \sin 2\pi f_1 t$

---

[4]The value of $1/\delta t$ must be more than twice the highest frequency contained in $y(t)$.

where

$$f_1 = \omega_1/2\pi = 1\,\text{Hz}$$
$$C(f_1 = 1\,\text{Hz}) = 10\,\text{V}$$
$$\phi(f_1) = 0$$
$$\delta t = 0.125\,\text{s}$$
$$N = 8$$

Table 2.2   Discrete Data Set for $y(t) = 10\sin 2\pi t$

r	$y(r\,\delta t)$	r	$y(r\,\delta t)$
1	7.071	5	−7.071
2	10.000	6	−10.000
3	7.071	7	−7.071
4	0.000	8	0.000

FIND $\{y(r\delta t)\}$

SOLUTION   The measurement of $y(t)$ every 0.125 s produces the discrete data set $\{y(r\delta t)\}$ given in Table 2.2. Note that the measurement produces four complete periods of the signal and that the signal duration is given by $N\delta t = 1$ s. The signal and its discrete representation as a series of impulses in a time domain are plotted in Figure 2.19. See Example 2.7 for more on how to create this discrete series by using common software.

EXAMPLE 2.7

Estimate the amplitude spectrum of the discrete data set in Example 2.6.

KNOWN   Discrete data set $\{y(r\delta t)\}$ of Table 2.2

$$\delta t = 0.125\,\text{s}$$
$$N = 8$$

FIND $C(f)$

SOLUTION   We can use either the Matlab® Fourier transform software (file *FunSpect*), the Labview® program WaveformGeneration, or a spreadsheet program to find $C(f)$ (see the Comment for details). For $N = 8$, a one-sided Fourier transform algorithm will return $N/2$ values for $C(f)$, with each successive amplitude corresponding to a frequency spaced at intervals of $1/N\delta t = 0.125$ Hz. The amplitude spectrum is shown in Figure 2.20 and has a spike of 10 V centered at a frequency of 1 Hz.

COMMENT   The discrete series and the amplitude spectrum are easily reproduced by using either the Matlab® program file called *FunSpect*, by using the Labview® program Waveform-Generation, or by using spreadsheet software. The Fourier analysis capability of this software can also be used on an existing data set, such as with the Matlab® program file called *DataSpect*.

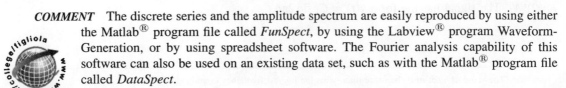

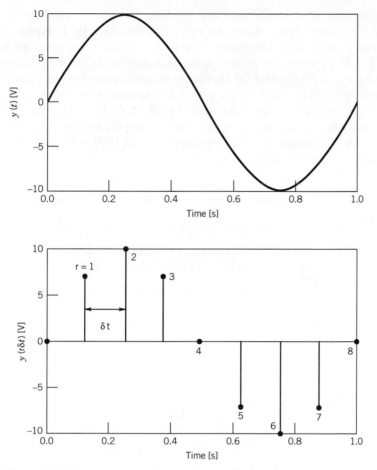

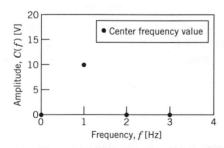

Figure 2.19 Representation of a simple periodic function as a discrete signal.

Figure 2.20 Amplitude as a function of frequency for a discrete representation of $10 \sin 2\pi t$ resulting from a discrete Fourier transform algorithm.

The following discussion of Fourier Analysis using a spreadsheet program makes specific references to procedures and commands from Microsoft® Excel*; similar functions are available in other engineering analysis software. When a spreadsheet is used, the $N = 8$ data point sequence $\{y(r\delta t)\}$ is created as in column 3. Under Tools/Data Analysis select Fourier Analysis. At the prompt, enter the cells containing $\{y(r\delta t)\}$ and define the cell destination. The analysis executes the DFT of equation (2.38) by using the

---

*Microsoft® and Excell are either registered trademarks or trademarks of Microsoft Coporation in the United States and/or other countries.

FFT algorithm, and it returns N complex numbers, the Fourier coefficients $Y(f)$ of equation (2.32). The analysis returns the two-sided transform [as in equation (2.30)], which contains N coefficients. However, a finite data set is one-sided, so we are interested in only the first $N/2$ Fourier coefficients, shown in column 4 (the second $N/2$ coefficients just mirror the first $N/2$). To find the coefficient magnitude as given in equation (2.33), compute or use the function IMABS $\left(= \sqrt{A^2 + B^2}\right)$ on each of the first $N/2$ coefficients, and then scale each coefficient magnitude by dividing by $N/2$. The $N/2$ scaled coefficients (column 5) now represent the discrete amplitudes corresponding to $N/2$ discrete frequencies (column 6) extending from $f = 0$ to $(N/2 - 1)/N\delta t$ Hz, with each frequency separated by $1/N\delta t$.

Spreadsheet

		Column			
1	2	3	4	5	6
r	$t(s)$	$y(r\delta t)$	$Y(f) = A - Bi$	$C(f)$	$f(\text{Hz})$
1	0.125	7.07	0	0	0
2	0.25	10	$28.8 - 28.28i$	10	1
3	0.375	7.07	0	0	2
4	0.5	0	0	0	3
5	0.625	−7.07			
6	0.75	−10			
7	0.875	−7.07			
8	1	0			

These same operations in Matlab[5] are as follows:

$t = 1/8 : 1/8 : 1$	defines time from 0.125 s to 1 s in increments of 0.125 s
$y = 10^* \sin(2^*\text{pi}^*t)$	creates the discrete time series with $N = 8$
$\text{ycoef} = \text{fft}(y)$	performs the Fourier analysis; returns N coefficients
$c = \text{coef}/4$	divides by $N/2$ to scale and determine the magnitudes

When signal frequency content is not known prior to conversion to a discrete signal, it is necessary to experiment with the parameters of frequency resolution, N and δt, to obtain an unambiguous representation of the signal. Techniques for this will be explored in Chapter 7.

2.6 SUMMARY

This chapter has provided a fundamental basis for the description of signals. The capabilities of a measurement system can be properly specified only if the nature of the input signal is known. The descriptions of general classes of input signals will be seen in Chapter 3 to allow universal descriptions of measurement system dynamic behavior.

Any signal can be represented by a static magnitude and a series of varying frequencies and amplitudes. As such, measurement system selection and design must consider the frequency content of the input signals the system is intended to measure. Fourier analysis was introduced to allow a precise definition of the frequencies and the

---

[5]Matlab is a registered trademark of Mathworks, Inc.

phase relationships among various frequency components within a particular signal. In Chapter 7, it will be shown as a tool for the accurate interpretation of discrete signals.

In conclusion, signal characteristics form an important basis for the selection of measurement systems and the interpretation of measurement system output. In Chapter 3 these ideas are combined with the concept of a generalized set of measurement system behaviors. The combination of generalized measurement system behavior and generalized descriptions of input waveforms provides for an understanding of a wide range of instruments and measurement systems.

REFERENCES

1. Monforte, J., The digital reproduction of sound, *Scientific American*, 251(6): 78, 1984.
2. Bracewell, R. N., *The Fourier Transform and Its Applications*, 3d ed., rev., McGraw Hill, New York, 1999.
3. Champeney, D. C., *Fourier Transforms and Their Physical Applications*, Academic, London, 1973.
4. Bracewell, R. N., The Fourier transform, *Scientific American*, 260(6): 86, 1989.
5. Cooley, J. W., and Tukey, J. W., *Math Comput.* 19:207, April 1965 (see also Special Issue on the fast Fourier transform, *IEEE Transactions on Audio and Electroacoustics* AU-2, June 1967).
6. Bendat, J. S., and A. G. Piersol, *Random Data: Analysis and Measurement Procedures*, 3d ed., Wiley, New York, 2000 (see also Bendat, J. S., and A. G. Piersol, *Engineering Applications of Correlation and Spectral Analysis*, 2d ed., Wiley, New York, 1993).
7. Cochran, W. T., et al., What is the fast Fourier transform?, *Proceedings of the IEEE* 55(10): 1664, 1967.

SUGGESTED READING

Halliday, D., and R. Resnick, *Fundamentals of Physics*, 6th ed., Wiley, New York, 2000.
Kreyszig, E., *Advanced Engineering Mathematics*, 8th ed., Wiley, New York, 1998.

NOMENCLATURE

f	frequency, in Hz $[t^{-1}]$	N	total number of discrete data points; integer
k	spring constant $[m - t^2]$	T	period $[t]$
m	mass $[m]$	U	unit step function
t	time $[t]$	Y	Fourier transform of y
y	dependent variable	α	angle [rad]
y_m	discrete data points	β	angle [rad]
A	amplitude	δf	frequency resolution $[t^{-1}]$
B	amplitude	δt	sample time increment $[t]$
C	amplitude	ϕ	phase angle [rad]
F	force $[m - l - t^{-2}]$	ω	circular frequency in rad/s $[t^{-1}]$

PROBLEMS

2.1   Define the term signal as it relates to measurement systems. Provide two examples of static and dynamic input signals to measurement systems.

2.2   List the important characteristics of input and output signals and define each.

2.3   Determine the average and rms values for the function

$$y(t) = 30 + 2\cos 6\pi t$$

over the time periods (a) 0 to 0.1 s (b) 0.4 to 0.5 s (c) 0 to 33 s (d) 0 to 20 s. Comment on the nature and meaning of the results in terms of analysis of dynamic signals.

2.4   The following values are obtained by sampling two time-varying signals once every 0.4 s:

t	$y_1(t)$	$y_2(t)$	t	$y_1(t)$	$y_2(t)$
0	0	0			
0.4	11.76	15.29	2.4	−11.76	−15.29
0.8	19.02	24.73	2.8	−19.02	−24.73
1.2	19.02	24.73	3.2	−19.02	−24.73
1.6	11.76	15.29	3.6	−11.76	−15.29
2.0	0	0	4.0	0	0

Determine the mean and the rms values for this discrete data. Discuss the significance of the rms value in distinguishing these signals.

2.5   A *moving average* is an averaging technique that can be applied to an analog, discrete time, or digital signal. A moving average is based on the concept of windowing, as illustrated in Figure 2.21. That portion of the signal that lies inside the window is averaged and the average values plotted as a function of time as the window moves across the signal. A 10-point moving average of the signal in Figure 2.21 is plotted in Figure 2.22.

a. Discuss the effects of employing a moving average on the signal depicted in Figure 2.21.

b. Develop a computer-based algorithm for computing a moving average, and determine the effect of the width of the averaging window on the signal described by

$$y(t) = \sin 5t + \cos 11t$$

This signal should be represented as a discrete time signal by computing the value of the function at equally spaced time intervals. An appropriate time interval for this signal would be 0.05 seconds. Examine the signal with averaging windows of 4 and 30 points.

2.6   Determine the value of the spring constant that would result in a spring-mass system that would execute one complete cycle of oscillation every 2.7 s, for a mass of 0.5 kg. What natural frequency does this system exhibit in radians/second?

2.7   A spring with $k = 5000$ N/cm supports a mass of 1 kg. Determine the natural frequency of this system in radians/second and Hertz.

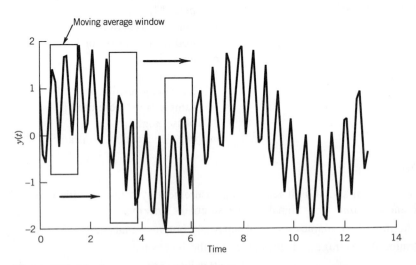

Figure 2.21 Moving average and windowing.

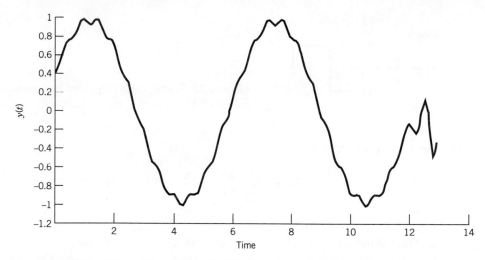

Figure 2.22 Effect of moving average on signal illustrated in Figure 2.21.

2.8 For the following sine and cosine functions determine the period, the frequency in hertz, and the circular frequency in radians/second and (note: t represents time in seconds.)
   a. $\sin 2\pi t/5$
   b. $5\cos 20t$
   c. $\sin 3n\pi t$ for $n = 1$ to ∞

2.9 Express the following function in terms of a cosine term only:

$$y(t) = 5\sin 4t + 3\cos 4t$$

2.10 Express the function

$$y(t) = 4\sin 2\pi t + 15\cos 2\pi t$$

in terms of (a) a cosine term only and (b) a sine term only.

2.11 Express the Fourier series given by

$$y(t) = \sum_{n=1}^{\infty} \frac{2\pi n}{6}\sin n\pi t + \frac{4\pi n}{6}\cos n\pi t$$

using only cosine terms.

2.12 If T is a period of $y(t)$, show that nT for $n = 2, 3, \ldots$ is a period of that function.

2.13 The nth partial sum of a Fourier series is defined as

$$A_0 + A_1\cos\omega_1 t + B_1\sin\omega_1 t + \cdots + A_n\cos\omega_n t + B_n\sin\omega_n t$$

For the third partial sum of the Fourier series given by

$$y(t) = \sum_{n=1}^{\infty} \frac{3n}{2}\sin nt + \frac{5n}{3}\cos nt$$

   a. What is the fundamental frequency and the associated period?
   b. Express this partial sum as cosine terms only.

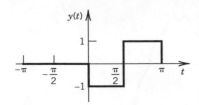

Figure 2.23 Function to be expanded in a Fourier series in Problem 2.15.

2.14 For the Fourier series given by

$$y(t) = 4 + \sum_{n=1}^{\infty} \frac{2n\pi}{10} \cos \frac{n\pi}{4} t + \frac{120n\pi}{30} \sin \frac{n\pi}{4} t$$

where t is time in seconds:
a. What is the fundamental frequency in hertz and radians/second?
b. What is the period T associated with the fundamental frequency?
c. Express this Fourier series as an infinite series containing sine terms only.

2.15 Find the Fourier series of the function shown in Figure 2.23, assuming the function has a period of 2π. Plot an accurate graph of the first three partial sums of the resulting Fourier series.

2.16 Determine the Fourier series for the function

$$y(t) = t \quad \text{for } -5 < t < 5$$

by expanding the function as an odd periodic function with a period of 10 units, as shown in Figure 2.24. Plot the first, second, and third partial sums of this Fourier series.

2.17 a. Show that $y(t) = t^2(-\pi < t < \pi)$, $y(t + 2\pi) = y(t)$ has the Fourier series

$$y(t) = \frac{\pi^2}{3} - 4\left(\cos t - \frac{1}{4}\cos 2t + \frac{1}{9}\cos 3t - + \cdots \right)$$

b. By setting $t = \pi$ in this series, show that a series approximation for π, first discovered by Euler, results as

$$\sum_{n=1}^{\infty} \frac{1}{n^2} = 1 + \frac{1}{4} + \frac{1}{9} + \frac{1}{16} + \cdots = \frac{\pi^2}{6}$$

2.18 Determine the Fourier series which represents the function $y(t)$ where

$$y(t) = t \quad \text{for } 0 < t < 1$$

and

$$y(t) = 2 - t \quad \text{for } 0 < t < 1$$

Clearly explain your choice for extending the function to make it periodic.

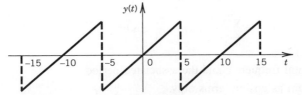

Figure 2.24 Sketch for Problem 2.16.

2.19 Classify the following signals as static or dynamic, and identify any that may be classified as periodic:

a. $\sin 10t$ V   **c.** $5t$ s

b. $5 + 2 \cos 2t$ m   **d.** 2 V

2.20 A particle executes linear harmonic motion around the point $x = 0$. At time zero the particle is at the point $x = 0$ and has a velocity of 5 cm/s. The frequency of the motion is 1 Hz. Determine: (a) the period (b) the amplitude of the motion (c) the displacement as a function of time, and (d) the maximum speed.

2.21 Define the following characteristics of signals: (a) frequency content, (b) amplitude, (c) magnitude, and (d) period.

2.22 Construct an amplitude spectrum plot for the Fourier series in Problem 2.16 for $y(t) = t$. Discuss the significance of this spectrum for measurement or interpretation of this signal. Hint: The plot can be done by inspection or by using software such as *DataSpect*.

2.23 Construct an amplitude spectrum plot for the Fourier series in Problem 2.17 for $y(t) = t^2$. Discuss the significance of this spectrum for selecting a measurement system. Hint: The plot can be done by inspection or by using software such as *DataSpect* or WaveformGeneration.vi.

2.24 Sketch representative waveforms of the following signals, and represent them as mathematical functions (if possible):

a. The output signal from the thermostat on a refrigerator.

b. The electrical signal to a spark plug in a car engine.

c. The input to a cruise control from an automobile.

d. A pure musical tone (e.g., 440 Hz is an A).

e. The note produced by a guitar string.

f. AM and FM radio signals.

2.25 Represent the function

$$e(t) = 5 \sin 31.4t + 2 \sin 44t$$

as a discrete set of $N = 128$ numbers separated by a time increment of $(1/N)$. Use an appropriate algorithm to construct an amplitude spectrum from this data set. (Hint: A spreadsheet program or the program file *DataSpect* will handle this task.)

2.26 Repeat Problem 2.25 using a data set of 256 numbers at $\delta t = (1/N)$ and $\delta t = (1/2N)$ s. Compare and discuss the results.

2.27 A particular strain sensor is mounted to an aircraft wing that is subjected to periodic wind gusts. The strain measurement system indicates a periodic strain that ranges from 3250×10^{-6} in./in. to 4150×10^{-6} in./in. at a frequency of 1 Hz. Determine:

a. The average value of this signal.

b. The amplitude and the frequency of this output signal when expressed as a simple periodic function.

c. A one-term Fourier series that represents this signal.

Construct an amplitude spectrum plot for the output signal.

2.28 For a dynamic calibration involving a force measurement system, a known force is applied to a sensor. The force varies between 100 and 170 N at a frequency of 10 rad/s. State the average (static) value of the input signal, its amplitude, and its frequency. Assuming that the signal may be represented by a simple periodic waveform, express the signal as a one-term Fourier series.

2.29 A displacement sensor is placed on a dynamic calibration rig known as a *shaker*. This device produces a known periodic displacement that serves as the input to the sensor. If the known displacement is set to vary between 2 and 5 mm at a frequency of 100 Hz, express the input signal as a one-term Fourier series. Plot the signal in the time domain, and construct an amplitude spectrum plot.

2.30 Consider the upward flow of water and air in a tube having a circular cross section, as shown in Figure 2.25. If the water and air flow rates are within a certain range, there are slugs of liquid and large gas bubbles flowing upward together. This type of flow is called "slug flow." The data file gas_liquid_data.txt with the companion software contains measurements of pressure made at the wall of a tube in which air and water were flowing. The data were acquired at a sample frequency of 300 Hz. The average flow velocity of the air and water is 1 m/sec.

  a. Construct an amplitude spectrum from the data, and determine the dominant frequency.

  b. Using the frequency information from part (a), determine the length L shown in the drawing in Figure 2.25. Assume that the dominant frequency is associated with the passage of the bubbles and slugs across the pressure sensor.

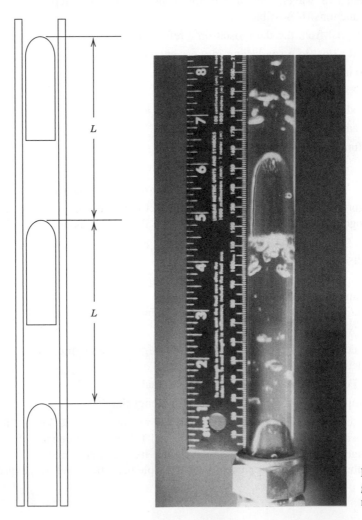

Figure 2.25 Upward gas–liquid flow: slug flow regime.

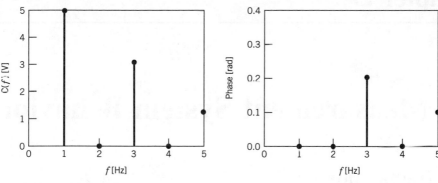

Figure 2.26 Spectrum for Problem 2.32.

2.31 Classify the following signals as completely as possible:

  a. Clock face having hands.

  b. Morse code.

  c. Musical score, as input to a musician.

  d. Flashing neon sign.

  e. Telephone conversation.

  f. Fax transmission.

2.32 Describe the signal defined by the amplitude spectrum and its phase shift of Figure 2.26 in terms of its Fourier series. What is the frequency resolution of these plots? What was the sample time increment used?

2.33 For the even-functioned triangle wave signal defined by

$$y(t) = (4C/T)t + C \quad -T/2 \le t \le 0$$
$$y(t) = (-4C/T)t + C \quad 0 \le t \le T/2$$

where C is an amplitude and T is the period:

  a. Show that this signal can be represented by the Fourier series

$$y(t) = \sum_{n=1}^{\infty} \frac{4C(1 - \cos n\pi)}{(\pi n)^2} \cos \frac{2\pi n t}{T}$$

  b. Expand the first three nonzero terms of the series. Identify the terms containing the fundamental frequency and its harmonics.

  c. Plot each of the first three terms on a separate graph. Then plot the sum of the three. For this, set $C = 1$ V and $T = 1$ s. Note: the program *FourCoef* or Waveform Generation is useful in this problem.

  d. Sketch the amplitude spectrum for the first three terms of this series, first separately for each term and then combined.

2.34 Figure 2.16 illustrates how the inclusion of higher frequency terms in a Fourier series refine the accuracy of the series representation of the original function. However, measured signals often display an amplitude spectrum for which amplitudes decrease with frequency. Using Figure 2.16 as a resource, discuss the effect of high-frequency, low-amplitude noise on a measured signal. What signal characteristics would be unaffected? Consider both periodic and random noise.

2.35 The program Sound.vi will sample the ambient room sounds using your laptop computer microphone and sound board and return the amplitude spectrum of the sounds. Experiment with different sounds (e.g. tapping, whistling, talking, humming) and describe your findings in a brief report.

Chapter 3

Measurement System Behavior

3.1 INTRODUCTION

Each measurement system will respond differently to different types of input signals. So a particular system may not be suitable for the measurement of certain input signals. Yet a measurement system will always output information regardless of how well (or poorly!) this information might reflect the actual input signal. To correctly interpret the information about an input signal, based on the information contained in the output signal, requires some understanding of how the system would respond to a variety of input signals. In this chapter, the concept of simulating the measurement system behavior through mathematical modeling is introduced. From this modeling, the important aspects of measurement system response pertinent to system design and specification are introduced and investigated.

Throughout this chapter, we use the term "measurement system" in a generic sense. We refer either to the response of the measurement system as a whole or to the response of any component or instrument that makes up that system. Either is important, and both are interpreted in similar ways. Each individual stage of the measurement system will have its own response to a given input. The overall system response will be affected by the response of each stage that makes up the complete system.

3.2 GENERAL MODEL FOR A MEASUREMENT SYSTEM

As pointed out in Chapter 2, all input and output signals can be broadly classified as being static, dynamic, or some combination of the two. For a static signal, only the signal magnitude is needed to reconstruct the input signal based on the indicated output signal. Consider the measurement of the length of a board using a ruler. Once the ruler is positioned, the indication (output) of the magnitude of length is immediately displayed because the board length (input) does not change over the time required to make the measurement. Thus, the board length represents a static input signal that is interpreted through the static magnitude output indicated by the ruler.

Dynamic Measurements

For dynamic signals, signal amplitude, frequency, and/or general waveform become necessary to reconstruct the input signal. Because dynamic signals vary with time, the measurement system must be able to respond fast enough literally to keep up with the input signal. We recognize the importance of measurement system response even when using the most common of measurement systems. Consider the time response of a typical bulb thermometer during the act of measuring body temperature. The thermometer, initially at approximately room temperature, is placed under the tongue. But even after several seconds, the thermometer does not indicate the expected value of body temperature and

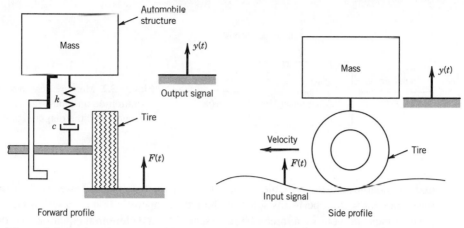

Figure 3.1 Model of an automobile suspension with input and output signals.

its display continually changes. What has happened? Surely your body temperature is not changing. Using the magnitude of the output signal after only several seconds might lead to false conclusions about one's health! Experience shows that within a few minutes, the correct body temperature will be indicated; so we wait. In this example, body temperature itself is a constant during the measurement, but the input signal to the thermometer is suddenly changed from room temperature to body temperature, i.e., mathematically, a step change. This is a dynamic event as the thermometer (the measurement system) sees it! The thermometer must gain energy from its new environment to reach thermal equilibrium and this takes a finite amount of time. The ability of any measuring system to follow dynamic signals is a characteristic of the design of the measuring system components.

Now consider the task of assessing the ride quality of an automobile suspension system. A simplified view for one wheel of this system is shown in Figure 3.1. As a tire moves along the road, the road surface provides the time-dependent input signal, $F(t)$, to the suspension at the tire contact point. The motion sensed by the passengers, $y(t)$, is a basis for the ride quality and can be described by a waveform that depends on the input from the road and the behavior of the suspension. An engineer must anticipate the form of input signals so as to design the suspension to attain a desirable output signal.

Measurement systems play a key role in documenting ride quality. But just as the road and car interact to provide ride quality, the input signal and the measurement system interact to form an output signal. In many situations, the goal of the measurement is actually to deduce the input signal based on the output signal. Either way, we see that it is important to understand how a measurement system responds to different forms of input signals.

The behavior of measurement systems to a few special inputs will, for the most part, define the input–output signal relationships necessary to enable the correct interpretation of the measured signal. We will show that only a few measurement system characteristics are needed to define the system response, so that dynamic calibrations can be restricted to one or two specific tests. The results of these tests can be used to judge the suitability of a particular system to interpret the signal information.

With the previous discussion in mind, consider that the primary task of a measurement system is to sense an input signal and to translate that information into a readily

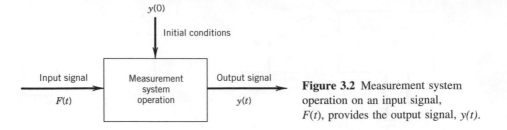

Figure 3.2 Measurement system operation on an input signal, *F(t)*, provides the output signal, *y(t)*.

understandable and quantifiable output form. With that in mind, we can reason that a measurement system performs some mathematical operation on a sensed input. In fact, a general measurement system can be represented by a differential equation that describes the operation that a measurement system performs on the input signal.

This concept of relating the input signal to the output signal is illustrated in Figure 3.2 with the mathematical operation that the measurement system performs represented within the box. For an input signal, $F(t)$, the system performs some operation that yields the output signal, $y(t)$. Then we must use $y(t)$ to infer $F(t)$. Therefore, at least a qualitative understanding of the operation that the measurement system performs is imperative to correctly interpret the input signal. We will propose a general mathematical model for a measurement system. Then, by representing a typical input signal as some function that acts as an input to the model, we can study just how the measurement system would behave by solving the model equation. In essence, we will perform the analytical equivalent of a system calibration. This information can then be used to determine those input signals for which a particular measurement system is best suited.

Measurement System Model

Consider the following model of a measurement system which consists of a general nth-order linear ordinary differential equation in terms of a general output signal, represented by variable $y(t)$, and subject to a general input signal, represented by the forcing function, $F(t)$:

$$a_n \frac{d^n y}{dt^n} + a_{n-1} \frac{d^{n-1} y}{dt^{n-1}} + \cdots + a_1 \frac{dy}{dt} + a_0 y = F(t) \tag{3.1}$$

The coefficients $a_0, a_1, a_2, \ldots, a_n$ represent physical system parameters whose properties and values will depend on the measurement system itself. The forcing function can be generalized into the mth-order form

$$F(t) = b_m \frac{d^m x}{dt^m} + b_{m-1} \frac{d^{m-1} x}{dt^{m-1}} + \cdots + b_1 \frac{dx}{dt} + b_0 x \qquad m \le n$$

where $b_0, b_1, \ldots, b_m$ also represent physical system parameters. Real measurement systems can be modeled this way by considering their governing system equations. These equations are generated by application of pertinent fundamental physical laws of nature to the measurement system. Our discussion will be limited to measurement system concepts, but a general treatment of systems can be found in text books dedicated to that topic (e.g., [1–3]).

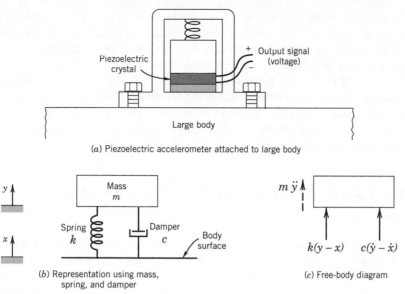

(a) Piezoelectric accelerometer attached to large body

(b) Representation using mass, spring, and damper

(c) Free-body diagram

Figure 3.3 Accelerometer of Example 3.1.

EXAMPLE 3.1

As an illustration, consider the seismic accelerometer depicted in Figure 3.3(a). Various configurations of this instrument are used in seismic and vibration engineering to determine the motion of large bodies to which the accelerometer is attached. Basically, as the small accelerometer mass reacts to motion it places the piezoelectric crystal into compression or tension, which causes a surface charge to develop on the crystal. The charge is proportional to the motion. As the large body moves, the mass of the accelerometer will move with an inertial response. The stiffness of the spring, k, will provide a restoring force to move the accelerometer mass back to equilibrium while internal frictional damping, c, will oppose any displacement away from equilibrium. A model of this measurement device in terms of ideal elements of stiffness, mass, and damping is given in Figure 3.3(b) and the corresponding free-body diagram in Figure 3.3(c). Let y denote the position of the small mass within the accelerometer and x denote the displacement of the body. Solving Newton's second law for the free body yields the second-order linear, ordinary differential equation

$$m\frac{d^2y}{dt^2} + c\frac{dy}{dt} + ky = c\frac{dx}{dt} + kx$$

Since the displacement y is the pertinent output from the accelerometer due to displacement x, the equation has been written such that all output terms, that is, all the y terms, are on the left side. All other terms are considered to be input signals and are placed on the right side. Comparing this to the general form for a second-order equation ($n = 2$; $m = 1$) from equation (3.1),

$$a_2\frac{d^2y}{dt^2} + a_1\frac{dy}{dt} + a_0y = b_1\frac{dx}{dt} + b_0x$$

we can see that $a_2 = m$, $a_1 = b_1 = c$, $a_0 = b_0 = k$, and that the forces developed due to the velocity and displacement of the body become the inputs to the accelerometer. If we could anticipate the waveform of x, e.g., $x(t) = x_0 \sin \omega t$, we could solve for $y(t)$, which is the measurement system response.

Fortunately, many measurement systems can be modeled by zero-, first-, or second-order linear, ordinary differential equations. More complex systems can usually be simplified to a lower order. Our intention here is to attempt to understand how systems behave and how response is closely related to the design features of a measurement system; it is not to simulate the exact system behavior. The exact input–output relationship is always found from calibration. But modeling will guide us in narrowing down our choice to specific instruments and determining the type, range, and specifics of calibration. Next, we examine several special cases of equation (3.1) that identify the most important concepts of measurement system behavior.

3.3   SPECIAL CASES OF THE GENERAL SYSTEM MODEL

Zero-Order Systems

A basic model for measurement systems used to measure static signals is the zero-order system model. It is the simplest model for a measurement system and is represented by this zero-order differential equation:

$$a_0 y = F(t)$$

with $F(t) = b_0 x$. Dividing through by a_0 gives

$$y = KF(t) \tag{3.2}$$

where $K = 1/a_0$. K is called the static sensitivity or steady gain of the system. This system property was introduced in Chapter 1 as the relation between the static output change associated with a change in input. In zero-order behavior, the system output is considered to respond to the input signal instantly. If an input signal of magnitude $F(t) = A$ were applied, the instrument would indicate KA, as modeled by equation (3.2). The scale of the measuring device would be calibrated to indicate A directly.

For real systems, the zero-order system concept is used to model measurement system response to static inputs. In fact, the zero-order concept appropriately models any system during a static calibration. When dynamic input signals are involved, a zero-order model is valid only at static equilibrium. This is because most real measurement systems possess inertial or storage capabilities that require higher-order differential equations to correctly model their time-dependent behavior to dynamic input signals.

Determination of K.   The static sensitivity is found from the static calibration of the measurement system. It is the slope of the calibration curve, $K = dy/dx$.

EXAMPLE 3.2

A pencil-type pressure gauge commonly used to measure tire pressure can be modeled at static equilibrium by considering the force balance on the gauge sensor, a piston that slides up and down a cylinder. In Figure 3.4(a), we model the piston motion[1] as being restrained by an internal spring of stiffness, k, so that at static equilibrium the absolute pressure force, F, bearing on the piston equals the force exerted on the piston by the spring, F_s, plus the atmospheric pressure force, F_{atm}. In this manner, the transduction of

---

[1]This is an idealization. In a common gauge there is no mechanical spring. Rather the piston motion is resisted by the motion of the air within the cylinder as it bleeds past the moving piston.

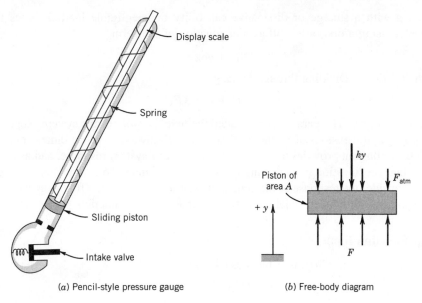

Display scale

Spring

Sliding piston

Intake valve

Piston of area A

ky

F_{atm}

$+y$

F

(a) Pencil-style pressure gauge

(b) Free-body diagram

Figure 3.4 Pressure gauge of Example 3.2.

pressure into displacement occurs. Considering the piston free-body at static equilibrium in Figure 3.4(b), the static force balance, $\Sigma F = 0$, yields

$$ky = F - F_{atm}$$

where y is measured relative to some static reference position marked as zero on the output display. Pressure is simply the force acting inward over the piston surface area, A. Dividing through by area provides the zero-order response equation between output displacement and input pressure

$$y = (A/k)(p - p_{atm})$$

The term $(p - p_{atm})$ represents the pressure relative to atmospheric pressure. It is the pressure indicated by this gauge. A direct comparison with equation (3.2) implies that the input pressure is translated into piston displacement through the static sensitivity, $K = A/k$. The system operates on pressure so as to bring about a relative displacement of the piston, the magnitude of which is used to indicate the pressure. The equivalent of spring stiffness and piston area affect the magnitude of this displacement, factors considered in its design. The exact static input–output relationship is found through calibration of the gauge. Because elements such as piston inertia and frictional dissipation were not considered, this model would not be appropriate for studying the dynamic response of the gauge.

First-Order Systems

Measurement systems that contain storage elements cannot respond instantaneously to changes in input. The bulb thermometer discussed in Section 3.2 is a good example. The bulb exchanges energy with its environment until the two are at the same temperature, storing energy during the exchange. The temperature of the bulb sensor will change with time until this equilibrium is reached, which accounts physically for its lag in response. The rate at which temperature changes with time can be modeled with a first-order derivative and the thermometer behavior modeled as a first-order equation. In general, systems with a storage or dissipative capability but negligible inertial forces may be modeled using a first-order differential equation of the form

$$a_1 \dot{y} + a_0 y = F(t) \tag{3.3}$$

with $\dot{y} = dy/dt$. Dividing through by a_0 gives

$$\tau \dot{y} + y = KF(t) \tag{3.4}$$

where $\tau = a_1/a_0$. The parameter τ is called the *time constant* of the system. Regardless of the physical dimensions of a_0 and a_1, their ratio will always have the dimensions of time. The time constant provides a measure of the speed of system response, and as such is an important specification in measuring dynamic input signals. To explore this concept more fully, consider the response of the general first-order system to the following two forms of an input signal: the step function and the simple periodic function.

Step Function Input

The step function, $AU(t)$, is defined as

$$AU(t) = 0 \quad t \leq 0^-$$
$$AU(t) = A \quad t \geq 0^+$$

where A is the amplitude of the step function and $U(t)$ is defined as the unit step function as depicted in Figure 3.5. Physically, this function describes a sudden change in the input signal from a constant value of one magnitude to a constant value of some other magnitude, such as a sudden change in loading, displacement, or temperature. When we apply a step function input to a measurement system, we obtain information about the speed at which a system will respond to a change in input signal. To illustrate this, let us apply a step function as an input to the general first-order system.

Setting $F(t) = AU(t)$ in equation (3.4) gives

$$\tau \dot{y} + y = KAU(t) = KF(t)$$

with an arbitrary initial condition denoted by, $y(0) = y_0$. Solving at $t \geq 0^+$ yields

$$\underbrace{y(t)}_{\text{time response}} = \underbrace{KA}_{\text{steady response}} + \underbrace{(y_0 - KA)e^{-t/\tau}}_{\text{transient response}} \tag{3.5}$$

The solution of the differential equation, $y(t)$, is the time response (or simply the response) of the system. Equation (3.5) describes the behavior of the system to a step change in input. This means that $y(t)$ is in fact the output indicated by the display stage of

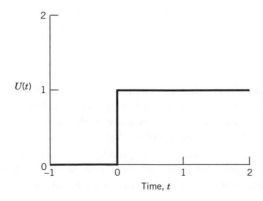

Figure 3.5 The unit step function, $U(t)$.

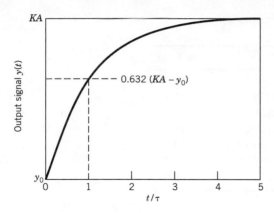

Figure 3.6 First-order system time response to a step function input: the time response, $y(t)$.

the system. It should represent the time variation of the output display of the measurement system if an actual step change were applied to the system. We have simply used mathematics to simulate this response.

The second term on the right side of equation (3.5) is known as the transient response of $y(t)$ because, as $t \to \infty$, the magnitude of this term eventually reduces to zero. The first term is known as the steady response because, as $t \to \infty$, the response of $y(t)$ approaches this steady value. The steady response is that portion of the output signal that remains after the transient response has decayed to zero.

For illustrative purposes only, let $y_0 < A$ so that the time response becomes as shown in Figure 3.6. Over time, the indicated output value rises from its initial value, at the instant the change in input is applied, to an eventual constant value, $y_\infty = KA$, at steady response. Compare this general time response to the recognized behavior of the bulb thermometer when measuring body temperature as discussed earlier. We see a qualitative similarity. In fact, the measurement of body temperature represents a real step function input to the thermometer, itself a real first-order measuring system.

Suppose we rewrite the response equation (3.5) in the form

$$\Gamma(t) = \frac{y(t) - y_\infty}{y_0 - y_\infty} = e^{-t/\tau} \tag{3.6}$$

The term $\Gamma(t)$ is called the *error fraction* of the output signal. Equation (3.6) is plotted in Figure 3.7 where the time axis has been nondimensionalized by the time constant. We see that the error fraction decreases from a value of 1 and approaches a value of 0 with

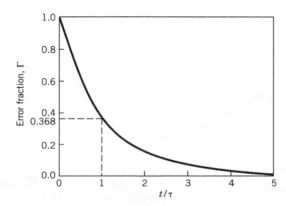

Figure 3.7 First-order system time response to a step function input: the error fraction, Γ.

Table 3.1   First-Order System Response and
Error Fraction

t/τ	% Response	Γ	% Error
0	0.0	1.0	100.0
1	63.2	0.368	36.8
2	86.5	0.135	13.5
2.3	90.0	0.100	10.0
3	95.0	0.050	5.0
5	99.3	0.007	0.7
∞	100.0	0.0	0.0

increasing t/τ. At the instant just after the step change in input is introduced, $\Gamma = 1.0$ so that the indicated output from the measurement system is 100% in error. This implies that the system has responded 0% to the input change. From Figure 3.7, it is apparent that as time moves forward the system responds with decreasing error in its indicated value. Let the percent response of the system to a step change be given as $(1 - \Gamma) \times 100$. Then by $t = \tau$, where $\Gamma = 0.368$, the system will have responded to 63.2% of the step change. Further, when $t = 2.3\tau$ the system will have responded ($\Gamma = 0.10$) to 90% of the step change; by $t = 5\tau$, 99.3%. Values for the percent response and the corresponding error as functions of $t = \tau$ are summarized in Table 3.1. The time required for a system to respond to a value that is 90% of the step input, $y_\infty - y_0$, is called the *rise time* of the system.

   Based on this behavior, we see that the time constant is in fact a measure of the speed of first-order measurement system response to a change in input value. A smaller time constant indicates a shorter time between the instant that an input is applied and when the system reaches an essentially steady output. The *time constant* will be defined as the time required for a first-order system to achieve 63.2% of the step change magnitude, $y_\infty - y_0$. The time constant is a system property.

Determination of τ.   From the development above, the time constant can be experimentally determined by recording the system's response to a step function input of a known magnitude. In practice, it is best to record that response from $t = 0$ until steady response is achieved. The data can then be plotted as error fraction versus time on a semilog plot, such as in Figure 3.8. This type of plot is equivalent to the transformation

$$\ln \Gamma = 2.3 \log \Gamma = -(1/\tau)t \tag{3.7}$$

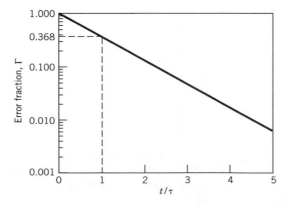

Figure 3.8 The error fraction plotted on semilog coordinates.

which is of the linear form, $Y = mX + B$ (where $Y = \ln \Gamma$, $m = -(1/\tau)$, $X = t$, and $B = 0$ here). A linear curve fit through the data will provide a good estimate of the slope, m, of the resulting plot. From equation (3.7), we see that $m = -1/\tau$, which yields the estimate for τ.

This method offers advantages over attempting to compute τ directly from the time required to achieve 63.2% of the step-change magnitude. First, real systems will deviate somewhat from perfect first-order behavior. On a semilog plot (Figure 3.8), such deviations are readily apparent as clear trends away from a straight line. Modest deviations do not pose any problem. But strong deviations provide a good indication that the system is not behaving as expected, thus requiring a closer examination either of the system operation, the step function experiment, or the assumed form of the system model. Second, acquiring data during the step function experiment is prone to some random error in each data point. The use of a data curve fit to determine τ utilizes all of the data over time so as to minimize the influence of an error in any one data point. Third, the method eliminates the need to determine the $\Gamma = 1.0$ and 0.368 points, which are difficult to establish in practice and so are prone to a systematic error.

EXAMPLE 3.3

Suppose a bulb thermometer originally indicating 20°C is suddenly exposed to a fluid temperature of 37°C. Develop a model that simulates the thermometer output response.

KNOWN

$$T(0) = 20°C$$
$$T_\infty = 37°C$$
$$F(t) = [T_\infty - T(0)]U(t)$$

ASSUMPTIONS   To keep things simple, assume the following:
No installation effects (neglect conduction and radiation effects)
Sensor mass is mass of liquid in bulb only
Uniform temperature within bulb (lumped mass)
Thermometer scale is calibrated to indicate temperature

FIND

$T(t)$

SOLUTION   Consider the energy balance developed in Figure 3.9. According to the first law of thermodynamics, the rate at which energy is exchanged between the sensor and its environment through convection, $\dot{Q}$, must be balanced by the storage of energy within the thermometer, dE/dt. This conservation of energy is written as

$$\frac{dE}{dt} = \dot{Q}$$

Energy storage in the bulb is manifested by a change in bulb temperature so that for a constant mass bulb, $dE(t)/dt = mc_v\, dT(t)/dt$. Energy exchange by convection between the bulb at T(t) and an environment at T_∞ has the form $\dot{Q} = hA_s\Delta T$. The first law can be written as

$$mc_v\frac{dT(t)}{dt} = hA_s[T_\infty - T(t)]$$

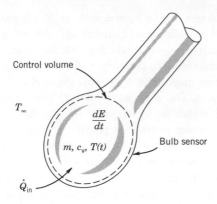

Control volume

T_∞

$\dfrac{dE}{dt}$

$m, c_v, T(t)$

Bulb sensor

$\dot{Q}_{in}$

Figure 3.9 Thermometer and energy balance of Example 3.3.

This equation can be written in the form

$$mc_v \frac{dT(t)}{dt} + hA_s[T(t) - T(0)] = hA_s F(t) = hA_s[T_\infty - T(0)]U(t)$$

with initial condition $T(0)$

where

m = mass of liquid within thermometer

c_v = specific heat of liquid within thermometer

h = convection heat transfer coefficient between bulb and environment

A_s = thermometer surface area

The term hA_s controls the rate at which energy can be transferred between a fluid and a body; it is analogous to an electrical conductance. By comparison with equation (3.3), $a_0 = hA_s$, $a_1 = mc_v$, and $b_0 = hA_s$. Rewriting for $t \geq 0+$ and simplifying yields

$$\frac{mc_v}{hA_s} \frac{dT(t)}{dt} + T(t) = T_\infty$$

From equation (3.4), this implies that the time constant and static sensitivity are

$$\tau = \frac{mc_v}{hA_s} \qquad K = \frac{hA_s}{hA_s} = 1$$

Direct comparison with equation (3.5) yields this thermometer response:

$$T(t) = T_\infty + [T(0) - T_\infty]e^{-t/\tau}$$
$$= 37 - 17e^{-t/\tau} \quad [^\circ\text{C}]$$

COMMENT   Two interactive examples of the thermometer problem (Examples 3.3–3.5) are available. In program *FirstOrd,* the user can choose input functions and study the system response. In Labview® program *Temperature_response*, the user can apply interactively a step change in temperature and study the first order system response of a thermal sensor.

Clearly the time constant, τ, of the thermometer can be reduced by decreasing its mass to area ratio or by increasing h (for example, increasing the fluid velocity around the sensor). Without modeling, such information could be ascertained only by trial and error, a time-consuming and costly method with no assurance of success. Also, it is significant that we found that the response of the temperature measurement system in this case will depend on the environmental conditions of the measurement that control h, because the magnitude of h affects the magnitude of τ. If h is not controlled during response tests (i.e., if it is an extraneous variable), ambiguous results are possible. For example, the curve of Figure 3.8 will become nonlinear and/or replications will not yield the same values for τ.

Review of this example should make it apparent that the results of a well-executed step calibration may not be indicative of an instrument's performance during a measurement if the measurement conditions differ from those existing during the step calibration.

EXAMPLE 3.4

For the thermometer of Example 3.3 subjected to a step change in input, calculate the 90% rise time in terms of t/τ.

KNOWN   Same as Example 3.3

ASSUMPTIONS   Same as Example 3.3

FIND   90% response time in terms of t/τ

SOLUTION   The percent response of the system is given by $(1 - \Gamma) \times 100$ with the error fraction, Γ, defined by equation (3.6). From equation (3.5), we note that at $t = \tau$, the thermometer will indicate $T(t) = 30.75°C$, which represents only 63.2% of the step change from 20 to 37 C°. The 90% rise time represents the time required for Γ to drop to a value of 0.10. Then

$$\Gamma = 0.10 = e^{-t/\tau}$$

or $t/\tau = 2.3$.

COMMENT   In general, a time equivalent to 2.3τ is required to achieve 90% of the applied step input for a first-order system.

EXAMPLE 3.5

A particular thermometer is subjected to a step change, such as in Example 3.3, in an experimental exercise to determine its time constant. The temperature data are recorded with time and presented in Figure 3.10. Determine the time constant for this thermometer. In the experiment, the heat transfer coefficient, h, is estimated to be 6 W/m^2-°C from engineering handbook correlations.

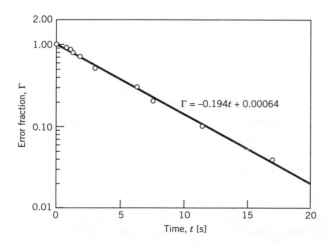

Figure 3.10 Temperature–time history of Example 3.5.

KNOWN   Data of Figure 3.10

$$h = 6\,\text{W/m}^2\text{-}°\text{C}$$

ASSUMPTIONS   First-order behavior and using the Model of Example 3.3
Constant properties

FIND

τ

SOLUTION   According to equation (3.7), the time constant should be the negative reciprocal of the slope of a line drawn through the data of Figure 3.10. Aside from the first few data points, the data appear to follow a linear trend, indicating a nearly first-order behavior and validating our model assumption. A first-order least-squares fit[2] yields the equation

$$2.3 \log \Gamma = (-0.194)t + 0.00064$$

With $m = -0.194 = -1/\tau$, the time constant can be estimated to be $\tau = 5.15$ s for a constant $h = 6$ W/m^2-°C.

COMMENT   If the experimental data were to deviate significantly from first-order behavior, we would want to examine the conduct of the experiment and/or the rigor of the model and its assumptions relative to the real system as possible causes.

Sine Function Input

Periodic signals are commonly encountered in many engineering processes, such as vibrating structures, vehicle suspension dynamics, and reciprocating pump flows. When periodic inputs are applied to a first-order system, the frequency of the input has an important influence on the measuring system time response and affects the output signal. This behavior can be studied effectively by applying a simple periodic waveform as an input to the system. Consider the first-order measuring system to which an input of the form of a simple periodic function, $F(t) = A \sin \omega t$, is applied for $t \geq 0^+$:

$$\tau \dot{y} + y = KA \sin \omega t$$

with initial conditions $y(0) = y_0$. Note that ω in [rad/s] $= 2\pi f$ with f in [Hz]. The general solution to this differential equation yields the measuring system output signal, the time response to the applied input, $y(t)$:

$$y(t) = Ce^{-t/\tau} + \frac{KA}{\sqrt{1 + (\omega\tau)^2}} \sin(\omega t - \tan^{-1}\omega\tau) \tag{3.8}$$

where the value for C will depend on the exact value of y_0.

So what has happened? The output signal, $y(t)$, consists of a transient and a steady response. The first term on the right side is the transient response. As t increases, this term decays to zero and no longer influences the output signal. Transient response is important only during the initial period following the application of the new input. We already have information about the system transient response from the step function study, so let us focus our attention on the second term, the steady response. This term persists for as long

---

[2]The least-squares approach to curve fitting is discussed in detail in Chapter 4.

as the periodic input is maintained. From equation (3.8), we see that the frequency of the steady response term remains the same as the input signal frequency, but note that the amplitude of the steady response depends on the value of the applied frequency, ω. Also, the phase angle of the periodic function has changed.

Equation (3.8) can be rewritten in a general form

$$y(t) = Ce^{-t/\tau} + B(\omega)\sin[\omega t + \Phi(\omega)]$$

$$B(\omega) = \frac{KA}{\sqrt{1 + (\omega\tau)^2}}$$

$$\Phi(\omega) = -\tan^{-1}\omega\tau$$

(3.9)

where $B(\omega)$ represents the amplitude of the steady response and the angle $\Phi(\omega)$ represents the *phase shift*. A relative illustration between the input signal and the system output response is given in Figure 3.11 for an arbitrary frequency and system time constant. From equation (3.9), both B and f are frequency dependent. Hence, the exact form of the output response will depend on the value of the frequency of the input signal. The steady response of any system to which a periodic input of frequency, ω, is applied is known as the frequency response of that system. The frequency dependence of a system affects the magnitude of amplitude B and brings about a time delay between when the input is applied and when the measuring system responds to the input signal. This time delay, β_1, is seen in the phase shift, $\Phi(\omega)$, of the steady response. For a phase shift given in radians, the time delay in units of time is

$$\beta_1 = \frac{\Phi(\omega)}{\omega}$$

that is, we can write

$$\sin(\omega t + \Phi) = \sin\left[\omega\left(t + \frac{\Phi}{\omega}\right)\right] = \sin[\omega(t + \beta_1)]$$

The value for β_1 will be negative, indicating a time delay between the output and input signals. Since equation (3.9) applies to all first-order measuring systems, the magnitude and phase shift by which the output signal differs from the input signal are predictable.

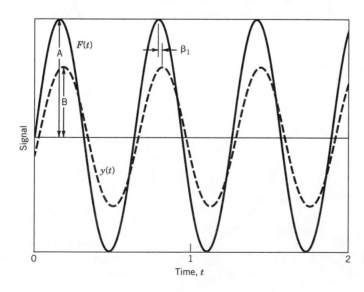

Figure 3.11 Relationship between a sinusoidal input and output: amplitude, frequency, and time lag.

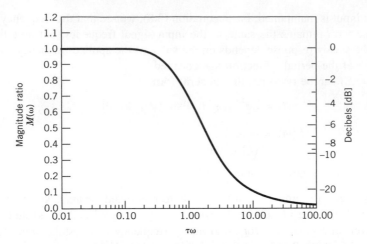

Figure 3.12 First-order system frequency response: magnitude ratio.

We define a magnitude ratio, $M(\omega)$, as the ratio of the output signal amplitude to the input signal amplitude, $M(\omega) = B/KA$. For a first-order system subjected to a simple periodic input, the magnitude ratio is

$$M(\omega) = \frac{B}{KA} = \frac{1}{\sqrt{1 + (\omega\tau)^2}} \qquad (3.10)$$

The magnitude ratio for a first-order system is plotted in Figure 3.12, and the corresponding phase shift is plotted in Figure 3.13. The effects of both system time constant and input signal frequency on frequency response are apparent in both figures. This behavior can be interpreted in the following manner. For those values of ωt for which the system responds with values of $M(\omega)$ near unity, the measurement system will transfer all or nearly all of the input signal amplitude to the output and with very little time delay; that is, B will be nearly equal to KA in magnitude and $\Phi(\omega)$ will be near $0°$. At large values of $\omega\tau$ the measurement system will essentially filter out the frequency

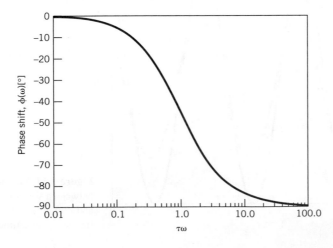

Figure 3.13 First-order system frequency response: phase shift.

information of the input signal by responding with very small amplitudes, as represented by small $M(\omega)$, and larger time delays, as evidenced by increasingly nonzero β_1.

Any combination of ω and τ will produce the same results. If one wants to measure signals with high-frequency content, then a system having an appropriately small τ will be necessary. On the other hand, systems of large τ may be adequate to measure signals of low-frequency content.

The *dynamic error*, $\delta(\omega)$, of a system can be defined as $\delta(\omega) = M(\omega) - 1$. It is a measure of the inability of a system to adequately reconstruct the amplitude of the input signal for a particular input frequency. Measurement systems having a magnitude ratio at or near unity over the anticipated frequency band of the input signal minimize $\delta(\omega)$. Perfect reproduction of the input signal is not possible, so some dynamic error is inevitable. For a first-order system, the *frequency bandwidth* is defined traditionally as the frequency band over which $M(\omega) \geq 0.707$; in terms of the decibel (plotted in Figure 3.12) and defined as

$$\text{dB} = 20 \log M(\omega) \tag{3.11}$$

it is the band of frequencies within which $M(\omega)$ remains above -3 dB.

The functions $M(\omega)$ and $\Phi(\omega)$ represent the frequency response of the measurement system to periodic inputs. These equations and universal curves are used for guidance in the selection of measurement systems and system components. They are not to be used to correct measured data. This can be understood by examining Figures 3.12 and 3.13. As $M(\omega)$ falls away from unity and as $\Phi(\omega)$ falls away from zero, these curves take on rather steep slopes. In these regions of the curves, small errors in the prediction of τ and deviations of the real systems from ideal first-order behavior can lead to errors of unpredictable magnitude.

Determination of Frequency Response.   The frequency response of a measurement system is found by a dynamic calibration. In this case, the calibration would entail applying a simple periodic waveform of known amplitude and frequency to the system sensor stage and measuring the corresponding output stage amplitude and phase shift. In practice, developing a method to produce a periodic input signal in the form of a physical variable may demand considerable ingenuity and effort. Hence, in many situations an engineer will elect to rely on modeling to infer system frequency response behavior.

EXAMPLE 3.6

A temperature sensor is to be selected to measure temperature within a reaction vessel. It is suspected that the temperature will behave as a simple periodic waveform with a frequency somewhere between 1 and 5 Hz. Several size sensors are available, each with a known time constant. Based on time constant, select a suitable sensor, assuming that a dynamic error of $\pm 2\%$ is acceptable.

KNOWN

$$1 \leq f \leq 5\,\text{Hz}$$
$$|\delta(\omega)| \leq 0.02$$

ASSUMPTIONS   First-order system

$$F(t) = A \sin \omega t$$

FIND   Time constant, τ

SOLUTION   With $|\delta(\omega)| \le 0.02$ we seek a magnitude ratio between $0.98 \le M \le 1.02$. From Figure 3.12, we see that first-order systems never exceed $M = 1$. So the constraint becomes $0.98 \le M \le 1$. Then,

$$0.98 \le M(\omega) = \frac{1}{\sqrt{1 + (\omega\tau)^2}} \le 1$$

From Figure 3.12, this constraint is maintained over the range $0 \le \omega\tau \le 0.2$. We can also see in this figure that for a system of fixed time constant, the smallest value of $M(\omega)$ will occur at the largest frequency. So with $\omega = 2\pi f = 2\pi(5)$ rad/s and solving for $M(\omega) = 0.98$ yields, $\tau \le 6.4$ ms. Accordingly, a sensor having a time constant of 6.4 ms or less will work.

Second-Order Systems

Systems that possess inertia will contain a second-derivative term in their model equation (e.g., see Example 3.1). A system that is modeled by a second-order differential equation is called a second-order system. Examples of second-order measurement systems include accelerometers and diaphragm pressure transducers (including microphones and loud-speakers).

In general, a second-order measurement system subjected to an arbitrary input, $F(t)$, can be described by an equation of the form

$$a_2\ddot{y} + a_1\dot{y} + a_0 y = F(t) \tag{3.12}$$

where a_0, a_1, and a_2 are physical parameters used to describe the system and $\ddot{y} = d^2 y / dt^2$. This equation can be rewritten as

$$\frac{1}{\omega_n^2}\ddot{y} + \frac{2\zeta}{\omega_n}\dot{y} + y = KF(t) \tag{3.13}$$

where

$$\omega_n = \sqrt{\frac{a_0}{a_2}} = \text{natural frequency of the system}$$

$$\zeta = \frac{a_1}{2\sqrt{a_0 a_2}} = \text{damping ratio of the system}$$

Consider the homogeneous solution to equation (3.13). Its form will depend on the roots of the characteristic equation of equation (3.13):

$$\frac{1}{\omega_n^2}\lambda^2 + \frac{2\zeta}{\omega_n}\lambda + 1 = 0$$

This quadratic equation has two roots,

$$\lambda_{1,2} = -\zeta\omega_n \pm \omega_n\sqrt{\zeta^2 - 1}$$

Depending on the value for ζ, three forms of homogeneous solution are possible:

$$0 \le \zeta < 1 \text{ (underdamped system solution)}$$

$$y_h(t) = Ce^{-\zeta\omega_n t}\sin\left(\omega_n\sqrt{1 - \zeta^2}t + \Theta\right) \tag{3.14a}$$

$$\zeta = 1 \text{ (critically damped system solution)}$$

$$y_h(t) = C_1 e^{\lambda_1 t} + C_2 t e^{\lambda_2 t} \tag{3.14b}$$

$$\zeta > 1 \text{ (overdamped system solution)}$$

$$y_h(t) = C_1 e^{\lambda_1 t} + C_2 e^{\lambda_2 t} \tag{3.14c}$$

The homogeneous solution determines the transient response of a system. For systems having $0 \leq \zeta \leq 1$, the transient response will be oscillatory, whereas for $\zeta \geq 1$, the transient response will not oscillate. The critically damped solution, $\zeta = 1$, denotes the demarcation between oscillatory and nonoscillatory behavior in the transient response. The damping ratio is a measure of system damping, a property of a system that enables it to dissipate energy internally.

Step Function Input

Again, the step function input is applied to determine the general behavior and speed at which the system will respond to a change in input. The response of a second-order measurement system to a step function input is found from the solution of equation (3.13), with $F(t) = AU(t)$, to be

$$y(t) = KA - KAe^{-\zeta \omega_n t}\left[\frac{\zeta}{\sqrt{1 - \zeta^2}}\sin\left(\omega_n\sqrt{1 - \zeta^2}t\right) + \cos\left(\omega_n\sqrt{1 - \zeta^2}t\right)\right] \quad 0 \leq \zeta < 1$$

$$(3.15a)$$

$$y(t) = KA - KA(1 + \omega_n t)e^{-\omega_n t} \quad \zeta = 1 \tag{3.15b}$$

$$y(t) = KA - KA\left[\frac{\zeta + \sqrt{\zeta^2 - 1}}{2\sqrt{\zeta^2 - 1}}e^{\left(-\zeta + \sqrt{\zeta^2 - 1}\right)\omega_n t} + \frac{\zeta - \sqrt{\zeta^2 - 1}}{2\sqrt{\zeta^2 - 1}}e^{\left(-\zeta - \sqrt{\zeta^2 - 1}\right)\omega_n t}\right] \quad \zeta > 1$$

$$(3.15c)$$

where the initial conditions, $y(0)$ and $\dot{y}(0)$, have been set to zero for convenience.

Equations (3.15a)–(3.15c) are plotted in Figure 3.14 for several values of ζ. The interesting feature is the transient response. For underdamped systems, the transient response is oscillatory about the steady value and occurs with a period

$$T_d = \frac{2\pi}{\omega_d} = \frac{1}{f_d} \tag{3.16}$$

$$\omega_d = \omega_n\sqrt{1 - \zeta^2} \tag{3.17}$$

where ω_d is called the *ringing frequency*. In instruments, this oscillatory behavior is called "ringing." The ringing phenomenon and the associated ringing frequency are properties

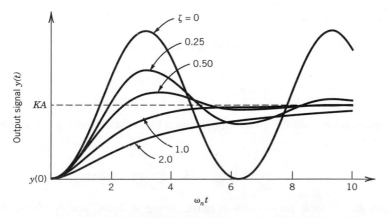

Figure 3.14 Second-order system time response to a step function input.

of the measurement system and are independent of the input signal. It is the free oscillation frequency of a system displaced from its equilibrium.

The duration of the transient response is controlled by the $\zeta\omega_n$ term. In fact, its influence is equivalent to that of a time constant in a first-order system, such that we could define a second-order time constant as $\tau = 1/\zeta\omega_n$. The system settles to KA more quickly when it is designed with a larger $\zeta\omega_n$ (i.e., smaller τ). Nevertheless, for all systems with $\zeta > 0$, the response will eventually indicate the steady value of $y_\infty = KA$ as $t \to \infty$.

Recall that the *rise time* is defined as the time required to achieve a value within 90% of the step input. For a second-order system, the rise time is the time required to first achieve 90% of $KA - y_0$. Rise time is reduced by decreasing the damping ratio, as seen in Figure 3.14. However, the severe ringing associated with very lightly damped systems can delay the time to achieve a steady value compared to systems of higher damping. This is demonstrated by comparing the response at $\zeta = 0.25$ with the response at $\zeta = 1$ in Figure 3.14. With this in mind, the time required for a measurement system's oscillations to settle to within $\pm 10\%$ of the steady value, KA, is defined as its *settling time*. The settling time is an approximate measure of the time to achieve a steady response. A damping ratio of about 0.7 appears to offer a good compromise between ringing and settling time. If an error fraction (Γ) of a few percent is acceptable, then a system with $\zeta = 0.7$ will achieve steady response in about one-half the time of a system having $\zeta = 1$. For this reason, most measurement systems intended to measure sudden changes in input signal are typically designed such that parameters a_0, a_1, and a_2 provide a damping ratio between 0.6 and 0.8.

Determination of Ringing Frequency and Rise and Settling Times.   The experimental determination of the ringing frequency associated with underdamped systems is performed by applying a step input to the second-order measurement system and recording the response with time. This type of calibration will also yield information concerning the time to steady response of the system, which includes rise and settling times. Example 3.8 describes such a test. Typically, measurement systems suitable for dynamic signal measurements have specifications that include 90% rise time and settling time.

EXAMPLE 3.7

Determine the physical parameters that affect the natural frequency and damping ratio of the accelerometer of Example 3.1.

KNOWN   Accelerometer shown in Figure 3.3

ASSUMPTIONS   Second-order system as modeled in Example 3.1

FIND

ω_n, ζ

SOLUTION   A comparison between the governing equation for the accelerometer in Example 3.1 and equation (3.13) gives

$$\omega_n = \sqrt{\frac{k}{m}} \qquad \zeta = \frac{c}{2\sqrt{km}}$$

Accordingly, the physical parameters of mass, spring stiffness, and frictional damping control the natural frequency and damping ratio of this measurement system.

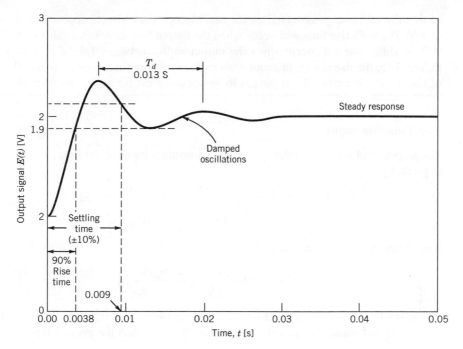

Figure 3.15 Pressure transducer time response to a step input for Example 3.8.

EXAMPLE 3.8

The curve given in Figure 3.15 was reproduced from an oscilloscope trace of the voltage signal output of a diaphragm pressure transducer subjected to a step change in input. During a static calibration using a pressure standard, it was found that the pressure–voltage relationship was linear over the range 1 atmosphere (atm) to 4 atm and that the static sensitivity of the measurement system was 1 V/atm. For the step test, it is known that the initial pressure was atmospheric pressure, p_a, and the final pressure was $2\,p_a$. Estimate the rise time, the settling time, and the ringing frequency of the measurement system.

KNOWN

$$p(0) = 1\,\text{atm}$$
$$p_\infty = 2\,\text{atm}$$
$$K = 1\,\text{V/atm}$$

ASSUMPTIONS   Oscilloscope does not affect transducer performance (see "loading error" in Chapter 6)
Second-order system

FIND   Rise and settling times; ω_d

SOLUTION   The ringing behavior of the system noted on the oscilloscope trace supports the assumption that the transducer can be described as having a second-order behavior. From the given information,

$$E(0) = Kp(0) = 1\,\text{V}$$
$$E_\infty = Kp_\infty = 2\,\text{V}$$

so that the step change observed on the oscilloscope trace should appear as a magnitude of 1 V. The 90% rise time will occur when the output first achieves a value of 1.9 V. The 90% settling time will occur when the output settles between $1.9 \leq E(t) \leq 2.1$ V. From Figure 3.15, the rise occurs in about 4 ms and the settling time is about 9 ms. The period of the ringing behavior, T_d, is judged to be about 13 ms for an $\omega_d \approx 485$ rad/s.

Sine Function Input

The response of a second-order system to a sinusoidal input of the form, $F(t) = A \sin \omega t$, is given by

$$y(t) = y_h + \frac{KA \sin[\omega t + \Phi(\omega)]}{\left\{ \left[1 - (\omega/\omega_n)^2 \right]^2 + [2\zeta\omega/\omega_n]^2 \right\}^{1/2}} \tag{3.18}$$

with frequency-dependent phase shift

$$\Phi(\omega) = \tan^{-1}\left(-\frac{2\zeta\omega/\omega_n}{1 - (\omega/\omega_n)^2} \right) \tag{3.19}$$

The exact form for y_h is found from equations (3.14a–c) and depends on the value of ζ. The steady response, the second term on the right side, has the general form

$$y_{\text{steady}}(t) = y(t \rightarrow \infty) = B(\omega)\sin[\omega t + \Phi(\omega)] \tag{3.20}$$

with amplitude $B(\omega)$. Comparing equations (3.18) and (3.20) shows that the amplitude of the steady response of a second-order system subjected to a sinusoidal input is dependent on the value of ω. So the amplitude of the output signal is frequency dependent. In general, we can define the magnitude ratio, $M(\omega)$, for a second-order system as

$$M(\omega) = \frac{B(\omega)}{KA} = \frac{1}{\left\{ \left[1 - (\omega/\omega_n)^2 \right]^2 + [2\zeta\omega/\omega_n]^2 \right\}^{1/2}} \tag{3.21}$$

The magnitude ratio-frequency dependence for a second-order system is plotted in Figure 3.16 for several values of damping ratio. A corresponding plot of the phase-shift dependency on input frequency and damping ratio is shown in Figure 3.17. For an ideal measurement system, $M(\omega)$ would equal unity and $\Phi(\omega)$ would equal zero for all values of measured frequency. Instead, $M(\omega)$ will approach zero and $\Phi(\omega)$ will approach $-\pi$ as ω/ω_n becomes large. Keep in mind that ω_n is a property of the measurement system, while ω is a property of the input signal.

System Characteristics

Several tendencies are apparent in Figures 3.16 and 3.17. For a system of zero damping, $\zeta = 0$, $M(\omega)$ will approach infinity and $\Phi(\omega)$ jumps to $-\pi$ in the vicinity of $\omega = \omega_n$. This behavior is characteristic of system resonance. Real systems possess some amount of damping, which modifies the abruptness and magnitude of resonance, but underdamped systems may still achieve resonance. This region on Figures 3.16 and 3.17 is called the *resonance band* of the system, referring to the range of frequencies over which the system is in resonance. Resonance in underdamped systems occurs at the *resonance frequency*, $\omega_R = \omega_n \sqrt{1 - 2\zeta^2}$. The resonance frequency is a property of the measurement system. Resonance is excited by a sinusoidal input signal frequency. The resonance frequency

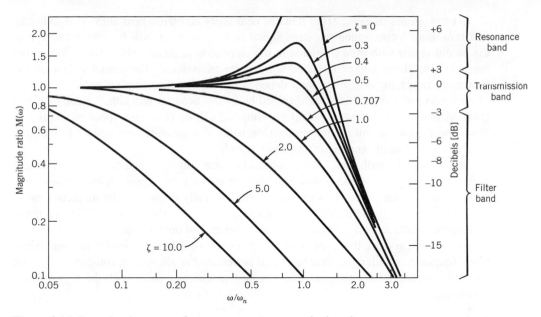

Figure 3.16 Second-order system frequency response: magnitude ratio.

differs from the ringing frequency of free oscillation. Resonance behavior results in values of $M(\omega) > 1$ and considerable phase shift. For most applications, operating at frequencies within the resonance band is undesirable, confusing, and could even be damaging to some delicate sensors. Resonance behavior is very nonlinear and results in distortion of the signal. On the other hand, systems with $\zeta > 0.707$ do not resonate.

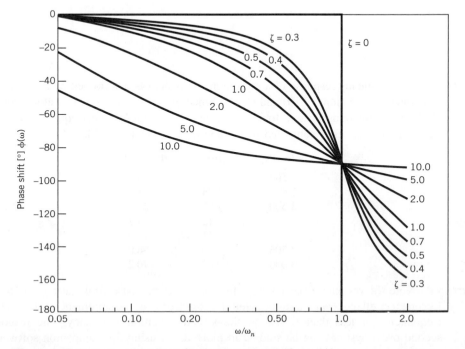

Figure 3.17 Second-order system frequency response: phase shift.

At low values of ω/ω_n, $M(\omega)$ remains near unity and $\Phi(\omega)$ near zero. This indicates that information concerning the input signal of frequency, ω, will be passed through to the output signal with little alteration in the amplitude or phase shift. This region on the frequency response curves is called the *transmission band*. The actual extent of the frequency range for near unity gain depends on the system damping ratio. The transmission band of a system is either specified by its frequency bandwidth, typically defined for a second-order system as $-3\,\mathrm{dB} \leq M(\omega) \leq 3\,\mathrm{dB}$, or otherwise specified explicitly. You need to operate within the transmission band of a measurement system to measure correctly the dynamic content of the input signal.

At large values of ω/ω_n, $M(\omega)$ approaches zero. In this region, the measurement system will attenuate the amplitude information in the input signal. A large phase shift occurs. This region is known as the *filter band*, typically defined as the frequency range over which $M(\omega) \leq -3\,\mathrm{dB}$. Most readers are familiar with the use of a filter to remove undesirable features from a desirable product. When you operate within the filter band of a measurement system, the amplitudes of the portion of dynamic signal corresponding to those frequencies within the filter band will be reduced or eliminated completely. So you need to match carefully the measurement system characteristics with the signal being measured.

EXAMPLE 3.9

Determine the frequency response of a pressure transducer that has a damping ratio of 0.5 and a ringing frequency (found by a step test) of 1200 Hz.

KNOWN

$\zeta = 0.5$

$\omega_d = 2\pi(1200\,\mathrm{Hz}) = 7540\,\mathrm{rad/s}$

ASSUMPTIONS   Second-order system

FIND

$M(\omega)$ and $\Phi(\omega)$

SOLUTION   The frequency response of a measurement system is determined by $M(\omega)$ and $\Phi(\omega)$ as defined in equations (3.19) and (3.21). Since $\omega_d = \omega_n\sqrt{1 - \zeta^2}$, the natural frequency of the pressure transducer is found to be $\omega_n = 8706$ rad/s. The frequency response at selected frequencies is computed from equations (3.19) and (3.21):

ω [rad/s]	$M(\omega)$	$\Phi(\omega)$ [°]
500	1.00	−3.3
2 600	1.04	−18.2
3 500	1.07	−25.6
6 155	1.15	−54.7
7 540	1.11	−73.9
8 706	1.00	−90.0
50 000	0.05	−170.2

COMMENT   Note the resonance behavior in the transducer response peaks at $\omega_R = 6155$ rad/s. Resonance effects can be minimized by operating a measurement system at input frequencies of less than $\sim$30% of the system's natural frequency. The response of second-order systems can be studied in more detail using the companion software. Try program *SecondOrd* and Labview® program *Second_order*.

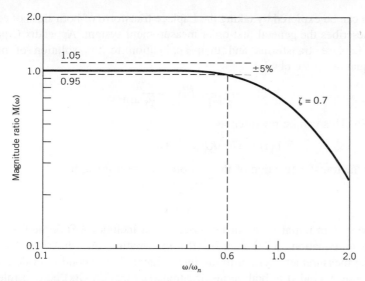

Figure 3.18 Magnitude ratio for second-order system with $\zeta = 0.7$ for Example 3.10.

EXAMPLE 3.10

An accelerometer is to be selected to measure a time-dependent motion. In particular, input signal frequencies below 100 Hz are of prime interest. Select a set of acceptable parameter specifications for the instrument assuming a dynamic error of ±5%.

KNOWN
  $f \leq 100\,\text{Hz}$ (i.e., $\omega \leq 628\,\text{rad/s}$)

ASSUMPTIONS   Second-order system
  Dynamic error of ±5% acceptable

FIND   Select ω_n and ζ

SOLUTION   To meet a ±5% dynamic error constraint, we will want $0.95 \leq M(\omega) \leq 1.05$ over the frequency range $0 \leq \omega \leq 628$ rad/s. This solution is open ended in that a number of instruments with different ω_n will do the task. So as one solution, let us set $\zeta = 0.7$ and then solve for the required ω_n using equation (3.21):

$$0.95 \leq M(\omega) = \frac{1}{\left\{ \left[1 - (\omega/\omega_n)^2 \right]^2 + (2\zeta\omega/\omega_n)^2 \right\}^{1/2}} \leq 1.05$$

With $\omega = 628$ rad/s, these equations yield $\omega_n \geq 1047$ rad/s. Alternatively, we could plot equation (3.21) with $\zeta = 0.7$, as shown in Figure 3.18. In Figure 3.18, we find that $0.95 \leq M(\omega) \leq 1.05$ over the frequency range $0 \leq \omega/\omega_n \leq 0.6$. Again, this puts $\omega_n \geq 1047$ rad/s as being acceptable. So as one solution, an instrument having $\zeta = 0.7$ and $\omega_n \geq 1047$ rad/s will meet the problem constraints.

3.4   TRANSFER FUNCTIONS

Consider the schematic representation in Figure 3.19. The measurement system operates on the input signal, $F(t)$, by some function, $G(s)$, so as to indicate the output signal, $y(t)$.

This operation can be explored by taking the Laplace transform of both sides of equation (3.4), which describes the general first-order measurement system. Appendix C provides a review of Laplace transforms and their application to the solution of ordinary differential equations. One obtains

$$Y(s) = \frac{1}{\tau s + 1} KF(s) + \frac{y_0}{\tau s + 1}$$

where $y_0 = y(0)$. This can be rewritten as

$$Y(s) = G(s)KF(s) + G(s)Q(s) \tag{3.22}$$

where $G(s)$ is the transfer function of the first-order system given by

$$G(s) = \frac{1}{\tau s + 1} \tag{3.23}$$

and $Q(s)$ is the system initial state function. Because it includes $KF(s)$, the first term on the right side of equation (3.22) contains the information that describes the steady response of the measurement system to the input signal. The second term describes its transient response. Included in both terms, the transfer function $G(s)$ plays a role in the complete time response of the measurement system. As indicated in Figure 3.19, the *transfer function* defines the mathematical operation that the measurement system performs on the input signal $F(t)$ to yield the time response (output signal) of the system.

The system frequency response, which has been shown to be given by $M(\omega)$ and $\Phi(\omega)$, can be found by finding the value of $G(s)$ at $s = i\omega$. This yields the complex number,

$$G(s = i\omega) = G(i\omega) = \frac{1}{\tau i\omega + 1} = M(\omega)e^{i\Phi(\omega)} \tag{3.24}$$

where $G(i\omega)$ is a vector on the real–imaginary plane having a magnitude, $M(\omega)$, and inclined at an angle, $\Phi(\omega)$, relative to the real axis as indicated in Figure 3.20. For the first-order system, the magnitude of $G(i\omega)$ is simply that given by $M(\omega)$ from equation (3.10) and the phase shift angle by $\Phi(\omega)$ from equation (3.9).

For a second-order or higher system, the approach is the same. The governing equation for a second-order system is defined by equation (3.13), with initial conditions $y(0) = y_0$ and $\dot{y}(0) = \dot{y}_0$. The Laplace transform yields

$$Y(s) = \frac{1}{(1/\omega_n^2)s^2 + (2\zeta/\omega_n)s + 1} KF(s) + \frac{s\dot{y}_0 + y_0}{(1/\omega_n^2)s^2 + (2\zeta/\omega_n)s + 1} \tag{3.25}$$

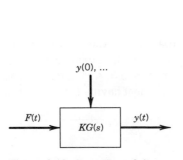

Figure 3.19 Operation of the transfer function. Compare with Figure 3.2.

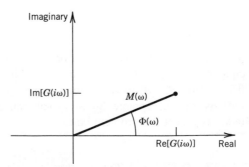

Figure 3.20 Complex plane approach to describing frequency response.

which can again be represented by

$$Y(s) = G(s)KF(s) + G(s)Q(s) \tag{3.26}$$

By inspection, the transfer function is given by

$$G(s) = \frac{1}{(1/\omega_n^2)s^2 + (2\zeta/\omega_n)s + 1} \tag{3.27}$$

Solving for $G(s)$ at $s = i\omega$, we obtain for a second-order system

$$G(s = i\omega) = \frac{1}{(i\omega)^2/\omega_n^2 + 2\zeta i\omega/\omega_n + 1} = M(\omega)e^{i\Phi(\omega)} \tag{3.28}$$

which yields the same magnitude ratio and phase shift relations as given by equations (3.19) and (3.21).

3.5   PHASE LINEARITY

We can see from Figures 3.16 and 3.17 that systems having a damping ratio near 0.7 possess the broadest frequency range over which $M(\omega)$ will remain at or near unity and that over this same frequency range the phase shift will essentially vary in a linear manner with frequency. Although it is not possible to design a measurement system without accepting some amount of phase shift, it is desirable to design a system such that the phase shift varies linearly with frequency. This is because operating outside this range will be accompanied with a significant distortion in the waveform of the output signal relative to the input signal. *Distortion* refers to a notable change in the shape of the waveform, as opposed to simply an amplitude alteration or relative phase shift. To minimize distortion, many measurement systems are designed with $0.6 \leq \zeta \leq 0.8$.

Signal distortion can be illustrated by considering a particular complex waveform represented by a general function, $u(t)$:

$$u(t) = \sum_{n=1}^{\infty} \sin n\omega t = \sin \omega t + \sin 2\omega t + \cdots \tag{3.29}$$

Suppose during a measurement a phase shift of this signal were to occur such that the phase shift remained linearly proportional to the frequency; that is, the measured signal, $v(t)$, could be represented by

$$v(t) = \sin(\omega t - \Phi) + \sin(2\omega t - 2\Phi) + \cdots \tag{3.30}$$

Or, by setting

$$\theta = (\omega t - \Phi) \tag{3.31}$$

we can write

$$v(t) = \sin \theta + \sin 2\theta + \cdots \tag{3.32}$$

We see that $v(t)$ in equation (3.22) is equivalent to the original signal, $u(t)$. If the phase shift were not linearly related to the frequency, this would not be so. This is demonstrated in Example 3.11.

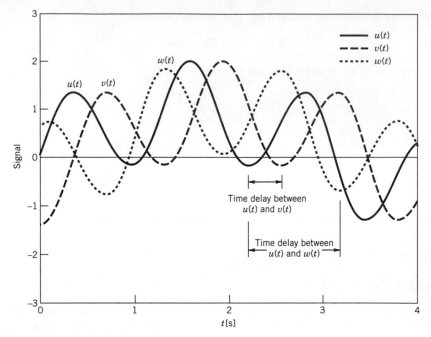

Figure 3.21 Waveforms for Example 3.11.

EXAMPLE 3.11

Consider the effect of the variations in phase shift with frequency on a measured signal by examination of the signal defined by the function

$$u(t) = \sin t + \sin 5t$$

Suppose this signal is measured in such a way that a phase shift that is linearly proportional to the frequency occurs in the form

$$v(t) = \sin(t - 0.35) + \sin[5t - 5(0.35)]$$

Both $u(t)$ and $v(t)$ are plotted in Figure 3.21. We can see that the two waveforms are identical except that $v(t)$ lags $u(t)$ by some time increment.

Now suppose this signal is measured in such a way that the relation between phase shift and frequency was nonlinear, such as in the signal output form

$$w(t) = \sin(t - 0.35) + \sin(5t - 5)$$

The $w(t)$ signal is also plotted in Figure 3.21 where we see it behaves differently than $u(t)$. This different behavior is signal distortion. In comparison of $u(t)$, $v(t)$, and $w(t)$, it should be apparent that distortion can be brought about by the nonlinear relation of phase shift with frequency.

3.6 MULTIPLE-FUNCTION INPUTS

So far we have discussed measurement system response to a signal containing only a single frequency. What about measurement system response to multiple input frequencies? Or to an input that consists of both a static and a dynamic part, such as a periodic strain or temperature signal? When using models that are linear, ordinary differential equations subjected to inputs that are linear in terms of the dependent variable, the principle of superposition of linear systems will apply to the solution of these equations. The *principle of superposition* states that a linear combination of input signals applied to a linear measurement system produces an output signal that is simply the linear addition of the separate output signals that would result if each input term had been applied separately. Because the form of the transient response is not affected by the input function, we can focus on the steady response. In general, we can write that if the forcing function of a form

$$F(t) = A_0 + \sum_{k=1}^{\infty} A_k \sin \omega_k t \qquad (3.33)$$

is applied to a system, then the combined steady response will have the form

$$KA_0 + \sum_{k=1}^{\infty} B(\omega_k) \sin[\omega_k t + \Phi(\omega_k)] \qquad (3.34)$$

where $B(\omega_k) = KA_k M(\omega_k)$. The development of the superposition principle can be found in basic texts on dynamic systems (e.g., [4]).

EXAMPLE 3.12

A second-order instrument having a $K = 1$ unit/unit, $\zeta = 2$, and $\omega_n = 628$ rad/s is to be used to measure an input signal of the form

$$F(t) = 5 + 10 \sin 25t + 20 \sin 400t$$

Predict its steady output signal.

KNOWN   Second-order system

$K = 1$ unit/unit

$\zeta = 2.0$

$\omega_n = 628$ rad/s

$F(t) = 5 + 10 \sin 25t + 20 \sin 400t$

ASSUMPTIONS   Linear system (superposition holds)

FIND $y(t)$

SOLUTION   Since $F(t)$ has a form consisting of a linear addition of multiple input functions, the steady response signal will have the form of equation (3.33) of $y_{steady}(t) = KF(t)$ or

$$y(t) = 5K + 10 KM(25 \, \text{rad/s}) \sin[25t + \Phi(25 \, \text{rad/s})]$$
$$+ 20 KM(400 \, \text{rad/s}) \sin[400t + \Phi(400 \, \text{rad/s})]$$

Using equations (3.19) and (3.21), or, alternatively, using Figures 3.16 and 3.17, with $\omega_n = 628$ rad/s and $\zeta = 2.0$, we find that

$$M(25 \, \text{rad/s}) = 0.99 \quad \Phi(25 \, \text{rad/s}) = -9.1°$$
$$M(400 \, \text{rad/s}) = 0.39 \quad \Phi(400 \, \text{rad/s}) = -77°$$

So that the steady output signal will have the form

$$y(t) = 5 + 9.9 \sin(25t - 9.1°) + 7.8 \sin(400t - 77°)$$

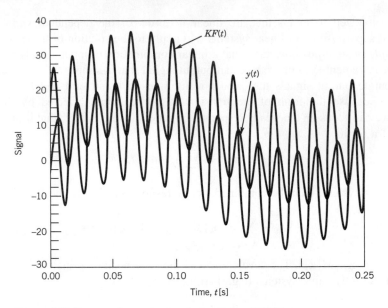

Figure 3.22 Input and output signals for Example 3.12.

The output signal is plotted against the input signal in Figure 3.22. The amplitude spectra for both the input signal and the output signal are shown in Figure 3.23(a) and (b). Spectra can also be generated by using the accompanying software using programs *FunSpect* and *DataSpect*.

COMMENT   We see that information concerning the average value and concerning the 25 rad/s component of the input signal will be passed along to the output signal. However, amplitude information in the output signal concerning the 400-rad/s component of the input signal has been severely reduced (down 61%). If information concerning the 400-rad/s component were important, then we would need to select an instrument that has an improved frequency response in the 400-rad/s range.

3.7   COUPLED SYSTEMS

As instruments in each stage of a measurement system are connected, the output from one stage becomes the input to the next stage and so forth. Such measurement systems will have a coupled output response to the original input signal that is some combination of the individual instrument responses to the input. However, the system concepts of zero-, first-, and second-order systems studied previously can still be used for a case-by-case study of the coupled measurement system. This is done by considering the input to each stage of the measurement system as the output of the previous stage.

This concept is easily illustrated by considering a first-order sensor that may be connected to a second-order output device (for example, a temperature sensor–transducer connected to a chart recorder). Suppose the input to the sensor is a simple periodic waveform, $F_1(t) = A \sin \omega t$. The transducer will respond with an output signal of the form of equation (3.8):

$$y_t(t) = Ce^{-t/\tau} + \frac{K_t A}{\sqrt{1 + (\omega\tau)^2}} \sin(\omega t + \Phi_t)$$

$$\Phi_t = -\tan^{-1}\omega\tau$$

(3.35)

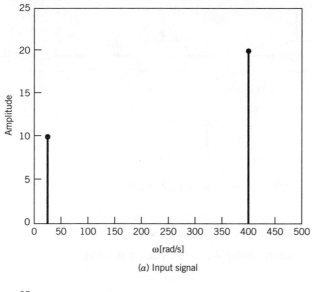

(a) Input signal

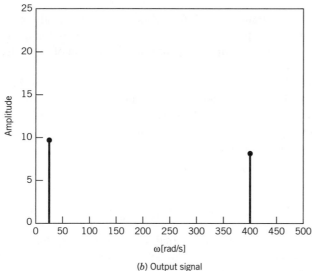

(b) Output signal

Figure 3.23 Amplitude spectrum of Example 3.12.

where the subscript t refers to the transducer. However, the transducer output signal now becomes the input signal, $F_2(t) = y_t$, to the second-order device. The output from the second-order device, $y_s(t)$, will be a second-order response appropriate for input $F_2(t)$,

$$y_s(t) = y_h(t) + \frac{K_t K_s A \sin[\omega t + \Phi_t + \Phi_s]}{\left[1 + (\omega\tau)^2\right]^{1/2} \left\{\left[1 - (\omega/\omega_n)^2\right]^2 + [2\zeta\,\omega/\omega_n]^2\right\}^{1/2}} \tag{3.36}$$

$$\Phi_s = -\tan^{-1} \frac{2\zeta\omega/\omega_n}{1 - (\omega/\omega_n)^2}$$

where subscript s refers to the chart recorder and $y_h(t)$ is the transient response. The output signal displayed on the recorder, $y_s(t)$, is the measurement system response to the

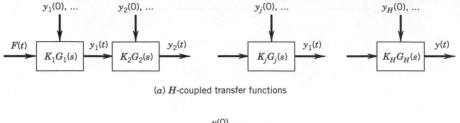

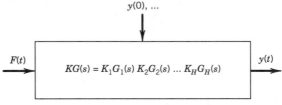

(a) H-coupled transfer functions

(b) Equivalent system transfer function

Figure 3.24 Coupled systems: describing the system transfer function.

original input signal to the transducer, $F_1(t) = A \sin \omega t$. The steady amplitude of the recorder output signal is the product of the static sensitivities and magnitude ratios of the first- and the second-order systems. The phase shift is the sum of the phase shifts of the two systems.

Based on equation (3.36) we can make a general observation. Consider the schematic representation in Figure 3.24, which depicts a measurement system consisting of H interconnected devices, $j = 1, 2, \ldots, H$, each device described by a linear system model. The overall transfer function of the combined system, $G(s)$, will be the product of the transfer functions of each of the individual devices, $G_j(s)$, such that

$$KG(s) = K_1G_1(s)K_2G_2(s)\ldots K_HG_H(s) \tag{3.37}$$

At $s = i\omega$, equation (3.37) becomes

$$KG(i\omega) = (K_1K_2\ldots K_H) \times [M_1(\omega)M_2(\omega)\ldots M_H(\omega)]e^{i[\Phi_1(\omega)+\Phi_2(\omega)+\cdots+\Phi_H(\omega)]} \tag{3.38}$$

According to equation (3.38), given an input signal to device 1, the system steady output signal at device H will be described by the system frequency response $G(i\omega) = M(\omega)e^{i\Phi(\omega)}$, with an overall system static sensitivity described by

$$K = K_1K_2\ldots K_H \tag{3.39}$$

The overall system magnitude ratio will be the product

$$M(\omega) = M_1(\omega)M_2(\omega)\ldots M_H(\omega) \tag{3.40}$$

and the overall system phase shift will be the sum

$$\Phi(\omega) = \Phi_1(\omega) + \Phi_2(\omega) + \cdots + \Phi_H(\omega) \tag{3.41}$$

This will hold true provided that significant loading effects do not exist, a situation discussed in Chapter 6.

3.8   SUMMARY

The response of a measurement system to a time-dependent input depends on several factors including the inherent rate of response of that system and the frequency response of that system. Modeling has enabled us to develop and to illustrate these concepts. Those system design parameters that affect system response are exposed through modeling, and this assists in instrument selection. Modeling has also suggested the methods by which measurement system specifications such as time constant, response time, frequency response, damping ratio, and resonance frequency can be determined both analytically and experimentally. Interpretation of these system properties and their effect on system performance was determined.

The rate of response of a system to a change in input is estimated by use of the step function input. The system parameters of time constant, for first-order systems, and natural frequency and damping ratio, for second-order systems, are used as indicators of system response rate. The magnitude ratio and phase shift define the frequency response of any system and are found by an input of a periodic waveform to a system. Figures 3.12, 3.13, 3.16, and 3.17 are universal frequency response curves for first- and second-order systems, respectively. These curves can be found in most engineering and mathematical handbooks and can be applied to any first- or second-order system, as appropriate.

REFERENCES

1. Close, C. M., and D. K. Frederick, *Modeling and Analysis of Dynamic Systems*, 2d ed., Wiley, Boston, 1994.
2. Doebelin, E. O., *System Modeling and Response, Theoretical and Experimental Approaches*, Wiley, New York, 1980.
3. Ogata, K., *System Dynamics*, 4th ed., Prentice-Hall, Englewood Cliffs, NJ, 2003.
4. Palm, W. J., III, *Modeling, Analysis and Control of Dynamic Systems*, 2d ed., Wiley, New York, 2000.

NOMENCLATURE

$a_0, a_1, \ldots, a_n$	physical coefficients	$M(\omega)$	magnitude ratio, B/KA
$b_0, b_1, \ldots, b_m$	physical coefficients	$T(t)$	temperature [°]
c	damping coefficient [$m\ t^{-1}$]	T_d	ringing period [t]
f	cyclical frequency ($f = \omega/2\pi$) [Hz]	$U(t)$	unit step function
k	spring constant or stiffness [$m\ t^{-2}$]	β_1	time lag [t]
m	mass [m]	$\delta(\omega)$	dynamic error
$p(t)$	pressure [$m\ l^{-1}\ t^{-2}$]	τ	time constant [t]
t	time[t]	$\Phi(\omega)$	phase shift
$x(t)$	independent variable	ω	circular frequency [t^{-1}]
$y(t)$	dependent variable	ω_n	natural frequency magnitude [t^{-1}]
y^n	nth time derivative of $y(t)$	ω_d	ringing frequency [t^{-1}]
$y^n(0)$	initial condition of y^n	ω_R	resonance frequency [t^{-1}]
A	input signal amplitude	ζ	damping ratio
$B(\omega)$	output signal amplitude	Γ	error fraction
C	constant		
$E(t)$	voltage [V] or energy [$m\ l\ t^{-3}$]	**Subscripts**	
F	force [$m\ l\ t^{-2}$]		
$F(t)$	forcing function	0	initial value
$G(s)$	transfer function	∞	final or steady value
K	static sensitivity	h	homogeneous solution

PROBLEMS

Although not required, the companion software can be used for solving many of these problems. We encourage the reader to explore the software provided.

3.1 A particular measurement system has a static sensitivity of 2 V/kg. An input range of 1 kg–10 kg needs to be measured. Determine the expected range of values for the output signal. What would be the significance of increasing the static sensitivity?

3.2 Determine the 75, 90, and 95% response time for each of the systems given (assume zero initial conditions):

   a. $0.4\dot{T} + T = 4U(t)$

   b. $\ddot{y} + 2\dot{y} + 4y = U(t)$

   c. $2\ddot{P} + 8\dot{P} + 8P = 2U(t)$

   d. $5\dot{y} + 5y = U(t)$

3.3 A special sensor is designed to sense the percent vapor present in a liquid–vapor mixture. If during a static calibration the sensor indicates 80 units when in contact with 100% liquid, 0 units with 100% vapor, and 40 units with a 50–50% mixture, determine the static sensitivity of the sensor.

3.4 A measurement system can be modeled by the equation

$$0.5\dot{y} + y = F(t)$$

Initially, the output signal is steady at 75 volts. The input signal is then suddenly increased to 100 volts

   a. Determine the response equation.

   b. On the same graph, plot both the input signal and the system time response until steady response is reached.

3.5 Suppose a thermometer similar to that of Example 3.3 is known to have a time constant of 30 s in a particular application. Plot its time response to a step change from 32 to 120°F. Determine its 90% rise time.

3.6 Referring back to Example 3.3, a student establishes the time constant of a temperature sensor by first holding it immersed in hot water and then suddenly removing it and holding it immersed in cold water. Several other students perform the same test with similar sensors. Overall, their results are inconsistent with each other with estimated time constants differing by as much as a factor of 1.2. Offer any suggestions as to why this might happen. Hint: Try this yourself and think about control of test conditions.

3.7 A thermocouple, which responds as a first-order instrument, has a time constant of 20 ms. Determine its 90% rise time.

3.8 During a step function calibration, a first-order instrument is exposed to a step change of 100 units. If after 1.2 s the instrument indicates 80 units, estimate the instrument time constant. Estimate the error in the indicated value after 1.5 s. $y(0) = 0$ units; $K = 1$ unit/ unit.

3.9 Estimate any dynamic error that could result from measuring a 2-Hz periodic waveform using a first-order system having a time constant of 0.7 s.

3.10 A first-order instrument having a time constant of 1 s is used to measure a signal that can be represented by $F(t) = 10 \cos 2.5t$. Write the expected indicated steady response output signal. What is the expected time lag between input and output signal? $y(0) = 0$; $K = 1$.

3.11 A first-order instrument with a time constant of 2 s is to be used to measure a periodic input. If a dynamic error of ±2% can be tolerated, determine the maximum frequency of periodic input that can be measured. What is the associated time lag (in seconds) at this frequency?

3.12   Determine the frequency response [$M(\omega)$ and $\phi(\omega)$] for an instrument having a time constant of 10 ms. Estimate the instrument's usable frequency range to keep its dynamic error within 10%.

3.13   A temperature measuring device with a time constant of 0.15 s outputs a voltage that is linearly proportional to temperature. The device is used to measure an input signal of the form $T(t) = 115 + 12 \sin 2t\,°C$. Plot the input signal and the predicted output signal with time assuming first-order behavior and a static sensitivity of 5 mV/°C. Determine the dynamic error and time lag in the steady response. $T(0) = 115°C$.

3.14   A first-order sensor is to be installed into a reactor vessel to monitor temperature. If a sudden rise in temperature greater than 100°C should occur, shutdown of the reactor will need to begin within 5 s after reaching 100°C. Determine the maximum allowable time constant for the sensor.

3.15   A single-loop *LR* circuit having a resistance of 1 MΩ is to be used as a low-pass filter between an input signal and a voltage measurement device. To attenuate undesirable frequencies above 1000 Hz by at least 50% select a suitable inductor size. The time constant for this circuit is given by *L/R*.

3.16   A measuring system has a natural frequency of 0.5 rad/s, a damping ratio of 0.5, and a static sensitivity of 0.5 m/V. Estimate its 90% rise time and settling time if $F(t) = 2\,U(t)$ and the initial condition is zero. Plot the response $y(t)$ and indicate its transient and steady responses.

3.17   Plot the frequency response, based on equations (3.19) and (3.21), for an instrument having a damping ratio of 0.6. Determine the frequency range over which the dynamic error remains within 5%. Repeat for a damping ratio of 0.9 and 2.0.

3.18   The output from a temperature system indicates a steady, time-varying signal having an amplitude which varies between 30 and 40°C with a single frequency of 10 Hz. Express the output signal as a waveform equation, $y(t)$. If the dynamic error is to be less than 1%, what must be the system's time constant?

3.19   A cantilever beam instrumented with strain gauges is used as a force scale. A step-test on the beam provides a measured damped oscillatory signal with time. If the signal behavior is second order, show how a data-reduction design plan could use the information in this signal to determine the natural frequency and damping ratio of the cantilever beam. (Hint: consider the shape of the decay of the peak values in the oscillation.)

3.20   A step test of a transducer brings on a damped oscillation decaying to a steady value. If the period of oscillation is 0.577 ms, what is the transducer ringing frequency?

3.21   If an instrument has a known damping ratio of 0.8, ringing frequency of 1000 Hz, and static sensitivity of 1.5 V/V, determine its expected steady response to an input signal that oscillates sinusoidally between 12 and 24 V with a frequency of 300 Hz.

3.22   An application demands that a sinusoidal pressure variation of 250 Hz be measured with no more than 2% dynamic error. In selecting a suitable pressure transducer from a vendor catalog, you note that a desirable line of transducers has a fixed natural frequency of 600 Hz but that you have a choice of transducer damping ratios of between 0.5 and 1.5 in increments of 0.05. Select a suitable transducer.

3.23   A compact disk player is to be isolated from room vibrations by placing it on an isolation pad. The isolation pad can be considered as a board of mass, *m*, a foam mat of stiffness, *k*, and a damping coefficient, *c*. For expected vibrations in the frequency range of between 2 and 40 Hz, select reasonable values for *m*, *k*, and *c* such that the room vibrations are attenuated by at least 50%. Assume that the only degree of freedom is in the vertical direction.

3.24   A single-loop *RCL* electrical circuit can be modeled as a second-order system in terms of current. Show that the differential equation for such a circuit subjected to a forcing function potential $E(t)$ is given by

$$L\frac{d^2I}{dt^2} + R\frac{dI}{dt} + \frac{I}{C} = E(t)$$

Determine the natural frequency and damping ratio for this system. For a forcing potential, $E(t) = 1 + 0.5 \sin 2000t$ V, determine the system steady response when $L = 2$ H, $C = 1$ μF, and $R = 10,000\,\Omega$. Plot the steady output signal and input signal versus time. $I(0) = \dot{I}(0) = 0$.

3.25   A transducer that behaves as a second-order instrument has a damping ratio of 0.7 and a natural frequency of 1000 Hz. It is to be used to measure a signal containing frequencies as large as 750 Hz. If a dynamic error of ±10% can be tolerated, show whether or not this transducer is a good choice.

3.26   A strain-gauge measurement system is mounted on an airplane wing to measure wing oscillation and strain during wind gusts. The strain system has a 90% rise time of 100 ms, a ringing frequency of 1200 Hz, and a damping ratio of 0.8. Estimate the dynamic error in measuring a 1-Hz oscillation. Also, estimate any time lag. Explain in words the meaning of this information.

3.27   An instrument having a resonance frequency of 1414 rad/s with a damping ratio of 0.5 is used to measure a signal of ∼ 6000 Hz. Estimate the expected dynamic error and phase shift.

3.28   Select one set of appropriate values for damping ratio and natural frequency for a second-order instrument used to measure frequencies up to 100 rad/s with no more than ±10% dynamic error. A catalog offers models with damping ratios of 0.4, 1, and 2 and natural frequencies of 200 and 500 rad/s. Explain your reasoning.

3.29   A signal of the form $F(t) = \sin t + 0.3 \sin 20t$ N is input to a first-order system that has a time constant of 0.2 s and a static sensitivity of 1 V/N. Determine the steady response (steady output signal) from the system. Discuss the transfer of information from the input to the output. Can the input signal be resolved based on the output?

3.30   A force transducer having a damping ratio of 0.5 and a natural frequency of 4000 Hz is available for use to measure a periodic signal of 2000 Hz. Show whether or not the transducer passes a ±10% dynamic error constraint. Estimate its resonance frequency.

3.31   An accelerometer, whose frequency response is defined by equations (3.19) and (3.21), has a damping ratio of 0.4 and a natural frequency of 18,000 Hz. It is used to sense the relative displacement of a beam to which it is attached. If an impact to the beam imparts a vibration at 4500 Hz, estimate the dynamic error and phase shift in the accelerometer output. Estimate its resonance frequency.

3.32   Derive the equation form for the magnitude ratio and phase shift of the seismic accelerometer of Example 3.1. Does its frequency response differ from that predicted by equations (3.19) and (3.21)? For what type of measurement would you suppose this instrument would be best suited?

3.33   Suppose the pressure transducer of Example 3.9 had a damping ratio of 0.6. Plot its frequency response $M(\omega)$ and $\phi(\omega)$. At which frequency is $M(\omega)$ a maximum?

3.34   The output stage of a first-order transducer is to be connected to a second-order display stage device. The transducer has a known time constant of 1.4 ms and static sensitivity of

2 V/°C while the display device has values of sensitivity, damping ratio and natural frequency of 1 V/V, 0.9, and 5000 Hz, respectively. Determine the steady response of this measurement system to an input signal of the form, $T(t) = 10 + 50 \sin 628t°C$.

3.35 The displacement of a solid body is to be monitored by a transducer (second-order system) with signal output displayed on a recorder (second-order system). The displacement is expected to vary sinusoidally between 2 and 5 mm at a rate of 85 Hz. Select appropriate design specifications for the measurement system for no more than 5% dynamic error (i.e., specify an acceptable range for natural frequency and damping ratio for each device).

3.36 A force measurement system has a resonance frequency of 82.5 rad/s, a damping ratio of 0.4, and static sensitivity of 2 V/N. If an input of the form

$$F(t) = 3 + \sin 8t + \sin 165t\,\text{N}$$

is to be applied, write the expected form of the output signal in volts. Comment on the suitability of the instrument for this measurement. Use a suitable software package to estimate and to plot its amplitude spectrum.

3.37 A signal suspected to be of the nominal form

$$y(t) = 5 \sin 1000t\,\text{mV}$$

is input to a first-order instrument having a time constant of 100 ms and $K = 1$ V/V. It is then to be passed through a second-order amplifier having a $K = 100$ V/V, a natural frequency of 15,000 Hz, and a damping ratio of 0.8. What is the expected form of the output signal, $y(t)$? Estimate the dynamic error and phase lag in the output. Is this system a good choice here? If not, do you have any suggestions?

3.38 A typical modern dc audio amplifier has a frequency bandwidth of 0 to 20,000 Hz ± 1 dB. Explain the meaning of this specification and its relation to music reproduction.

3.39 The displacement of a rail vehicle chassis as it rolls down a track is measured using a transducer ($K = 10$ mV/mm, $\omega_n = 10\,000$ rad/s, $\zeta = 0.6$) and a recorder ($K = 1$ mm/mV, $\omega_n = 700$ rad/s, $\zeta = 0.7$). The resulting amplitude spectrum of the output signal consists of a spike of 90 mm at 2 Hz and a spike of 50 mm at 40 Hz. Are the measurement system specifications suitable for the displacement signal? (If not, suggest changes.) Estimate the actual displacement of the chassis. State any assumptions.

3.40 The amplitude spectrum of the time-varying displacement signal from a vibrating U-shaped tube is expected to have a spike at 85, 147, 220, and 452 Hz. Each spike is related to an important vibrational mode of the tube. Displacement transducers available for this application have a range of natural frequencies from 500 Hz through 1000 Hz, with a fixed damping ratio of about 0.5. The transducer output is to be monitored on a DFT-based spectrum analyzer that has a frequency bandwidth extending from 0.1 Hz to 250 kHz. Within the stated range of availability, select a suitable transducer for this measurement from the given range.

3.41 A sensor mounted to a cantilever beam indicates beam motion with time. When the beam is deflected and released (step test), the sensor signal indicates that the beam oscillates as an underdamped second-order system with a ringing frequency of 10 rad/s. The maximum displacement amplitudes are measured at three different times corresponding to the 1st, 16th, and 32nd cycles and found to be 17, 9, and 5 mV, respectively. Estimate the damping ratio and natural frequency of the beam based on this measured signal, $K = 1$ mm/mV.

3.42 Write a short essay on how system properties of static sensitivity, natural frequency, and damping ratio affect the output information from a measurement system. Be sure to discuss the relative importance of the transient and steady aspects of the resulting signal.

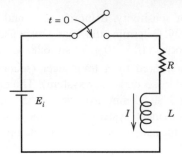

Figure 3.25 Problem 3.43.

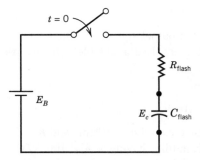

Figure 3.26 Problem 3.44.

3.43 The starting transient of a dc motor can be modeled as an *RL* circuit, with the resistor and inductor in series (Figure 3. 25). Let $R = 4\,\Omega$, $L = 0.1$H, and $E_i = 50$ V with $t(0) = 0$. Find the current draw with time for $t > 0^+$.

3.44 A camera flash light is driven by the energy stored in a capacitor. Suppose the flash operates off a 6-V battery. Determine the time required for the capacitor stored voltage to reach 90% of its maximum energy ($= 1/2\ CE_B^2$). Model this as an *RC* circuit (Figure 3.26). For the flash: $C = 1000\ \mu$F, $R = 1\ k\Omega$, and $E_c(0) = 0$.

3.45 Run program *Thermal Response*. The program allows the user to vary time constant τ and signal amplitude. To impose a step change move the input value up/down. Discuss the effect of time constant on the time required to achieve a steady signal following a new step change.

Chapter 4

Probability and Statistics

4.1 INTRODUCTION

Suppose we had a large box containing thousands of similar round bearings. To get an idea of the size of the bearings, we might measure two dozen. The resulting diameter values form a data set, which we then use to imply something about the size of the bearings in the box as a whole, such as average size and size variation. But how close are these values taken from our data set to the actual average size and variation of all the bearings in the box? If we took another data set, should we expect the values to be exactly the same? These are the same questions that surround engineering measurements. What is the mean value of the set? Do the variations meet the tolerances? How good are these results? These have answers in probability and statistics.

Just as engineering processes show variations, engineering measurements taken repeatedly under seemingly identical conditions will show variations in measured values. Sources that contribute to variations include:

Measurement System

- Resolution
- Repeatability

Measurement Procedure and Technique

- Repeatability

Measured Variable

- Temporal variation
- Spatial variation

For a given set of measurements, we want to be able to quantify (1) a single representative value that best characterizes the average of the data set; (2) a representative value that provides a measure of the variation in the measured data set; and (3) we need to know how well this single average value represents the true average value of the variable measured, which is done by establishing an interval about the representative value in which the true value is expected to lie.

This chapter presents an introduction to the concepts of probability and statistics at a level sufficient to provide information for a large class of engineering judgments. Such material allows for the reduction of raw data into results. Some of this material, such as the concepts of mean value and standard deviations, should already be familiar to an engineering student and are presented along with their relations with probability to provide for a correct interpretation of data.

4.2   STATISTICAL MEASUREMENT THEORY

A *sample* of data refers to a set of data obtained during repeated measurements of a variable under fixed operating conditions. This variable is known as the *measured variable* or the *measurand*. Fixed operating conditions imply that the external conditions that control the process from which the measured value is obtained are held at fixed values while obtaining the sample. In actual engineering practice, the ability to control the operating conditions at truly fixed conditions may be impossible and the term "fixed operating conditions" should be considered in a nominal sense. That is, the process conditions are maintained as closely as possible.

In this chapter, we consider the effects of random errors and how to quantify them. For now, we will assume that systematic error in the measurement is negligible.[1] Recall from Chapter 1 that this is the case where the average error in a data set is zero. We begin by considering the measurement problem of estimating the true value, x', based on the information derived from the repeated measurement of x. The true value of x is the mean value of all possible values of x. This is the value that we want to estimate from the measurement. A sample of the variable x under controlled, fixed operating conditions renders a finite number of data points. We use these data to infer x' based on the mean value of the sample, $\bar{x}$. We can imagine that if the number of data points, N, is very small, then our estimation of x' from the data set could be heavily influenced by the value of any one data point. If a single data point showed a large variation from x' relative to the other data, then the estimate could show a large error. If the data set were larger, then the influence of any one data point would be offset by the larger influence of the other data. As $N \to \infty$, all the possible variations in x would be included in the data set. From a practical view, only finite-sized data sets are possible, in which case the measured data can provide only an estimate of the true value.

From a statistical analysis of the data set and an analysis of sources of error that influence these data, we can estimate x' as

$$x' = \bar{x} \pm u_x \quad (P\%) \tag{4.1}$$

where $\bar{x}$ represents the most probable estimate of x' based on the available data and $\pm u_x$ represents the *uncertainty interval* in that estimate at some probability level, $P\%$. The uncertainty interval combines the estimates of the random error and of the systematic error in the measurement of x.[2] Since statistical methods are used to estimate the random error only, we neglect systematic error here in Chapter 4. In Chapter 5, we find the uncertainty interval by combining both random and systematic errors. In this chapter, we seek methods that will estimate x' and the random error in x due to the variation in the data set.

Probability Density Functions

Regardless of the care taken in obtaining a set of data from independent measurements under identical conditions, random scatter in the data values will occur. As such, the measured variable behaves as a *random variable*. If the variable is continuous in time or space, then it is said to be a *continuous random variable*. A variable represented only by

---

[1]Systematic error does not vary with repeated measurements and so will not affect the statistics of a measurement. Systematic error is considered in Chapter 5.

[2]Statistics texts that ignore systematic error entirely refer to this uncertainty interval as a "confidence interval."

Table 4.1   Sample of Random Variable x

i	x_i	i	x_i
1	0.98	11	1.02
2	1.07	12	1.26
3	0.86	13	1.08
4	1.16	14	1.02
5	0.96	15	0.94
6	0.68	16	1.11
7	1.34	17	0.99
8	1.04	18	0.78
9	1.21	19	1.06
10	0.86	20	0.96

discrete values while time or space is continuous, such as a discrete set of data points, is called a *discrete random variable*. During repeated measurements of a variable under fixed operating conditions, each data point may tend to assume one preferred value or lie within some interval about this value more often than not, when all the data are compared. This tendency toward one central value about which all the other values are scattered is known as a *central tendency* of a random variable.[3] Probability deals with the concept that certain values for a variable will be measured with some frequency of occurrence relative to other values.

The central value and those values scattered about it can be determined from the probability density of the measured variable. The frequency with which the measured variable assumes a particular value or interval of values is described by its probability density. Consider a sample of x shown in Table 4.1, which consists of N individual measurements, x_i, where $i = 2, \ldots, N$, each measurement taken at random but under identical test operating conditions. The measured values of this variable are plotted on a single axis as shown in Figure 4.1.

In Figure 4.1, there exists a region on the axis where the data points tend to clump; this region contains the central value. Such behavior is typical of most engineering measured variables. We might expect that the true mean value of x is contained somewhere in this clump.

This description for variable x can be extended. Suppose we replot the data of Table 4.1. The abscissa will be divided between the maximum and minimum measured values of x into K small intervals. Let the number of times, n_j, that a measured value assumes a value within an interval defined by $x - \delta x \leq x \leq x + \delta x$ be plotted on the ordinate. For small N, K should be conveniently chosen but such that $n_j \geq 5$ for at least

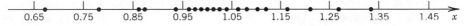

Figure 4.1   Concept of density in reference to a measured variable (from Example 4.1).

---

[3]Not all random variables display a central tendency, e.g., the value in a fair roll of a die could show equal values from 1 through 6 with equal probability of 1/6. But the fair roll of two dice will show a central tendency of the value of 7.

one interval. An estimate of the number of intervals K required [1] for a viable statistical analysis can be found from

$$K = 1.87(N - 1)^{0.40} + 1 \qquad (4.2)$$

This correlation is derived from the estimates provided in reference [2]. As N becomes large, a value of $K \approx N^{1/2}$ works reasonably well. The resulting plot of n_j versus x is called a *histogram* of the variable. The histogram is just another way of viewing both the tendency and the probability density of a variable. If the ordinate were nondimensionalized by dividing n_j by the total number of measurements of the variable, N, a *frequency distribution* of the variable results. For any value of the variable, the frequency, f_j, at which that value of the variable occurred is found from its frequency distribution. This concept is illustrated in Example 4.1.

EXAMPLE 4.1

Compute the histogram and frequency distribution for the data of Table 4.1.

KNOWN   Data of Table 4.1

   $N = 20$

ASSUMPTIONS   Fixed operating conditions

FIND   Histogram and frequency distribution

SOLUTION   To develop the histogram, compute a reasonable number of intervals for this data set. For $N = 20$, a convenient estimate of K is found from (4.2) to be

$$K = 1.87(N - 1)^{0.40} + 1 = 7$$

Next, determine the maximum and minimum values of the data set and divide this range into K intervals. For a minimum of 0.68 and a maximum of 1.34, a value of $\delta x = 0.05$ is chosen. The intervals are shown below.

j	Interval	n_j	$f_j = n_j/N$
1	$0.65 \le x_i < 0.75$	1	0.05
2	$0.75 \le x_i < 0.85$	1	0.05
3	$0.85 \le x_i < 0.95$	3	0.15
4	$0.95 \le x_i < 1.05$	7	0.35
5	$1.05 \le x_i < 1.15$	4	0.20
6	$1.15 \le x_i < 1.25$	2	0.10
7	$1.25 \le x_i < 1.35$	2	0.10

Since at least one interval has an $n_j \ge 5$, the interval number is adequate. The results are plotted in Figure 4.2. The plot displays a definite central tendency at the maximum frequency of occurrence within the interval 0.95 to 1.05.

COMMENT   The sum of the number of occurrences,

$$\sum_{j=1}^{K} n_j$$

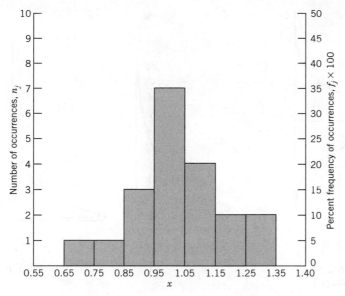

Figure 4.2 Histogram and frequency distribution for data in Table 4.1.

must equal the total number of measurements, N. Likewise, the area under the percent frequency distribution curve will always equal the total frequency of occurrence of 100%; that is,

$$100 \times \sum_{j=1}^{K} f_j = 100\%$$

Probability-density.vi and *Running-histogram.vi* demonstrate the influence of population size and interval numbers on the histogram.

The *probability density function*, $p(x)$, results from the frequency distribution, in the limit as $N \to \infty$ and $\delta x \to 0$, by

$$p(x) = \lim_{N \to \infty, \, \delta x \to 0} \frac{n_j}{N(2\delta x)} \tag{4.3}$$

The probability density function defines the probability that a measured variable might assume a particular value upon any individual measurement. It also provides the central tendency of the variable. This central tendency is the desired representative value that gives the best estimate of the true mean value.

The actual shape that the probability density function takes depends on the variable it represents and the circumstances affecting the process from which the variable is obtained. There are a number of standard distribution shapes that suggest how a variable will be distributed on the probability density plot. The specific values of the variable and the width of the distribution depend on the actual process but the overall shape of the plot will most likely fit some standard distribution. A number of standard distributions that engineering data are likely to follow along with specific comments regarding the types of processes from which these are likely to be found are given in Table 4.2. Generally, the experimentally determined histograms are used to determine which standard distribution the measured variable tends to follow. In turn, the standard distribution is used to interpret the data. Of course, the list in Table 4.2 is not all-inclusive and the reader is referred to more complete treatments of this subject [3, 5].

Table 4.2 Standard Statistical Distributions and Relations to Measurements

Distribution	Applications	Mathematical Representation	Shape
Normal	Most physical properties that are continuous or regular in time or space. Variations due to random error.	$p(x) = \dfrac{1}{\sigma(2\pi)^{1/2}} \exp\left[-\dfrac{1}{2}\dfrac{(x-x')^2}{\sigma^2}\right]$	
Log normal	Failure or durability projections; events whose outcomes tend to be skewed toward the extremity of the distribution.	$p(x) = \dfrac{1}{\pi\sigma(2\pi)^{1/2}} \exp\left[-\dfrac{1}{2}\ln\dfrac{(x-x')^2}{\sigma^2}\right]$	
Poisson	Events randomly occurring in time; $p(x)$ refers to probability of observing x events in time t. Here λ refers to x'.	$p(x) = \dfrac{e^{-\lambda}\lambda^x}{x!}$	

Weibull

Fatigue tests; similar to log normal applications.

See [4]

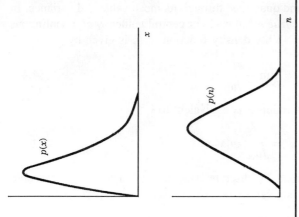

$p(x)$ x

Binomial

Situations describing the number of occurrences, n, of a particular outcome during N independent tests where the probability of any outcome, P, is the same.

$$p(n) = \left[\frac{N!}{(N-n)!n!}\right] P^n (1-P)^{N-n}$$

$p(n)$ n

Regardless of the type of distribution assumed by a variable, a variable that shows a central tendency can be described and quantified through its mean value and variance. In the absence of systematic errors, the *true mean value* or central tendency of a continuous random variable, $x(t)$, having a probability density function $p(x)$, is given by

$$x' = \lim_{T \to \infty} \frac{1}{T} \int_0^T x(t)dt \tag{4.4a}$$

which for any continuous random variable x is equivalent to

$$x' = \int_{-\infty}^{\infty} xp(x)dx \tag{4.4b}$$

If the measured variable is described by discrete data, the mean value of measured variable, x_i, where $i = 1, 2, \ldots, N$, is given by

$$x' = \lim_{N \to \infty} \frac{1}{N} \sum_{i=1}^{N} x_i \tag{4.5}$$

Physically, the width of the density function reflects the data variation. For a continuous random variable, the *true variance* is given by

$$\sigma^2 = \lim_{T \to \infty} \frac{1}{T} \int_0^T [x(t) - x']^2 dt \tag{4.6a}$$

which is equivalent to

$$\sigma^2 = \int_{-\infty}^{\infty} (x - x')^2 p(x)dx \tag{4.6b}$$

or for discrete data, the variance is given by

$$\sigma^2 = \lim_{N \to \infty} \frac{1}{N} \sum_{i=1}^{N} (x_i - x')^2 \tag{4.7}$$

The *standard deviation*, σ, a commonly used statistical parameter, is defined as the square root of the variance.

The fundamental difficulty in using equations (4.3)–(4.7) becomes immediately obvious as they assume an infinite number of measurements. But what if the data set is finite? Real data set sizes may range from one value to a large but always finite number. For now we will study "infinite statistics" to introduce the connection between probability and statistics. Then we will turn our attention to the practical treatment of finite data sets.

4.3   INFINITE STATISTICS

A common distribution found in measurements is the *normal* or *Gaussian distribution*.[4] Many measurands common to engineering measurements are described by this distribution which predicts that the scatter seen in a measured data set will be distributed

---

[4]Actually, this distribution was independently suggested in the 18th century by Gauss, LaPlace and DeMoivre. However, Gauss retains the honor.

symmetrically about some central tendency. The shape of the normal distribution is the familiar bell curve.

The probability density function for a random variable, x, having a normal distribution is defined as

$$p(x) = \frac{1}{\sigma\sqrt{2\pi}}\exp\left[-\frac{1}{2}\frac{(x-x')^2}{\sigma^2}\right] \tag{4.8}$$

where x' is defined as the true mean value of x and σ^2 as the true variance of x. Hence, the exact form of $p(x)$ depends on the specific values for x' and σ^2. Note that a maximum in $p(x)$ will occur at $x = x'$, the true mean value. This implies that in the absence of systematic error the central tendency of a random variable having a normal distribution is toward its true mean value. The value most expected from any single measurement, that is, the most probable value, would be the true mean value.

How can we predict the probability that any future measurement will fall within some stated interval? The probability, given by $P(x)$, that a random variable, given by x, will assume a value within the interval $x' \pm \delta x$ is given by the area under $p(x)$. This area is found by integration over this interval. Thus, this probability is given by

$$P(x' - \delta x \leq x \leq x' + \delta x) = \int_{x'-\delta x}^{x'+\delta x} p(x)dx \tag{4.9}$$

Integration of equation (4.9) is made easier through the following transformations. Begin by defining the terms $\beta = (x - x')/\sigma$, as the standardized normal variate for any value x, and $z_1 = (x_1 - x')/\sigma$, as the z variable which specifies an interval on $p(x)$. It follows that

$$dx = \sigma d\beta \tag{4.10}$$

so that equation (4.9) becomes

$$P(-z_1 \leq \beta \leq z_1) = \frac{1}{\sqrt{2\pi}}\int_{-z_1}^{z_1} e^{-\beta^2/2}d\beta \tag{4.11}$$

Since for a normal distribution, $p(x)$ is symmetrical about x', one can write

$$\frac{1}{\sqrt{2\pi}}\int_{-z_1}^{z_1} e^{-\beta^2/2}d\beta = 2\left[\frac{1}{\sqrt{2\pi}}\int_{0}^{z_1} e^{-\beta^2/2}d\beta\right] \tag{4.12}$$

Known as the normal error function, the value in brackets in equation (4.12) provides one-half of the probability sought from equation (4.9) or its equivalent equation (4.11). This half value is tabulated in Table 4.3 for the interval defined by z_1 shown in Figure 4.3.

It should now be clear that the statistical terms defined by equations (4.4)–(4.7) are actually statements of probability. The area under the portion of the probability density function curve, $p(x)$, defined by the interval $x' - z_1\sigma \leq x \leq x' + z_1\sigma$ provides the probability that a measurement will assume a value within that interval. Direct integration of $p(x)$ for a normal distribution between the limits $x' \pm z_1\sigma$ yields that for $z_1 = 1.0$, 68.26% of the area under $p(x)$ lies within $\pm 1.0\sigma$ of x'. This means that there is a 68.26% chance that a measurement of x will have a value within the interval $x' \pm 1.0\sigma$. As the interval defined by z_1 is increased, the probability of occurrence increases. For

$z_1 = 2.0$: 95.45% of the area under $p(x)$ lies within $\pm z_1\sigma$ of x'.

$z_1 = 3.0$: 99.73% of the area under $p(x)$ lies within $\pm z_1\sigma$ of x'.

This concept is illustrated in Figure 4.4.

Table 4.3   Probability Values for Normal Error Function

One-Sided Integral Solutions for $p(z_1) = \dfrac{1}{(2\pi)^{1/2}} \displaystyle\int_0^{z_1} e^{-\beta^2/2} d\beta$

$z_1 = \dfrac{x_1 - x'}{\sigma}$	0.00	0.01	0.02	0.03	0.04	0.05	0.06	0.07	0.08	0.09
0.0	0.0000	0.0040	0.0080	0.0120	0.0160	0.0199	0.0239	0.0279	0.0319	0.0359
0.1	0.0398	0.0438	0.0478	0.0517	0.0557	0.0596	0.0636	0.0675	0.0714	0.0753
0.2	0.0793	0.0832	0.0871	0.0910	0.0948	0.0987	0.1026	0.1064	0.1103	0.1141
0.3	0.1179	0.1217	0.1255	0.1293	0.1331	0.1368	0.1406	0.1443	0.1480	0.1517
0.4	0.1554	0.1591	0.1628	0.1664	0.1700	0.1736	0.1772	0.1809	0.1844	0.1879
0.5	0.1915	0.1950	0.1985	0.2019	0.2054	0.2088	0.2123	0.2157	0.2190	0.2224
0.6	0.2257	0.2291	0.2324	0.2357	0.2389	0.2422	0.2454	0.2486	0.2517	0.2549
0.7	0.2580	0.2611	0.2642	0.2673	0.2704	0.2734	0.2764	0.2794	0.2823	0.2852
0.8	0.2881	0.2910	0.2939	0.2967	0.2995	0.3023	0.3051	0.3078	0.3106	0.3133
0.9	0.3159	0.3186	0.3212	0.3238	0.3264	0.3289	0.3315	0.3340	0.3365	0.3389
1.0	0.3413	0.3438	0.3461	0.3485	0.3508	0.3531	0.3554	0.3577	0.3599	0.3621
1.1	0.3643	0.3665	0.3686	0.3708	0.3729	0.3749	0.3770	0.3790	0.3810	0.3830
1.2	0.3849	0.3869	0.3888	0.3907	0.3925	0.3944	0.3962	0.3980	0.3997	0.4015
1.3	0.4032	0.4049	0.4066	0.4082	0.4099	0.4115	0.4131	0.4147	0.4162	0.4177
1.4	0.4192	0.4207	0.4222	0.4236	0.4251	0.4265	0.4279	0.4292	0.4306	0.4319
1.5	0.4332	0.4345	0.4357	0.4370	0.4382	0.4394	0.4406	0.4418	0.4429	0.4441
1.6	0.4452	0.4463	0.4474	0.4484	0.4495	0.4505	0.4515	0.4525	0.4535	0.4545
1.7	0.4554	0.4564	0.4573	0.4582	0.4591	0.4599	0.4608	0.4616	0.4625	0.4633
1.8	0.4641	0.4649	0.4656	0.4664	0.4671	0.4678	0.4686	0.4693	0.4699	0.4706
1.9	0.4713	0.4719	0.4726	0.4732	0.4738	0.4744	0.4750	0.4758	0.4761	0.4767
2.0	0.4772	0.4778	0.4783	0.4788	0.4793	0.4799	0.4803	0.4808	0.4812	0.4817
2.1	0.4821	0.4826	0.4830	0.4834	0.4838	0.4842	0.4846	0.4850	0.4854	0.4857
2.2	0.4861	0.4864	0.4868	0.4871	0.4875	0.4878	0.4881	0.4884	0.4887	0.4890
2.3	0.4893	0.4896	0.4898	0.4901	0.4904	0.4906	0.4909	0.4911	0.4913	0.4916
2.4	0.4918	0.4920	0.4922	0.4925	0.4927	0.4929	0.4931	0.4932	0.4934	0.4936
2.5	0.4938	0.4940	0.4941	0.4943	0.4945	0.4946	0.4948	0.4949	0.4951	0.4952
2.6	0.4953	0.4955	0.4956	0.4957	0.4959	0.4960	0.4961	0.4962	0.4963	0.4964
2.7	0.4965	0.4966	0.4967	0.4968	0.4969	0.4970	0.4971	0.4972	0.4973	0.4974
2.8	0.4974	0.4975	0.4976	0.4977	0.4977	0.4978	0.4979	0.4979	0.4980	0.4981
2.9	0.4981	0.4982	0.4982	0.4983	0.4984	0.4984	0.4985	0.4985	0.4986	0.4986
3.0	0.49865	0.4987	0.4987	0.4988	0.4988	0.4988	0.4989	0.4989	0.4989	0.4990

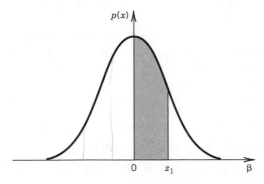

Figure 4.3 Integration terminology for the normal error function.

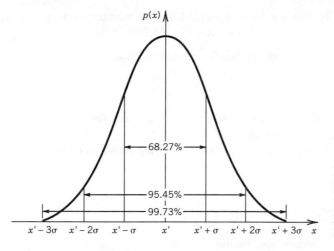

Figure 4.4 Relationship between the probability density function and its statistical parameters, x' and σ, for a normal (Gaussian) distribution.

It follows directly that the representative value that characterizes the variation of a measured data set is the standard deviation. The probability that the ith measured value of x will have a value between $x' \pm z_1 \sigma$ is $2P(z_1) \times 100 = P\%$. This is written as

$$x_i = x' \pm z_1 \sigma \quad (P\%) \tag{4.13}$$

Thus, simple statistical analyses can provide useful quantification of a measured variable in terms of probability. This in turn can be useful in engineering situations where the probable outcome of a measured variable needs to be predicted or specified. These ideas are demonstrated in the Examples 4.2 and 4.3.

EXAMPLE 4.2

Using the probability values in Table 4.3, show that the probability that a measurement will yield a value within $x' \pm \sigma$ is 0.6826 or 68.26%.

KNOWN Table 4.3

$z_1 = 1$

ASSUMPTIONS Data follow a normal distribution.

FIND

$P(x' - \sigma \leq x \leq x' + \sigma)$

SOLUTION To estimate the probability that a single measurement will yield a value within some interval, we need to solve the integral

$$\frac{1}{\sqrt{2\pi}} \int_0^{z_1 = 1.0} e^{-\beta^2/2} d\beta$$

over the interval defined by z_1. Table 4.3 lists the solutions for this integral. Using Table 4.3 for $z_1 = 1.0$, we find $P(z_1) = 0.3413$. However, since $z_1 = (x_1 - x')/\sigma$ then the probability that any measurement of x will produce a value within the interval $0 \leq x \leq x' + \sigma$ is 34.13%. Since the normal distribution is symmetric about x', the probability that x will fall within the interval defined between $-z_1 \sigma$ and $+z_1 \sigma$ for $z_1 = 1.0$ is $(2)(0.3413) = 0.6826$ or 68.26%. Accordingly, if a measurement of x were made, the probability that the value indicated by the measurement system would lie within the interval $x' - \sigma \leq x \leq x' + \sigma$ would be 68.26%.

COMMENT   Similarly, for $z_1 = 1.96$, the probability would be 95.0%.

EXAMPLE 4.3

It is known that the statistics of a well-defined voltage signal are given by $x' = 8.5$ V and $\sigma^2 = 2.25$ V^2. If a single measurement of the voltage signal is made, determine the probability that the measured value indicated is between 10.0 and 11.5 V.

KNOWN

$$x' = 8.5 \text{ V}$$

$$\sigma^2 = 2.25 \text{ V}^2$$

ASSUMPTIONS   Signal has a normal distribution about x'

FIND

$$P(10.0 \leq x \leq 11.5)$$

SOLUTION   The standard deviation of the variable is $\sigma = \sqrt{\sigma^2} = 1.5$ V. The probability that a value will fall between $8.5 \leq x \leq 10.0$ is given in Table 4.3 for $z_1 = (10.0 - 8.5)/1.5 = 1$. This yields $P(8.5 \leq x \leq 10.0) = P(z_1) = 0.3413$. Similarly, for the interval defined by $8.5 \leq x \leq 11.5$, we find $P(8.5 \leq x \leq 11.5) = P(z_1 = 2) = 0.4772$. The probability that x will fall into the interval $10.0 \leq x \leq 11.5$ is given by the area under $p(x)$ over this interval. This area is just the overlap of the two intervals

$$P(10.0 \leq x \leq 11.5) = P(8.5 \leq x \leq 11.5) - P(8.5 \leq x \leq 10.0)$$

$$= 0.4772 - 0.3413 = 0.1359$$

So there is a 13.59% probability that the measurement will yield a value between 10.0 and 11.5 V.

COMMENT   In general, the probability that a measured value will lie within an interval defined by any two values of z_1, such as z_a and z_b, is found by integration of $p(x)$ between z_a and z_b. For a normal density function, this probability is identical to the operation, $P(z_b) - P(z_a)$.

4.4   FINITE STATISTICS

If we recall the box of bearings discussed in Section 4.1, some two dozen bearings were measured, each having been randomly selected from a population containing thousands. Can we use the resulting statistics from this sample to characterize the mean size and variance of the bearings within the box? And if so, how? Within the constraints imposed by probability, it is possible to estimate the true mean and true variance of the box of bearings from such a finite sample size. The method is now discussed.

Suppose we examine the case where we measure random variable x. Assume for the moment that N measurements (that is, N repetitions) of x have been made, each measurement represented by x_i, where $i = 1, 2, \ldots, N$, and N is a finite value. In cases where N is not infinite, statistical values obtained from such finite data sets should be regarded only as estimates of the true statistics of the measurand. Such statistics will be called *finite statistics*. An important point: Whereas *infinite statistics describe the true behavior of a variable, finite statistics describe only the behavior of the finite data set.*

Finite-sized data sets can provide the statistical estimates known as the sample mean value ($\bar{x}$), the sample variance (S_x^2), and its outcome, the sample standard deviation (S_x), defined by

$$\bar{x} = \frac{1}{N}\sum_{i=1}^{N} x_i \tag{4.14a}$$

$$S_x^2 = \frac{1}{N-1}\sum_{i=1}^{N}(x_i - \bar{x})^2 \tag{4.14b}$$

$$S_x = \sqrt{S_x^2} = \left(\frac{1}{N-1}\sum_{i=1}^{N}(x_i - \bar{x})^2\right)^{1/2} \tag{4.14c}$$

where $(x_i - \bar{x})$ is called the deviation of x_i. The degrees of freedom in the data set, ν, is given as $N - 1$, as seen in denominator of equations (4.14b) and (4.14c). The sample mean value provides a most probable estimate of the true mean value, x'. The sample variance represents a probable measure of the variation found in a data set. The *degrees of freedom* in a statistical estimate equate to the number of data points minus the number of previously determined statistical parameters used in estimating that value. These equations are robust and provide reasonable statistical estimates regardless of the probability density function of the measurand.

The predictive utility of infinite statistics can be extended to data sets of finite sample size with only some modification. When sample sizes are finite, the z variable of infinite statistics does not provide a reliable weight estimate of the true probability. However, the sample variance can be weighted in a similar manner so as to compensate for the difference between the finite statistical estimates and the expected infinite statistics for a measured variable. For a normal distribution of x about some sample mean value, $\bar{x}$, one can state that statistically

$$x_i = \bar{x} \pm t_{\nu,P}S_x \quad (P\%) \tag{4.15}$$

where the variable $t_{\nu,P}$ is obtained from a new weighting function used for finite data sets and which replaces the z variable. This new variable is referred to as the t estimator. The interval $\pm t_{\nu,P}S_x$ represents a precision interval, given at probability $P\%$, within which one should expect any measured value to fall.

The value for the t estimator is a function of the probability, P, and the degrees of freedom in the data set, ν. These t values can be obtained from Table 4.4, which is a tabulation of the *Student's t distribution* as developed by William S. Gosset[5] (1876–1937). Gossett recognized that the use of the z variable with S_x in place of σ did not yield accurate estimates of the precision interval at small degrees of freedom. Careful inspection of the Table 4.4 shows that the t value inflates the size of the interval required to attain a percent probability, $P\%$. That is, it has the effect of increasing the magnitude of $t_{\nu,P}S_x$ relative to $z_1\sigma(P\%)$ when N has a finite value. As the value of N increases, t approaches those values given by the z variable just as the value of S_x must approach σ. It should be understood that for very small sample sizes ($N \leq 10$), sample statistics can be misleading. In that situation other information regarding the measurement may be required, including additional measurements.

---

[5]At the time, Gosset was employed as a brewer and statistician by a well-known Irish Brewery. You might pause to reflect on his multifarious contributions.

Table 4.4   Student-t Distribution

ν	t_{50}	t_{90}	t_{95}	t_{99}
1	1.000	6.314	12.706	63.657
2	0.816	2.920	4.303	9.925
3	0.765	2.353	3.182	5.841
4	0.741	2.132	2.770	4.604
5	0.727	2.015	2.571	4.032
6	0.718	1.943	2.447	3.707
7	0.711	1.895	2.365	3.499
8	0.706	1.860	2.306	3.355
9	0.703	1.833	2.262	3.250
10	0.700	1.812	2.228	3.169
11	0.697	1.796	2.201	3.106
12	0.695	1.782	2.179	3.055
13	0.694	1.771	2.160	3.012
14	0.692	1.761	2.145	2.977
15	0.691	1.753	2.131	2.947
16	0.690	1.746	2.120	2.921
17	0.689	1.740	2.110	2.898
18	0.688	1.734	2.101	2.878
19	0.688	1.729	2.093	2.861
20	0.687	1.725	2.086	2.845
21	0.686	1.721	2.080	2.831
30	0.683	1.697	2.042	2.750
40	0.681	1.684	2.021	2.704
50	0.680	1.679	2.010	2.679
60	0.679	1.671	2.000	2.660
∞	0.674	1.645	1.960	2.576

Standard Deviation of the Means

If we were to measure another two dozen of the bearings discussed in Section 4.1, we would expect the statistics from this new sample of randomly selected bearings to differ somewhat from the previous sample. This is simply due to the combined effects of a finite sample size and random variation in bearing size from manufacturing tolerances. So if this is expected behavior, how can we quantify how good the estimate is of the true mean based on a sample mean? That method is now discussed.

Suppose we were to measure a variable N times under fixed operating conditions. If we duplicated this procedure M times, somewhat different estimates of the sample mean value and sample variance would be obtained for each of the M data sets. Why? The chance occurrence of events in any finite sample will affect the estimate of sample statistics. In fact, in experimental situations it is quite straightforward to demonstrate the variation of the sample mean value among different finite-sized samples of the same variable, even under identical operating conditions. After M replications of the N measurements, a set of mean values would be obtained, which are themselves each normally distributed about some central value. In fact, regardless of the probability density function of the measurand, the mean values obtained from M replications will

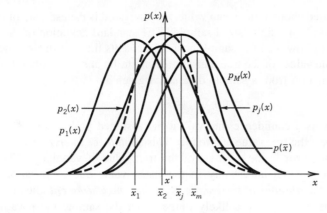

Figure 4.5 The normal distribution tendency of the sample means about a true value in the absence of systematic error.

follow a normal distribution.[6] This process is visualized in Figure 4.5. The amount of variation possible in the sample means would depend on two values: the sample variance, S_x^2, and sample size, N. The discrepancy tends to increase with variance and decrease with $N^{1/2}$.

This tendency for finite sample sets to have somewhat different statistics should not be surprising, for that is precisely the problem inherent to a finite data set. The variation in the sample statistics will be characterized by a normal distribution of the sample mean values about the true mean. The variance of the distribution of mean values that could be expected can be estimated from a single finite data set through the *standard deviation of the means*, $S_{\bar{x}}$:

$$S_{\bar{x}} = \frac{S_x}{\sqrt{N}} \tag{4.16}$$

An illustration of the relation between the standard deviation of a data set and the standard deviation of the means is given in Figure 4.6. The standard deviation of the means is a property of a finite data set. It reflects an estimate of how the sample mean

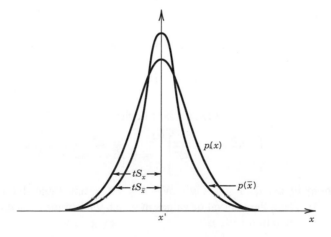

Figure 4.6 Relationships between S_x and a distribution of x and between $S_{\bar{x}}$ and the true value x'.

---

[6]This is a consequence of the *central limit theorem* [3, 5].

values may be distributed about a true mean value. So how good is the estimate of the true mean of a variable based on a finite-sized sample? The standard deviation of the means represents a measure of how well the sample mean represents the true mean. The range over which the possible values of the true mean value might lie at some probability level, P, based on the information from a sample data set is given as

$$\bar{x} \pm t_{v,P}S_{\bar{x}} \quad (P\%) \tag{4.17}$$

where $\pm t_{v,P}S_{\bar{x}}$ represents a confidence interval, at the assigned probability, $P\%$, within which one should expect the true value of x to fall. This *confidence interval* is a quantified measure of the random error in the estimate of the true value of variable x. Therefore, $\pm t_{v,P}S_{\bar{x}}$ represents the random uncertainty in the mean value due to variation in the measured data set. **In the absence of systematic error in a measurement**, the confidence interval states the true value within a likely range about the sample mean value. The estimate of the true mean value based on a finite data set is then stated as

$$x' = \bar{x} \pm t_{v,P}S_{\bar{x}} \quad (P\%) \tag{4.18}$$

Equation (4.18) is an important and powerful equation in engineering measurements.

EXAMPLE 4.4

Consider the data of Table 4.1. (a) Compute the sample statistics for this data set. (b) Estimate the interval of values over which 95% of the measurements of the measurand should be expected to lie. (c) Estimate the true mean value of the measurand at 95% probability based on this finite data set.

KNOWN   Table 4.1

$\quad$ $N = 20$

ASSUMPTIONS   Data set follows a normal distribution.
$\qquad\qquad\qquad$ No systematic errors.

FIND

$\quad$ $\bar{x}$, $\bar{x} \pm tS_x$, and $\bar{x} \pm tS_{\bar{x}}$

SOLUTION   The sample mean value is computed for the $N = 20$ values by the relation

$$\bar{x} = \frac{1}{20} \sum_{i=1}^{20} x_i = 1.02$$

This, in turn, is used to compute the sample standard deviation

$$S_x = \left[\frac{1}{19} \sum_{i=1}^{20} (x_i - \bar{x})^2 \right]^{1/2} = 0.16$$

The degrees of freedom in the data set is $v = N - 1 = 19$. From Table 4.4 at 95% probability, $t_{19,95}$ is 2.093. Then, the interval of values in which 95% of the measurements of x should lie is given by equation (4.15):

$$x_i = 1.02 \pm (2.093 \times 0.16) = 1.02 \pm 0.33 \quad (95\%)$$

Accordingly, if a 21st data point were to be taken, there is a 95% probability that its value would lie between 0.69 and 1.35.

The true mean value is estimated by the sample mean value. However, the confidence interval at 95% probability for this estimate is $\pm t_{19,95} S_{\bar{x}}$, where

$$S_{\bar{x}} = \frac{S_x}{\sqrt{N}} = \frac{0.16}{\sqrt{20}} = 0.04$$

Then, in the absence of systematic errors, we write from equation (4.18)

$$x' = \bar{x} \pm t_{\nu,P} S_{\bar{x}} = 1.02 \pm 0.08 \quad (95\%)$$

Program *Finite-population.vi* demonstrates the effect of sample size on the histogram and finite statistics.

Pooled Statistics

As discussed in Chapter 1, a good test plan uses replication, as well as repetition, in the measurement. Since replications are independent estimates of the same measured value, their data represent separate data samples that can be combined to provide a better statistical estimate of a measured variable than are obtained from a single sample. Samples that are grouped in a manner so as to determine a common set of statistics are said to be pooled.

Consider M replicates of a measurement of variable x, each of N repeated readings so as to yield the data set x_{ij}, where $i = 1, 2, \ldots, N$ and $j = 1, 2, \ldots, M$. The *pooled mean* of x is defined by

$$\langle \bar{x} \rangle = \frac{1}{MN} \sum_{j=1}^{M} \sum_{i=1}^{N} x_{ij} \tag{4.19}$$

The *pooled standard deviation* of x is defined by

$$\langle S_x \rangle = \sqrt{\frac{1}{M(N-1)} \sum_{j=1}^{M} \sum_{i=1}^{N} (x_{ij} - \bar{x}_j)^2} = \sqrt{\frac{1}{M} \sum_{j=1}^{M} S_{x_j}^2} \tag{4.20}$$

with degrees of freedom, $\nu = M(N-1)$. The *pooled standard deviation of the means* of x is defined by

$$\langle S_{\bar{x}} \rangle = \frac{\langle S_x \rangle}{\sqrt{MN}} \tag{4.21}$$

If the number of measurements of x are not the same between replications, then it is appropriate to weight each replication by its particular degrees of freedom. The pooled mean is then defined by its weighted mean

$$\langle \bar{x} \rangle = \frac{\sum_{j=1}^{M} N_j \bar{x}_j}{\sum_{j=1}^{M} N_j} \tag{4.22}$$

where subscript j refers to a particular data set. The pooled standard deviation is given by

$$\langle S_x \rangle = \sqrt{\frac{\nu_1 S_{x_1}^2 + \nu_2 S_{x_2}^2 + \cdots + \nu_M S_{x_M}^2}{\nu_1 + \nu_2 + \cdots + \nu_M}} \tag{4.23}$$

with degrees of freedom $v = \sum_{j=1}^{M} v_j = \sum_{j=1}^{M} (N_j - 1)$ and the pooled standard deviation of the means by

$$\langle S_{\bar{x}} \rangle = \frac{\langle S_x \rangle}{\sqrt{\sum_{j=1}^{M} N_j}} \tag{4.24}$$

4.5   CHI-SQUARED DISTRIBUTION

In the previous section, we discussed how different finite-sized data sets of the same measured variable would have somewhat different statistics. We used this argument to develop the concept of the standard deviation of the means as a precision indicator in the mean value. Similarly, we can estimate how well S_x^2 predicts σ^2. If we plotted the sample standard deviation for many data sets, each having N data points, we would generate the probability density function, $p(\chi^2)$. The $p(\chi^2)$ follows the so-called *Chi-squared* (χ^2) *distribution* depicted in Figure 4.7.

For the normal distribution, the χ^2 statistic is defined by (see [1, 3, or 5])

$$\chi^2 = v S_x^2 / \sigma^2 \tag{4.25}$$

with degrees of freedom $v = N - 1$.

Precision Interval in a Sample Variance

A precision interval for the sample variance can be formulated by the probability statement

$$P(\chi_{1-\alpha/2}^2 \leq \chi^2 \leq \chi_{\alpha/2}^2) = 1 - \alpha \tag{4.26}$$

with a probability of $P(\chi^2) = 1 - \alpha$. The term α is called the level of significance. Combining equation (4.25) with equation (4.26) gives

$$P\left(v S_x^2 / \chi_{\alpha/2}^2 \leq \sigma^2 \leq v S_x^2 / \chi_{1-\alpha/2}^2\right) = 1 - \alpha \tag{4.27}$$

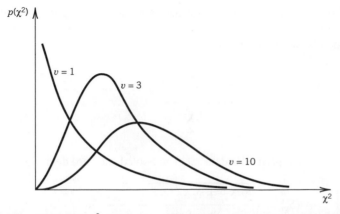

Figure 4.7 The χ^2 distribution with its dependency on degrees of freedom.

Table 4.5 Values for χ_α^2

ν	$\chi_{0.99}^2$	$\chi_{0.975}^2$	$\chi_{0.95}^2$	$\chi_{0.90}^2$	$\chi_{0.50}^2$	$\chi_{0.05}^2$	$\chi_{0.025}^2$	$\chi_{0.01}^2$
1	0.000	0.000	0.000	0.016	0.455	3.84	5.02	6.63
2	0.020	0.051	0.103	0.211	1.39	5.99	7.38	9.21
3	0.115	0.216	0.352	0.584	2.37	7.81	9.35	11.3
4	0.297	0.484	0.711	1.06	3.36	9.49	11.1	13.3
5	0.554	0.831	1.15	1.61	4.35	11.1	12.8	15.1
6	0.872	1.24	1.64	2.20	5.35	12.6	14.4	16.8
7	1.24	1.69	2.17	2.83	6.35	14.1	16.0	18.5
8	1.65	2.18	2.73	3.49	7.34	15.5	17.5	20.1
9	2.09	2.70	3.33	4.17	8.34	16.9	19.0	21.7
10	2.56	3.25	3.94	4.78	9.34	18.3	20.5	23.2
11	3.05	3.82	4.57	5.58	10.3	19.7	21.9	24.7
12	3.57	4.40	5.23	6.30	11.3	21.0	23.3	26.2
13	4.11	5.01	5.89	7.04	12.3	22.4	24.7	27.7
14	4.66	5.63	6.57	7.79	13.3	23.7	26.1	29.1
15	5.23	6.26	7.26	8.55	14.3	25.0	27.5	30.6
16	5.81	6.91	7.96	9.31	15.3	26.3	28.8	32.0
17	6.41	7.56	8.67	10.1	16.3	27.6	30.2	33.4
18	7.01	8.23	9.39	10.9	17.3	28.9	31.5	34.8
19	7.63	8.91	10.1	11.7	18.3	30.1	32.9	36.2
20	8.26	9.59	10.9	12.4	19.3	31.4	34.2	37.6
30	15.0	16.8	18.5	20.6	29.3	43.8	47.0	50.9
60	37.5	40.5	43.2	46.5	59.3	79.1	83.3	88.4

For example, the 95% precision interval by which S_x^2 estimates σ^2, is given by

$$\nu S_x^2 / \chi_{0.025}^2 \le \sigma^2 \le \nu S_x^2 / \chi_{0.975}^2 \quad (95\%) \tag{4.28}$$

Note that this interval is bounded by the 2.5% and 97.5% levels of significance (for 95% coverage).

The χ^2 distribution estimates the discrepancy expected due to random chance. Values for χ_α^2 are tabulated in Table 4.5 as a function of the degrees of freedom. The $P(\chi^2)$ value equals the area under $p(\chi^2)$ as measured from the left, and the α value is the area as measured from the right, as noted in Figure 4.8. The total area under $p(\chi^2)$ is equal to unity.

EXAMPLE 4.5

Ten steel tension specimens are tested from a large batch, and a sample variance of $40,000 \, (\text{kN/m}^2)^2$ is found. State the true variance expected at 95% confidence.

KNOWN

$$S_x^2 = 40,000 \, (\text{kN/m}^2)^2$$
$$N = 10$$

FIND Precision interval for σ^2

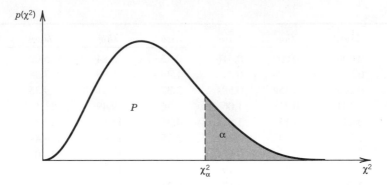

Figure 4.8 The χ^2 distribution as it relates to probability P and to the level of significance, $\alpha\,(= 1 - P)$.

SOLUTION With $\nu = N - 1 = 9$, we find from Table 4.5, $\chi^2 = 19.0$ at $\alpha = 0.025$ written $\chi^2_{.025} = 19.0$ and $\chi^2 = 2.7$ at $\alpha = 0.975$ written $\chi^2_{.975} = 2.7$. Thus, from equation (4.27),

$$(9)(40000)/19.0 \le \sigma^2 \le (9)(40000)/2.7 \quad (95\%)$$

or, the precision interval for the variance is

$$18,947 \le \sigma^2 \le 133,333(\text{kN}/\text{m}^2)^2 \quad (95\%)$$

This is the precision interval about σ^2 due to random chance. As N becomes larger, the precision interval will narrow as $S^2 \to \sigma^2$.

EXAMPLE 4.6

A manufacturer knows from experience that the variance in the diameter of the roller bearings used in their bearings is 3.15 μm^2. Rejecting bearings drives up the unit cost. However, they reject any batch of roller bearings if the sample variance of 20 pieces selected at random exceeds 5 μm^2. Assuming a normal distribution, what is the probability that any given batch will be rejected even though its true variance is actually within the tolerance limits?

KNOWN

$$\sigma^2 = 3.15 \ \mu\text{m}^2$$
$$S_x^2 = 5 \ \mu\text{m}^2 \text{ based on } N = 20$$

ASSUMPTIONS Variations between bearings fit a normal distribution.

FIND

$$\chi^2_\alpha$$

SOLUTION This problem could be reposed as follows: What is the probability that S_x^2 based on 20 measurements will not predict σ_x^2 for the entire batch?

For $\nu = N - 1 = 19$, and using equation (4.25), the χ^2 value is

$$\chi^2_\alpha(\nu) = \nu S_x^2/\sigma^2 = 30.16$$

Inspection of Table 4.5 shows $\chi^2_{.05}(19) = 30.1$, so we can take $\alpha \approx 0.05$. Taking χ^2_α as a measure of discrepancy due to random chance, we interpret this result as a 5% chance that a batch actually within tolerance will be rejected. So there is a probability, $P = 1 - \alpha$, of 95% that S^2_x does predict the σ^2 for the batch. Rejecting a batch on this basis is a good decision.

Goodness-of-Fit Test

Just how well does a set of measurements follow an assumed distribution function? For example, in Example 4.4, we assumed that the data of Table 4.1 followed a normal distribution based only on the rough form of its histogram (Figure 4.2). A more rigorous approach would apply the chi-squared test using the chi-squared distribution. The chi-squared test provides a measure of the discrepancy between the measured variation of a data set and the variation predicted by the assumed distribution function.

To begin, construct a histogram of K intervals from a data set of N measurements. This establishes the measured number of occurrences, n_j, that the measured value lies within the jth interval. Then calculate the degrees of freedom in the variance for the data set, $v = N - m$, where m is the number of restrictions imposed. From v, estimate the predicted number of occurrences, n'_j, to be expected from the distribution function. For this test, the χ^2 value is calculated from the entire histogram by

$$\chi^2 = \frac{\sum_j (n_j - n'_j)^2}{n'_j} \quad j = 1, 2, \ldots, K \tag{4.29}$$

For a given degree of freedom, the better a data set fits the assumed distribution function, the lower its χ^2 value will be, whereas the higher the χ^2 value, the more dubious is the fit.

The χ^2_α table, given in Table 4.5, can be interpreted as a measure of the discrepancy expected due to random chance alone. For example, a value for α of 0.95 implies that 95% of the discrepancy between the histogram and the assumed distribution is due only to random variation. With $P(\chi^2) = 1 - \alpha$, this leaves only 5% of the discrepancy as caused by a systematic tendency, such as a different distribution or other external influence. In general, a value of $P(\chi^2) < 0.05$ confers a very strong measure of a good fit to the assumed distribution, an unequivocal result. On the other hand, values within the range

$$5\% \le P(\chi^2) \le 95\%$$

suggest only that the data *could* be described by the distribution function assumed (an equivocal or ambiguous result). In that case, the use of the assumed distribution may be as reasonable as some other. Lastly, values of $P(\chi^2) > 95\%$ provide a strong measure against the assumed distribution (a rejected result). The data do not fit the distribution, and a different distribution should be attempted.

As with all finite data sets, statistics can only be used to suggest what is probable. Conclusions are left to the user. On the other side of the interpretations, data that fit a distribution too well could be suspected of being "constructed" or "fixed," such as might be the outcome from a loaded pair of dice (or other games of chance).

EXAMPLE 4.7

Test the hypothesis that the variable x as given by the measured data of Table 4.1 is described by a normal distribution.

KNOWN Table 4.1 and Histogram of Figure 4.2

From Example 4.4: $\bar{x} = 1.02$ $S_x = 0.16$ $N = 20$ $K = 7$

Table 4.6 χ^2 Test for Example 4.7

j	n_j	n'_j	$(n_j - n'_j)^2/n'_j$
1	1	0.92	0.07
2	1	1.96	0.47
3	3	3.76	0.15
4	7	4.86	0.94
5	4	4.36	0.03
6	2	2.66	0.16
7	2	1.51	1.16
Totals	20	20	$\chi^2_\alpha = 1.98$

FIND   Apply the chi-squared test to the data set to test for normal distribution.

SOLUTION   Figure 4.2 (Example 4.1) provides a histogram for the data of Table 4.1 giving the values for n_j for $K = 7$ intervals. To evaluate the hypothesis, we must find the predicted number of occurrences, n'_j, for each of the seven intervals based on a normal distribution. To do this, substitute $x' = \bar{x}$ and $\sigma = S_x$ and compute the probabilities based on the corresponding z values.

For example, consider the second interval $(j = 2)$. Using Table 4.3, the predicted probabilities are

$$P(0.75 \le x_i \le 0.85) = P(0.75 \le x_i < x') - P(0.85 \le x_i < x')$$
$$= P(z_a) - P(z_b)$$
$$= P(1.6875) - P(1.0625)$$
$$= 0.454 - 0.356 = 0.098$$

So for a normal distribution, we should expect the measured value to lie within the second interval for 9.8% of the measurements. With

$$n'_2 = N \times P(0.75 \le x_i < 0.85) = 20 \times 0.098 = 1.96$$

that is, 1.96 occurrences are expected out of 20 measurements in this second interval. The actual measured data set shows $n_2 = 1$.

The results are summarized in Table 4.6 with χ^2 based on equation (4.29). Because two calculated statistical values ($\bar{x}$ and S_x) are used in the computations, the degrees of freedom in χ^2 are restricted by 2. So, $\nu = K - 2 = 7 - 2 = 5$ and from Table 4.5, for $\chi^2_\alpha(\nu) = 1.98$, $\alpha \approx 0.85$ or $P(\chi^2) \approx 0.15$ (Note: $\alpha = 1 - P$ is found here by interpolation between columns). While there is a high probability that the discrepancy between the histogram and the normal distribution is due only to random variation of a finite data set, there is a 15% chance the discrepancy is by some other systematic tendency. We should consider this result as being equivocal. The hypothesis that x is described by a normal distribution is neither proven nor disproven.

4.6   REGRESSION ANALYSIS

A measured variable is often a function of one or more independent variables that are controlled during the measurement. When the measured variable is sampled, these variables are controlled, to the extent possible, as are all the other operating conditions. Following the sample, one of these variables is changed and a sample is made under the new operating conditions. This is the procedure used to document the relationship between the measured variable and an independent process variable. A regression

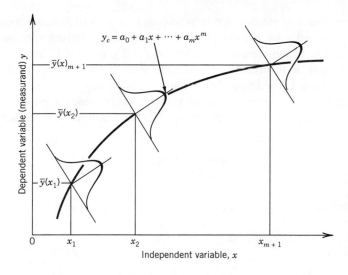

Figure 4.9 Distribution of measured value y about each fixed value of independent variable x. The curve y_c represents a possible functional relationship.

analysis can be used to establish a functional relationship between the dependent variable and the independent variable, which will hold on the average. It is used in those situations where the relationship anticipated is either polynomial in form or can be approximated by a Fourier series. This discussion pertains directly to polynomial fits. More information on Fourier series fits can be found in [4].

A regression analysis assumes that the variation found in the dependent measured variable follows a normal distribution about each fixed value of the independent variable. Such behavior is illustrated in Figure 4.9 by considering the dependent variable $y_{i,j}$ consisting of N measurements, $i = 1, 2, \ldots, N$, of y at each of n values of independent variable, $x_j, j = 1, 2, \ldots, n$. This type of behavior is most common during calibrations, where the input to the measuring system, x, is held nominally fixed while the measurement of y occurs. It is also frequent in many types of measurements in which the dependent variable is measured under fixed operating conditions. Repeated measurements of y will yield a normal distribution with variance $S_y^2(x_j)$, about some mean value, $\bar{y}(x_j)$.

Most spreadsheet and engineering software packages can perform a regression analysis on a data set. Regardless of the package used, the engineer remains responsible for selecting the appropriate analysis for a given data set and interpreting the results of such analyses. The following discussion presents the concepts of regression analysis, its interpretation, and its limitations.

Least-Squares Regression Analysis

The regression analysis for a single variable of the form $y = f(x)$ provides an mth-order polynomial fit of the data in the form

$$y_c = a_0 + a_1 x + a_2 x^2 + \cdots + a_m x^m \tag{4.30}$$

where y_c refers to the value of y predicted by the polynomial equation for a given value of x. For n different values of the independent variable included in the analysis, the highest order, m, of the polynomial that can be determined is restricted to $m \leq n - 1$. The values of the m coefficients $a_0, a_1, \ldots, a_m$ are determined by the analysis. The most common form for regression analysis for engineering applications is the method of least-squares. The *method of least-squares* attempts to minimize the sum of the squares of the deviations between the actual data and the polynomial fit of a stated order by adjusting the values of the coefficients.

An mth-order polynomial relationship is to be found for a set of N data points of the form (x, y) in which x and y are the independent and dependent variables, respectively. Consider the situation in which there are N values of x and y, referred to as x_i, y_i, where $i = 1, 2, \ldots, N$. The task is to find the $m + 1$ coefficients, $a_0, a_1, \ldots, a_m$, of the polynomial of equation (4.30). Let's define the deviation between any dependent variable y_i and the polynomial as $y_i - y_{c_i}$, where y_{c_i} is the value of the polynomial evaluated at the data point (x_i, y_i). The sum of the squares of this deviation for all values of y_i, $i = 1, 2, \ldots, N$, is

$$D = \sum_{i=1}^{N} (y_i - y_{c_i})^2 \tag{4.31}$$

The goal of the method of least-squares is to reduce D to a minimum for a given order of polynomial.

Combining equations (4.30) and (4.31), we write

$$D = \sum_{i=1}^{N} [y_i - (a_0 + a_1 x + \cdots + a_m x^m)]^2 \tag{4.32}$$

Now the total differential of D is dependent on the $m + 1$ coefficients through

$$dD = \frac{\partial D}{\partial a_0} da_0 + \frac{\partial D}{\partial a_1} da_1 + \cdots + \frac{\partial D}{\partial a_m} da_m$$

To minimize the sum of the squares of the deviations, we want dD to be zero. This is accomplished by setting each partial derivative equal to zero:

$$\frac{\partial D}{\partial a_0} = 0 = \frac{\partial}{\partial a_0} \left\{ \sum_{i=1}^{N} [y_i - (a_0 + a_1 x + \cdots + a_m x^m)]^2 \right\}$$

$$\frac{\partial D}{\partial a_1} = 0 = \frac{\partial}{\partial a_1} \left\{ \sum_{i=1}^{N} [y_i - (a_0 + a_1 x + \cdots + a_m x^m)]^2 \right\}$$

$$\vdots \tag{4.33}$$

$$\frac{\partial D}{\partial a_m} = 0 = \frac{\partial}{\partial a_m} \left\{ \sum_{i=1}^{N} [y_i - (a_0 + a_1 x + \cdots + a_m x^m)]^2 \right\}$$

This yields $m + 1$ equations that are solved simultaneously to yield the unknown regression coefficients, $a_0, a_1, \ldots, a_m$.

In general, the polynomial found using a regression analysis will not pass through every data point (x_i, y_i) exactly, so there will be some deviation, $y_i - y_{c_i}$, between each data point and the polynomial. We compute a standard deviation based on these differences by

$$S_{yx} = \sqrt{\frac{\sum_{i=1}^{N} (y_i - y_{c_i})^2}{\nu}} \tag{4.34}$$

where ν is the degrees of freedom of the fit and $\nu = N - (m + 1)$. S_{yx} is referred to as the *standard error of the fit* and is related to how closely a polynomial fits the data set.

The best order of polynomial fit to a particular data set is that *lowest* order of fit that maintains a logical physical sense between the dependent and independent variables and reduces S_{yx} to an acceptable value. This first point is important. Do not rely only on the S_{yx} value to determine the best fit. If the underlying physics of a problem implies that a certain order relationship should exist between dependent and independent variables, there is no sense in forcing the data to fit any other order of polynomial regardless of the value of S_{yx}. Higher-order fits generally will have lower values of S_{yx} but often do not reflect the behavior of the data set very well. Because the method of least-squares tries to minimize the sum of the squares of the deviations, it forces inflections in the curve fit that may not be real. In any event, it is a good rule to have at least two independent data points for each order of polynomial attempted, i.e., at least two data points for a first-order curve fit, four for a second-order, etc.

If we consider variability in both the independent and dependent variables, then the confidence interval of curve fit y_c due to random data scatter about the curve fit at any value of x is estimated by [1, 5]:

$$\pm t_{v,P} S_{yx} \left[\frac{1}{N} + \frac{(x - \bar{x})^2}{\sum_{i=1}^{N} (x_i - \bar{x})^2} \right]^{1/2} \quad (P\%) \tag{4.35}$$

where

$$\bar{x} = \sum_{i=1}^{N} x_i$$

and x is the value used to estimate y_c in equation (4.30). Often in engineering measurements, the independent variable is a known and controlled value. In such cases, we assume that the principal source of variation in the curve fit is due to the random error in the dependent (measured) variable. It is then reasonable to simplify equation (4.35) to a confidence interval due to random data scatter about the fit as

$$\pm t_{v,P} \frac{S_{yx}}{\sqrt{N}} \quad (P\%) \tag{4.36}$$

We then can state that the curve fit is best described by

$$y_c \pm t_{v,P} \frac{S_{yx}}{\sqrt{N}} \quad (P\%) \tag{4.37}$$

where y_c is defined by equation (4.30). A simple comparison of the values of equations (4.35) and (4.36) will determine if the simplified form of equation (4.36) can be used. Either equation provides a measure of the random uncertainty in a curve fit. The effect of the second term in the brackets of equation (4.35) is to increase the confidence interval toward the outer limits of the polynomial.

There is no rule that can be used to estimate which order fit is best and still yield an acceptable value of S_{yx} without trial and error. This is the attractive feature of having a least-squares software package available. The choice of the actual order of fit used is always a compromise between reducing the random error of the fit, the problem physics, and the convenience of using a lower-order polynomial.

Regression analysis of multiple variables of the form $y = f(x_1, x_2, \ldots)$ is also possible. It will not be discussed here, but the concepts generated for the single-variable analysis are carried through for multiple-variable analysis. The interested reader is referred to references [3, 5].

The Labview® program *Polynomial_Fit* performs a least-squares regression analysis. It allows the user to enter data points manually or to read data from a file.

EXAMPLE 4.8

The following data are suspected to follow a linear relationship. Find an appropriate equation of the first-order form.

x [cm]	y [V]
1.0	1.2
2.0	1.9
3.0	3.2
4.0	4.1
5.0	5.3

KNOWN

Independent variable, x
Dependent measured variable, y
$N = 5$

ASSUMPTIONS   Linear relation

FIND

$y_c = a_0 + a_1 x$

SOLUTION   We seek a polynomial of the form $y_c = a_0 + a_1 x$, which minimizes the term

$$D = \sum_{i=1}^{N} (y_i - y_{c_i})^2$$

setting the derivatives to zero:

$$\frac{\partial D}{\partial a_0} = 0 = -2 \left\{ \sum_{i=1}^{N} [y_i - (a_0 + a_1 x)] \right\}$$

$$\frac{\partial D}{\partial a_1} = 0 = -2 \left\{ \sum_{i=1}^{N} [y_i - (a_0 + a_1 x)] \right\}$$

solving simultaneously for the coefficients a_0 and a_1 yields

$$a_0 = \frac{\sum x_i \sum x_i y_i - \sum x_i^2 \sum y_i}{\left(\sum x_i \right)^2 - N \sum x_i^2}$$

$$a_1 = \frac{\sum x_i \sum y_i - N \sum x_i y_i}{\left(\sum x_i \right)^2 - N \sum x_i^2}$$

(4.38)

Substituting the data set into equations (4.38) yields: $a_0 = 0.02$ and $a_1 = 1.04$. Hence,

$$y_c = 0.02 + 1.04x \text{ V}$$

COMMENT   Although the polynomial described by y_c is the linear curve fit for this data set, we still have no idea of how well this curve fits this data set or even if a first-order fit is appropriate. These questions are addressed below and in Example 4.9.

Linear Polynomials

For linear polynomials a correlation coefficient, r, can be found by

$$r = \sqrt{1 - \frac{S_{yx}^2}{S_y^2}} \tag{4.39}$$

where

$$S_y^2 = \frac{1}{N-1} \sum_{i=1}^{N} (y_i - \bar{y})^2$$

with

$$\bar{y} = \frac{1}{N} \sum_{i=1}^{N} y_i$$

The correlation coefficient represents a quantitative measure of the linear association between x and y. It is bounded by ± 1, which represents perfect correlation; the sign indicates that y increases or decreases with x. For $\pm 0.9 < r \leq \pm 1$, a linear regression can be considered a reliable relation between y and x. Alternatively, the value r^2 is often reported, which is indicative of how well the variance in y is accounted for by the fit. It is a ratio of the variation assumed by the linear fit to the actual measured variations in the data. However, the correlation coefficient and the r^2 value are only indicators of the hypothesis that y and x are linearly related. They are not effective estimators of the random error in y_c. The S_{yx} value is used for that purpose, as discussed earlier.

The precision estimate in the slope of the fit can be estimated by

$$S_{a_1} = S_{yx} \sqrt{\frac{N}{N \sum_{i=1}^{N} x_i^2 - \left(\sum_{i=1}^{N} x_i \right)^2}} \tag{4.40}$$

For example, S_{a_1}, would provide a measure of the static sensitivity error of a measurement system when using a linear fit of the calibration data.

The precision estimate of the zero intercept can be estimated by

$$S_{a_0} = S_{yx} \sqrt{\frac{N \sum_{i=1}^{N} x_i^2}{N \left[N \sum_{i=1}^{N} x_i^2 - \left(\sum_{i=1}^{N} x_i \right)^2 \right]}} \tag{4.41}$$

An error in a_0 would offset a calibration curve from its y intercept. The derivation and further discussion on equations (4.39)–(4.41) can be found in [1, 3, 5].

EXAMPLE 4.9

Compute the correlation coefficient and the standard error of the fit for the data in Example 4.8. Estimate the random uncertainty associated with the fit. State the correlation with its 95% confidence interval.

KNOWN

$$y_c = 0.02 + 1.04x \, [\text{V}]$$

ASSUMPTIONS Errors are normally distributed.
No systematic errors

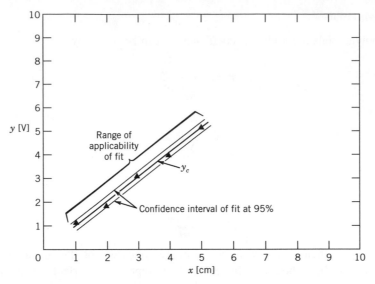

Figure 4.10 Results of the regression analysis of Example 4.9.

FIND r and S_{yx}

SOLUTION Direct application of equation (4.39) with the data set yields for the correlation coefficient $r = 0.996$. An equivalent estimator is r^2. Here $r^2 = 0.99$, which indicates that 99% of the variance in y is accounted for by the fit, whereas only 1% is unaccountable. These values suggest that a linear fit is a reliable relation between x and y.

The random error between the data and this fit is quantified through S_{yx}. Using equation (4.34),

$$S_{yx} = \sqrt{\frac{\sum_{i=1}^{N} (y_i - y_{c_i})^2}{\nu}} = 0.16$$

with degrees of freedom, $\nu = N - (m+1) = 3$. The t estimator, $t_{3,95} = 3.18$, establishes a random uncertainty about the fit of $\pm(t_{3,95}S_{yx}/\sqrt{N}) = \pm 0.23$. Accordingly, using equation (4.37) and the estimate for the random error, the polynomial fit can be stated at 95% confidence as

$$y_c = 1.04x + 0.02 \pm 0.23 \text{ V } (95\%)$$

This curve is plotted in Figure 4.10 with its 95% confidence interval. The regression polynomial with its confidence interval is necessary when reporting a curve fit to a data set.

EXAMPLE 4.10

A velocity probe provides a voltage output that is related to velocity, U, by the form $E = a + bU^m$. A calibration is performed, and the data $(N = 5)$ are recorded below. Estimate an appropriate curve fit.

U[m/s]	E[V]
0.0	3.19
10.0	3.99
20.0	4.30
30.0	4.48
40.0	4.65

KNOWN $N = 5$

ASSUMPTIONS   Data related by $E = a + bU^m$

FIND a and b

SOLUTION   The equation can be transformed into

$$\log(E - a) = \log b + m \log U$$

which has the linear form

$$Y = B + mX$$

Because at $U = 0$ m/s, $E = 3.19$ V, the value of a must be 3.19 V. The values for Y and X are computed below with the corresponding deviations from the resulting fit:

U	Y	X	$E_i - E_{c_i}$
0.0	—	—	0.0
10.0	−0.097	1.0	−0.01
20.0	0.045	1.30	0.02
30.0	0.111	1.48	0.0
40.0	0.164	1.60	−0.01

Substituting the values for Y and X into equation (4.38) results in: $a_0 = B = -0.525$ and $a_1 = m = 0.43$. The standard error of the fit is found from equation (4.34) to be $S_{yx} = 0.01$. From Table 4.4, $t_{3,95} = 3.18$ so that the confidence interval $\pm\left(t_{3,95}S_{yx}/\sqrt{N}\right)$ is estimated to be ± 0.014 V. This gives

$$Y = -0.525 + 0.43X \pm 0.014 \text{ V} \quad (95\%)$$

The polynomial is transformed back to the form $E = a + bU^m$:

$$E = 3.19 + 0.30\, U^{0.43} \quad \text{V}$$

The curve fit with its 95% confidence interval is shown on Figure 4.11.

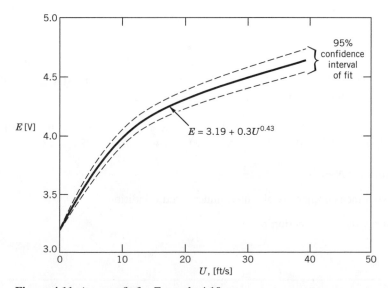

Figure 4.11 A curve fit for Example 4.10.

4.7   DATA OUTLIER DETECTION

It is not uncommon to find a spurious data point that does not appear to be related to the expected tendency of the data set. Data that lie outside the probability of normal variation will incorrectly offset the sample mean value estimate, inflate the random error estimates, and influence a least-squares correlation. Statistical techniques can be used to detect such data points, which are known as *outliers* because their measured values fall outside the probable range of values for a measurand. Once detected, a decision as to whether to remove the data point from the data set can be made. Histograms are particularly effective for visual detection of suspected outliers. If outliers are removed from a data set, the data set statistics are recomputed using the remaining data. But such outlier treatment can only be done once!

The specific methods presented here were chosen for their ease of use. The methods detect those data points that fall outside the normal range of variation expected in a data set; the normal range of variation is based on the variance of that data set. This range is defined by some multiple of the standard deviation of the data set. Since a multiple of the standard deviation is related to the area under the probability density function, the basis for the outlier detection lies with the probability of occurrence for the data value in question.

A simple approach is to label data points that lie outside the range of 99.8% probability of occurrence, $\bar{x} \pm t_{\nu,99.8}S_x$, as outliers. This *three-sigma*[7] *test* works well with data sets of 10 or more points.

For large data sets, the modified three-sigma test for outliers can be used. The sample mean and standard deviations are estimated and a modified z variable is computed for each data point by

$$z_0 = \left| \frac{x_i - \bar{x}}{S_x} \right| \tag{4.42}$$

The probability that x lies outside the one-sided range defined by 0 and z_0 is $[0.5 - P(z_0)]$ where $P(z_0)$ is found from the one-sided z chart of Table 4.3. For N data points, if $N[0.5 - P(z_0)] \leq 0.1$, the data point can be considered an outlier. Other methods of outlier detection are discussed elsewhere (e.g., [3, 6]).

EXAMPLE 4.11

Consider the data given below for 10 measurements of tire pressure made using an inexpensive hand-held gauge (Note: 14.5 psi = 1 bar). Compute the statistics of the data set; then test for outliers using the modified three-sigma test.

i	1	2	3	4	5	6	7	8	9	10
x_i[psi]	28	31	27	28	29	24	29	28	18	27

KNOWN   Data values for $N = 10$

ASSUMPTIONS   Each measurement is obtained under fixed conditions.

FIND $\bar{x}$ and S_x; apply outlier detection tests

---

[7]So named since, as $\nu \rightarrow \infty$, $t \rightarrow 3$ for 99.8% probability.

SOLUTION   Based on the 10 data points, the sample mean and sample standard deviation are found from equations (4.14a) and (4.14c) to be $\bar{x} = 27$ psi with an $S_x = 3.8$. But the tire pressure should not vary much between two readings beyond the precision capabilities of the measurement system and technique, so data point 9 looks suspect as an outlier and is probably the result of a measurement blunder. Apply the modified three-sigma test to this data point.

For $x = 18$, $z_0 = 2.34$ and, from Table 4.3, $P(z_0) = 0.4904$, so we find $[0.5 - P(z_0)] = 0.0096$. Hence, $N[0.5 - P(z_0)] = 0.0096$, a value that is much less than 0.1. So data point 9 can be considered an outlier. The data point is rejected and removed. The remaining data variations reflect the random errors in the measurement (variable, procedure, and instrument). The data set statistics become $N = 9$, $\bar{x} = 28$ psi (or $\bar{x} = 1.9$ bar) with $S_x = 2.0$ psi ($S_x = 0.1$ bar). We see that the outlying data point offset the sample mean and significantly inflated the data variation.

4.8   NUMBER OF MEASUREMENTS REQUIRED

Statistics can be used to assist in the design and planning of a test program. For example, how many measurements, N, are required to reduce the random error due to variation in the data set to an acceptable level? To answer this question, begin with equation (4.18), which expresses the true value based on a sample mean and its confidence interval:

$$x' = \bar{x} \pm t_{v,P}S_{\bar{x}} \quad (P\%) \tag{4.18}$$

We can express the 95% confidence interval in (4.18) as CI, that is,

$$\mathrm{CI} = \pm t_{v,95}S_{\bar{x}} = \pm t_{v,95}\frac{S_x}{\sqrt{N}} \quad (95\%) \tag{4.43}$$

To evaluate equation (4.43), we must assign a value to S_x. S_x should be a conservative estimate based on previous test data, prior experience, or manufacturer's information.

The confidence interval is two sided about the mean, defining a range from $-t_{v,95}\frac{S_x}{\sqrt{N}}$ to $+t_{v,95}\frac{S_x}{\sqrt{N}}$. We introduce the one-sided precision value d as

$$d = \frac{\mathrm{CI}}{2} = t_{v,95}\frac{S_x}{\sqrt{N}} \tag{4.44}$$

Then, it follows that the required number of measurements is estimated by

$$N \approx \left(\frac{t_{v,95}S_x}{d}\right)^2 \quad (95\%) \tag{4.45}$$

The approximation serves as a reminder that this expression is based on an assumed value for S_x and finite statistics. The accuracy of equation (4.45) will depend on how well the assumed value of S_x approximates σ. Although we work at 95% in this example, the approach works at any probability level.

A shortcoming of this method is the need for an estimate of S_x. One way around this is to make a preliminary small number of measurements, N_1, to obtain an estimate of the sample variance, S_1, to be expected. Then S_1 is used to estimate the number of measurements required. The total number of measurements, N_T, will be estimated by

$$N_T \approx \left(\frac{t_{N-1,95}S_1}{d}\right)^2 \quad (95\%) \tag{4.46}$$

This establishes that $N_T - N_1$ additional measurements will be required.

EXAMPLE 4.12

Determine the number of measurements required to reduce the 95% confidence interval of the mean value of a variable to within 1 unit, if the variance of the variable is estimated to be about 64 units.

KNOWN

$$\text{CI} = 1 \text{ unit} \qquad P = 95\%$$
$$d = 1/2 \qquad \sigma^2 = 64 \text{ units}$$

ASSUMPTIONS

$$\sigma^2 \approx S_x^2$$

FIND N required

SOLUTION   Because equation (4.45) has two unknowns, begin this problem by guessing at some value for N. Then, using this guessed value, compute the t variable at the probability level desired. An updated value for N can then be found from the formulation

$$N \approx \left(\frac{t_{v,95}S_x}{d}\right)^2 \quad (95\%)$$

Then use trial and error iteration to converge on a value for N.

Suppose we begin by guessing that $N = 500$ so that $t_{v,95} \approx 1.96$ is reasonable. Then

$$v = 499 \quad t_{499,95} = 1.96 \Rightarrow N = 983$$

So, now guess $N = 983$. Then

$$v = 982 \quad t_{982,95} = 1.96 \Rightarrow N = 983$$

We have converged on $N = 983$. Thus, at least 983 measurements must be made to achieve the desired confidence interval in the measured variable. Analyze the results after 983 measurements to ensure that the variance level used was representative of the actual data set.

COMMENT   Since the confidence interval is reduced as $N^{1/2}$, the procedure of increasing N to decrease this interval becomes one of diminishing returns.

EXAMPLE 4.13

From 51 measurements of a variable, S_1 is found to be 160. For a 95% confidence interval of 60 units in the mean value, estimate the total number of measurements required.

KNOWN

$$S_1 = 160 \text{ units} \qquad N_1 = 51$$
$$d = \text{CI}/2 = 30 \qquad t_{50,95} = 2.01$$

ASSUMPTIONS

$$\sigma^2 \approx S_1^2$$

FIND

$$N_T$$

Table 4.7 Summary Table for a Sample of N Data Points

Sample mean	$\bar{x} = \dfrac{1}{N} \sum_{i=1}^{N} x_i$
Sample standard deviation	$S_x = \sqrt{\dfrac{1}{N-1} \sum_{i=1}^{N} (x_i - \bar{x})^2}$
Standard deviation of the means[1]	$S_{\bar{x}} = \dfrac{S_x}{\sqrt{N}}$
Precision interval for a single data point, x_i	$\pm t_{v,P} S_x \quad (P\%)$
Confidence interval[2,3] for a mean value, $\bar{x}$	$\pm t_{v,P} S_{\bar{x}} \quad (P\%)$
Confidence interval[2,4] for curve fit, $y = f(x)$	$\pm t_{v,P} \dfrac{S_{yx}}{\sqrt{N}} \quad (P\%)$

[1]Measure of standard random uncertainty in $\bar{x}$.
[2]In the absence of systematic errors.
[3]Measure of random uncertainty in $\bar{x}$.
[4]Measure of random uncertainty in curve fit.

SOLUTION   The total number of measurements required is estimated by equation (4.46):

$$N_T \approx \left(\frac{t_{N-1,95} S_1}{d} \right)^2 = \left(\frac{2.01 \times 160}{30} \right)^2 = 115 \quad (95\%)$$

Thus, as a first guess, a total of 115 measurements are estimated to be necessary. This means that an additional 64 measurements should be taken. The N_T data are then analyzed to be certain that the constraint is met.

4.9   SUMMARY

The normal scatter of data about some central mean value is brought about through several contributing factors, including the process variable's own temporal unsteadiness and spatial distribution under nominally fixed operating conditions, as well as random errors in the measurement system and in the measurement procedure. Data scatter introduces an additional element, an uncertain vagueness, into the measurement scheme, which requires statistical methods to sort out. Further, a finite number of measurements can only go so far in estimating the true value of the measurand. Statistics, then, becomes a powerful tool used to interpret and present data. In this chapter, we developed the most basic methods used to understand and quantify finite data sets. Methods to estimate the true mean value based on a limited number of data points and the random error in such estimates were presented along with treatment of data curve fitting. A summary table of these statistical estimators is given as Table 4.7. Of course, throughout this chapter we have assumed negligible systematic error and random errors from other sources. In the next chapter, we will include these other error estimates in our estimate of the true value of a measurement.

REFERENCES

1. Kendal, M.G., and A. Stuart, *Advanced Theory of Statistics*, Vol. 2, Griffin, London, 1961.
2. Bendat, J., and A. Piersol, *Random Data Analysis*, Wiley, New York, 1971.

3. Lipson, C., and N.J. Sheth, *Statistical Design and Analysis of Engineering Experiments*, McGraw-Hill, New York, 1973.

4. James, M.L., G.M. Smith, and J.C. Wolford, *Applied Numerical Methods For Digital Computation*, 2d ed., Harper & Row, New York, 1977.

5. Miller, I., and J.E. Freund, *Probability and Statistics for Engineers*, 3d ed., Prentice Hall, Englewood Cliffs, NJ, 1985.

6. ASME/ANSI Power Test Codes, *Test Uncertainty PTC 19.1-1998*, American Society of Mechanical Engineers, New York, 1998.

NOMENCLATURE

$a_0, a_1, \ldots, a_m$	polynomial regression coefficients	$P\%$	percent probability
f_j	frequency of occurrence of a value	S_x	sample standard deviation x
$p(x)$	probability density function of x	$S_{\bar{x}}$	sample standard deviation of the means of x
r	correlation coefficient	S_x^2	sample variance of x
t	Student's-t variable	$\langle S_x \rangle$	pooled sample standard deviation of x
x	measured variable; measurand	$\langle S_{\bar{x}} \rangle$	pooled standard deviation of the means of x
x_i	ith measured value in a data set		
x'	true mean value of the measured value x	$\langle S_x \rangle^2$	pooled sample variance of x
$\bar{x}$	sample mean value of x	S_{yx}	standard error of the (curve) fit between y and x
$\langle \bar{x} \rangle$	pooled sample mean of x	β	normalized standard variate
y	variable	σ	true standard deviation of the measured value
z	z variable		
N	total number of measurements	σ^2	true variance of the measured value
$P(x)$	probability of measuring a specific value of x	χ^2	chi-squared value
		ν	degrees of freedom

PROBLEMS

4.1 A very large data set $(N > 10{,}000)$ has a mean value of 9.2 units and a standard deviation of 1.1 units. Determine the range of values in which 50% of the data set should be found assuming a normal probability density.

4.2 For a very large data set $(N > 10{,}000)$, a mean value of 204 units with a standard deviation of 18 units has been determined. Determine the range of values in which 90% of the data fall.

4.3 At a fixed operating setting, the pressure in a line downstream of a reciprocating compressor is found to have a mean value of 121.6 psi with a standard deviation of 14 psi based on a very large data set obtained from continuous monitoring. What is the probability that the line pressure will exceed 150 psi during any measurement?

4.4 Consider the toss of four coins. There are 2^4 possible outcomes of a single toss. Develop a histogram of the number of heads (one side of the coin) that can appear on any toss. Does it look like a Gaussian distribution? Should this be expected? What is the probability that three heads will appear on any toss?

4.5 As a game, slide a matchbook across a table, trying to make it stop at some predefined point on each attempt. Measure the distance from the starting point to the stopping point. Repeat this 10, 20, through 50 times. Plot the frequency distribution from each set. Would you expect them to look like a Gaussian distribution? What statistical outcomes would distinguish a better player?

Table 4.8 Measured Force Data for Exercise Problems

F [N] Set 1	F [N] Set 2	F [N] Set 3
51.9	51.9	51.1
51.0	48.7	50.1
50.3	51.1	51.4
49.6	51.7	50.5
51.0	49.9	49.7
50.0	48.8	51.6
48.9	52.5	51.0
50.5	51.7	49.5
50.9	51.3	52.4
52.4	52.6	49.5
51.3	49.4	51.6
50.7	50.3	49.4
52.0	50.3	50.8
49.4	50.2	50.8
49.7	50.9	50.2
50.5	52.1	50.1
50.7	49.3	52.3
49.4	50.7	48.9
49.9	50.5	50.4
49.2	49.7	51.5

Problems 6 through 15 refer to the three measured data sets in Table 4.8. Assume that the data have been recorded from three replications from the same process under a fixed operating condition.

4.6 Develop a histogram for the data listed in column 1. Discuss each axis and describe its overall shape.

4.7 Develop a frequency distribution for the data given in column 3. Discuss each axis and describe its overall shape.

4.8 Develop and compare the histograms of the three data sets represented under columns 1, 2, and 3. If these are taken from the same process, why might the histograms vary? Do they appear to show a central tendency?

4.9 For the data in each column, determine the sample mean value, standard deviation, and standard deviation of the means. State the degrees of freedom in each.

4.10 Explain the concept of "central tendency" by comparing the range of the measured values and the sample mean values from each of the three data sets.

4.11 From the data in column 1, estimate the range of values for which you would expect 95% of all possible measured values for this operating condition to fall. Repeat for columns 2 and then 3. Discuss these outcomes in terms of what you might expect from finite statistics.

4.12 From the data in column 1, determine the best estimate of the mean value at a 95% probability level. How does this estimate differ from the estimates made in problem 4.11? Repeat for columns 2 and then 3. Why do the estimates vary for each data set? Discuss these outcomes in terms of what you might expect from finite statistics if these are measuring the same measured variable during the same process.

4.13 For the data in column 3, if one additional measurement were made, estimate the interval of values in which the value of this measurement would fall with a 95% probability.

4.14 Compute a pooled sample mean value for the process. State the range for the best estimate in force at 95% probability based on these data sets. Discuss whether this pooled sample mean value is reasonable given the sample mean values for the individual data sets. Which sample mean gives a better estimate of the true value of force? Write a short essay explanation in terms of the limitations of sample statistics, the number of measurements, variations in data sets, and statistical estimators.

4.15 Apply the χ^2 Goodness-of-Fit test to the data in column 1 and test the assumption of a normal distribution.

4.16 Consider a process in which the applied measured load has a known true mean of 100 N with variance of 400 N^2. An engineer takes 16 measurements at random. What is the probability that this sample will have a mean value between 90 and 110?

4.17 A professor grades students on a normal curve. For any grade x, based on a course mean and standard deviation developed over years of testing, the following applies:

$$A: \quad x > \bar{x} + 1.6\,\sigma$$
$$B: \quad \bar{x} + 0.4\,\sigma < x \le \bar{x} + 1.6\,\sigma$$
$$C: \quad \bar{x} - 0.4\,\sigma < x \le \bar{x} + 0.4\,\sigma$$
$$D: \quad \bar{x} - 1.6\,\sigma < x \le \bar{x} - 0.4\,\sigma$$
$$F: \quad x \le \bar{x} - 1.6\,\sigma$$

How many A, C, and D grades are given per 100 students?

4.18 The production of a certain polymer fiber follows a Gaussian distribution with a true mean diameter of 20 µm and a standard deviation of 30 µm. Compute the probability of a measured value greater than 80 µm. Compute the probablility of a measured value between 50 and 80 µm.

4.19 An automotive manufacturer removes the friction linings from the clutch plates of drag race cars following test runs. A sample of 10 linings is studied. Measurements for wear showed the following values (in µm): 204.5, 231.1, 157.5, 190.5, 261.6, 127.0, 216.6, 172.7, 243.8, and 291.0. Estimate the average wear and its variance. Based on this sample, how many clutch plates out of a large set will be expected to show wear of more than 203 µm?

4.20 Determine the mean value of the life of an electric light bulb if

$$p(x) = 0.001\, e^{-0.001x} \quad x \ge 0$$

and $p(x) = 0$ otherwise. Here x is the life in hours.

4.21 Compare the reduction in the random error in estimating x' by taking a sample of 16 measurements as opposed to only four measurements. Then compare 100 measurements to 25. Explain "diminishing returns" as it applies to using larger sample sizes to reduce random error in estimating the true mean value.

4.22 The variance in the strength test values of 270 bricks is 6.89 $(MN/m^2)^2$ with a mean of 6.92 MN/m^2. What is the random error in the mean value and the confidence interval at 95%?

4.23 A sample of 61 data points of force indicates a sample mean value of 44.20 N with sample variance of 4.0 N^2. Estimate the probability that an additional measurement would indicate a value between 45.56 and 48.20 N.

4.24 A batch of rivets is tested for shear strength. A sample of 36 rivets shows a mean strength of 924.2 MPa with a standard deviation of 18 MPa. State the estimate of the mean shear strength for the batch at 95% probability.

4.25 Suppose three sets of data are collected of some variable during a similar process operating condition. The statistics are found to be

$$N_1 = 16 \qquad \bar{x}_1 = 32 \qquad S_{x_1} = 3 \, \text{units}$$
$$N_2 = 21 \qquad \bar{x}_1 = 30 \qquad S_{x_2} = 2 \, \text{units}$$
$$N_3 = 9 \qquad \bar{x}_1 = 34 \qquad S_{x_3} = 6 \, \text{units}$$

Determine the degrees of freedom in the pooled data. Neglecting systematic errors and random errors other than the variation in the measured data set, compute an estimate of the weighted mean value of this variable and the range in which the true mean should lie with 95% confidence.

4.26 Eleven core samples of fresh concrete are taken by a county engineer from the loads of 11 concrete trucks used in pouring a structural footing. After curing, the engineer tests to find a mean compression strength of 3027 lb/in.^{2} with a standard deviation of 53 lb/in.^{2}. State codes require a minimum strength of 3000 lb/in.^{2}. Should the footing be repoured based on a 95% confidence interval of the test data?

4.27 The following data were collected during the repeated measurement of the force load acting on a small area of a beam under "fixed" conditions:

Reading Number	Output [N]	Reading Number	Output [N]
1	923	6	916
2	932	7	927
3	908	8	931
4	932	9	926
5	919	10	923

Determine if any of these data points can be considered outliers. If so, reject the data point. Estimate the true mean value from this data set assuming that the only error is from variations in the data set.

4.28 An engineer measures the diameter of 20 tubes selected at random from a large shipment. The sample yields: $\bar{x} = 47.5 \, \text{mm}$ and $S_x = 8.4 \, \text{mm}$. The manufacturer of the tubes claims that $x' = 42.1 \, \text{mm}$. What can the engineer conclude about this claim?

4.29 In manufacturing a particular set of motor shafts, only shaft diameters of between 38.10 and 37.58 mm are usable. If the process mean is found to be 37.84 mm with a standard deviation of 0.13 mm, what percentage of the manufactured shafts are usable?

For Problems 4.30 through 4.33, it would be helpful to use a least-squares software package, such as Polynomial_fit, or the least-squares analytical capability of spreadsheet software.

4.30 Determine the static sensitivity for the following data by using a least-squares regression analysis. Plot the data and fit including the 95% confidence interval due to random scatter about the fit. Compare the uncertainty interval estimated by equation (4.36) with that of equation (4.35) and discuss the differences.

Y:	2.7	3.6	4.4	5.2	9.2
X:	0.4	1.1	1.9	3.0	5.0

4.31 Using the following data, determine a suitable least-squares fit of the data. Which order polynomial best fits this data set? Plot the data and the fit with 95% confidence interval on an appropriate graph.

Y:	1.4	4.7	17.3	82.9	171.6	1227.1
X:	0.5	1.1	2.0	2.9	5.1	10.0

4.32 The following data have the form $y = ax^b$. Plot the data and their best fit with 95% confidence interval. Estimate the static sensitivity at each value of X.

Y:	0.14	2.51	15.30	63.71
X:	0.5	2.0	5.0	10.0

4.33 A fan performance test yields the following data:

Q:	2000	6000	10000	14000	18000	22000
h:	5.56	5.87	5.73	4.95	3.52	1.08

where Q is flow rate in m^3/s and h is static head in $cm-H_2O$. Find the lowest-degree polynomial that best fits the data as $h = f(Q)$. Note: physically, a second-order polynomial is a reasonable fit. Explain your choice.

4.34 A camera manufacturer purchases glass to be ground into lenses. From experience it is known that the variance in the refractive index of the glass is 1.25×10^{-4}. The company will reject any shipment of glass if the sample variance of 30 lenses measured at random exceeds 2.10×10^{-4}. What is the probability that such a batch of glass will be rejected even though it is actually within the normal variance for refractive index?

4.35 The set-up for grinding a type of bearing is considered under control if the bearings have a mean diameter of 5.000 mm. Normal procedure is to measure 30 bearings each 10 minutes to monitor production. Production rates are 1000 per minute. What action do you recommend if such a sample shows a mean of 5.060 mm with a standard deviation of 0.0025 mm?

4.36 Referring to the information in Problem 4.35, suppose the mean diameter of the bearings must be held to within 0.2%. Based on given information, how many bearings should be measured at each 10-minute interval?

4.37 The manufacturer of general aircraft vacuum pumps wishes to estimate the mean failure time of its product at 95% confidence. Six pumps are tested to failure with these results (in hours of operation): 1272, 1384, 1543, 1465, 1250, and 1319. Estimate the sample mean and the 95% confidence interval of the true mean. How many more data points would be needed to improve the confidence value to 50 hours?

4.38 Strength tests on a batch of cold-drawn steel yield the following:

Strength (MPa)	Occurrences
421–480	4
481–540	8
541–600	12
601–660	6

Test the hypothesis that the data are a random sample that follows a normal distribution.

4.39 Referring to the strength test of Problem 4.38, how many test measurements should actually be made if the mean strength estimate should be accurate to within ±3%?

4.40 Estimate the number of specimens that should be tested to determine the mean failure strength of an alloy to within a confidence interval with a range no greater than 5% of the mean using ASTM standard procedures. A sample run of six alloy specimens suggests a mean strength of 71,327 lb/in.^{2} with standard deviation of 8345 lb/in.^{2}. What sources contribute to this variation in "measured" strength?

4.41 Estimate the number of measurements of a time-dependent acceleration signal obtained from a vibrating vehicle that would lead to an acceptable confidence interval about the mean of 0.1 g, if the standard deviation of the signal is expected to be 2 g.

4.42 The sample standard deviation of a time-dependent electrical signal, based on 60 measurements, is estimated to be 1.52 V. How many more measurements would be required to provide a confidence interval in the mean of 0.28 V?

4.43 Estimate the probability with which an observed $\chi^2 = 19.0$ will be exceeded if there are 10 degrees of freedom.

4.44 A conductor is insulated using an enameling process. It is known that the number of insulation breaks per meter of wire is 0.07. What is the probability of finding x breaks in a piece of wire 5 m long? Use the Poisson distribution.

4.45 We know that 2% of the screws made by a certain machine are defective with the defects occurring randomly. The screws are packaged 100 to a box. Estimate the probability that a box will contain x defective screws. (i) Use the Binomial distribution. (ii) Use the Poisson distribution.

4.46 An optical velocity measuring instrument provides an updated signal on the passage of small particles through its focal point. Suppose the average number of particles passing in a given time interval is 4. Estimate the probability of observing x particle passages in a given time. Use the Poisson distribution for $x = 1$ through 10.

4.47 Run program *Finite Population*. Describe the evolution of the measured histogram as N increases. Why do the statistical values keep changing? Is there a tendency in these values?

4.48 Run program *Running Histogram*. Study the influence of the number of data points and the number of intervals on the shape of the histogram and magnitude of each interval. Describe this influence. Why are some values "out of range."

4.49 Run program *Probability Density*. Vary the sample size and number of intervals used to create the density function. Why does it change shape with each new data set?

Chapter 5

Uncertainty Analysis

5.1 INTRODUCTION

Whenever we plan a test or later report a test result, we need to know something about the quality of the results. Uncertainty analysis provides a methodical approach to estimating the accuracy of the results. It can be used to report the quality of the results from an anticipated test or from a completed test. This chapter focuses on how to estimate the "± what?" in a planned test or in a stated test result.

Suppose the competent dart thrower of Chapter 1 tossed several practice rounds of darts at a bull's-eye. This would give us a good idea of the thrower's throw tendencies and accuracy. Then let the thrower toss another round. Without looking, can you guess where the darts hit? Test measurements that include systematic and random error components are much like this. We can calibrate a measurement system to get a good idea of its behavior and accuracy. However, from the calibration we can only estimate how well a measured value might estimate the actual "true" value in a subsequent measurement. So we need to quantify, on average, how closely the measured value agrees with some known true value and then base our interpretation of subsequent measured results on that information.

Errors are a property of the measurement. Measurement errors are introduced from various elements, e.g., the individual instruments, the data set finite statistics, the approach used. We have defined measurement as the process of assigning a value to a physical variable. The error in a measurement is simply the difference between the value assigned by our measurement and the true value of the variable. But we do not know the true value, only the measured value. So while we cannot estimate the actual error, we draw from what we do know about the measurement to estimate a range of probable error in the result from that measurement. It is this estimate that is called the *uncertainty* in the reported measured value. The uncertainty describes an interval about the measured value within which we suspect the true value must fall. It is the process of identifying, quantifying, and combining the errors that we call *uncertainty analysis*.

Uncertainty is a property of the result. In this chapter, we present a systematic approach for identifying, quantifying, and combining the estimates of the errors of a measurement in a way that estimates the uncertainty in the final result. Such uncertainty analysis provides a powerful design tool for evaluating different measurement systems and methods, designing a test plan, and for reporting on the quality of the determined result. While the chapter stresses the methodology of analyses, we have tried to illustrate the concomitant need for an equal application of critical thinking and professional judgment in applying the analyses. The quality of an uncertainty analysis will depend on the engineer's knowledge of the test, the measured variables, the equipment, and the measurement procedures [1].

There are two accepted professional documents on uncertainty analysis. The ANSI/ASME International's PTC 19.1 Test Uncertainty [2] is the United States engineering test standard, and our approach favors that method. The International Organization on Standardization's "Guide to the Expression of Uncertainty in Measurement" (ISO GUM) [1] is an international metrology standard. The two differ in some terminology and how errors are cataloged. For example, PTC 19.1 refers to random and systematic errors, whereas ISO GUM refers to Type A and Type B errors. PTC 19.1 favors classifying errors by how they manifest themselves in the measurement, while ISO GUM favors classifying errors by how they were originally estimated. These differences are real but they are not significant to the outcome. Once past the classifications, the two methods are quite similar. The important point is: The end result of an uncertainty analysis by either method will yield a similar result!

5.2   MEASUREMENT ERRORS

In the discussion that follows, errors will be grouped into two very general categories: systematic error and random error. We will not consider measurement blunders that result in obviously fallacious data—such data should be discarded.

Consider the repeated measurement of a variable under conditions that are expected to produce the same value of the measured variable. The relationship between the true value and the measured data set, containing both systematic and random errors, can be illustrated as in Figure 5.1. The total error in a set of measurements obtained under seemingly fixed conditions can be described by the systematic errors and a statistical estimate of the random errors in those measurements. The systematic errors will shift the sample mean away from the true mean of the measured variable by a fixed amount; and within a sample of many measurements, the random errors bring about a distribution of measured values about the sample mean. Even a so-called accurate measurement contains small amounts of systematic and random errors.

Measurement errors enter during all aspects of a test and obscure our ability to ascertain the information that we desire: the true value of the variable measured. If the result depends on more than one measured variable, these errors will further propagate to

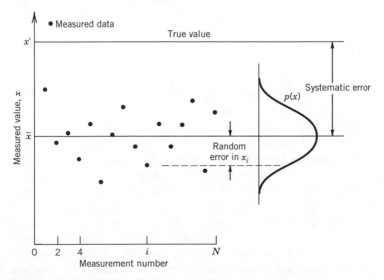

Figure 5.1. Distribution of errors upon repeated measurements.

the result. In Chapter 4, we stated that the best estimate of the true value sought in a measurement is provided by its sample mean value and the uncertainty in that value,

$$x' = \bar{x} \pm u_x \quad (P\%) \tag{4.1}$$

Uncertainty analysis is the method used to quantify the u_x term.

Certain assumptions are implicit in an uncertainty analysis:

1. The test objectives are known and the measurement itself is a clearly defined process.

2. Any known corrections for systematic error are already applied, in which case the systematic uncertainty assigned is the uncertainty of the correction. Random errors cannot be corrected.

3. Unless stated, the systematic errors are assumed to be independent (uncorrelated) of each other. Random errors are uncorrelated by their nature.

4. Data are obtained under fixed operating conditions.

5. The engineers have some "experience" with the system components.

When systematic errors are independent of each other, the value of one error could be high when the other is low and so forth. We say that these errors are uncorrelated. But some systematic errors may be related in a way that they are not independent. Examples and treatment of such correlated errors are discussed in Section 5.9.

In regards to item (5), we use the term "experience" to mean that the engineer can make an estimate of systematic and random errors based on some evidence or judgment, such as manufacturer's performance specifications, previous use, a professional test code, or performance information discussed in the technical literature.

5.3   DESIGN-STAGE UNCERTAINTY ANALYSIS

We might begin the design of a measurement system with an idea and some catalogs, and end the project after data have been obtained and analyzed. As with any part of the design process, the uncertainty analysis will evolve as the design of the measurement system and process matures. We will study uncertainty analysis for the following measurement situations: (1) design stage, (2) advanced stage and/or single measurement, and (3) multiple measurement.

Design-stage uncertainty analysis refers to the initial analysis performed prior to the measurement. It is useful for selecting instruments, selecting measurement techniques, and obtaining an approximate estimate of the uncertainty likely to exist in the measured data. At this point, the measurement system and associated procedures are but a concept. Often little is known about the instruments, which in many cases might still be just pictures in a catalog. Major facilities may need to be built and equipment ordered with a considerable lead time. Uncertainty analysis should be used to assist in the selection of equipment and procedures based on their relative performance and cost. In the design stage, distinguishing between systematic and random errors might be too difficult to be of concern. So for this discussion, consider only sources of uncertainty in general.

The goal in a design-stage analysis is to estimate the magnitude of uncertainty in the measured value that would result from the measurement. But a measurement system will usually consist of sensors and instruments each with their respective contributions to system uncertainty. So let's first talk about individual contributions to uncertainty.

Even when all errors are otherwise zero, a measured value must be affected by our ability to resolve the information provided by the instrument. We call this the *zero-order uncertainty* of the instrument, u_0. At *zero-order uncertainty*, we assume that the variation expected in the measured values will be that amount due to instrument resolution and that all other aspects of the measurement are perfectly controlled. Essentially, u_0 is an estimate of the expected random uncertainty caused by the data scatter due to reading the instrument.

As an arbitrary rule, assign a numerical value to u_0 of one-half of the instrument resolution[1] or equal to its digital least count with a probability of 95%:

$$u_0 = +\frac{1}{2}\text{resolution} \quad (95\%) \tag{5.1}$$

At 95% probability, we assume that only one measured value in 20 (20:1 odds) would have a value exceeding the interval defined by u_0.

The second piece of information that is usually available is the manufacturer's statement concerning instrument error. In a catalog, the value stated will be some typical value for that type of instrument under ideal conditions. We can assign this stated value as the *instrument uncertainty*, u_c. Essentially, u_c is an estimate of the systematic error for the instrument.

Sometimes the instrument errors will be stated in parts, each part due to some contributing factor (e.g., Table 1.1). A probable estimate in u_c can be made by combining all known errors in some reasonable manner. The accepted manner of combining errors is termed the root-sum-squares (RSS) method.

Combining Elemental Errors: RSS Method

Each individual measurement error will combine in some manner with other errors to increase the uncertainty of the measurement. We will refer to each individual error as an "elemental error." For example, the sensitivity error and linearity error of a transducer are two elemental errors. Consider a measurement of x which is subject to some K elements of error, e_k, where $k = 1, 2, \ldots, K$. A realistic estimate of the uncertainty in the measurement, u_x, due to these elemental errors can be computed using the *root-sum-squares method (RSS) method*:

$$
\begin{aligned}
u_x &= \pm\sqrt{e_1^2 + e_2^2 + \cdots + e_K^2} \\
&= \pm\sqrt{\sum_{k=1}^{K} e_k^2} \quad (P\%)
\end{aligned}
\tag{5.2}
$$

It is imperative to maintain consistency in the units of each error. Ideally, each error should be estimated at the same probability level.

A general, albeit somewhat arbitrary, rule is to use a 95% probability level ($P\% = 95\%$) throughout all uncertainty calculations. Engineers tend to follow this 95% rule, and it is equivalent to assuming the probability covered by two standard deviations. However, some prefer to use a 68% probability level ($P\% = 68\%$), which is equivalent to a spread of one standard deviation. We use the 95% level in our calculations but point out that other probability levels may be substituted, provided they are applied consistently, without any effect on the procedures.

---

[1]In some situations, it may be possible to assign a value for u_0 that is smaller than 1/2 the scale resolution. Common sense and discretion should be used.

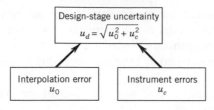

Figure 5.2. Design-stage uncertainty procedure in combining uncertainties.

Design-Stage Uncertainty

The RSS method of combining errors is based on the assumption that the possible variations in the values of an error encountered over repeated measurements will tend to follow a Gaussian distribution. As such, the RSS estimate of contributing errors should provide a "probable" measure of the uncertainty due to these in any estimates of error. Although the simple arithmetic sum of the elemental errors would provide a larger uncertainty estimate, that particular approach assumes that all errors can occur in their worst possible way for each and all measurements. Such a method has no basis in probability and would be appropriate in very few applications (such as in catastrophic prediction).

The *design-stage uncertainty*, u_d, for the instrument is found by combining the instrument uncertainty with the zero-order uncertainty,

$$u_d = \sqrt{u_0^2 + u_c^2} \quad (P\%) \tag{5.3}$$

This procedure for estimating the design-stage uncertainty is outlined in Figure 5.2.

The design-stage uncertainty estimate is not intended to be a final statement of the total uncertainty in a measurement. It is to be used only as a guide for selecting equipment and procedures before a test. For example, design-stage uncertainty assumes perfect control of test conditions and measurement procedure. As presented, it includes only instrument errors and may be interpreted as a reasonable estimate of the minimum uncertainty that can be achieved with a particular test strategy. *If additional information about other measurement errors is known at the design stage, then these errors should be used to adjust equation (5.3).*

A measurement system typically consists of a chain of sensors and instruments. The design-stage uncertainty for the measurement system is arrived at by combining each of the design-stage uncertainties for each component in the system using the RSS method. When using the RSS method for several components, take care to use consistent units between each error or uncertainty!

EXAMPLE 5.1

Consider the force measuring instrument described by the following catalog data. Provide an estimate of the uncertainty attributable to this instrument and the instrument design-stage uncertainty.

Resolution:	0.25 N
Range:	0 to 100 N
Linearity:	within 0.20 *N* over range
Hysteresis:	within 0.30 *N* over range

KNOWN   Catalog specifications

ASSUMPTIONS   Values representative of instrument

FIND

$$u_c, u_d$$

SOLUTION   We will follow the procedure outlined in Figure 5.2. An estimate of the instrument uncertainty depends on the uncertainty assigned to each of the contributing elemental errors of linearity, e_1, and hysteresis, e_2:

$$u_1 = 0.20\,\text{N} \quad u_2 = 0.30\,\text{N}$$

Then using equation (5.2) with $K = 2$ yields

$$u_c = \pm\sqrt{(0.2)^2 + (0.3)^2}$$
$$= \pm 0.36\,\text{N} \quad (95\%)$$

The instrument resolution is given as 0.25 N, from which $u_0 = \pm 0.125$ N. From equation (5.3), the design-stage uncertainty of this instrument would be

$$u_d = \pm\sqrt{u_0^2 + u_c^2} = \pm\sqrt{(0.36)^2 + (0.125)^2}$$
$$= \pm 0.38\,\text{N} \quad (95\%)$$

COMMENT   The design-stage uncertainty for this instrument is simply an estimate based on the "experience" at hand. Additional information, such as calibrations or prior use of the instrument, would provide justification for modifying this estimate by substituting more relevant values for each error or by including additional elemental errors in the analysis.

EXAMPLE 5.2

A voltmeter is to be used to measure the output from a pressure transducer as an electrical signal. The nominal pressure is expected to be about 3 psi (3 lb/in.^{2}). Estimate the design-stage uncertainty in this combination. The following information is available:

Voltmeter

    Resolution: 10 μV

    Accuracy: within 0.001% of reading

Transducer

    Range: ± 5 psi

    Sensitivity: 1 V/psi

    Input power: 10 VDC $\pm 1\%$

    Output: ± 5 V

    Linearity: within 2.5 mV/psi over range

    Sensitivity: within 2 mV/psi over range

    Resolution: negligible

KNOWN   Instrument specifications

ASSUMPTIONS   Values representative of instrument at 95% probability

FIND u_c for each device and u_d for the measurement system

SOLUTION   The procedure in Figure 5.2 will be used for both instruments to estimate the design-stage uncertainty in each. The resulting uncertainties will then be combined using the RSS approximation to estimate the system u_d.

The uncertainty in the voltmeter at the design stage is given by equation (5.3) as

$$(u_d)_E = \pm\sqrt{(u_o)_E^2 + (u_c)_E^2}$$

From the information available,

$$(u_0)_E = \pm 5\,\mu V \quad (95\%)$$

For a nominal pressure of 3 psi, we expect to measure an output of 3 V. Then,

$$(u_c)_E = \pm(3\,V \times 0.00001) = \pm 30\,\mu V \quad (95\%\ \text{assumed})$$

so that the design-stage uncertainty in the voltmeter is

$$(u_d)_E = \pm 30.4\,\mu V \quad (95\%)$$

The uncertainty in the pressure transducer output at the design stage is also given by equation (5.17). Assuming that we operate within the input power range specified, the instrument output uncertainty can be estimated by considering the uncertainty in each of the instrument elemental errors of linearity, e_1, and sensitivity, e_2:

$$(u_c)_p = \sqrt{e_1^2 + e_2^2} \quad (95\%\ \text{assumed})$$

$$= \pm\sqrt{(2.5\,mV/psi \times 3\,psi)^2 + (2\,mV/psi \times 3\,psi)^2}$$

$$= \pm 9.61\,mV \quad (95\%)$$

Since $(u_0) \approx 0$ V/psi, then the design-stage uncertainty in the transducer in terms of indicated voltage is $(u_d)_p = \pm 9.61$ mV (95%).

Finally, u_d for the combined system is found by use of the RSS method using the design-stage uncertainties of the two devices. The design-stage uncertainty in pressure as indicated by this measurement system is estimated to be

$$u_d = \pm\sqrt{(u_d)_E^2 + (u_d)_p^2}$$

$$= \pm\sqrt{(0.030\,mV)^2 + (9.61\,mV)^2}$$

$$= \pm 9.61\,mV \quad (95\%)$$

But since the sensitivity is 1 V/psi, the uncertainty in pressure is better stated as

$$u_d = \pm 0.0096\,psi \quad (95\%)$$

COMMENT   Note that essentially all of the uncertainty is due to the transducer. A better voltmeter will not improve the uncertainty in this measurement! Design-stage uncertainty analysis shows us that if better pressure results are desired, then a better transducer will need to be employed.

5.4   ERROR SOURCES

Design-stage uncertainty provides essential information to assess methodology and instrument selection. But it does not address all of the possible errors that influence a measured result. Here we look at potential sources of errors in measurements and provide a helpful checklist of common errors. It is not necessary to classify error sources as we do here, but it is a good bookkeeping practice.

Table 5.1   Calibration Error Source Group

Element	Error Source[a]
1	Primary to interlab standard
2	Interlab to transfer standard
3	Transfer to lab standard
4	Lab standard to measurement system
5	Calibration and technique
etc.	

[a]Systematic error and/or random error in each element.

Consider the measurement process as consisting of three distinct steps: calibration, data acquisition, and data reduction. Errors that enter during each of these steps will be grouped under their respective error source heading: (1) calibration errors, (2) data-acquisition errors, and (3) data-reduction errors. Within each of these three *error source groups*, list the types of errors encountered. Such errors are the elemental errors of the measurement. Do not become preoccupied with these groupings. They are used as a guide to identify errors. If you place an error in an "incorrect" group, it is okay. The final uncertainty will not be changed!

Calibration Errors

Calibration in itself does not eliminate measurement system errors; it merely reduces the uncertainty in the calibration result to more acceptable values. Calibration errors include those elemental errors that enter the measuring system during its calibration. *Calibration errors* tend to enter through two principal sources: (1) the systematic and random errors inherent to the standard used in the calibration and (2) the manner in which the standard is applied to the measuring system or system component. For example, the laboratory standard used for calibration contains some inherent uncertainty. Accordingly, there may be a difference between the value of the standard used for the calibration and the primary standard that it represents. Hence, there will be an uncertainty in the input value on which the calibration is based. In addition, there can be a difference between the value supplied by the standard used and the calibration value actually sensed by the measuring system. Either of these effects will be built into the calibration data. In Table 5.1, we list some common elemental errors contributing to this error source group.

Data-Acquisition Errors

All errors that arise during the actual act of measurement are referred to as data-acquisition errors. These errors include sensor and instrument errors, uncontrolled variables, such as changes or unknowns in measurement system operating conditions, and sensor installation effects on the measured variable. In addition, the measured variable temporal and spatial variations contribute errors, and these may be quantified through the application of finite statistics. We list some elemental errors common to this source in Table 5.2.

Data-Reduction Errors

The use of curve fits and correlations with their associated unknowns (Section 4.5) introduces data-reduction errors into the reported test results. Also, the resolution of

Table 5.2   Data-Acquisition Error Source Group

Element	Error Source[a]
1	Measurement system operating conditions
2	Sensor-transducer stage (instrument error)
3	Signal conditioning stage (instrument error)
4	Output stage (instrument error)
5	Process operating conditions
6	Sensor installation effects
7	Environmental effects
8	Spatial variation error
9	Temporal variation error
etc.	

[a]Systematic error and/or random error in each element.
Note: A total input-to-output measurement system calibration will combine elements 2, 3, 4, and possibly 1 within this error source group.

computational operations required to reduce the data into some desired result contributes error. We list elemental errors typical of this error source group in Table 5.3.

5.5   SYSTEMATIC AND RANDOM ERRORS

In general, whether an error element is classified as containing a systematic error, a random error, or both can be simplified by considering the methods used to quantify these errors. Treat an error as a random error if it is estimated using the statistics from an available data set; otherwise, treat the error as a systematic error.

Systematic Error

A systematic error (also known as a bias error) remains constant in repeated measurements under fixed operating conditions. Thus, each repeated measurement would contain the same amount of systematic error. Being a fixed value, the systematic error cannot be directly discerned by statistical means alone. It can be difficult to estimate its value or in many cases even recognize its presence. A systematic error may cause either a high or a low estimate of the true value. Accordingly, an estimate of the range of systematic error is represented by an interval, defined as $\pm B$, within which the true value is expected to lie. This range of suspected systematic error is called the systematic uncertainty.

The reader is probably familiar with some systematic errors in measurements. For example, if you report the barefoot height of a person based on a measurement taken while the person was wearing heeled shoes, the reported value will be in error. This is one

Table 5.3   Data-Reduction Error Source Group

Element	Error Source[a]
1	Calibration curve fit
2	Truncation error
etc.	

[a]Systematic error and/or random error in each element.

clear systematic error in a measurement. In this case this systematic error, a data-acquisition error, is the height of the heels. But that is too simple!

Consider a home bathroom scale, which may have a systematic error. Does it? How might you estimate the magnitude of the error in an indicated weight? Perhaps you can calibrate the scale using calibrated masses, account for local gravitational acceleration, and correct the output, thereby estimating the systematic error of the measurement (direct calibration against a local standard). Or perhaps you can compare it to a measurement taken in a physician's office or one at the gym and can compare each reading (a sort of interlaboratory comparison). Or perhaps you can carefully measure your volume displacement in water and compare the results to estimate differences (concomitant methodology). Without any of the above, what value would you assign? Would you even suspect a systematic error?

But let us think about this. An insidious aspect of systematic error has been revealed. Why should one doubt a measurement indication and suspect a systematic error? Figuratively speaking, there will be no shoe heels staring at you. Experience teaches us to think through each measurement carefully because systematic error is always present at some magnitude. We see that it is difficult to estimate systematic error without comparison, so a good design should include some means to estimate it.

Systematic error can be estimated by comparison. Various methodologies can be utilized: (1) calibration, (2) concomitant methodology, (3) interlaboratory comparisons, or (4) experience. When available, the most direct method is by calibration using a suitable standard or alternatively, using calibration methods of inherently negligible or small systematic error. Another procedure is the use of concomitant methodology, which is using different methods of estimating the same thing and comparing the results. Concomitant methods that depend on different physical measurement principles are preferable, as are methods that rely on calibrations that are independent of each other. In this regard, even accepted analytical methods could be used for comparison[2] or at least to estimate the range of systematic error due to influential sources such as environmental conditions, instrument response errors, and loading errors. Lastly, an elaborate but good approach is through interlaboratory comparisons of similar measurements, an excellent replication method. This approach introduces different instruments, facilities, and personnel into an otherwise similar measurement procedure in an effort to compare the systematic errors between measuring facilities [1]. In lieu of the above, a value based on experience may have to be assigned.

As a note, calibration cannot eliminate all systematic error; it can only reduce it. Consider the calibration of a temperature transducer against an NIST standard certified to be correct to within 0.01°C. If the calibration data show that the transducer output has a systematic error of 0.2°C relative to the standard then we would just correct all the data obtained with this transducer by 0.2°C. Simple enough! But the standard itself still has an intrinsic systematic uncertainty of ±0.01°C, and this uncertainty remains in the calibrated transducer, as would the uncertainty in the correction value applied.

Random Error

Random errors vary randomly in repeated measurements. When repeated measurements are made under nominally fixed operating conditions, random errors manifest themselves

---

[2]Reference [3] provides examples for determining jet engine thrust, and several complementary measurements are used with an energy balance to estimate systematic error.

as scatter of the measured data. Random error, which may also be called precision error, is affected by the repeatability and resolution of the measurement system components, the measured variable's own temporal and spatial variations, the variations in the process operating and environmental conditions from which measurements are taken, and the repeatability of the measurement procedure and technique.

Because of the varying nature of random errors, exact values cannot be given, but probable estimates of the error can be made through statistical analyses. For example, the sample mean, $\bar{x}$, from a data set has a confidence interval estimated by $\pm t_{v,P} S_{\bar{x}}$, and this provides a probable estimate for the range of random error contained in the mean value. This range of random error is called the *random uncertainty*.

In summary, the most direct method of both reducing and estimating systematic error is through a comparison, such as a calibration. Random errors are identified and quantified through repeated measurements and statistical analysis of the results.

5.6   UNCERTAINTY ANALYSIS: ERROR PROPAGATION

Suppose we wanted to determine how long it would take to fill a swimming pool from a garden hose. We might time the filling of a bucket of known volume to estimate the flow rate from the garden hose. Then with an estimate or measurement of the volume of the pool, the time to fill the pool could be estimated. Clearly, small errors in estimating the flow rate from the garden hose would translate into large differences in the time required to fill the pool! Here we are using measured values, the flow rate and volume, to estimate a result, the time required to fill the pool.

Very often in engineering, results are determined through a functional relationship with measured values. For example, we just calculated a flow rate above by measuring time, t, and bucket volume, $\forall$, since $Q = f(t, \forall)$. But how do uncertainties in either quantity contribute to uncertainty in flow rate? Is the uncertainty in Q more sensitive to uncertainty in volume or in time? In general, how are errors propagated to a calculated result? These issues are now explored.

Propagation of Error

A general relationship between some dependent variable y and a measured variable x, that is $y = f(x)$, is illustrated in Figure 5.3. Now suppose we measure x a number of times at some operating condition so as to establish its sample mean value and the uncertainty in the random error in this mean value, $t S_{\bar{x}}$. This implies that, neglecting other random and systematic errors, the true value for x lies somewhere within the interval $\bar{x} \pm t S_{\bar{x}}$. It is reasonable to assume that the true value of y, which is determined from the measured values of x, falls within the interval defined by

$$\bar{y} \pm \delta y = f(\bar{x} \pm t S_{\bar{x}}) \tag{5.4}$$

Expanding this as a Taylor series yields

$$\bar{y} \pm \delta y = f(\bar{x}) \pm \left[\left(\frac{dy}{dx} \right)_{x=\bar{x}} t S_{\bar{x}} + \frac{1}{2} \left(\frac{d^2 y}{dx^2} \right)_{x=\bar{x}} (t S_{\bar{x}})^2 + \cdots \right] \tag{5.5}$$

By inspection, the mean value for y must be $f(\bar{x})$ so that the term in brackets estimates $\pm \delta y$. A linear approximation for δy can be made, which is valid when $t S_{\bar{x}}$ is small and neglects the higher order terms in equation (5.5), as

$$\delta y \approx \left(\frac{dy}{dx} \right)_{x=\bar{x}} t S_{\bar{x}} \tag{5.6}$$

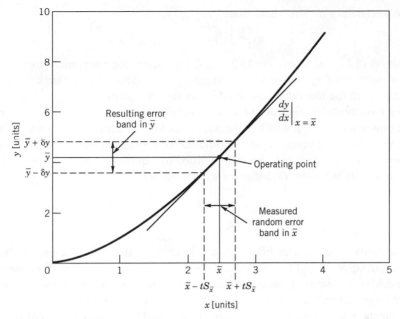

Figure 5.3. Relationship between a measured variable and a resultant calculated using the value of that variable.

The derivative term, $(dy/dx)_{x=\bar{x}}$, defines the slope of a line that passes through the point specified by $\bar{x}$. For small deviations from the value of $\bar{x}$, this slope predicts an acceptable, approximate relationship between $tS_{\bar{x}}$ and δy. The derivative term is a measure of the sensitivity of y to changes in x. Since the slope of the curve can be different for different values of x, it is important to evaluate the slope using a representative value. The width of the interval defined by $\pm tS_{\bar{x}}$ corresponds to an interval about y, that is, $\pm \delta y$, within which we should expect the true value of y to lie. Figure 5.3 illustrates the concept that errors in a measured variable are propagated through to a resultant variable in a predictable way. In general, we apply this analysis to the errors that contribute to the uncertainty in x, written as u_x. The uncertainty in x will be related to the uncertainty in the resultant y by

$$u_y = \left(\frac{dy}{dx}\right)_{x=\bar{x}} u_x \tag{5.7}$$

Compare equations (5.6) and (5.7) with Figure 5.3. This idea can be extended to multivariable relationships. Consider a result, R, which is determined through some functional relationship between independent variables, $x_1, x_2, \ldots, x_L$, defined by

$$R = f_1\{x_1, x_2, \ldots, x_L\} \tag{5.8}$$

where L is the number of independent variables involved. Each variable will contain some measure of uncertainty that will affect the result. The best estimate of the true mean value, R', would be stated as

$$R' = \overline{R} \pm u_R \quad (P\%) \tag{5.9}$$

where the sample mean of R is found from

$$\overline{R} = f_1\{\bar{x}_1, \bar{x}_2, \ldots \bar{x}_L\} \tag{5.10}$$

and the uncertainty in $\overline{R}$ is found from

$$u_R = f_1\{u_{\overline{x}_1}, u_{\overline{x}_2}, \ldots, u_{\overline{x}_L}\} \tag{5.11}$$

In equation (5.11), each $u_{\overline{x}_i}$, $i = 1, 2, \ldots, L$ represents the uncertainty associated with the best estimate of x_1 and so forth through x_L. The value of u_R reflects the individual contributions of the individual uncertainties as they are propagated through to the result.

The most probable estimate of u_R is generally accepted as that value given by the second power relation [5], which is the square root of the sum of the squares (RSS). The RSS form can be derived from the linearized approximation of the Taylor series expansion of the multivariable function defined by equation (5.8). A general sensitivity index, θ_i, results from the Taylor series expansion and is given by

$$\theta_i = \frac{\partial R}{\partial x_i}\bigg|_{x=\overline{x}} \quad i = 1, 2, \ldots, L \tag{5.12}$$

The sensitivity index relates how changes in each x_i affect R. The use of the partial derivative in equation (5.12) is necessary when R is a function of more than one variable. This index is evaluated using either the mean values or, lacking these estimates, the expected nominal values of the variables. From a Taylor series expansion, the contribution of the uncertainty in x to the result, R, is estimated by the term $\theta_i u_{\overline{x}_i}$. The propagation of uncertainty in the variables to the result will yield an uncertainty estimate given by

$$u_R = \pm \left[\sum_{i=1}^{L} (\theta_i u_{\overline{x}_i})^2 \right]^{1/2} \quad (P\%) \tag{5.13}$$

Sequential Perturbation

A numerical approach can also be used to estimate the propagation of uncertainty through to a result [6]. Often referred to as sequential perturbation, it is generally the preferred method when direct partial differentiation is too cumbersome or intimidating, or the number of variables involved is large. It is easily programmed using spreadsheet software to reduce data already stored in discrete form, but it can also be executed on a hand calculator.

The method is straightforward and uses a finite difference method to approximate the derivatives:

1. Based on measurements for the independent variables under some fixed operating condition, calculate a result R_o where $R_o = f(x_1, x_2, \ldots, x_L)$. This value fixes the operating point for the numerical approximation (e.g., see Figure 5.3).

2. Next, increase the independent variables by their respective uncertainties and recalculate the result based on each of these new values. Call these values R_i^+. That is,

$$R_1^+ = f(x_1 + u_{x_1}, x_2, \ldots, x_L),$$
$$R_2^+ = f(x_1, x_2 + u_{x_2}, \ldots, x_L), \ldots \tag{5.14}$$
$$R_L^+ = f(x_1, x_2, \ldots, x_L + u_{x_L}),$$

3. Next, in a similar manner, decrease the independent variables by their respective uncertainties and recalculate the result based on each of these new values. Call these values R_i^-.

4. Calculate the differences δR_i^+ and δR_i^- for $i = 1, 2, \ldots, L$

$$\delta R_i^+ = R_i^+ - R_o$$
$$\delta R_i^- = R_i^- - R_o \tag{5.15}$$

5. Finally, evaluate the approximation of the uncertainty contribution from each variable,

$$\delta R_i = \frac{\delta R_i^+ - \delta R_i^-}{2} \approx \theta_i u_i \tag{5.16}$$

Then, the uncertainty in the result is

$$u_R = \pm \left[\sum_{i=1}^{L} (\delta R_i)^2 \right]^{1/2} \quad (\text{P\%}) \tag{5.17}$$

Equations 5.13 and 5.17 provide two methods for estimating the propagation of uncertainty to a result. In most cases, use of either equation will yield nearly identical results, and the choice of method is left to the user. We point out that sometimes unreasonable estimates of u_R may be generated by either method. When this happens the cause can be traced to a sensitivity index that changes rapidly with small changes in the independent variable, x_i, coupled with a large value of the uncertainty u_{x_i}. In these situations, the engineer should examine the cause and extent of the variation in sensitivity and to use a more accurate approximation for the sensitivity.

In subsequent sections, we develop methods to estimate the uncertainty values from available information.

EXAMPLE 5.3

For a displacement transducer having a calibration curve $y = KE$, estimate the uncertainty in displacement y for $E = 5.00$ V, if $K = 10.10$ mm/V with $u_K = \pm 0.10$ mm/V and $u_E = \pm 0.01$V at 95% confidence.

KNOWN

$y = KE$

$E = 5.00\,\text{V} \quad u_E = \pm 0.01\,\text{V}$

$K = 10.10\,\text{mm/V} \quad u_K = \pm 0.10\,\text{mm/V}$

FIND

u_y

SOLUTION   Based on equations (5.10) and (5.11), respectively,

$$\bar{y} = f(\bar{E}, \bar{K}) \quad \text{and} \quad u_y = f(u_E, u_K)$$

From equation (5.13), the uncertainty in the displacement at $y = KE$ is

$$u_y = \pm \left[(\theta_E u_E)^2 + (\theta_K u_K)^2 \right]^{1/2}$$

where the sensitivity indices are evaluated from equation (5.12) as

$$\theta_E = \frac{\partial y}{\partial E} = K \quad \text{and} \quad \theta_K = \frac{\partial y}{\partial K} = E$$

or we can write equation (5.13) as

$$u_y = \pm\left[(Ku_E)^2 + (Eu_K)^2\right]^{1/2}$$

The operating point occurs at the nominal or the mean values of $E = 5.00$ V and $y = 50.50$ mm. With $E = 5.00$ V and $K = 10.10$ mm/V and substituting for u_E and u_K, evaluate u_y at its operating point:

$$u_y|_{y=50.5} = \pm\left[(0.10)^2 + (0.5)^2\right]^{1/2} = \pm 0.51 \text{ mm} \quad (95\%)$$

Alternatively, we can use sequential perturbation. The operating point for the perturbation will again be $y = R_o = 50.5$ mm. Using equations (5.14)–(5.16) gives

i	x_i	R_i^+	R_i^-	δR_i^+	δR_i^-	δR_i
1	E	50.60	50.40	0.1	−0.1	0.1
2	K	51.00	50.00	0.5	−0.5	0.5

Then, using equation (5.17) gives

$$u_y|_{y=50.5} = \pm\left[(0.10)^2 + (0.5)^2\right]^{1/2} = \pm 0.51 \text{ mm}$$

The two methods give the identical result. We state the calculated displacement in the form of equation (5.9) as

$$y' = 50.50 \pm 0.51 \text{ mm} \quad (95\%)$$

5.7 ADVANCED-STAGE UNCERTAINTY ANALYSIS

In designing a measurement system a pertinent question is, how would it affect the result if this particular aspect of the technique or equipment were changed? In design-stage uncertainty analysis, we only considered the errors due to a measurement system's resolution and estimated instrument calibration errors. But if additional information is available, we can get a better idea of the uncertainty in a measurement. An advanced-stage uncertainty analysis permits taking design-stage analysis further by considering procedural and test control errors that affect the measurement. We consider it as a method for a thorough uncertainty analysis when a large data set is not available. This is often the case in the early stages of a test program or for certain tests where repeating measurements may not be possible. Such an advanced-stage analysis, also known as *single-measurement uncertainty analysis* [6],[3] is to be used: (1) In the advanced design stage of a test to estimate the expected uncertainty, beyond the initial design stage estimate; and (2) To report the results of a test program that involved measurements over a range of one or more parameters but with no or relatively few repeated measurements of the pertinent variables at each test condition. Essentially, the method assesses different aspects of the test to quantify potential errors.

In this section, the goals will be either to estimate the uncertainty in some measured value, x, or in some general result, R, through an estimation of the uncertainty in each of the factors which may affect x or R. If all factors that influence a measurement were held constant, we might expect the measured value to remain constant upon repeated measurements. However, measurement errors will replicate; that is, they will affect the measured value somewhat differently with each measurement. The overall uncertainty in

---

[3]It is called replication level analysis in [3, 6].

any measurement is affected by this replication of errors and is an estimate of the possible value of these errors. A technique that uses a step-by-step approach for identifying and estimating errors is presented. We seek the combined value of the estimates at each step.

Zero-Order Uncertainty

At zero-order uncertainty, all variables and parameters that affect the outcome of the measurement, including time, are assumed to be fixed except for the physical act of observation itself. Under such circumstances, any data scatter introduced upon repeated observations of the output value will be the result of instrument resolution alone. The value u_0 estimates the extent of variation expected in the measured value when all influencing effects are controlled. An estimate of u_0 is found using equation (5.1).

At this level, the uncertainty value calculated would be the minimum possible as the use of zero-order analysis provides an estimate only of the effect of instrument resolution on the measurement. Obviously, a zero-order uncertainty analysis is inadequate for the reporting of test results.

Higher-Order Uncertainty

Higher-order uncertainty estimates consider the controllability of the test operating conditions and the variability of all measured variables. For example, at the first-order level, the effect of time as an extraneous variable in the measurement might be considered. That is, what would happen if we started the test, set the operating conditions, and sat back and watched? If a variation in the measured value is observed then time is a factor in the test, presumably due to some extraneous influence or simply inherent in the behavior of the variable being measured.

In practice, the uncertainty at this first level would be evaluated for each particular measured variable by operating the test facility at some single operating condition that would be within the range of conditions to be used during the actual tests. A set of data (say, $N \geq 30$) would be obtained under some set operating condition. The first-order uncertainty of our ability to estimate the true value of the measured value could be estimated as

$$u_1 = \pm t_{\nu,p} S_{\bar{x}} \tag{5.18}$$

The uncertainty at u_1 includes the effects of resolution, u_o. So only when $u_1 = u_0$ is time not a factor in the test. In itself, the first-order uncertainty is inadequate for reporting of test results.

At each successive order, each factor identified as affecting the measured value is introduced into the analysis, thus giving a higher but more realistic estimate of the uncertainty. For example, at the second level it might be appropriate to assess the limits of the ability to duplicate the exact operating conditions and the consequences on the test outcome. Or perhaps spatial variations that affect the outcome are assessed, such as when a value from a point measurement is assigned to quantify a larger volume.

Nth-Order Uncertainty

At the Nth-order estimate, instrument calibration characteristics are entered into the scheme through the instrument uncertainty, u_c. A practical estimate of the Nth-order

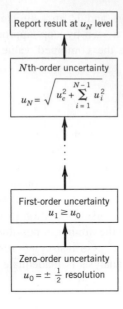

Figure 5.4. Advanced-stage and single-measurement uncertainty procedure in combining uncertainties.

uncertainty, u_N, is given by

$$u_N = \left[u_c^2 + \left(\sum_1^{N-1} u_i^2 \right) \right]^{1/2} \quad (P\%) \tag{5.19}$$

Uncertainty estimates at the Nth order allow for the direct comparison between results of similar tests obtained either using different instruments or at different test facilities. The procedure for a single-measurement analysis is outlined in Figure 5.4.

Note that as a minimum design-stage analysis includes only the effects found as a result of u_0 and u_c. It is those in-between levels that allow measurement procedure and control effects to be considered in the uncertainty analysis scheme. Only when $u_N = u_d$ are the test conditions under complete control. As such, the Nth-order uncertainty estimate provides the uncertainty value sought in advanced design stage or in single-measurement analyses. It is the only appropriate value to be used in the reporting of results from single-measurement tests.

EXAMPLE 5.4

As an exercise, obtain and examine a dial oven thermometer. How would you assess the zero- and first-order uncertainty in the measurement of the temperature of a kitchen oven using the device?

KNOWN   Dial thermometer

ASSUMPTION   Negligible systematic error in the instrument

FIND   Estimate u_0 and u_1 in oven temperature

SOLUTION   The zero-order uncertainty would be that contributed by the interpolation error of the measurement system only. For example, most gauges of this type have a resolution of 10°C. So at the zero order from equation (5.1) we might estimate

$$u_0 = \pm 5°C \quad (95\%)$$

At the first order, the uncertainty would be affected by any time variation in the measured value (temperature) and variations in the operating conditions. If we placed the thermometer in the center of an oven, set the oven to some relevant temperature, allowed the oven to preheat to a steady condition, and then proceeded to record the temperature indicated by the thermometer at random intervals, we could estimate our ability to control the oven temperature. For J measurements, the first-order uncertainty of the measurement in this oven's mean temperature using this technique and this instrument would be from equation (5.18)

$$u_1 = \pm t_{J-1, 95} S_{\overline{T}} \quad (95\%)$$

COMMENT   Let's suppose we are interested in how accurately we could set the oven mean temperature. Then the idea of setting the operating condition, the temperature, becomes important. We could estimate our ability to repeatedly set the oven to a desired mean temperature (time-averaged temperature). This could be done by changing the oven setting and then resetting the thermostat back to the original operating setting. If we were to repeat this sequence M times (reset thermostat, measure a set of data, reset thermostat, etc.), we could compute the different u_1 by

$$u_1 = \pm t_{M(J-1), 95} \langle S_{\overline{T}} \rangle \quad (95\%)$$

Note that the variation in oven temperature with time at any setting is included in this new pooled estimate. The effects of instrument calibration would enter at the Nth order through u_c. The uncertainty in oven temperature at some setting would be well approximated by the estimate from equation (5.19)

$$u_N = \pm \left(u_1^2 + u_c^2 \right)^{1/2} \quad (95\%)$$

By inspection of the different values u_0, u_c, and u_i, where $i = 1, 2, \ldots, N - 1$, single-measurement analysis provides a capability to pinpoint those aspects of a test that contribute most to the overall uncertainty in the measurement, as well as provide a reasonable estimate of the uncertainty in a single measurement.

EXAMPLE 5.5

A stopwatch is to be used to estimate the time between the start of an event and the end of this event. Event duration might range from several seconds to 10 min. Estimate the probable uncertainty in a time estimate using a hand-operated stopwatch that claims an accuracy of 1 min/month and a resolution of 0.01 s.

KNOWN

$u_0 = \pm 0.005$ s   (95%)

$u_c = \pm 60$ s/month   (95% assumed)

FIND

u_d, u_N

SOLUTION   The design-stage uncertainty will give an estimate of the suitability of an instrument for a measurement. At 60 s/month, the instrument accuracy works out to about 0.01 s/10 min of operation. This gives a design-stage uncertainty of

$$u_d = \pm \left(u_o^2 + u_c^2 \right)^{1/2} = \pm 0.01 \text{ s} \quad (95\%)$$

for an event lasting 10 min versus ± 0.005 s (95%) for an event lasting 10 s. Note that instrument calibration error controls the longer duration measurement, whereas instrument resolution controls the short duration measurement.

But do instrument resolution and calibration error actually control the uncertainty in this measurement? The design-stage analysis does not include the data-acquisition error involved in the act of physically turning the watch on and off. But a first-order analysis might be run to estimate the uncertainty that enters through the procedure of using the watch. Suppose a typical trial run of 20 tries of simply turning a watch on and off suggests that the uncertainty in determining the duration of an occurrence is estimated by

$$u_1 = \pm 0.15 \, \text{s} \quad (95\%)$$

The uncertainty in measuring the duration of an event would then be better estimated by equation (5.19).

$$u_N = \pm \left(u_1^2 + u_c^2 \right)^{1/2} = \pm 0.15 \, \text{s} \quad (95\%)$$

This estimate will hold for periods of up to about 2 h. Clearly, procedure controls the uncertainty, not the watch. Now a better decision as to whether the watch and procedure are suitable for the intended test can be made.

EXAMPLE 5.6

A flow meter can be calibrated by providing a known flow rate to the meter and measuring the meter output. One method of calibration with liquid systems is the use of a catch and time technique whereby a volume of liquid, after passing through the meter, is diverted to a tank for a measured period of time from which the flow rate, volume/time, is computed. There are two procedures that can be used to determine the known flow rate, Q, in, say, ft^3/min:

1. The volume of liquid, $\forall$, caught in known time, t, can be measured.
   Suppose we arbitrarily set $t = 6$ s and assume that our available facilities can determine volume (at Nth order) to ± 0.001 ft^3. Note: The chosen time value depends on how much liquid we can accommodate in the tank.

2. The time, t, required to collect 1 ft^3 of liquid can be measured.
   Based on Example 5.5, assume an Nth-order uncertainty in time of ± 0.15 s. This uncertainty includes our ability to control the on/off time and watch accuracy, but assumes that the diversion of flow is instantaneous. In either case the same instruments are used. Determine which method is better to minimize uncertainty over a range of flow rates based on these preliminary estimates.

KNOWN

$$u_N = \pm 0.001 \, \text{ft}^3$$

$$u_t = \pm 0.15 \, \text{s}$$

$$Q = f(\forall, t) = \forall / t$$

ASSUMPTIONS Flow diversion is instantaneous.

FIND Preferred method

SOLUTION From the available information, the propagation of probable uncertainty to the result, Q, is estimated from equation (5.13):

$$u_Q = \pm \left[\left(\frac{\partial Q}{\partial \forall} u_\forall \right)^2 + \left(\frac{\partial Q}{\partial t} u_t \right)^2 \right]^{1/2}$$

or dividing through by Q, the percent uncertainty in flow rate, u_Q/Q, is

$$\frac{u_Q}{Q} = \pm \left[\left(\frac{u_\forall}{\forall} \right)^2 + \left(\frac{u_t}{t} \right)^2 \right]^{1/2}$$

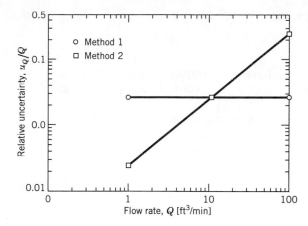

Figure 5.5. Uncertainty plot for the design analysis of Example 5.6.

Representative values of Q are needed to solve this relation. Consider a range of flow rates, say, 1, 10, and 100 ft³/min; the results are listed in the table for both methods.

Q [ft³/min]	t [s]	$\forall$ [ft³]	$u_\forall/\forall$	u_t/t	u_Q/Q
Method 1					
1	6	0.1	0.01	0.025	0.027
10	6	1.0	0.001	0.025	0.025
100	6	10.0	0.0001	0.025	0.025
Method 2					
1	60.0	1.0	0.001	0.003	0.003
10	6.0	1.0	0.001	0.025	0.025
100	0.6	1.0	0.001	0.250	0.250

In method 1, it is clear the uncertainty in time contributes the most to the relative uncertainty in Q, provided that the flow diversion is instantaneous. But in method 2, uncertainty in measuring either time or volume can contribute more to the uncertainty in Q, depending on the time sample length. The results for both methods are compared in Figure 5.5. It is apparent that for the conditions selected and preliminary uncertainty values used, method 2 would be a better procedure for flow rates up to 10 ft³/min. At higher flow rates, method 1 would be better. However, the minimum uncertainty in method 1 is limited to 2.5%. The engineer may be able to reduce this uncertainty by improvements in the time procedure.

COMMENT These results are without consideration of some other elemental errors that are present in the experimental procedure. For example, the diversion of the flow is assumed to occur instantaneously. A first-order uncertainty estimate could be used to estimate the added uncertainty to flow rate intrinsic to the time required to divert the liquid to and from the catch tank. Operator influence should be randomized in actual tests. Accordingly, the above calculations may be used as a guide in procedure selection but should not be used as the final estimate in the overall uncertainty in any results obtained from the actual calibration.

EXAMPLE 5.7

Repeat Example 5.6, using sequential perturbation for the operating conditions $\forall = 1$ ft³ and $t = 6$ s.

SOLUTION   The operating point is $R_o = Q = \forall/t = 0.1667$ ft³/s. Then, solving equations (5.14)–(5.17) gives:

i	x_i	R_i^+	R_i^-	δR_i^+	δR_i^-	δR_i
1	$\forall$	0.1668	0.1665	0.000167	−0.000167	0.000167
2	t	0.1626	0.1709	−0.00407	0.00423	0.00415

Applying equation (5.17) gives the uncertainty about this operating point

$$u_Q = \pm\left[0.00415^2 + 0.000167^2\right]^{1/2} = \pm 4.15 \times 10^{-3} \text{ ft}^3/\text{s}$$

or

$$u_Q/Q = \pm 0.025$$

COMMENT   These last two examples give similar results for the uncertainty aside from insignificant differences due to round-off in the computations.

5.8   MULTIPLE-MEASUREMENT UNCERTAINTY ANALYSIS

This section develops a method for estimating the uncertainty in the value assigned to a variable based on a set of measurements obtained under fixed operating conditions. The method parallels the uncertainty standards approved by professional societies and by NIST in the United States [2, 8] and is in harmony with international guidelines [1]. The procedures assume that the errors follow a normal probability function, although the procedures are actually quite insensitive to deviations away from such behavior [5]. Sufficient repetitions must be present in the measured data to assess random error; otherwise, some estimate of the magnitude of expected variation should be provided in the analysis.

Propagation of Elemental Errors

The procedures for multiple-measurement uncertainty analysis consist of the following steps:

- Identify the elemental errors in the measurement. As an aid, consider the errors in each of the three source groups (calibration, data acquisition, and data reduction).
- Estimate the magnitude of systematic and random error in each of the elemental errors.
- Calculate the uncertainty estimate for the result.

Considerable guidance can be obtained from Tables 5.1–5.3 and from the design-stage analysis for identifying the elemental errors. In multiple-measurement analysis, it is possible to divide the estimates for elemental errors into random and systematic errors. Statistics are used to estimate the random uncertainty in each error, and other means are used to estimate the systematic errors, as discussed earlier.

Consider the measurement of variable x, which is subject to elemental random errors each estimated by their standard random uncertainty, P_k, and systematic errors each estimated by their systematic uncertainty, B_k. Let subscript k, where $k = 1, 2, \ldots, K$, refer to each of up to any K elements of error, e_k. A method to estimate the uncertainty in x based on the uncertainties in each of the elemental random and systematic errors is given below and outlined in Figure 5.6.

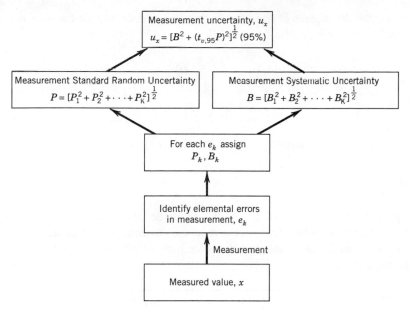

Figure 5.6. Multiple-measurement uncertainty procedure for combining uncertainties.

The *propagation of random uncertainty* due to the K random errors is given by the standard random uncertainty, P. P is estimated by the RSS method of equation (5.2):

$$P = \left(P_1^2 + P_2^2 + \cdots + P_K^2\right)^{1/2} \tag{5.20}$$

The *measurement standard random uncertainty*, P, represents a basic measure of the elemental errors affecting the variations found in the overall measurement of variable x.

The *propagation of elemental systematic errors*, B_k, is treated in a similar manner. The measurement systematic uncertainty, B, is given by

$$B = \left(B_1^2 + B_2^2 + \cdots + B_K^2\right)^{1/2} \tag{5.21}$$

The *measurement systematic uncertainty*, B, represents a basic measure of the elemental systematic errors that affect the measurement of variable x.

The *uncertainty in x*, written as u_x, is reported as a combination of the systematic uncertainty and standard random uncertainty in x at a desired confidence level, as indicated in Figure 5.6:

$$u_x = \pm(B^2 + (t_{v,95}P)^2)^{1/2} \quad (95\%) \tag{5.22}$$

Equation (5.22) represents the uncertainty in the measured value x at the 95% confidence level.[4] In this case, we evaluate the random uncertainties at 95% confidence through the use of the appropriate t value at 95%, usually a value of or near 2. Equation (5.22), as written, assumes that the systematic uncertainties are evaluated at 95% confidence.

The degrees of freedom, v, in the standard random uncertainty, P, requires some discussion since P is based on equation (5.20), which is composed of errors elements that

---

[4]This is referred to as the *expanded uncertainty in x* in [2].

may have different degrees of freedom. In this case, the degrees of freedom is estimated using the Welch–Satterthwaite formula [2]:

$$\nu = \frac{\left(\sum\limits_{k=1}^{K} P_k^2\right)^2}{\sum\limits_{k=1}^{K} \left(P_k^4/\nu_k\right)} \tag{5.23}$$

where k refers to each elemental error with its own $\nu_k = N_k - 1$.

EXAMPLE 5.8

Ten repeated measurements of force, F, are made over time under fixed operating conditions. The data are listed below. Estimate the standard random uncertainty due to the elemental error in the mean value of the force that is introduced into the measured data through the data scatter.

n	$F[N]$	n	$F[N]$
1	123.2	6	119.8
2	115.6	7	117.5
3	117.1	8	120.6
4	125.7	9	118.8
5	121.1	10	121.9

KNOWN   Measured data set

   $N = 10$

ASSUMPTIONS   Error due to data scatter only

FIND   Estimate P_1 [equation (5.20)]

SOLUTION   The mean value of the force based on this finite data set is computed from equation (4.14) as $\overline{F} = 120.1$ N. A random error is associated with the estimate of the mean value because of data scatter. This error enters the measurement during data acquisition (Table 5.2). The standard random uncertainty in this error can be computed through the standard deviation of the means, equation (4.16):

$$P_1 = S_{\overline{F}} = \frac{S_F}{\sqrt{N}}$$

$$= \frac{3.2}{\sqrt{10}} = 1.01 \text{N}$$

COMMENT   It is common to call this elemental error, which arises due to data scatter, a temporal variation error as noted in Table 5.2.

EXAMPLE 5.9

Suppose that the force measuring device of Example 5.1 was used to generate the data set of Example 5.8. Estimate the systematic uncertainty in the measurement instrument.

KNOWN

$$e_1 = B_1 = 0.20\,\text{N}$$
$$e_2 = B_2 = 0.30\,\text{N}$$

ASSUMPTIONS   Manufacturer specifications reliable at 95% probability

FIND B

SOLUTION   Elemental errors due to instruments are considered as data-acquisition source errors (Table 5.2). And since we have no information as to the statistics used to generate the numbers e_1 and e_2, we consider them as systematic errors. Accordingly, the transducer systematic uncertainty in the measurement device, B, is estimated by

$$B = \left(B_1^2 + B_2^2\right)^{1/2}$$
$$= 0.36\,\text{N}$$

COMMENT   Lacking specific calibration data points or information such as degrees of freedom in the error estimates, we consider the manufacturer specifications as a listing of systematic errors. Note that the estimate for systematic error due to the instrument is equal in value to an estimate of u_c that would have been arrived at by a design-stage analysis.

EXAMPLE 5.10

In Examples 4.8 and 4.9, a set of measured data was used to generate a polynomial curve fit equation between variables x and y. If during a test, variable x were to be measured and y computed from x through the polynomial, estimate the uncertainty due to the data-reduction error involved.

KNOWN   Data set of Example 4.8
Polynomial fit of Example 4.9

ASSUMPTIONS   Data fits curve $y = 1.04x + 0.02$.

FIND P_1

SOLUTION   The curve fit was found to be given by $y = 1.04x + 0.02$ with a standard error of the fit, $S_{yx} = 0.16$. Accordingly, the standard random uncertainty due to the curve fit, noted as P_1 (see Table 5.3), is

$$P_1 = S_{yx}/\sqrt{N} = 0.072$$

COMMENT   Likewise, the standard random uncertainty in the power law equation fit of Example 4.10 can be computed. For example, at $E = 4.3$ V, $P_1 = S_{yx}/\sqrt{N} = 0.013$ V, which is equivalent to 0.58 ft/s.

EXAMPLE 5.11

The measurement of x is found to contain three elemental random errors from data acquisition. Each elemental error is evaluated with the following conclusions:

$$P_1 = 0.60\,\text{units}, \; v_1 = 29$$
$$P_2 = 0.80\,\text{units}, \; v_2 = 9$$
$$P_3 = 1.10\,\text{units}, \; v_3 = 19$$

Estimate the standard random uncertainty due to data-acquisition errors.

KNOWN P_k with $k = 1, 2, 3$

FIND P [using equation (5.20)]

SOLUTION   Using equation (5.20), the standard random uncertainty due to data-acquisition errors, P, is found by

$$P = \left(P_1^2 + P_2^2 + P_3^2\right)^{1/2}$$
$$= \left(0.60^2 + 0.80^2 + 1.10^2\right)^{1/2} = 1.49\,\text{units}$$

From equation (5.23), the degrees of freedom in P is determined by

$$v = \frac{\left(\sum\limits_{k=1}^{3} P_k^2\right)^2}{\sum\limits_{k=1}^{3} \left(P_k^4/v_k\right)} = \frac{\left(0.6^2 + 0.8^2 + 1.1^2\right)^2}{\left(0.6^4/29\right) + \left(0.8^4/9\right) + \left(1.1^4/19\right)}$$

or $v = 38.8 \approx 39$.

EXAMPLE 5.12

After an experiment to measure stress in a loaded beam, an uncertainty analysis reveals the following values of uncertainty in stress measurement whose magnitudes were computed from elemental errors using equations (5.20) and (5.21).

$$
\begin{array}{lll}
B_1 = 1.0\,\text{N/cm}^2 & B_2 = 2.1\,\text{N/cm}^2 & B_3 = 0\,\text{N/cm}^2 \\
P_1 = 4.6\,\text{N/cm}^2 & P_2 = 10.3\,\text{N/cm}^2 & P_3 = 1.2\,\text{N/cm}^2 \\
v_1 = 14 & v_2 = 37 & v_3 = 8
\end{array}
$$

If the mean value of the stress in the measurement is $\overline{\sigma} = 223.4\,\text{N/cm}^2$, determine the best estimate of the stress at a 95% confidence level, assuming all errors are accounted for.

KNOWN   Experimental errors

ASSUMPTIONS   All elemental errors ($K = 3$) have been included.

FIND P, B, and u_σ (using equations 5.20–5.23)

SOLUTION   We seek values for the statement, $\sigma' = \overline{\sigma} \pm u_\sigma(95\%)$, given that $\overline{\sigma} = 223.4\,\text{N/cm}^2$. The uncertainty estimate in the measurement is obtained through equations (5.20) through (5.23). The measurement standard random uncertainty is given by equation (5.20) as

$$P = \left(P_1^2 + P_2^2 + P_2^2\right)^{1/2} = 11.3\,\text{N/cm}^2$$

The measurement systematic uncertainty is given by equation (5.21) as

$$B = \left(B_1^2 + B_2^2 + B_2^2\right)^{1/2} = 2.3\,\text{N/cm}^2$$

The degrees of freedom in P is found from equation (5.23) to be

$$v = \frac{\left(\sum\limits_{k=1}^{3} P_k^2\right)^2}{\sum\limits_{k=1}^{3} P_k^4/v_k} \approx 49$$

Therefore, the t estimator is $t_{49,95} \sim 2.0$. The uncertainty estimate is found using equation (5.22) to be

$$u_\sigma = \pm \left[B^2 + \left(t_{v,95} P \right)^2 \right]^{1/2} = \pm \left[\left(2.3\,\text{N/cm}^2 \right)^2 + \left(2 \times 11.3\,\text{N/cm}^2 \right)^2 \right]^{1/2}$$
$$= \pm 22.7\,\text{N/cm}^2$$

The best estimate is given in the form of equation (4.1) as

$$\sigma' = 223.4 \pm 22.7\,\text{N/cm}^2 \quad (95\%)$$

Propagation of Uncertainty to a Result

Consider the result, R, which is determined through the functional relationship between the measured independent variables x_i, $i = 1, 2, \ldots, L$ as defined by equation (5.8). Again, L is the number of independent variables involved and each x_i has an associated systematic uncertainty, B, given by the measurement systematic uncertainty determined for that variable by equation (5.21), and a measurement standard random uncertainty, P, determined using equation (5.20). The best estimate of the true value, R', is given as

$$R' = \overline{R} \pm u_R \quad (P\%) \tag{5.24}$$

where the mean value of the result is determined by

$$\overline{R} = f_1(\overline{x}_1, \overline{x}_2, \ldots, \overline{x}_L) \tag{5.25}$$

and where the uncertainty in the result u_R is given by

$$u_R = f_2\{B_{x_1}, B_{x_2}, \ldots, B_{x_L}; P_{x_1}, P_{x_2}, \ldots, P_{x_L}\} \tag{5.26}$$

where subscripts x_1 through x_L refer to the measurement systematic uncertainties and measurement random uncertainties in each of these L variables.

The propagation of random uncertainty through the variables to the result will yield a result standard random uncertainty given by

$$P_R = \left(\sum_{i=1}^{L} [\theta_i P_{x_i}]^2 \right)^{1/2} \tag{5.27}$$

where θ_i is the sensitivity index defined by equation (5.12). The propagation of systematic uncertainty of the variables to the result will yield a result systematic uncertainty given by

$$B_R = \left(\sum_{i=1}^{L} [\theta_i B_{x_i}]^2 \right)^{1/2} \tag{5.28}$$

The terms $\theta_i P_{xi}$ and $\theta_i B_{xi}$ represent the individual contributions of the ith variable to the uncertainty in R. A comparison between the magnitudes of each individual contribution identifies how the uncertainty terms affect the result.

The measurement standard random uncertainty and the systematic uncertainty are then combined to yield an estimate of the *uncertainty in the result*,[5] written as u_R, by

$$u_R = \pm \left[B_R^2 + \left(t_{v,95} P \right)^2 \right]^{1/2} \quad (95\%) \tag{5.29}$$

---

[5]This is referred to as the *expanded uncertainty in the result* in [2].

Once again the t estimator is used to provide a reasonable weight to the interval defined by the random uncertainty to achieve the desired confidence, which here is 95%. If the degrees of freedom in each variable is large ($N \geq 30$), then a reasonable approximation is to take $t_{\nu,95} = 2$.

When the degrees of freedom in each of the variables, x_i, is not the same and are small, the Welch–Satterthwaite formula is used to estimate the degrees of freedom in the result expressed as

$$\nu_R = \frac{\left\{ \sum\limits_{i=1}^{L} (\theta_i P_{x_i})^2 \right\}^2}{\sum\limits_{i=1}^{L} \left\{ (\theta_i P_{x_i})^4 \Big/ \nu_{x_i} \right\}} \tag{5.30}$$

EXAMPLE 5.13

The density of a gas, ρ, which is believed to follow the ideal gas equation of state, $\rho = p/RT$, is to be estimated through separate measurements of pressure, p, and temperature, T. The gas is housed within a rigid impermeable vessel. The literature accompanying the pressure measurement system states an instrument uncertainty to within 1% of the reading, and that accompanying the temperature measuring system suggests $0.6°R$. Twenty measurements of pressure, $N_p = 20$, and ten measurements of temperature, $N_T = 10$, are made with the following statistical outcome:

$$\bar{p} = 2253.91 \text{ psfa} \quad S_p = 167.21 \text{ psfa}$$
$$\bar{T} = 560.4°R \quad\quad S_T = 3.0°R$$

where psfa refers to lb/ft^2 absolute. Determine a best estimate of the density. The gas constant is $R = 54.7$ ft-lb/lb$_m$-°R.

KNOWN

$$\bar{p}, S_p, \bar{T}, S_T$$
$$\rho = p/RT; \quad R = 54.7 \text{ ft-lb/lb}_m\text{-°R}$$

ASSUMPTIONS   Gas behaves as an ideal gas

FIND

$$\rho' = \bar{\rho} \pm u_\rho \quad (95\%)$$

SOLUTION   Our measurement objective is to determine the density of an ideal gas through temperature and pressure measurements. The independent and dependent variables are related through the ideal gas law, $\rho = p/RT$. Equation (5.25) provides the mean value of density as

$$\bar{\rho} = \frac{\bar{p}}{R\bar{T}} = 0.074 \text{ lb}_m/\text{ft}^3$$

The next step must be to identify and estimate the errors and determine how they contribute to the uncertainty in the mean value of density. Since no calibrations are performed and the gas is considered to behave as an ideal gas in an exact manner, the measured values of pressure and temperature are subject only to elemental errors within the data-acquisition error source group (Table 5.2): instrument errors and temporal variation errors.

The tabulated value of the gas constant is not without error. However, estimating the possible error in a tabulated value is sometimes difficult. According to [7] the uncertainty (systematic) in the evaluation of the gas constant is on the order of $\pm(0.33$ J/kg-K)/(gas molecular weight) or ± 0.06 (ft-lb/lb$_\text{m}$-°R)/(gas molecular weight). Since this yields a small value for a reasonable gas molecular weight, here we will assume a zero (negligible) systematic error in the gas constant.

Consider the pressure measurement. The temporal variation (data scatter) error will be based on the variation in the measured data obtained during presumably fixed operating conditions. The instrument error will be estimated as a systematic uncertainty and based on the manufacturer's statement:

$$(B_1)_p = 22.5 \text{ psfa} \quad (P_1)_p = 0$$

where the subscript simply keeps track of the error. The temporal variation causes a random uncertainty in the mean value of pressure and is estimated directly from statistics:

$$(P_2)_p = S_{\bar{p}} = \frac{S_p}{\sqrt{N}} = \frac{167.21 \text{ psfa}}{\sqrt{20}} = 37.4 \text{ psfa} \quad v_p = 19$$

and assigning no systematic uncertainty gives

$$(B_2)_p = 0$$

In a similar manner the data-acquisition source error in temperature is computed. Elemental errors assessed from manufacturer instrument error are considered as systematic only:

$$(B_1)_T = 0.6°\text{R} \quad (P_1)_T = 0$$

The elemental error due to temporal variation is considered as random uncertainty only and estimated by

$$(P_2)_T = S_{\bar{T}} = \frac{S_T}{\sqrt{N}} = \frac{3.0°\text{R}}{\sqrt{10}} = 0.9°\text{R} \quad v_p = 9$$

$$(B_2)_T = 0$$

The uncertainties in the data-acquisition source errors in pressure and temperature are estimated from equations (5.20) and (5.21):

$$(B)_p = \left[(22.5)^2 + (0)^2 \right]^{1/2} = 22.5 \text{ psfa}$$

$$(P)_p = \left[(0)^2 + (37.4)^2 \right]^{1/2} = 37.4 \text{ psfa}$$

similarly,

$$(B)_T = 0.6°\text{R}$$
$$(P)_T = 0.9°\text{R}$$

with degrees of freedom determined from equation (5.23) to be

$$(v)_p = N_p - 1 = 19$$
$$(v)_T = 9$$

The propagation of systematic and random uncertainties through to the result, the density, will be estimated using the RSS of equations (5.27) and (5.28):

$$P_\rho = \left[\left(\frac{\partial \rho}{\partial T} P_T \right)^2 + \left(\frac{\partial \rho}{\partial p} P_p \right)^2 \right]^{1/2}$$

$$= \left[(1.2 \times 10^{-4})^2 + (1.2 \times 10^{-3})^2 \right]^{1/2}$$

$$= 0.0012 \text{ lb}_\text{m}/\text{ft}^3$$

(5.31)

and

$$
\begin{aligned}
B_\rho &= \left[\left(\frac{\partial \rho}{\partial T} B_T \right)^2 + \left(\frac{\partial \rho}{\partial p} B_p \right)^2 \right]^{1/2} \\
&= \left[\left(8 \times 10^{-5} \right)^2 + \left(7 \times 10^{-4} \right)^2 \right]^{1/2} \\
&= 0.0007 \, \text{lb}_\text{m}/\text{ft}^3
\end{aligned}
\tag{5.32}
$$

The degrees of freedom in the density is determined from equation (5.30):

$$
\nu = \frac{\left[\left(\frac{\partial \rho}{\partial T} P_T \right)^2 + \left(\frac{\partial \rho}{\partial p} P_p \right)^2 \right]^2}{\left(\frac{\partial \rho}{\partial T} P_T \right)^4 \Big/ \nu_T + \left(\frac{\partial \rho}{\partial p} P_p \right)^4 \Big/ \nu_p} = 23
$$

From Table 4.4, $t_{23,95} = 2.06$.

The uncertainty in the mean value of density is estimated from equation (5.29):

$$
\begin{aligned}
u_\rho &= \left[B_\rho^2 + \left(t_{23,95} P_\rho \right)^2 \right]^{1/2} \\
&= 0.0025 \, \text{lb}_\text{m}/\text{ft}^3 \quad (95\%)
\end{aligned}
$$

The best estimate of the density is given in the form of equation (5.24):

$$
\rho' = 0.074 \pm 0.0025 \, \text{lb}_\text{m}/\text{ft}^3 \quad (95\%)
$$

This measurement of density has an uncertainty of about 3.4%.

COMMENT   If we examine the magnitude of each contribution in equations (5.31) and (5.32), we see that the pressure contributes more to the random uncertainty and to the systematic uncertainty of density than does temperature. However, the systematic uncertainty is small compared to the random uncertainty. The analysis reveals that the uncertainty in density could be reduced by actions that reduce the random uncertainty in the pressure measurement.

EXAMPLE 5.14

Consider determining the mean diameter of a shaft using a hand-held micrometer. The shaft was manufactured on a lathe presumably to a constant diameter. Identify possible elements of error that can contribute to the uncertainty in the estimated mean diameter.

SOLUTION   In the machining of the shaft, possible run-out during shaft rotation can bring about an eccentricity in the shaft cross-sectional diameter. Further, as the shaft is machined to size along its length possible run-out along the shaft axis can bring about a difference in the machined diameter. To account for such deviations, the usual measurement procedure is to repeatedly measure the diameter at one location of the shaft, rotate the shaft, and repeatedly measure again. Then the micrometer is moved to a different location along the axis of the shaft and the above procedure repeated.

It is unusual to calibrate a micrometer in a working machine shop, although an occasional offset error check against accurate gauge blocks is normal procedure. Let us assume that the micrometer is used as is without calibration. Data-acquisition errors are introduced from at least several elements:

1. Since the micrometer is not calibrated, the reading during any measurement can be affected by a possible systematic error in the micrometer markings. This can be labeled as an uncertainty due to instrument error, B_1 (see Table 5.2). Experience

shows that this value will be on the order of the resolution of the micrometer at 95% confidence.

2. The random uncertainty on repeated readings will be affected by the resolution of the readout, eccentricity of the shaft, and the exact placement of the micrometer on any cross section along the shaft. It is not possible to separate these errors, so they are grouped as variation errors, with standard random uncertainty, P_2. This value can be discerned from the statistics of the measurements made at any cross section (replication).

3. Spatial variations in the diameter along its length will introduce scatter between the statistical values for each cross section. Since this affects the spatial precision of the overall mean diameter, its effect must be accounted for. Label this error as a spatial error with standard random uncertainty, P_3. It can be estimated by the pooled statistical standard deviation of the mean values at each measurement location (replication).

Since a diameter is a direct outcome from any reading, there will be no data-reduction errors introduced.

COMMENT   If we were interested in the uncertainty associated with the production of these shafts, note that machinist, machine, micrometer, and possibly day of the week would be extraneous variables in need of a randomized test plan.

EXAMPLE 5.15

The mean temperature in an oven is to be estimated by using the information obtained from a temperature probe. The manufacturer offers a statement suggesting an uncertainty within 0.6°C with this probe. Determine the oven temperature.

The measurement process is defined such that the probe is to be positioned at several strategic locations within the oven. The measurement strategy is to make four measurements within each of two equally spaced cross-sectional planes of the oven. The positions are chosen so that each measurement location corresponds to the centroid of an area of value equal for all locations so that spatial variations in temperature throughout the oven are accounted. The measurement locations within the cubical oven are shown in Figure 5.7. In this example, 10 measurements are made at each position, with the results given in Table 5.4. Assume that a temperature readout device with a resolution of 0.1°C is used.

KNOWN   Data of Table 5.4

FIND
$$T' = \overline{T} \pm u_T \quad (95\%)$$

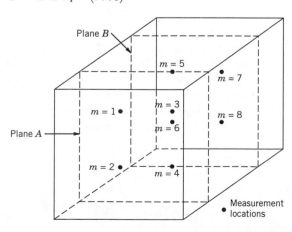

Figure 5.7. Measurement locations for Example 5.15.

Table 5.4   Example 5.15: Oven Temperature Data, $N = 10$

Location	$\overline{T}_m$	S_{T_m}	Location	$\overline{T}_m$	S_{T_m}
1	342.1	1.1	5	345.2	0.9
2	344.2	0.8	6	344.8	1.2
3	343.5	1.3	7	345.6	1.2
4	343.7	1.0	8	345.9	1.1

Note: $\overline{T}_m = \frac{1}{N}\sum\limits_{n=1}^{N} T_{mn}$; $S_{T_m} = \left[\frac{1}{N-1}\sum\limits_{n=1}^{N}\left(T_{mn} - \overline{T}_m\right)^2\right]^{1/2}$

SOLUTION   The mean values at each of the locations will be averaged to yield a mean oven temperature using pooled averaging:

$$\langle\overline{T}\rangle = \frac{1}{8}\sum_{m=1}^{8}\overline{T}_m$$

Elemental errors in this test are found in the data-acquisition group (Table 5.2) and due to (1) the temperature probe system (instrument error), (2) spatial variation errors, and (3) temporal variation errors. Consider first the elemental error in the probe system. A statement of 0.6°C was provided by the manufacturer. Because no statistical information is provided, the total uncertainty due to the probe system will be considered to be a systematic uncertainty. Thus,

$$B_1 = 0.6°C \qquad P_1 = 0$$

Consider next the spatial error contribution to the estimate in mean temperature, $\overline{T}$. This error arises from the use of discrete measurement locations within the oven coupled with the nonuniformity in the oven temperatures. An estimate of spatial temperature distribution within the oven can be made by examining the mean temperatures at each measured location. These temperatures show that the oven is not uniform in temperature. The mean temperatures within the oven display a standard deviation of

$$S_T = \sqrt{\frac{\sum\limits_{m=1}^{8}\left(\overline{T}_m - \langle\overline{T}\rangle\right)^2}{7}} = 1.26°C$$

Thus, the random uncertainty of the oven mean temperature is $S_{\overline{T}}$ or

$$P_2 = \frac{S_T}{\sqrt{7}} = 0.45°C$$

with degrees of freedom, $n = 7$. We assume no systematic uncertainty in this elemental error:

$$B_2 = 0$$

During each of the 10 measurements of temperature at each location, time variations in the probe output cause data scatter, as evidenced by the respective S_{T_m} values. Such time variations are caused by random local temperature variations as measured by the probe, probe system resolution, and oven temperature control variations during fixed operating conditions. It is rarely possible to separate these, so they are estimated together as a single error. A random uncertainty can be found. Using the pooled standard deviation, equation (4.20),

$$\langle S_T\rangle = \sqrt{\frac{\sum\limits_{m=1}^{8}\sum\limits_{n=1}^{10}\left(\overline{T}_{mn} - \langle\overline{T}\rangle\right)^2}{MN}} = 1.08°C$$

and so the random uncertainty is

$$P_3 = \frac{\langle S_T \rangle}{\sqrt{80}} = 0.12°C$$

with degrees of freedom, $\nu = 72$. We assign $B_3 = 0$.

The measurement systematic uncertainty is estimated from equation (5.22) to be

$$B = \left(B_1^2 + B_2^2 + B_3^2\right)^{1/2} = 0.60°C$$

and the measurement standard random uncertainty is estimated from equation (5.20) to be

$$P = \left(P_1^2 + P_2^2 + P_3^2\right)^{1/2} = 0.47°C$$

with degrees of freedom estimated from equation (5.23) to be $\nu = 8$.

The uncertainty in the mean oven temperature is estimated from equation (5.22).

$$u_T = \left[B^2 + \left(t_{8,95}P\right)^2\right]^{1/2}$$

with $t_{8,95} = 2.306$. The best estimate of the mean oven temperature is given by equation (4.1) as

$$T' = \overline{T} \pm u_T = 344.4 \pm 1.24°C \quad (95\%)$$

5.9   CORRECTION FOR CORRELATED SYSTEMATIC ERRORS

So far, we have assumed that all of the different elements of systematic error in a test are uncorrelated. This means that each error is independent of the others and the errors are not related. But it is quite common to have some errors that are not independent between instruments used in a test. Instruments that share a common element that contributes a systematic error to the measurement will have a dependent error. We say that such errors are "correlated."

For example, when an instrument is calibrated against a standard, we assign the systematic error from the standard to the instrument. But all instruments calibrated against the same standard would have the same amount of systematic error from that standard. The offset in each instrument reading resulting from that error would have the same magnitude and sign. Hence, these errors would be correlated. But the method we use to propagate systematic uncertainties, the RSS method, assumes that the errors are independent (i.e., uncorrelated). The effect of correlated errors on the uncertainty depends on the functional relationship between the variables and the magnitudes of the elemental systematic errors that are correlated. We now introduce a correction for treating the correlated systematic errors. Remember that *random errors by their nature are not correlated*.

Consider the result, R, which is determined through the functional relationship between the measured independent variables x_i, $i = 1, 2, \ldots, L$ where L is the number of independent variables involved. Each x_i is subject to elemental systematic errors, B_k, where $k = 1, 2, \ldots, K$, refer to each of up to any of K elements of error. Now allow that H number of these K elemental errors are correlated between variables while the rest $(K - H)$ are uncorrelated. The systematic uncertainty in a result, equation (5.28), is to be estimated by

$$B_R = \left[\sum_{i=1}^{L} \left(\theta_i B_{x_i}\right)^2 + 2\sum_{i=1}^{L-1}\sum_{j=i+1}^{L} \theta_i \theta_j B_{x_i x_j}\right]^{1/2} \tag{5.33}$$

where index j is a counter equal to $i + 1$ and

$$\theta_i = \frac{\partial R}{\partial x_{i_{x=\bar{x}}}} \tag{5.12}$$

$$B_{x_i x_j} = \sum_{h=1}^{H} (B_{x_i})_h (B_{x_j})_h \tag{5.34}$$

where H is the number of elemental errors that are correlated between variables x_i and x_j and h is a counter for each correlated error. Note that equation (5.33) reduces to equation (5.28) when no errors are correlated (i.e., when $H = 0$, $B_{x_i x_j} = 0$). The term $B_{x_i x_j}$ is called the covariance. References 1 and 9 discuss correlated systematic errors in more detail.

EXAMPLE 5.16

Suppose a result is a function of three variables, X_1, X_2, X_3. There are four systematic errors associated with both x_1 and x_2 and five associated with x_3. The first and second systematic elemental errors associated with the second and third variable (x_2 and x_3) are determined to be correlated because these errors arise from common sources. Express the systematic uncertainty in the result and express the covariance term.

SOLUTION   Equation (5.33) is expanded to the form

$$B_R = [(\theta_1 B_{x_1})^2 + (\theta_2 B_{x_2})^2 + (\theta_3 B_{x_3})^2 + 2\theta_1 \theta_2 B_{x_1 x_2} + 2\theta_1 \theta_3 B_{x_1 x_3} + 2\theta_2 \theta_3 B_{x_2 x_3}]^{1/2}$$

For the two correlated errors ($H = 2$) associated with variables 2 and 3, the covariance term is

$$B_{x_2 x_3} = \sum_{h=1}^{2} (B_{x_2})_h (B_{x_3})_h = (B_{x_2})_1 (B_{x_3})_1 + (B_{x_2})_2 (B_{x_3})_2$$

The covariance related to the other errors and variables is zero because they are uncorrelated. Hence, the systematic uncertainty in the result is written as

$$B_R = [(\theta_1 B_{x_1})^2 + (\theta_2 B_{x_2})^2 + (\theta_3 B_{x_3})^2 + 2\theta_2 \theta_3 B_{x_2 x_3}]^{1/2}$$

EXAMPLE 5.17

Suppose a result R is a function of two variables, X and Y, such that $R = X + Y$, and each variable having one elemental error. If $\bar{X} = 10.1$ V with $B_x = 1.1$ V and $\bar{Y} = 12.2$ V with $B_y = 0.8$ V, estimate the systematic uncertainty in the result if the systematic errors are (1) uncorrelated and (2) correlated.

SOLUTION   From the stated information, $\bar{R} = 10.1 + 12.2 = 22.3$ V. For the uncorrelated case,

$$B_{R_{\text{unc}}} = [(\theta_X B_X)^2 + (\theta_Y B_Y)^2]^{1/2} = [(1 \times 1.1)^2 + ((1 \times 0.8)^2)^2]^{1/2} = 1.36 \text{ V}$$

For the correlated case,

$$B_{R_{\text{cor}}} = [(\theta_X B_X)^2 + (\theta_Y B_Y)^2 + 2\theta_X \theta_Y B_{XY}]^{1/2}$$
$$= [(1 \times 1.1)^2 + (1 \times 0.8)^2 + 2(1)(1)(1.1)(0.8)]^{1/2} = 3.12 \text{ V}$$

where

$$B_{XY} = \sum_{h=1}^{1} (B_X)_1 (B_Y)_1 = B_X B_Y$$

COMMENT   In this case, the correlated systematic errors have a large impact on the systematic uncertainty through the covariance term. This is not always the case as it depends on the functional relationship itself. We leave it to the reader to show that if $R = X/Y$, then the covariance term in this problem would have little impact on the systematic uncertainty in the result.

5.10   HARMONIZATION

Equations (5.22) and (5.29) can readily be standardized to other confidence levels. The ISO GUM lists errors as Type A when uncertainties are estimated through standard deviations, S_{x_i} and the number of degrees of freedom v_i. Type B errors are estimated by other means and are quantified by an amount b_k, the standard systematic uncertainty. The quantities b_k^2 are to be treated like variances, the existence of which is assumed. The combined uncertainty should be characterized by the numerical value obtained by applying the usual method for the combination of variances, such as followed in equations (5.22) and (5.29). The confidence level and t value should be stated.

For example, we can state the uncertainty at the 68% confidence level preferred by ISO GUM (1) guidelines as follows. Define the elemental *standard systematic uncertainty* as

$$b_k = \frac{B_k}{2} \qquad (5.35)$$

where the denominator value of 2 represents the expected dispersion of the estimated systematic errors, B_k, at 95% confidence. Then, by treating the propagation of the standard systematic uncertainty as done in equations (5.21) and (5.28), as appropriate, we combine with the standard random uncertainty

$$u_{68} = \left[b^2 + P^2\right]^{1/2} \quad (68\%) \qquad (5.36)$$

Equation (5.36) is known as the *combined standard uncertainty* [1, 2] and is evaluated at the 68% confidence level (i.e., equivalent to an interval with width of one standard deviation and such that $t = 1$). The approach is the same whether calculating the uncertainty in variable x through equations (5.20) to (5.22) or in result R through equations (5.27) to (5.29).

The combined standard uncertainty value can be expressed at any confidence level by weighting the b and P values by the appropriate t value, such as we have done in equations (5.22) and (5.29) for 95% confidence. For example, if we accept that the standard systematic uncertainty has a very large degrees of freedom (as is assumed in equations 5.21 and 5.33) and that the standard random uncertainty has a known degrees of freedom, then

$$u = \left[(2b)^2 + \left(t_{v,95}P\right)^2\right]^{1/2} = \left[B^2 + \left(t_{v,95}P\right)^2\right]^{1/2} \quad (95\%) \qquad (5.37)$$

which returns us to equation (5.29). For very large data sets, the t value can be set to 2 and the expanded uncertainty is estimated as [2]

$$u = \left[B^2 + (2P)^2\right]^{1/2} \quad (95\%) \qquad (5.38)$$

Reference [2] contains advanced procedures for harmonization for other confidence levels and for estimating a common t value based on varying degrees of freedom in both b and P.

5.11   SUMMARY

Uncertainty analysis provides the "± what" to a test result or the anticipated result from a proposed test plan. In this chapter, we have discussed the manner in which various errors can enter into a measurement with emphasis on their influences on the test result. Both random errors and systematic errors are considered. Random errors differ on repeated measurements, lead to data scatter, and can be quantified by statistical methods. Systematic errors remain constant in repeated measurements under fixed operating conditions. Errors are quantified by uncertainties. Procedures of uncertainty analysis were introduced as a means of estimating uncertainty propagation within a measurement and the propagation of uncertainty among independent measured variables that are used to determine a result through some functional relationship. We have discussed the role and use of uncertainty analysis in both the design of measurement systems, through selection of equipment and procedures, as well as in the interpretation of measured data. We have presented procedures for estimating uncertainty at each of these stages. The procedures will provide reasonable estimates of the uncertainty to be expected in measurements.

We advise the reader that test uncertainty by its very nature evades exact values. But the methodology presented will give reasonable estimates provided that the engineer has used sound, impartial judgment in its implementation. The important thing is that the uncertainty analysis be performed during the design and development stages of a test, as well as at its completion to assess the stated results. Only then can the engineer use the results with appropriate confidence and professional and ethical responsibility.

REFERENCES

1. International Organization for Standardization (ISO), *Guide to the Expression of Uncertainty in Measurement*, Geneva, Switzerland, 1993.
2. ANSI/ASME Power Test Codes-PTC 19.1, *Test Uncertainty*, American Society of Mechanical Engineers, New York, 2005.
3. Smith, R. E., and S. Wenhofer, From measurement uncertainty to measurement communications, credibility and cost control in propulsion ground test facilities, *Journal of Fluids Engineering* 107: 165–172, 1985.
4. Dieck, R., *Measurement Uncertainty: Methods and Applications*, 3d ed., Instrumentation, Systems and Automation Society (ISA), Research Triangle Park, NC, 2002.
5. Kline, S. J., and F. A. McClintock, Describing uncertainties in single sample experiments, *Mechanical Engineering* 75, 1953.
6. Moffat, R. J., Uncertainty analysis in the planning of an experiment, *Journal of Fluids Engineering* 107: 173–181, 1985.
7. Kestin, J., *A Course in Thermodynamics*, revised printing, Hemisphere, New York, 1979.
8. Taylor, B.N., and C. Kuyatt, *Guidelines for Evaluating and Expressing the Uncertainty of NIST Measurement Results*, NIST Technical Note 1297, Gaithersburg, MD, 1994.
9. Coleman, H., and W.G. Steele, *Experimentation and Uncertainty Analysis for Engineers*, 2d ed., Wiley, New York, 1999.

NOMENCLATURE

e_k	kth elemental error	u_0	zero-order uncertainty
$f(\ )$	general functional relation	u_i	ith-order uncertainty
p	pressure $[m\text{-}l^{-1}\text{-}t^{-2}]$	u_N	Nth-order uncertainty
$t_{v,95}$	t variable (at 95% probability)	x	measured variable
t	time $[t]$	x'	true value of x
u_x	uncertainty of a measured variable	$\bar{x}$	sample mean value of x
u_d	design-stage uncertainty	b_x	standard systematic uncertainty in x
u_c	instrument calibration uncertainty	B	systematic uncertainty (at 95% probability)

B_{x_i}	elemental systematic uncertainty in x	$(P\%)$	percent probability
F	force $[m\text{-}l\text{-}t^{-2}]$	Q	flow rate $[l^3 - t^{-1}]$
N	number of measured values	T	temperature $[°]$
R	resultant value	$\forall$	volume $[l^{-3}]$
S_x	sample standard deviation of x	θ_i	sensitivity index
$S_{\bar{x}}$	sample standard deviation of the means;	ρ	gas density $[m - l^{-3}]$
	standard random uncertainty	ν	degrees of freedom
P	standard random uncertainty	$\langle\rangle$	pooled statistic
P_{x_i}	elemental random error in x		

PROBLEMS

5.1 In Chapter 1, the development of a test plan is discussed. Clearly discuss how a test plan should account for the presence of systematic and random errors. You will want to include calibration, randomization, and repetition in your discussion.

5.2 Discuss how a systematic uncertainty can be estimated for a measured value. How is random uncertainty estimated? What is the difference in the terms: error and uncertainty?

5.3 Explain what is meant by the terms true value, best estimate, mean value, uncertainty, and confidence interval.

5.4 Consider a common tire pressure gauge. How would you estimate the uncertainty in a measured pressure at the design stage and then at the Nth order? Should the estimates differ? Explain.

5.5 A micrometer has graduations scribed at 0.001-in. (0.025-mm) intervals. Estimate the uncertainty due to resolution at a 95% probability.

5.6 A tachometer has an analog display dial graduated in 5-revolutions-per-minute (rpm) increments. The user manual states an accuracy of 1% of reading. Estimate the design-stage uncertainty in the reading at 10, 500, and 5000 rpm.

5.7 An automobile speedometer is graduated in 5-mph (8-kph) increments and has an accuracy rated to be within $\pm4\%$. Estimate the uncertainty in indicated speed at 60 mph (90 kph).

5.8 A temperature measurement system is composed of a sensor and a readout device. The readout device has a claimed accuracy of 0.8°C with a resolution of 0.1°C. The sensor has an off-the-shelf accuracy of 0.5°C. Estimate a design-stage uncertainty in the temperature indicated by this combination.

5.9 Two resistors are to be combined to form an equivalent resistance of 1000 Ω. Readily available are two common resistors rated at 500 $\pm$ 50 Ω and two common resistors rated at 2000 Ω $\pm$ 5%. What combination of resistors (series or parallel) would provide the smaller uncertainty in an equivalent 1000 Ω resistance?

5.10 An equipment catalog boasts that a pressure transducer system comes in 3½-digit (e.g., 19.99) or 4½-digit (e.g., 19.999) displays. The 4½-digit model costs substantially more. Both units are otherwise identical. The specifications are:

> Linearity error: 0.15% FSO
>
> Hysteresis error: 0.20% FSO
>
> Sensitivity error: 0.25% FSO

For a full-scale output (FSO) of 20 kPa, select a readout based on appropriate uncertainty calculations. Explain.

5.11 The shear modulus, G, of an alloy can be determined by measuring the angular twist, θ, resulting from a torque applied to a cylindrical rod made from the alloy. For a rod of radius R and a torque applied at a length L from a fixed end, the modulus is found by $G = 2LT/\pi R^4 \theta$. Examine the effect of the relative uncertainty of each measured variable on the shear modulus. If during test planning all of the uncertainties are set at 1%, what is the uncertainty in G?

5.12 An ideal heat engine operates in a cycle and produces work as a result of heat transfer from a thermal reservoir at an elevated temperature, T_h, and by rejecting energy to a thermal sink at T_c. The efficiency for such an ideal cycle, termed a "Carnot cycle," is

$$\eta = 1 - (T_c/T_h).$$

Determine the required uncertainty in the measurement of temperature to yield an uncertainty in efficiency of 1%. Use $T_h = 1000$ K and $T_c = 300$ K.

5.13 Heat transfer from a rod of diameter D immersed in a fluid can be described by the Nusselt number, $Nu = hD/k$, where h is the heat-transfer coefficient and k is the thermal conductivity of the fluid. If h can be measured to within $\pm 7\%$ (95%), estimate the uncertainty in Nu for the nominal value of $h = 150$ W/m2-K. Let $D = 20 \pm 0.5$ mm and $k = 0.6 \pm 2\%$ W/m-K.

5.14 Estimate the design-stage uncertainty in determining the voltage drop across an electric heating element. The device has a nominal resistance of 30 Ω and power rating of 500 W. Available is an ohmmeter (accuracy: within 0.5%; resolution: 1 Ω) and ammeter (accuracy: within 0.1%; resolution: 100 mA). Recall $E = IR$.

5.15 Explain the critical difference(s) between a design-stage uncertainty analysis and an advanced-stage uncertainty analysis.

5.16 From an uncertainty analysis perspective, what important information does replication provide that is not found by repetition alone? How is this information included in an uncertainty analysis?

5.17 A displacement transducer has the following specifications:

Linearity:	$\pm 0.25\%$ reading
Drift:	$\pm 0.05\%/°C$ reading
Sensitivity:	$\pm 0.25\%$ reading
Excitation:	10–25 V dc
Output:	0–5 V dc
Range:	0–5 cm

The transducer output is to be indicated on a voltmeter having a stated accuracy of $\pm 0.1\%$ reading with a resolution of 10 μV. The system is to be used at room temperature, which can vary by $\pm 10°C$. Estimate an uncertainty in a nominal displacement of 2 cm at the design stage.

5.18 The displacement transducer of Problem 5.17 is used in measuring the displacement of a body impacted by a mass. Twenty measurements are made which yield

$$\bar{x} = 17.2 \, \text{mm} \quad S_x = 1.7 \, \text{mm}$$

Determine a best estimate for the mass displacement at 95% probability based on all available information.

5.19 A pressure transducer outputs a voltage to a readout device that converts the signal back to pressure. The device specifications are:

Resolution:	0.1 psi
Sensitivity:	0.1 psi
Linearity:	within 0.1% of reading
Drift:	less than 0.1 psi/6 months (32–90°F)

The transducer has a claimed accuracy of within 0.5% of reading. For a nominal pressure of 100 psi at 70°F, estimate the design-stage uncertainty in a measured pressure.

5.20 For a thin-walled pressure vessel of diameter, D, and wall thickness, t, subjected to an internal pressure, p, the tangential stress is given by $\sigma = pD/2t$. During one test, 10 measurements of pressure yielded a mean of 8610 lb/ft^2 with a standard deviation of 273.1. Cylinder dimensions are to be based on a set of 10 measurements which yielded: $\bar{D} = 6.2$ in., $S_D = 0.18$ and $\bar{t} = 0.22$ in., $S_t = 0.04$. Determine the best estimate of stress. Pressure measurements and dimensions have a systematic uncertainty of 1% of the reading.

5.21 Suppose a measured pressure contains three elemental random errors in its calibration:

$$P_1 = 0.90\,\text{N/m}^2 \quad P_2 = 1.10\,\text{N/m}^2 \quad P_3 = 0.09\,\text{N/m}^2$$
$$v_1 = 21 \quad v_2 = 10 \quad v_3 = 15$$

Estimate the random uncertainty due to calibration errors.

5.22 An experiment to determine force acting at a point on a member contains three elemental errors with resulting uncertainties:

$$
\begin{array}{lll}
B_1 = 2\,\text{N} & B_2 = 4.5\,\text{N} & B_3 = 3.6\,\text{N} \\
P_1 = 0 & P_2 = 6.1\,\text{N} & P_3 = 4.2\,\text{N} \\
& v_2 = 17 & v_3 = 19
\end{array}
$$

If the mean value for force is estimated to be 200 N, determine the uncertainty in the mean value at 95% confidence.

5.23 The area of a flat, rectangular parcel of land is computed from the measurement of the length of two adjacent sides, X and Y. Measurements are made using a scaled chain accurate to within 0.5% over its indicated length. The two sides are measured several times with the following results:

$$
\begin{array}{ll}
\bar{X} = 556\,\text{m} & \bar{Y} = 222\,\text{m} \\
S_x = 5.3\,\text{m} & S_y = 2.1\,\text{m} \\
v = 8 & v = 7
\end{array}
$$

Estimate the area of the land and state the confidence interval of that measurement at 95%.

5.24 Estimate the random uncertainty in the measured value of stress for a data set of 23 measurements if $\bar{\sigma} = 1061$ kPa and $S_\sigma = 22$ kPa.

5.25 Estimate the uncertainty at 95% confidence in the strength of a metal alloy. Six separate specimens are tested with a standard deviation of 1.23 MPa in the results. A systematic uncertainty of 1.48 MPa is estimated by the operator.

5.26 A pressure measuring system outputs a voltage that is proportional to pressure. It is calibrated against a transducer standard (certified error: within ± 0.5 psi) over its 0–100-psi range with the results given below. The voltage is measured with a voltmeter (instrument error: within ± 10 μV; resolution: 1 μV). The engineer intending to use this system estimates that installation effects can cause the indicated pressure to be off by another ±0.5 psi. Estimate the uncertainty at 95% confidence in using the system based on the known information.

E[mV]:	0.004	0.399	0.771	1.624	2.147	4.121
p[psi]:	0.1	10.2	19.5	40.5	51.2	99.6

5.27 The density of a metal composite is to be determined from the mass of a cylindrical ingot. The volume of the ingot is determined from diameter and length measurements. It is estimated that mass, m, can be determined to within 0.1 lb$_m$ using an available balance scale; length, L, can be determined to within 0.05 in. and diameter, D, to within 0.0005 in.

Instrumentation for each variable has a known calibration systematic uncertainty of 1% of its reading. Estimate the design-stage uncertainty in the determination of the density. Which measurement contributes most to the uncertainty in the density? Which measurement method should be improved first if the uncertainty in density is unacceptable? Use the nominal values of $m = 4.5\,\text{lb}_\text{m}$, $L = 6$ in., and $D = 4$ in.

5.28 For the ingot of Problem 5.27, the diameter of the ingot is measured 10 independent times at each of three cross sections. For the results below, give a best estimate in the diameter with its uncertainty:

$$\bar{d}_1 = 3.990\,\text{in.}\qquad \bar{d}_2 = 3.9892\,\text{in.}\qquad \bar{d}_3 = 3.9961\,\text{in.}$$
$$S_{d_1} = 0.005\,\text{in.}\qquad S_{d_2} = 0.001\,\text{in.}\qquad S_{d_3} = 0.0009\,\text{in.}$$

5.29 For the ingot of Problem 5.27, the following measurements are obtained from the ingot:

$$\bar{m} = 4.4\,\text{lb}_\text{m}\qquad \bar{L} = 5.85\,\text{in.}$$
$$S_m = 0.1\,\text{lb}_\text{m}\qquad S_L = 0.1\,\text{in.}$$
$$N = 21\qquad\qquad N = 11$$

Based on the results of Problem 5.28, estimate the density and the uncertainty in this value. Compare your answer with that of 5.27. Explain any difference.

5.30 A temperature measurement system is calibrated against a standard system certified to an uncertainty of $\pm 0.05°C$ at 95%. The system sensor is immersed alongside the standard within a temperature bath so that the two are separated by about 10 mm. The temperature uniformity of the bath is estimated at about 5°C/m. The temperature system sensor is connected to a readout that indicates the temperature in terms of voltage. The following are calibration results between the temperature indicated by the standard and the indicated voltage:

T[°C]	E[mV]	T[°C]	E[mV]
0.1	0.004	40.5	1.624
10.2	0.399	51.2	2.147
19.5	0.771	99.6	4.121

a. Compute the calibration curve fit.

b. Estimate the uncertainty in using the output from the temperature measurement system for temperature measurements.

5.31 The power usage of a strip heater is to be determined by measuring heater resistance and heater voltage drop simultaneously. The resistance is to be measured using an ohmmeter having a resolution of 1 Ω and an error stated to be within 1% of its reading, and voltage is to be measured using a voltmeter having a resolution of 1 V and a stated error of 1% of reading. It is expected that the heater will have a resistance of 100 Ω and use 100 W of power. Determine the uncertainty in power determination to be expected with this equipment at the design stage.

5.32 The power usage of a dc strip heater can be determined in either of two ways: (1) heater resistance and voltage drop can be measured simultaneously and power computed, or (2) heater voltage drop and current measured simultaneously and power computed. Instrument manufacturer specifications are listed:

Instrument	Resolution	Error (% reading)
Ohmmeter	1 Ω	0.5%
Ammeter	0.5 A	1%
Voltmeter	1 V	0.5%

For loads of 10 W, 1 kW, and 10 kW, determine the best method based on an appropriate uncertainty analysis. Assume nominal values as necessary for resistance, current, and voltage.

5.33 A set of tests is to be done to study how the cure temperature of a composite material will affect its strength. Tests will be done over a range of temperatures (20–60°C). Temperature is to be measured at each of four quadrants within the cure chamber using temperature equipment for which the accuracy can be estimated. The test specimens are expensive, so only one test at any four temperatures can be performed. Suppose prior to testing any composite, the chamber thermostat is set to, say, 30°C, and 20 measurements over time are taken. This exercise is then replicated five times so that the statistics can be computed. Compare this method to one in which a composite is placed in the chamber and cured while 100 measurements of chamber temperature are made (25 at each of the four locations). Would the uncertainty estimates for the cure temperature based on the two methods include the same effects? Explain.

5.34 Time variations in a signal require that the signal be measured often in time to determine its mean value. A preliminary sample is obtained by measuring the signal 50 times. The statistics from this sample show a mean value of 2.112 V with a standard deviation of 0.387 V. Systematic errors are negligible. If it is desired to state the uncertainty of the mean value to within 0.100 V at 95% confidence, how many measurements of the signal will be required based on the data variations?

5.35 The diameter of a shaft is estimated in a manner consistent with Example 5.14. A hand-held micrometer (resolution: 0.001 in.; accuracy: $< .001$ in.) is used to make measurements about four selected cross-sectional locations. Ten measurements of diameter are made around each location with the results noted below. Provide a reasonable statement as to the diameter of the shaft and the uncertainty in that estimate.

Location:	1	2	3	4
$\bar{d}$ (in.):	4.494	4.499	4.511	4.522
S_d:	0.006	0.009	0.010	0.003

5.36 The pressure in a large vessel is to be maintained at some set pressure for a series of tests. A compressor is to supply air through a regulating valve that is set to open and close at the set pressure. A dial gauge (resolution: 1 psi; accuracy: 0.5 psi) is used to monitor pressure in the vessel. Thirty trial replications of pressurizing the vessel to a set pressure of 50 psi are attempted to estimate pressure controllability and a standard deviation in set pressure of 2 psi is found. Estimate the uncertainty to be expected in the vessel set pressure.

5.37 The linear displacement of a vehicle due to an applied impact force is measured with a transducer. Transducer specifications are:

Input range:	0–5 m
Output range:	0–5 V
Linearity:	$\pm 0.25\%$ reading
Drift:	$\pm 0.05\%/°C$ FSO
Hysteresis:	$\pm 0.25\%$ reading
Sensitivity:	$\pm 0.10\%$ FSO

The transducer output is indicated on a voltmeter (accuracy: within $\pm 0.1\%$ reading; resolution: 10 mV). The expected nominal displacement of the vehicle is to be 4 m for an impact force of 2000 ± 100 N (95%). The force-displacement relation can be assumed linear. Four replications consisting of 10 measurements are made over the course of one day. The results are given with the ambient temperature for each run.

Test Run	N	Mean Value [m]	S_x [m]	$T_{ambient}$ [°C]
1	10	4.3	0.22	21
2	10	3.8	0.27	24
3	10	4.2	0.31	27
4	10	4.0	0.15	22

Report the mean displacement of the vehicle. Include a complete uncertainty estimate. Compare your uncertainty estimate to that from a simple design-stage analysis. Discuss why these differ.

5.38 The cooling of a thermometer (e.g., Examples 3.3 and 3.4) can be modeled as a first-order system with $\Gamma = e^{-t/\tau}$. If Γ can be measured to within 2% and time within 1%, determine the uncertainty in τ over the range $0 \le \Gamma \le 1$.

5.39 A J-type thermocouple monitors the temperature of air flowing through a duct. Its signal is measured by a thermostat. The air temperature is maintained constant by an electric heater whose power is controlled by the thermostat. To test the control system, 20 measurements of temperature were taken over a reasonable time period during steady operation. This was repeated three times with the results as:

Run	N	$\overline{T}$[°C]	S_T [°C]
1	20	181.0	3.01
2	20	183.1	2.84
3	20	182.1	3.08

The thermocouple itself has an error within 1°C (95%). It has a 90% rise time of 20 ms. Thermocouple insertion errors are estimated to be less than 1.2°C (95%). What information is found by performing replications? Identify the elemental errors that affect the system's control of the air temperature. What is the uncertainty in the set temperature?

5.40 Based on everyday experience, give an estimate of the systematic uncertainty you might expect in the following measuring instruments: bathroom scale, plastic ruler scale, micrometer, kitchen window bulb thermometer, and automobile speedometer.

5.41 A tank is pressurized with air at $25 \pm 2°C$ (95%). Determine the relative uncertainty in the tank's air density if tank pressure is known to within $\pm 1\%$ (95%). Assume ideal gas behavior.

5.42 The density of air must be known to within $\pm 0.5\%$. If the air temperature at 25°C can be determined to within $\pm 1°C$ (95%), what uncertainty can be tolerated in the pressure measurement if air behaves as an ideal gas?

5.43 In pneumatic conveying, solid particles such as flour or coal are carried through a duct by a moving air stream. Solids density at any duct location can be measured by passing a laser beam of known intensity I_o through the duct and measuring the light intensity transmitted to the other side, I. A transmission factor is found by

$$T = I/I_o = e^{-KEW} \quad 0 \le T \le 1$$

Here W is the width of the duct, K is the solids density, and E is a factor taken as 2.0 ± 0.04 (95%) for spheroid particles. Determine how u_K/K is related to the relative uncertainies of the other variables. If the transmission factor and duct width can be measured to within $\pm 1\%$, can solids density be measured to within 5%? 10%? Discuss your answer remembering that the transmission factor varies from 0 to 1.

5.44 A step test is run to determine the time constant of a first-order instrument (see Chapter 3). If the error fraction, $\Gamma(t)$, can be estimated to within $\pm 2\%$ (95%) and time, t, can be estimated in seconds to within $\pm 0.5\%$ (95%), plot u_τ/τ versus $\Gamma(t)$ over its range, $0 \leq \Gamma(t) \leq 1$.

5.45 The acceleration of a cart down a plane inclined at an angle α to horizontal can be determined by measuring the change in speed of the cart at two points, separated by a distance s, along the inclined plane. Suppose two photocells are fixed at the two points along the plane. Each photocell measures the time for the cart, which has a length, L, to pass it. If $L = 5 \pm 0.5$ cm, $s = 100 \pm 0.2$ cm, $t_1 = 0.054 \pm 0.001$s, and $t_2 = 0.031 \pm 0.001$s, all (95%), estimate the uncertainty in acceleration:

$$a = \frac{L^2}{2s}\left(\frac{1}{t_2^2} - \frac{1}{t_1^2}\right)$$

Compare as a concomitant check $a = g \sin \alpha$ using values for $\alpha = 30°$ and $90°$. What uncertainty in α is needed to make this the better method?

5.46 Golf balls are often tested using a mechanical player called an "Iron Byron" because the robotic golfer's swing was patterned after that of Byran Nelson, a famous golf professional. It is proposed to use initial velocity and launch angle from such testing to predict carry distance (how far the ball travels after impact). The following data represent test results for a particular golf ball when struck with a particular driver (golf club).

Initial Velocity (mph)	Launch Angle (deg)	Carry Distance (yd)
165.5	8	254.6
167.8	8	258.0
170	8	261.4
172.2	8	264.8
165.5	10	258.2
167.8	10	261.6
170	10	264.7
172.2	10	267.9
165.5	12	260.6
167.8	12	263.7
170	12	266.8
172.2	12	269.8

If the initial velocity can be measured to within a systematic uncertainty of 1 mph, and the launch angle to within $0.1°$, estimate the resulting uncertainty in the carry distance as a function of initial velocity and launch angle. Over this range of initial velocity and launch angle, can a single value of uncertainty be assigned?

5.47 A particular flow meter allows the volumetric flow rate, Q, to be inferred from a measured pressure drop, Δp. Theory predicts $Q \propto \sqrt{\Delta p}$, and a calibration provides the following data:

Q (m³/min)	Δp (Pa)
10	1000
20	4271
30	8900
40	16023

What uncertainty in the measurement of Δp is required to yield an uncertainty of 0.25% in Q over the range of flow rates from 10 to 40 m³/min?

5.48 Devise a simple experiment to estimate the true value of a variable through measurements. List the errors that enter into your measurement. Execute the experiment and estimate the uncertainty in the result. Discuss how you assigned values to each uncertainty at the design stage and after the experiment.

5.49 Devise a simple experiment to estimate the true value of a result that is a function of at least two variables that must be measured. List the errors that enter into your measurements. Execute the experiment and estimate the uncertainty in the result. Discuss how you assigned values to each uncertainty at the design stage and after the experiment.

5.50 A steel cantilever beam is fixed at one end and free to move at the other. A load (F) of 980 N with a systematic error (95%) of 10 N is applied at the free end. The beam geometry, as determined by a metric ruler having 1-mm increments, is 100 mm long (L), 30 mm wide (w), and 10 mm thick (t). Estimate the maximum stress at the fixed end and its uncertainty. The maximum stress is given by: $\sigma = \frac{Mc}{I} = \frac{FL(t/2)}{wt^3/12}$.

5.51 The heat flux in a reaction is estimated by $\dot{Q} = 5(T_1 - T_2)$ kJ/s. Two thermocouples are used to measure T_1 and T_2. Each thermocouple has a systematic uncertainty of 0.2°C. Based on a large sample of measurements, $\overline{T}_1 = 180°C$, $\overline{T}_2 = 90°C$, and the standard random uncertainty in each temperature measurement is 0.1°C. Compare the uncertainty in heat flux if the thermocouple errors are (i) uncorrelated, and (ii) correlated. Explain.

5.52 A comparative test uses the relationship, $R = p_2/p_1$, where pressure p_2 is measured to be 54.7 MPa and pressure p_1 is measured to be 42.0 MPa. Each pressure measurement has a single systematic error of 0.5 MPa. Compare the systematic uncertainty in R if these errors are (i) correlated and (ii) uncorrelated. Explain.

5.53 A sensitive material is to be contained within a glovebox. Codes require that the transparent panels, which make up the walls of the glovebox, withstand the impact of a 22-kg mass falling at 8 m/s. An impact test is performed where the 22 kg mass is dropped a known variable height onto a panel. The mass motion is sensed by photocells from which an impact velocity is estimated. The mean kinetic energy for failure was 717 N-m with a standard deviation of 60.7 N-m based on 24 failures. The systematic error in mass is 0.001 kg. The velocity at impact is 8 m/s with a systematic error in velocity estimated as 0.27 m/s. Estimate the combined standard uncertainty and the expanded uncertainty at 95% in the kinetic energy for failure.

5.54 A geometric stress concentration factor, K_t, is used to relate the actual maximum stress to a well-defined nominal stress in a structural member. The maximum stress is given by $\sigma_{max} = K_t \sigma_o$ where σ_{max} represents the maximum stress and σ_o represents the nominal stress. The nominal stress is most often stated at the minimum cross section.

Consider the axial loading shown in Figure 5.8, where the structural member is in axial tension and experiences a stress concentration as a result of a transverse hole. In this geometry, $\sigma_o = F/A$ where area $A = (w - d)t$. If $d = 0.5w$, then $K_t = 2.2$. Suppose $F = 10,000 \pm 500$ N, $w = 1.5 \pm 0.02$ cm, $t = 0.5 \pm 0.02$ cm, and the uncertainty in the value for d is 3%. Neglecting the uncertainty in stress concentration factor, determine the uncertainty in the maximum stress experienced by this part.

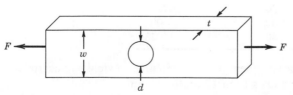

Figure 5.8. Structural member discussed in Problem 5.54.

Chapter **6**

Analog Electrical Devices
and Measurements

6.1 INTRODUCTION

This chapter provides an introduction into some basic electrical analog devices used with analog signals or to display signals in an analog form. We include analog output meters and the more common solid-state devices used in signal conditioning. Information is often transferred between stages of a measurement system as an analog electrical signal. This signal typically originates from the measurement of a physical variable using a fundamental electromagnetic or electrical phenomenon and then propagates from stage to stage of the measurement system (see Figure 1.1). Accordingly, we discuss some signal conditioning and output methods.

While analog devices have been supplanted by their digital equivalents in many applications, they are still widely used and remain engrained in engineered devices. An analog output format is often ergonomically superior in monitoring, as evidenced by modern car speedometer dials and by wristwatches. Too often we qualify a digital device by its digital readout, but internal analog circuits form the foundation for both analog and digital measuring systems. In fact, most systems that we interface with are actually analog and digital hybrids. Within a signal chain, it is common to find digital and analog electrical devices being used together and so requiring special signal conditioning. An understanding of analog device function provides insight, as well as a historical reference point, into the advantages and disadvantages of digital counterparts; such issues are discussed in this chapter.

6.2 ANALOG DEVICES: CURRENT MEASUREMENTS

Direct Current

A simple way to measure a dc electrical current is to use an analog device that responds to the force exerted on a current-carrying conductor in a magnetic field. Because an electric current is a series of moving charges, a magnetic field will exert a force on a current-carrying conductor. So this force can be used as a measure of the flow of current in a conductor by moving a pointer on a display.

Consider a straight length of a conductor through which a current, I, flows, as shown in Figure 6.1. The magnitude of force on this conductor due to a magnetic field strength $\mathscr{B}$ is

$$F = Il\mathscr{B} \tag{6.1}$$

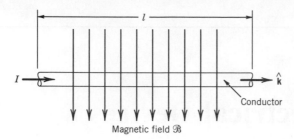

Figure 6.1 Current-carrying conductor in a magnetic field.

This relation is valid only when the direction of the current flow and the magnetic field are at right angles. In the general case

$$\mathbf{F} = Il\hat{k} \times \mathscr{B} \tag{6.2}$$

where $\hat{k}$ is a unit vector along the direction of the current flow, and the force $\mathbf{F}$ and the magnetic field $\mathscr{B}$ are also vector quantities. Equation 6.2 provides both the magnitude of the developed force $\mathbf{F}$, and, by the right-hand rule, the direction in which the force on the conductor acts.

Similarly, a current loop in a magnetic field will experience a torque if the loop is not aligned with the magnetic field, as illustrated in Figure 6.2. The torque[1] on a loop composed of N turns is given by

$$T_\mu = NIA\mathscr{B} \sin \alpha \tag{6.3}$$

where

$A = $ cross-sectional area defined by the perimeter of the current loop

$\mathscr{B} = $ magnetic field strength (magnitude)

$I = $ current

$\alpha = $ angle between the normal, the cross-sectional area of the current loop, and the magnetic field

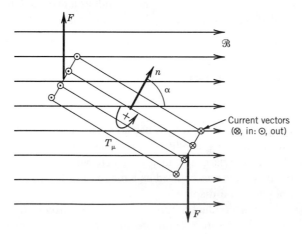

Figure 6.2 Forces and resulting torque on a current loop in a magnetic field.

---

[1]In vector form, the torque on a current loop can be written $\mathbf{T}_\mu = \mu \times \mathscr{B}$ where μ is the magnetic dipole moment.

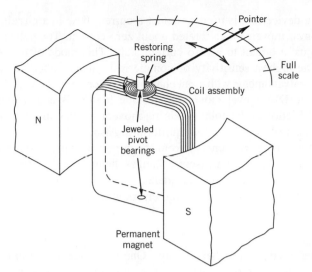

Figure 6.3 Basic D'Arsonval meter movement.

So how do we use equation (6.3) in a working unit? One approach is the D'Arsonval movement shown in Figure 6.3. In this arrangement, the uniform radial magnetic field and torsional spring result in a steady angular deflection of the coil which corresponds to the existing current through the coil. The coil and fixed permanent magnet are arranged in the normal direction to the current loop, i.e., $\alpha = 90°$. For extreme sensitivity, the pointer can be replaced by a mirror and light beam arrangement.

Most devices that use the D'Arsonval movement employ a pointer whose deflection increases with the magnitude of current applied. This mode of operation is called the *deflection mode*. Most electrical pointer meters use this mode of operation. A typical circuit for an analog current measuring device, an ammeter, is shown in Figure 6.4. Here the deflection of the pointer indicates the magnitude of the current flow. The range of current that can be measured is determined by selection of the combination of shunt resistor and the internal resistance of the meter movement. The shunt resistor provides a bypass for current flow, reducing the current that flows through the movement. A make-before-break switch prevents current overload through the meter coil.

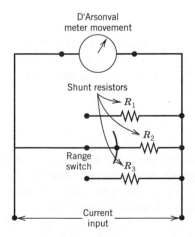

Figure 6.4 Simple multirange ammeter (with make-before-break selector switch). Shunt resistors determine meter range.

A *galvanometer* is a device that is used to detect a current flow in a circuit. It is a highly sensitive D'Arsonval movement calibrated about zero current. The galvanometer indication is used to adjust a circuit to a zero current state. This mode of operation is called the *null mode*. The highest sensitivity of commercial galvanometers is approximately 0.1 μA/div for a galvanometer with a mechanical pointer.

Errors inherent to the D'Arsonval movement include hysteresis and repeatability errors due to mechanical friction in the pointer-bearing movement, and linearity errors in the spring that provides the restoring force for equilibrium. Also, in developing a torque, the D'Arsonval movement must extract energy from the current flowing through it. This draining of energy from the signal being measured changes the measured signal, and such an effect is called a *loading error*. This is a consequence of all instruments that operate in deflection mode. A quantitative analysis of loading errors will be provided later.

Alternating Current

An ac current can be measured any number of ways. One technique, found in common deflection meters, uses diodes to form a rectifier that converts the time-dependent ac current into a dc current. This current then can be measured with a calibrated D'Arsonval movement meter as previously described. This is the same technique used in those ubiquitous small transformers used to convert ac wall current into a dc current to power electronic devices. Some meters employ electromagnets whose magnetic field strength increases with increasing current flow. An *electrodynamometer* is basically a D'Arsonval movement modified for use with ac current by replacing the permanent magnet with an electromagnet in series with the current coil. These ac meters have upper limits on the frequency of the alternating current that they can effectively measure; most common instruments are calibrated for use with standard line frequency.

An accurate measuring approach for large ac current is based on the Hall effect. The approach can be recognized by a probe that is clamped over the current-carrying wire (conductor) to measure its unknown current flow. To understand its use, let us mention two phenomena. The first is the *Hall effect*, which is a voltage that is developed from any current-carrying conductor placed perpendicular to a magnetic field. For a known current, the magnitude of this voltage directly depends on the magnitude of the magnetic field. The second is that a current passing through a wire will generate a magnetic field. So in practice, a Hall-effect probe is realized by coupling these two processes concurrently: use the unknown current within a wire to generate a magnetic field that develops a measurable voltage across the Hall-effect sensor.

The Hall-effect sensor is a thin conducting semiconductor wafer driven by a known current; this current is unrelated to the unknown current being measured and is provided by a separate source, such as a battery. The Hall-effect probe is an iron-core ring, which is placed around the wire of unknown current and used to concentrate the magnetic field, and a Hall-effect sensor, which is attached to the iron core so as to be aligned parallel to the wire of unknown current being measured.

EXAMPLE 6.1

A galvanometer consists of N turns of a conductor wound about a core of length l and radius r that is situated perpendicular to a magnetic field of uniform flux density $\mathscr{B}$. A dc current passes through the conductor due to an applied potential, $E_i(t)$. The output of the device is the rotation of the core and pointer, θ, as shown in Figure 6.5. Develop a model relating pointer rotation and current.

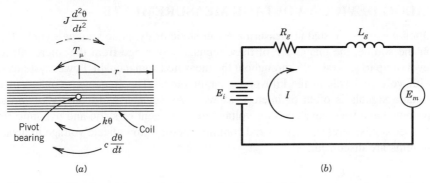

Figure 6.5 Circuit and free-body diagram for Example 6.1.

FIND A dynamic model relating θ and I.

SOLUTION The galvanometer is a rotational system consisting of a torsional spring, a pointer, and core which are free to deflect; the rotation is subject to frictional damping. The mechanical portion of the device consists of a coil having moment of inertia, J, bearings that provide damping from friction, with a damping coefficient, c, and a torsional spring of stiffness, k. The electrical portion of the device consists of a coil with a total inductance, L_g, and a resistance, R_g.

We apply Newton's second law with the mechanical free-body diagram shown in Figure 6.5(a).

$$J\frac{d^2\theta}{dt^2} + c\frac{d\theta}{dt} + k\theta = T_\mu \tag{6.4}$$

As a current passes through the coil, a force is developed that produces a torque on the coil

$$T_\mu = 2N\mathscr{B}lrI \tag{6.5}$$

This torque, which tends to rotate the coil, is opposed by an electromotive force, E_m, due to the Hall effect

$$E_m = \left(r\frac{d\theta}{dt} \times \mathscr{B}\right)l\hat{k} = 2N\mathscr{B}r\frac{d\theta}{dt} \tag{6.6}$$

which produces a current in the direction opposite to that produced by E_i. Using Kirchhoff's law to the electrical free body of Figure 6.5(b) gives

$$L_g\frac{dI}{dt} + R_gI = E_i - E_m \tag{6.7}$$

Equations (6.5) and (6.6) define the coupling relations between the mechanical equation (6.4) and the electrical equation (6.7). From these, we see that the current due to potential, E_i, brings about developed torque, T_m, that moves the galvanometer pointer, and that this motion is opposed by the mechanical restoring force of the spring and by the development of an opposing electrical potential, E_m. The system damping allows the pointer to settle to an equilibrium position.

COMMENT The system output, which is the pointer movement here, is governed by the second-order system response described in Chapter 3. However, the torque input to the mechanical system is a function of the electrical response of the system. In this measurement system, the input signal, which is the applied voltage here, is transformed into a mechanical torque to provide the measurable output deflection.

6.3   ANALOG DEVICES: VOLTAGE MEASUREMENTS

Often we are interested in measuring either static or dynamic voltage signals. Depending on the source, the magnitude of these signals may range from the microvolt level to a level of up to several volts throughout the measured signal chain. Power systems may deal with voltage levels in the kilovolt or larger range. The frequency content of dynamic voltage signals is often of interest, as well. As such, a wide variety of measurement systems have been developed for voltage measurement of static and dynamic signals. In this section, several convenient and common methods to indicate voltage in measurement systems are discussed.

Analog Voltage Meters

A dc voltage can be measured in through the analog circuit shown in Figure 6.6, where a D'Arsonval movement is used in series with a resistor. Although fundamentally sensitive to current flow, the D'Arsonval movement can be calibrated in terms of voltage by using an appropriate known fixed resistor. This basic circuit is employed in the construction of analog voltage dials and volt-ohmmeters (VOM), which for many years served as the common measurement device for current, voltage, and resistance.

An ac voltage can be measured either through rectification of the ac signal, or through use of an electromagnet, either in an electrodynamometer or with a movable iron vane. These instruments are sensitive to the *rms* (root-mean-squared) value of a simple periodic ac current, and can be calibrated in terms of voltage; shunt resistors can be used to establish the appropriate scale. The circuit shown in Figure 6.6 can also be used to measure an ac voltage if the input voltage is rectified prior to input to this circuit. The waveform output of the rectifier must be considered, if a steady meter reading is to be obtained. An ac meter indicates a true *rms* value for a simple periodic signal only, but a *true rms* ac voltmeter performs the signal integration, e.g., equation (2.4), required to accurately determine the *rms* value in a signal-conditioning stage and indicates true signal *rms* regardless of waveform.

Oscilloscope

The *oscilloscope* is a practical graphical display device providing an analog representation of a measured signal. It is used to measure voltage magnitude versus time for dynamic signals over a wide range of frequencies with a signal bandwidth extending well into the megahertz range and, with some units, into the gigahertz ranges [1, 2]. A useful diagnostic tool, the oscilloscope provides a visual output of signal magnitude, frequency, distortion, and a delineation of the dc and ac components. The visual image provides a direct means to detect the superposition of noise and interference on a measured signal,

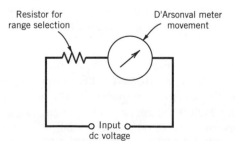

Figure 6.6 A dc voltmeter circuit.

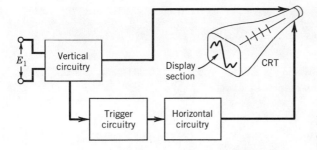

Figure 6.7 Schematic of basic cathod-ray tube oscilloscope.

something nonvisual metering devices cannot do. Extraneous information can be detected, their sources diagnosed, and appropriate action taken to reduce their effects. In addition to signal versus time, a typical unit can also display two or more signals [$X(t)$ and $Y(t)$], perform addition and subtraction of signals ($X + Y, X - Y$), and display amplitude vs. amplitude (XY) plots and other features. Some digital oscilloscopes have significant internal storage so as to mimic a data-logging signal recorder. Others have internal FFT circuitry to provide for direct spectral analysis of a signal (Chapter 2).

Although seen increasingly less today, the cathode ray oscilloscope is interesting in that the cathode tube operates as a time-based voltage transducer. Its schematic is shown in Figure 6.7. A beam of electrons is emitted by the cathode ray tube. The output of the oscilloscope is a signal trace on the screen of the instrument, created by the impact of the electrons on a phosphorescent coating on the screen. Because an electron beam is composed of charged particles, it can be guided by an electrical field. In the case of the oscilloscope, pairs of plates are oriented horizontally and vertically, to control the location of the impact of the electron beam on the screen. The beam sweeps horizontally across the screen at a known speed or frequency. Input voltages to the oscilloscope result in vertical deflections of the beam, and produce a trace of the voltage variations (vertical or y axis) versus time (horizontal or x axis) on the screen. The horizontal sweep frequency can be varied over a wide range, and the operation of the oscilloscope is such that high-frequency waveforms can be resolved.

A digital oscilloscope, such as shown in Figure 6.8(a), also provides an analog representation of the measured signal. But it does so by first sampling the signal to convert it into a digital form and then reconstructing the signal on a phosphorous or LCD display screen as an analog output, just as does a common LCD computer monitor. In this way, the digital oscilloscope measures and stores voltage in a digital manner but then displays it in an analog format as a sequence of measured points, as shown in Figure 6.8(b). The amplifier and time base controls are interpreted as with the analog oscilloscope. Sampling rates of over 60 GHz are available. These common lab devices can be very portable and can be packaged as a component of an analog-to-digital data acquisition system. Analog to digital sampling techniques used to measure voltages are discussed in Chapter 7.

The Labview® interactive program (virtual instrument) called *Oscilloscope* is available with the companion software. It features a basic two-channel oscilloscope. The user can vary channel displayed and the time sweep and gain using two built-in signals (sine wave and square wave) and an active signal trigger.

Potentiometer

The *potentiometer* is a device used to measure dc voltages that are in the microvolt to millivolt range. Equivalent to a balance scale, a potentiometer balances an unknown input

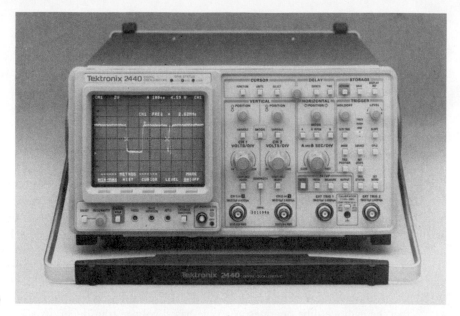

(a)

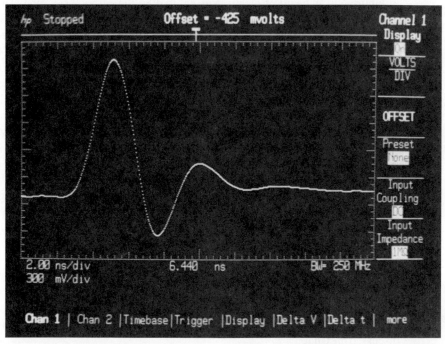

(b)

Figure 6.8 (*a*) Digital oscilloscope. (Photograph courtesy of Tektronix, Inc.) (*b*) Oscilloscope output. (Photograph courtesy of Hewlett-Packard Company.)

voltage against a known internal voltage until both sides are equal. A potentiometer is a null balance instrument in that it drives the loading error to essentially zero. Potentiometers have been largely supplanted by digital voltmeters, which are deflection mode devices but have very high input impedances so as to keep loading errors small, even at low-level voltages.

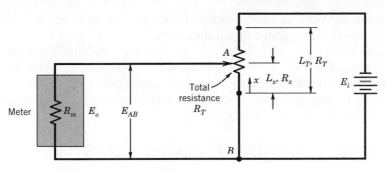

Figure 6.9 Voltage divider circuit.

Voltage Divider Circuit

An important, but general purpose, component found in the potentiometer circuit is the *voltage divider circuit* shown in Figure 6.9. The point labeled A in Figure 6.9 represents a sliding contact, which makes an electrical connection with the resistor R. The resistance between points A and B in the circuit is a linear function of the distance from A to B. So the output voltage sensed by the meter is given by

$$E_o = \frac{L_x}{L_T} E_i = \frac{R_x}{R_T} E_i \qquad (6.8)$$

which holds so long as the internal resistance of the meter, R_m, is very large relative to R_T.

Potentiometer Circuit

A simple potentiometer[2] circuit can be derived from the voltage divider circuit as shown in Figure 6.10. In this circuit a galvanometer is used to detect current flow; any current flow through the galvanometer, G, would be a result of an imbalance between the measured voltage, E_m, and the voltage imposed across points A to B, E_{AB}. The voltage E_{AB} is adjusted by moving the sliding contact A; if this sliding contact is calibrated in terms of a known supply voltage, E_i, the circuit may be used to measure voltage by creating a balanced condition, indicated by a zero current flow through the galvanometer. A null balance, corresponding to zero current flow through G, will occur only when $E_m = E_{AB}$.

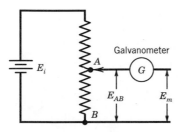

Galvanometer

Figure 6.10 Basic potentiometer circuit.

---

[2]The term "potentiometer" has several common uses: a sliding contact precision variable resistor, the divider circuit in Figure 6.9, and the voltage measuring instrument described.

With a known and constant supply voltage E_i, the position of A can be calibrated to indicate E_m directly as suggested by equation (6.8).

A practical potentiometer includes a means for setting the supply voltage, E_i. One way uses a separate divider circuit to adjust the output from a cheap, replaceable battery to equal the fixed, known output from a standard voltage cell. This way, the replaceable battery is used for subsequent measurements. Potentiometers can have systematic errors of $< 1\mu V$ and random errors $< 0.2\mu V$.

6.4   ANALOG DEVICES: RESISTANCE MEASUREMENTS

Resistance measurements may range from simple continuity checks, to determining changes in resistance on the order of $10^{-6}\Omega$, to determining absolute resistance ranging from 10^{-5} to 10^{15} Ω. As a result of the tremendous range of resistance values that have practical application, numerous measurement techniques that are appropriate for specific ranges of resistance or resistance changes have been developed. Some of these measurement systems and circuits provide necessary protection for delicate instruments, allow compensation for changes in ambient conditions, or accommodate changes in transducer reference points. In addition, the working principle of many transducers is a change in resistance relative to a change in the measured variable. We will discuss the measurement of resistance using basic techniques of voltage and current measurement in conjunction with Ohm's law.

Ohmmeter Circuits

A simple way to measure resistance is by imposing a voltage across the unknown resistance and measuring the resulting current flow. Clearly, from Ohm's law the value of resistance can be determined in a circuit such as in Figure 6.11, which forms the basis of a common analog ohmmeter. Analog ohmmeters use circuits similar to those shown in Figure 6.12, which use shunt resistors and a D'Arsonval mechanism for measuring a wide range of resistance while limiting the flow of current through the meter movement. In this technique, the lower limit on the measured resistance, R_1 is determined by the upper limit of current flow through it, i.e., I_1. A practical limit to the maximum current flow through a resistance is imposed by the ability of the resistance element to dissipate the power generated by the flow of current (I^2R heating). For example, the limiting ability of a thin

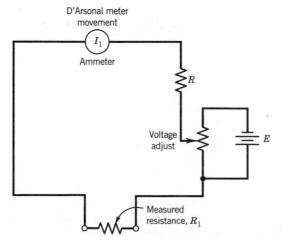

Figure 6.11 Basic analog ohmmeter (voltage is adjusted to yield full-scale deflection with terminals shorted).

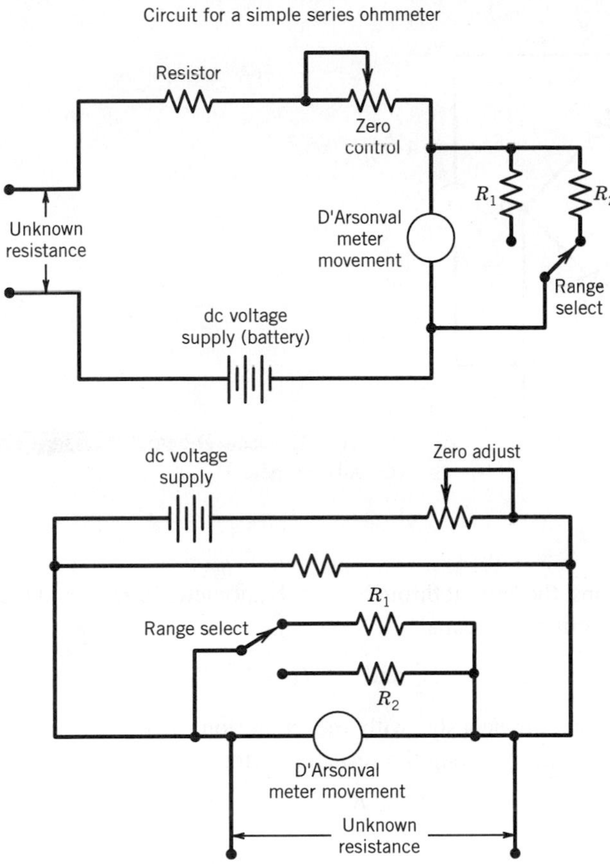

Circuit for a simple series ohmmeter

Figure 6.12 Multirange ohmmeter circuits.

metallic conductor to dissipate heat is the principle on which fuses are based. At too large a value of I_1, they melt. The same is true for delicate, small resistance sensors!

Bridge Circuits

A variety of bridge circuits have been devised for measuring capacitance, inductance, and most often for measuring resistance. A purely resistive bridge, called a *Wheatstone bridge*, provides a means for accurately measuring resistance, and for detecting small changes in resistance. Figure 6.13 shows the basic arrangement for a bridge circuit, where R_1 may be a transducer that experiences a change in resistance associated with a change in some physical variable. A dc voltage is applied as an input across nodes A to D, and the bridge forms a parallel circuit arrangement between these two nodes. The currents flowing through the resistors R_1 to R_4 are I_1 to I_4, respectively. Under the condition that the current flow through the galvanometer, I_g, is zero, the bridge is in a balanced condition (alternatively, a current sensing digital voltmeter might be used in place of the galvanometer). A specific relationship exists between the resistances that form the bridge at balanced conditions. To find this relationship, a circuit analysis is performed with $I_g = 0$. Under this balanced condition, there is no voltage drop from B to C and

$$I_1 R_1 - I_3 R_3 = 0$$
$$I_2 R_2 - I_4 R_4 = 0$$

$$(6.9)$$

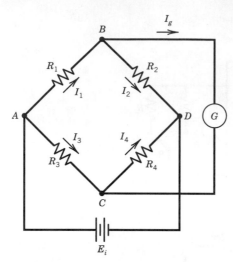

Figure 6.13 Basic Wheatstone bridge circuit (*G*, galvanometer).

Under balanced conditions, the current through the galvanometer is zero, and the currents through the arms of the bridge are equal:

$$I_1 = I_2 \quad \text{and} \quad I_3 = I_4 \tag{6.10}$$

Solving equations (6.9) simultaneously, with the condition stated in equation (6.10), yields the necessary relationship among the resistances for a balanced bridge:

$$\frac{R_2}{R_1} = \frac{R_4}{R_3} \tag{6.11}$$

Resistance and resistance change can be measured in one of two ways with this bridge circuit. If the resistor R_1 varies with changes in the measured physical variable, one of the other arms of the bridge can be adjusted to null the circuit and determine resistance. Another method uses a voltage measuring device to measure the voltage unbalance in the bridge as an indication of the change in resistance. Both of these methods will be analyzed further.

Null Method

Consider the circuit shown in Figure 6.13, where R_2 is an adjustable variable resistance. If the resistance R_1 changes due to a change in the measured variable, the resistance R_2 can be adjusted to compensate so that the bridge is once again balanced. In this null method of operation, the resistance R_2 must be a calibrated variable resistor, such that adjustments to R_2 directly indicate the value of R_1. The balancing operation may be accomplished either manually or automatically through a closed loop controller circuit. An advantage of the null method is that the input voltage need not be known, and changes in the input voltage do not affect the accuracy of the measurement. In addition, the current detector or controller need only detect if there is a flow of current, not measure its value.

However, even null methods are limited. Earlier in this section, we assumed that the galvanometer current was exactly zero when the bridge was balanced. In fact, because the resolution of the galvanometer is limited, the current cannot be set exactly to zero. Consider a bridge that has been balanced within the sensitivity of the meter, such that I_g is smaller than the smallest current detectable. This current flow due to the meter resolution is a loading error with an associated (systematic) uncertainty in the measured resistance, u_R. A basic analysis of the circuit with a current flow through the galvanometer yields

$$\frac{u_R}{R_1} = \frac{I_g(R_1 + R_g)}{E_i} \tag{6.12}$$

where R_g is the internal resistance of the galvanometer. Alternatively, if a current sensing digital voltmeter is used, then the input bias current from this meter remains and a small offset voltage will be present as a loading error.

Equation (6.12) can serve to guide the choice in a galvanometer or digital meter, and a battery voltage for a particular application. Clearly, the error is reduced by increased input voltages. However, the input voltage is limited by the power dissipating capability of the resistance device, R_1. The power that must be dissipated by this resistance is $I_1^2 R_1$.

Deflection Method

In an unbalanced condition, the magnitude of the current or voltage drop for the meter portion of the bridge circuit is a direct indication of the change in resistance of one or more of the arms of the bridge. Consider first the case where the voltage drop from node B to node C in the basic bridge is measured by a meter with infinite internal impedance, so that there is no current flow through the meter, as shown in Figure 6.14. Knowing the conditions for a balanced bridge given in equation (6.10), the voltage drop from B to C can be determined, since the current I_1 must equal the current I_2, as

$$E_o = I_1 R_1 - I_3 R_3 \tag{6.13}$$

Under these conditions, substituting equations (6.9–6.11) into equation (6.13) yields

$$E_o = E_i \left(\frac{R_1}{R_1 + R_2} - \frac{R_3}{R_3 + R_4} \right) \tag{6.14}$$

The bridge is usually initially balanced at some reference condition. Any transducer resistance change, as a result of a change in the measured variable, would then cause a deflection in the bridge voltage away from the balanced condition. Assume that from an

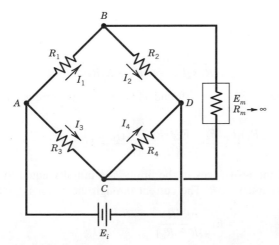

Figure 6.14 Voltage-sensitive Wheatstone bridge.

initially balanced condition where $E_o = 0$, a change in R_1 occurs to some new value, $R_1' = R_1 + \delta R$. The output from the bridge becomes

$$E_o + \delta E_o = E_i \left(\frac{R_1'}{R_1' + R_2} - \frac{R_3}{R_3 + R_4} \right) = E_i \frac{R_1' R_4 - R_3 R_2}{(R_1' + R_2)(R_3 + R_4)} \tag{6.15}$$

In many designs the bridge resistances are initially equal. Setting $R_1 = R_2 = R_3 = R_4 = R$ allows equation (6.15) to be reduced to

$$\frac{\delta E_o}{E_i} = \frac{\delta R / R}{4 + 2(\delta R / R)} \tag{6.16}$$

In contrast to the null method of operation of a Wheatstone bridge, the deflection bridge requires a meter capable of accurately indicating the output voltage, as well as a stable and known input voltage. But the bridge output should follow any resistance changes over any frequency input, up to the frequency limit of the detection device! So a deflection mode is often used to measure time-varying signals.

If the high-impedance voltage measuring device in Figure 6.14 is replaced with a relatively low-impedance current measuring device and the bridge is operated in an unbalanced condition, a current-sensitive bridge circuit results. Consider Kirchhoff's laws applied to the Wheatstone bridge circuit for a current sensing device with resistance R_g. The input voltage is equal to the voltage drop in each arm of the bridge,

$$E_i = I_1 R_1 + I_2 R_2 \tag{6.17}$$

but $I_2 = I_1 - I_g$, which implies

$$E_i = I_1 (R_1 + R_2) - I_g R_2 \tag{6.18}$$

If we consider the voltage drops in the path through R_1, R_g, and R_3, the total voltage drop must be zero

$$I_1 R_1 + I_g R_g - I_3 R_3 = 0 \tag{6.19}$$

For the circuit formed by R_g, R_4, and R_2,

$$I_g R_g + I_4 R_4 - I_2 R_2 = 0 \tag{6.20}$$

or with $I_2 = I_1 - I_g$ and $I_4 = I_3 + I_g$,

$$I_g R_g + (I_3 + I_g) R_4 - (I_1 - I_g) R_2 = 0 \tag{6.21}$$

Equations (6.18)–(6.21) form a set of three simultaneous equations in the three unknowns I_1, I_g, and I_3. Solving these three equations for I_g yields

$$I_g = \frac{E_i (R_2 R_3 - R_1 R_4)}{R_3 (R_1 + R_2)(R_g + R_2 + R_4) + R_1 R_2 R_4 - R_3 R_2^2 + R_g R_4 (R_1 + R_2)} \tag{6.22}$$

Then, the change in resistance of R_1 can be found in terms of the bridge deflection voltage, E_o, by

$$\frac{\delta R}{R_1} = \frac{(R_3 / R_1)[E_o / E_i + R_2 / (R_2 + R_4)]}{1 - E_o / E_i - R_2 / (R_2 + R_4)} - 1 \tag{6.23}$$

Consider the case when all of the resistances in the bridge are initially equal to R, and subsequently R_1 changes by an amount δR. The current through the meter is given by

$$I_g = E_i \frac{\delta R / R}{4(R + R_g)} \tag{6.24}$$

and the output voltage is given by $E_o = I_g R_g$,

$$E_o = E_i \frac{\delta R/R}{4(1 + R/R_g)} \tag{6.25}$$

The bridge impedance can affect the output from a constant voltage source having an internal resistance R_s. The effective bridge resistance, based on a Thévenin equivalent circuit analysis, is given by

$$R_B = \frac{R_1 R_3}{R_1 + R_3} + \frac{R_2 R_4}{R_2 + R_4} \tag{6.26}$$

such that for a power supply of voltage E_s

$$E_i = \frac{E_s R_B}{R_s + R_B} \tag{6.27}$$

In a similar manner, the bridge impedance can affect the voltage indicated by the voltage measuring device. For a voltage measuring device of internal impedance R_g, the actual bridge deflection voltage, relative to the indicated voltage, E_m, is

$$E_o = \frac{E_m}{R_g} \left(\frac{R_1 R_2}{R_1 + R_2} + \frac{R_3 R_4}{R_3 + R_4} + R_g \right) \tag{6.28}$$

The difference between the measured voltage, E_m, and the actual voltage, E_o, is a loading error, in this case due to the bridge impedance load. Loading errors are discussed next.

EXAMPLE 6.2

A certain temperature sensor experiences a change in electrical resistance with temperature according to the equation

$$R = R_o[1 + \alpha(T - T_o)] \tag{6.29}$$

where

R = sensor resistance [Ω]

R_0 = sensor resistance at the reference temperature, T_0 [Ω]

T = temperature [$°C$]

T_0 = reference temperature, $0°C$

α = constant: $0.00395°C^{-1}$

This temperature sensor is connected in a Wheatstone bridge like the one shown in Figure 6.13, where the sensor occupies the R_1 location, and R_2 is a calibrated variable resistance. The bridge is operated using the null method. The fixed resistances R_3 and R_4 are each equal to 500 Ω. If the temperature sensor has a resistance of 100 Ω at $0°C$, determine the value of R_2 that would balance the bridge at $0°C$.

KNOWN

$R_1 = 100\,\Omega$

$R_3 = R_4 = 500\,\Omega$

FIND R_2 for null balance condition

SOLUTION   From equation (6.11), a balanced condition for this bridge would be achieved when $R_2 = R_1R_4/R_3 = R_1 = 100\,\Omega$. Thus, a value of 100 Ω for R_2 creates a balanced condition. Notice that to be a useful circuit, R_2 must be adjustable and provide an indication of its resistance value at any setting.

EXAMPLE 6.3

Consider a deflection bridge, which initially has all arms of the bridge equal to 100 Ω, with the temperature sensor described in Example 6.2 again as R_1. The input or supply voltage to the bridge is 10 V. If the temperature of R_1 is changed such that the bridge output is 0.569 V, what is the temperature of the sensor? How much current flows through the sensor and how much power must it dissipate?

KNOWN

$E_i = 10$ V

Initial (reference) state:

$R_1 = R_2 = R_3 = R_4 = 100\,\Omega$

$E_o = 0$ V

Deflection state:

$E_o = 0.569$ V

ASSUMPTION   Voltmeter has infinite input impedance but source has negligible impedance $(E_s = E_i)$.

FIND

T_1, I_1, P_1

SOLUTION   The change in the sensor resistance can be found from equation (6.16) as

$$\frac{\delta E_o}{E_i} = \frac{\delta R/R}{4 + 2(\delta R/R)} \Rightarrow \frac{0.569}{10} = \frac{\delta R}{400 + 2\delta R}$$

$$\delta R = 25.67\,\Omega$$

This gives a sensor resistance of $R_1 + \delta R = 125.7\,\Omega$, which equates to a sensor temperature of $T_1 = 65°C$.

To determine the current flow through the sensor, consider first the balanced case where all the resistances are equal to 100 Ω. The equivalent bridge resistance, R_B, is simply 100 Ω, and the total current flow from the supply, $E_i/R_B = 100$ mA. Thus, at the initially balanced condition, through each arm of the bridge and through the sensor the current flow is 50 mA. If the sensor resistance changes to 125.67 Ω, the current will be reduced. If the output voltage is measured by a high-impedance device, such that the current flow to the meter is negligible, the current flow through the sensor is given by

$$I_1 = E_i \frac{1}{(R_1 + \delta R) + R_2} \tag{6.30}$$

This current flow is then 44.3 mA.

The power, $P_1 = I_1^2(R_1 + \delta R)$, that must be dissipated from the sensor is 0.25 W, which, depending on the surface area and local heat transfer conditions, may cause a change in the temperature of the sensor.

COMMENT   The current flow through the sensor results in a sensor temperature higher than would occur with zero current flow, due to I^2R heating. This creates a loading error in the indicated temperature. There is a tradeoff between the increased sensitivity, dE_o/dR_1, and the correspondingly increased current associated with an increased E_1. The input voltage must be chosen appropriately for a given application.

6.5   LOADING ERRORS AND IMPEDANCE MATCHING

In an ideal sense, an instrument or measurement system should not in itself affect the variable being measured. Any such effect will alter the variable and be considered as a "loading" that the measurement system exerts on the measured variable. A *loading error* is a difference between the value of the measurand and the indicated value brought on by the act of measurement. Loading effects can be of any form: mechanical, electrical, or optical. When the insertion of a sensor into a process somehow changes the physical variable being measured, a loading error occurs. A loading error can occur anywhere along the signal path of a measurement system. If the output from one system stage is in any way affected by the subsequent stage, then the signal is affected by *interstage loading error*. Good measurement system design minimizes loading errors.

Consider trying to measure the temperature of a volume of a high-temperature liquid using a mercury-in-glass thermometer. Some finite quantity of energy must flow from the liquid to the thermometer to achieve thermal equilibrium between the thermometer and the liquid (i.e., the thermometer may cool down or heat up the liquid). As a result of this energy flow, the liquid temperature is changed, and the measured value will not correspond to the initial liquid temperature. Because the goal of the measurement was to measure the temperature of the liquid in its initial state, this measurement has introduced a loading error.

Or consider the current flow that drives the galvanometer in the Wheatstone bridge of Figure 6.13. Under deflection conditions, some energy must be removed from the circuit to deflect the pointer. This reduces the current in the circuit, bringing about a loading error in the measured resistance. Under null balance conditions, there is no pointer deflection, to within the resolution of the meter, and so a negligible amount of current is removed from the circuit. There is then negligible loading error.

In general, null balance techniques will minimize the magnitude of loading error, usually to negligible levels. Deflection methods derive energy from the process being measured. Therefore deflection methods need careful consideration to make sure the resulting loading errors are kept to an acceptable level.

Loading Errors for Voltage Dividing Circuit

Consider the voltage dividing circuit shown in Figure 6.9 for the case where R_m is finite. Under these conditions, the circuit can be represented by the equivalent circuit shown in Figure 6.15. As the sliding contact at point A moves, it divides the full-scale deflection resistance, R, into R_1 and R_2, such that $R_1 + R_2 = R_T$. The resistances R_m and R_1 form a parallel loop, yielding an equivalent resistance, R_L, given by

$$R_L = \frac{R_1 R_m}{R_1 + R_m} \tag{6.31}$$

The equivalent resistance for the entire circuit, as seen from the voltage source, is R_{eq}:

$$R_{eq} = R_2 + R_L = R_2 + \frac{R_1 R_m}{R_1 + R_m} \tag{6.32}$$

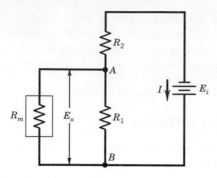

Figure 6.15 Instruments in parallel to signal path form an equivalent voltage dividing circuit.

The current flow from the voltage source is then

$$I = \frac{E_i}{R_{eq}} = \frac{E_i}{R_2 + R_1 R_m/(R_1 + R_m)} \tag{6.33}$$

and the output voltage is given by

$$E_o = E_i - IR_2 \tag{6.34}$$

This result can be expressed as

$$\frac{E_o}{E_i} = \frac{1}{1 + (R_2/R_1)(R_1/R_m + 1)} \tag{6.35}$$

The limit of this expression as R_m tends to infinity is

$$\frac{E_o}{E_i} = \frac{R_1}{R_1 + R_2} \tag{6.36}$$

which is just equation (6.8). Also, as R_2 approaches zero, the output voltage approaches the supply voltage, as expected. Expressing the value found from equation (6.8) as the true value $(E_o/E_i)'$, the loading error, e_I, may be given by

$$e_I = E_i \left[\left(\frac{E_o}{E_i} \right) - \left(\frac{E_o}{E_i} \right)' \right]$$

$$= E_i \frac{R_2 - R_2 \left[(R_1/R_m) + 1 \right]}{R_T + (R_2 R_T/R_1)\left[(R_1/R_m) + 1 \right]} \tag{6.37}$$

The loading error goes to zero as $R_m \to \infty$.

Interstage Loading Errors

Consider the common situation of Figure 6.16, in which the output voltage signal from one measurement system device provides the input to the following device. The open circuit potential, E_1, is present at the output terminal of device 1 with output impedance, Z_1. However, the output signal from device 1 provides the input to a second device, which, at its input terminals, has an input impedance, Z_m. As shown, the Thévenin equivalent circuit of device 1 consists of a voltage generator, of open circuit voltage E_1, with internal series impedance, Z_1. Connecting device 2 to the output terminals of device 1 is equivalent to placing Z_m across the Thévenin equivalent circuit. The finite impedance

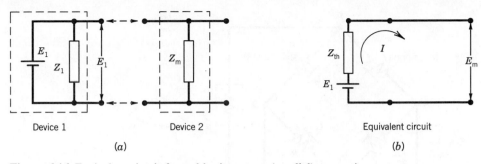

Figure 6.16 Equivalent circuit formed by interstage (parallel) connections.

of the second device causes a current, I, to flow in the loop formed by the two terminals and Z_m acts as a load on the first device. The potential sensed by device 2 will be

$$E_m = IZ_m = E_1 \frac{1}{1 + Z_1/Z_m} \tag{6.38}$$

The original potential has been changed due to the interstage connection which has caused a loading error, $e_I = E_m - E_1$,

$$e_I = E_1 \left(\frac{1}{1 + Z_1/Z_m} - 1 \right) \tag{6.39}$$

For a maximum in the voltage potential between stages, it is required that $Z_m \gg Z_1$ such that $e_I \to 0$. As a practical matter, this can be difficult to achieve at reasonable cost and some design compromise is inevitable.

When the signal is current driven, the maximum current transfer between devices 1 and 2 is desirable. Consider the circuit shown in Figure 6.17 in which device 2 is current sensitive, such as with a current-sensing galvanometer. The current through the loop indicated by the device 2 is given by

$$I_m = \frac{E_1}{Z_1 + Z_m} \tag{6.40}$$

However, if the measurement device is removed from the circuit, the current is given by the short-circuit value

$$I' = \frac{E_1}{Z_1} \tag{6.41}$$

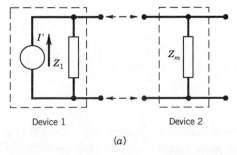
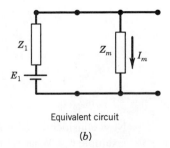

Figure 6.17 Instruments in series with signal path.

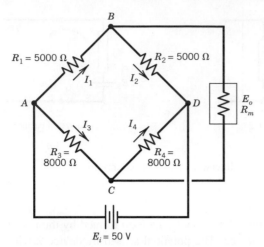

Figure 6.18 Bridge circuit for Example 6.4.

The loading error, $e_I = I_m - I'$, is

$$e_I = E_1 \frac{-Z_m}{Z_1^2 + Z_1 Z_m} \tag{6.42}$$

From equation (6.42) or equations (6.40) and (6.41), it is clear that to reduce current driven loading error then $Z_m \ll Z_1$, such that $e_1 \to 0$.

EXAMPLE 6.4

For the Wheatstone bridge shown in Figure 6.18, find the open circuit output voltage (when $R_m \to \infty$) if the four resistances change by

$$\delta R_1 = +40\ \Omega \quad \delta R_2 = -40\ \Omega \quad \delta R_3 = +40\ \Omega \quad \delta R_4 = -40\ \Omega$$

KNOWN   Measurement device resistance given by R_m

ASSUMPTIONS

$$R_m \to \infty$$

Negligible source impedance

FIND

$$E_o$$

SOLUTION   From equation (6.14)

$$E_o = E_i \left(\frac{R_1}{R_1 + R_2} - \frac{R_3}{R_3 + R_4} \right)$$

$$= 50V \left(\frac{5040\ \Omega}{5040\ \Omega + 4960\ \Omega} - \frac{8040\ \Omega}{8040\ \Omega + 7960\ \Omega} \right) = +0.075\ \text{V}$$

EXAMPLE 6.5

Now consider the case where a meter with internal impedance, $R_m = 20{,}000\ \Omega$ is used in Example 6.4. What is the output voltage with this particular meter in the circuit? Estimate the loading error.

KNOWN

$$R_m = 20,000 \, \Omega$$

FIND

$$E_o, e_I$$

SOLUTION   Because the output voltage, E_o, is equal to $I_m R_m$, equation (6.22) can be used for I_m. Then,

$$E_o = I_m R_m = \frac{E_i R_m (R_2 R_3 - R_1 R_4)}{R_3(R_1 + R_2)(R_m + R_2 + R_4) + R_1 R_2 R_4 - R_3 R_2^2 + R_m R_4 (R_1 + R_2)}$$

which for $R_m = 20,000 \, \Omega$, gives $E_o = -0.0566 \, \text{V}$.

The loading error, e_I, will be the difference between the output voltage assuming infinite impedance in the meter (Example 6.4) and the output voltage for the finite impedance value. This gives $e_I = 18.4 \, \text{mV}$.

The bridge output may also be expressed as a ratio of the output voltage with $R_m \to \infty$, E_o', and the output voltage with a finite meter resistance, E_o

$$\frac{E_o}{E_o'} = \frac{1}{1 + R_e/R_m}$$

where $R_e = \dfrac{R_1 R_2}{R_1 + R_2} + \dfrac{R_3 R_4}{R_3 + R_4}$ which yields for the present example

$$\frac{E_o}{E_o'} = \frac{1}{1 + 6500/20,000} = 0.755$$

The percent loading error, $100 \times \left(\dfrac{E_o}{E_o'} - 1 \right)$, is -24.5%.

6.6   ANALOG SIGNAL CONDITIONING: AMPLIFIERS

An amplifier is a device that scales the magnitude of an analog input signal according to the relation

$$E_o(t) = h\{E_i(t)\}$$

where $h\{E_i(t)\}$ defines a mathematical function. The simplest amplifier is the linear scaling amplifier in which

$$h\{E_i(t)\} = G E_i(t) \tag{6.43}$$

where the gain G is a constant that may be any positive or negative value. Many other types of operation are possible, including the "base x" logarithmic amplifier in which

$$h\{E_i(t)\} = G \log_x [E_i(t)] \tag{6.44}$$

Amplifiers have a finite frequency response and limited input voltage range.

The most widely used type of amplifier is the solid-state operational amplifier. This device is characterized by a high input impedance $(Z_i > 10^7 \, \Omega)$, a low output impedance $(Z_o < 100 \, \Omega)$, and a high internal gain $(A \approx 10^5 \text{ to } 10^6)$. As shown in the general diagram of Figure 6.19(a), an operational amplifier has two input ports, a noninverting and an inverting input, and one output port. The signal at the output port will be in phase with a signal passed through the noninverting input port but will be 180° out of phase with a signal passed through the inverting input port. The amplifier requires dual polarity dc

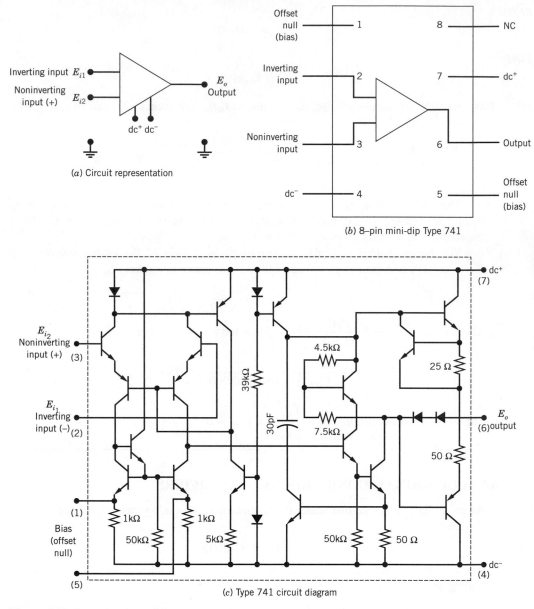

(a) Circuit representation

(b) 8–pin mini-dip Type 741

(c) Type 741 circuit diagram

Figure 6.19 Operational amplifier.

excitation power ranging from ± 5 V to ± 15 V. In addition, two dc voltage offset null (bias) input ports provide a means to zero out any output offset signal at zero input; usually a variable 10 kΩ resistor is placed across these inputs to adjust offset null.

As an example, the pin connection layout of a common operational amplifier circuit, the type 741, is shown in Figure 6.19(b) in its eight-pin, dual-in-line ceramic package form. This is the familiar rectangular black integrated circuit package seen on circuit boards. Each pin port is numbered and labeled as to function. An internal schematic diagram is shown in Figure 6.19(c) with the corresponding pin connections labeled. As shown, each input port (i.e., 2 and 3) is attached to the base of an npn transistor.

The high internal open loop gain, A, of an operational amplifier is given as

$$E_o = A[E_{i_2}(t) - E_{i_1}(t)] \tag{6.45}$$

The magnitude of A, flat at low frequencies, falls off rapidly at high frequencies, but this intrinsic gain curve is overcome by using external input and feedback resistors which will actually set the circuit gain, G, and circuit response. Some possible amplifier configurations using an operational amplifier are shown in Figure 6.20.

Because the amplifier has a very high internal gain and negligible current draw, resistors R_1 and R_2 are used to form a feedback loop and control the overall amplifier

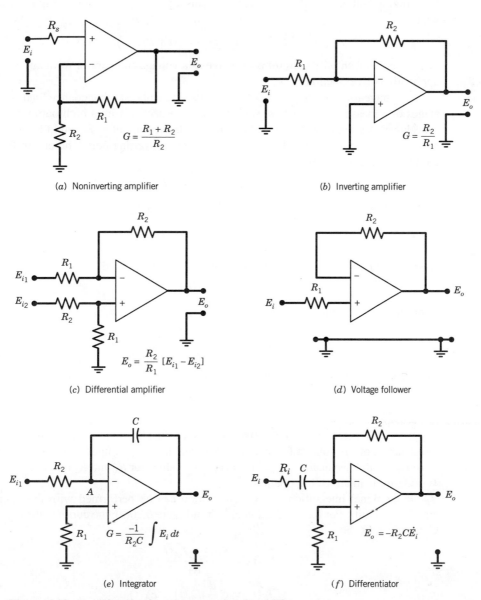

$$G = \frac{R_1 + R_2}{R_2}$$

(a) Noninverting amplifier

$$G = \frac{R_2}{R_1}$$

(b) Inverting amplifier

$$E_o = \frac{R_2}{R_1}[E_{i_1} - E_{i_2}]$$

(c) Differential amplifier

(d) Voltage follower

$$G = \frac{-1}{R_2 C}\int E_i \, dt$$

(e) Integrator

$$E_o = -R_2 C \dot{E}_i$$

(f) Differentiator

Figure 6.20 Amplifier circuits using operational amplifiers.

circuit gain, called the *closed loop gain*, G. The noninverting linear scaling amplifier circuit of Figure 6.20(a) has a closed loop gain of

$$G = \frac{E_o(t)}{E_i(t)} = \frac{R_1 + R_2}{R_2} \tag{6.46}$$

Resistor R_s does not affect the gain but is used to balance out potential variation problems at small currents. Its value is selected such that $R_s \approx R_1 R_2 / (R_1 + R_2)$. The inverting linear scaling amplifier circuit of Figure 6.20(b) provides a gain of

$$G = \frac{E_o(t)}{E_i(t)} = \frac{R_2}{R_1} \tag{6.47}$$

By utilizing both inputs, the arrangement forms a differential amplifier, Figure 6.20(c), in which

$$E_o(t) = [E_{i_2}(t) - E_{i_1}(t)](R_2/R_1) \tag{6.48}$$

The differential amplifier circuit is effective as a voltage comparator for many instrument applications (see Section 6.7).

A voltage follower circuit is commonly used to isolate an impedance load from other stages of a measurement system, such as might be needed with a high output impedance transducer. A schematic diagram of a voltage follower circuit is shown in Figure 6.20(d). Note that the feedback signal is sent directly to the inverting port. The output for such a circuit, using equation (6.45), is

$$E_o(t) = A[E_i(t) - E_o(t)]$$

or writing in terms of the circuit gain, $G = E_o(t)/E_i(t)$,

$$G = \frac{A}{1 + A} \approx 1 \tag{6.49}$$

Using Kirchhoff's law about the noninverting input loop yields

$$I_i(t)R_1 + E_o(t) = E_i(t)$$

Then the circuit input resistance, R_i, is

$$R_i = \frac{E_i(t)}{I_i(t)} = \frac{E_i(t)R_1}{E_i(t) - E_o(t)} \tag{6.50}$$

Likewise the output resistance is found to be

$$R_o = \frac{R_2}{1 + A} \tag{6.51}$$

Because A is large, equations (6.49) to (6.51) show that the input impedance, developed as a resistance, of the voltage follower can be large, its output impedance can be small, and the circuit will have near unity gain. Acceptable values for R_1 and R_2 range from 10 kΩ to 100 kΩ to maintain stable operation.

Input signal integration and differentiation can be performed with the operational amplifier circuits. For the integration circuit in Figure 6.20(e), the currents through R_2 and C are given by

$$I_{R_2}(t) = \frac{E_i(t) - E_A(t)}{R_2}$$

$$I_C(t) = C \frac{d}{dt}[E_o(t) - E_A(t)]$$

Summing currents at node A yields the integrator circuit operation

$$E_o(t) = -\frac{1}{R_2 C} \int E_i(t) dt \qquad (6.52)$$

As a special note, if the input voltage is a constant positive dc voltage, the output signal will be a negative linear ramp, $E_0 = -E_i t / RC$ with t in seconds, a feature used in ramp converters and integrating digital voltmeters (see Chapter 7.5). Integrator circuits are relatively unaffected by high-frequency noise as integration averages out noise.

Following a similar analysis, the differentiator circuit shown in Figure 6.20(f), performs the operation

$$E_o(t) = -R_2 C \frac{dE_i(t)}{dt} \qquad (6.53)$$

Differentiator circuits will amplify high-frequency noise in signals to the point of masking the true signal. Adding resistor R_i limits high-frequency gain to a -3 dB cutoff frequency of $f_c = 1/2\pi R_i C$. In noisy environments a passive RC differentiator may simply perform better.

6.7   ANALOG SIGNAL CONDITIONING: SPECIAL PURPOSE CIRCUITS

Analog Voltage Comparator

A voltage comparator provides an output that is proportional to the difference between two input voltages. As shown in Figure 6.21(a), a basic comparator consists of an operational amplifier operating in a high-gain differential mode. In this case, any difference between inputs E_{i_1} and E_{i_2} is amplified at the large open-loop gain A. Accordingly, the output from the comparator will saturate when $E_{i_1} - E_{i_2}$ is either slightly positive or negative, the actual value set by the threshold voltage E_T. The saturation output value, E_{sat}, is nearly the supply voltage E_s. For example, a 741 op-amp driven at ± 15 V might saturate at ± 13.5 V. The value for E_T can be adjusted by the amplifier bias (offset null) voltage, E_{bias}. The comparator output is given by

$$\begin{aligned} E_o &= A(E_{i_1} - E_{i_2}) &&\text{for} && |E_{i_1} - E_{i_2}| \leq E_T \\ &= +E_{sat} &&\text{for} && E_{i_1} - E_{i_2} > E_T \\ &= -E_{sat} &&\text{for} && E_{i_1} - E_{i_2} < -E_T \end{aligned} \qquad (6.54)$$

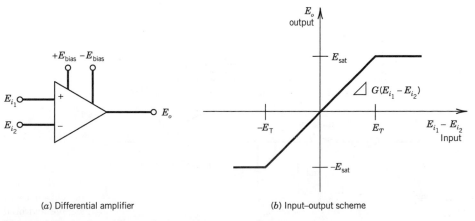

(a) Differential amplifier   (b) Input–output scheme

Figure 6.21 Analog voltage comparator.

This input/output relation is shown in Figure 6.21(*b*). For a ±15 V supply and a gain of about 200,000, the comparator might saturate with a voltage difference of only ~ 68 μV.

Often E_{i_2} will be a known reference voltage. This allows the comparator output to be used for control circuits to decide if E_{i_1} is less than or greater than E_{i_2}; e.g., a positive difference gives a positive output. One frequent use of the comparator is in an analog-to-digital converter (Chapter 7). A zener-diode connected between the +input and the output will provide a TTL compatible output signal for digital system use.

Sample and Hold Circuit

The sample and hold circuit (SHC) is used to take a narrow band measurement of a time-changing signal and to hold that measured value until reset. It is widely used in data acquisition systems using analog-to-digital converters. The circuit tracks the signal until it is triggered to sample the signal and hold it. This is illustrated in Figure 6.22(*a*) in which the track and hold logic provides the appropriate trigger.

The basic circuit for sample and hold is shown in Figure 6.22(*b*). The switch, S_1, is a fast analog device. When the switch is closed, the "hold" capacitor, C, is charged through

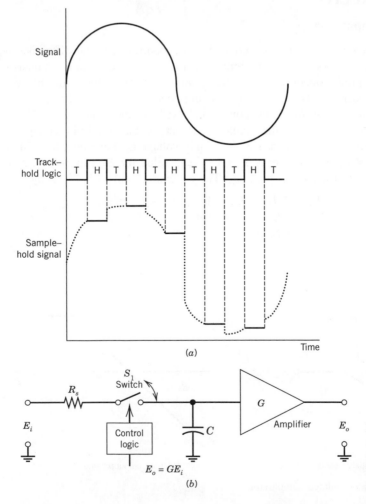

Figure 6.22 Sample and hold technique: (*a*) Original signal and sample and hold signal, (*b*) Circuit.

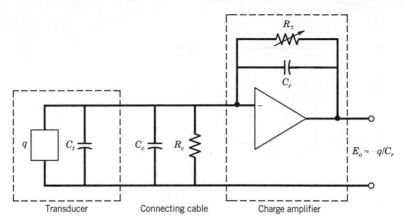

Figure 6.23 Charge amplifier circuit shown connected to a transducer.

the source resistor R_s. When the capacitor is charged, the switch is opened. The amplifier presents a very high input impedance and very low current which, together with a very low leakage capacitor, allows for a long hold time. The typical SHC is noninverting with a unit gain $(G = 1)$.

Charge Amplifier

A *charge amplifier* is used to convert a high-impedance charge, q, into an output voltage, E_o. The circuit consists of a high gain, inverting voltage operational amplifier such as shown in Figure 6.23. These circuits are commonly used with transducers which utilize piezoelectric crystals. A piezoelectric crystal will develop a time-dependent surface charge under varying mechanical load.

The circuit output voltage is determined by

$$E_o = -q/[C_r + (C_T/A)] \qquad (6.55)$$

with $C_T = C_t + C_c + C_r$ where C_t, C_c, and C_r represent the transducer, cable, and feedback capacitances, respectively, R_c is the cable and transducer resistances, and A is the amplifier open-loop gain. Because the open-loop gain of operational amplifiers is very large, equation (6.55) can be simplified to

$$E_o \approx -q/C_r \qquad (6.56)$$

A variable resistance, R_τ, and fixed feedback capacitor are used to average out low frequency fluctuations in the signal.

4–20 mA Current Loop

Low-level voltage signals below about 100 mV are quite vulnerable to noise. Examples of transducers having low-level signals include strain gauges, pressure gauges, and thermocouples. One means of transmitting such signals over long distances is by signal boosting through amplification. But amplification also boosts noise. A practical alternative method that is well suited to an industrial environment is a *4–20 mA-current loop* (read as "4 to 20"). In this approach, the low-level voltage is converted into a standard current loop signal of between 4 and 20 mA, the lower value for the minimum voltage and the higher value for the maximum voltage in the range. The 4–20 mA-current loop can be

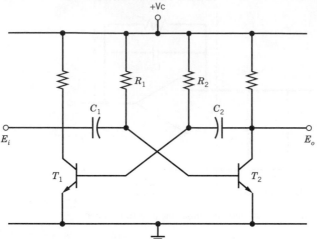

Figure 6.24 Basic multivibrator circuit.

transmitted over several hundred meters without degradation. The 4–20 mA output is a common option for measuring devices.

At the output display end, a receiver converts the current back to a low-level voltage. This can be as simple as a single resistor in parallel with the loop. For example, a 4–20 mA signal can be converted back to a 1–5 V signal by terminating the loop with a 250 Ω resistor in parallel and measuring across it.

Multivibrator and Flip-Flop Circuits

Multivibrator and *flip-flop* circuits are analog circuits that also form the basis of digital signals. They are useful for system control and triggering of events. The *astable multivibrator* is a switching circuit that toggles on and off continuously between a high- and low-voltage state in response to an applied time-dependent input voltage. It is used to generate a square waveform on prompt, where typically amplitudes will vary between 0 V (low) and 5 V (high) in a manner known as a *TTL signal*. Because the circuit continuously switches between high and low state, it is astable. The heart of this circuit, as shown in Figure 6.24, is the two transistors T_1 and T_2, which conduct alternately; the

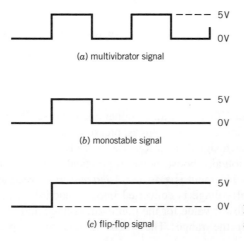

Figure 6.25 Circuit output response to an applied input signal.

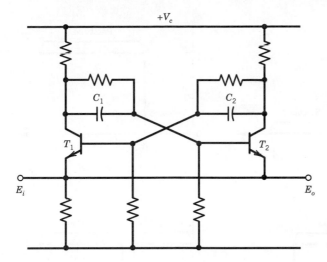

Figure 6.26 Basic flip-flop circuit.

circuit is symmetric around the transistors. This generates a square wave signal of fixed period and amplitude at the output as shown in Figure 6.25(a). The working mechanism is straightforward. The two transistors change state as the currents through capacitors C_1 and C_2 increase and decrease due to the applied input. For example, when T_2 turns on and its collector switches from V_c toward 0, the base of T_1 is driven negative which turns T_1 off. While C_2 discharges, C_1 charges. As the voltage across C_1 increases, the current across it decreases. But so long as the current through C_1 is large enough, T_2 remains on. It eventually falls to a value which turns T_2 off. This causes C_2 to charge, which turns T_1 on. The period of the resulting square wave is proportional to R_1C_1.

A useful variation of the multivibrator circuit is the *monostable*. In this arrangement T_2 stays on until a positive external pulse or change in voltage is applied to the input. At that moment, T_2 will turn off for a brief period of time providing a jump in the output voltage. The monostable circuit will cycle just once awaiting another pulse as shown in Figure 6.25(b). Hence, it is often referred to as a *one-shot*. Because it fires just once and resets itself, the monostable is an effective trigger.

Another variation of this circuit is the *flip-flop* or *bistable multivibrator* in Figure 6.26. This circuit is an effective electronic switch. Its operation is analogous to that of a light switch; it is either in an "on" (high) or an "off" (low) state. In practice, transistors T_1 and T_2 will change state every time a pulse is applied. If T_1 is on with T_2 off, a pulse will turn T_2 off, turning T_1 on, producing the output shown in Figure 6.25(c). So the flip-flop output level changes from low to high voltage or high to low voltage on command. The flip-flop is also the basic circuit of computer memory chips as it is capable of holding a single bit of information (high or low state) at any instant.

6.8   ANALOG SIGNAL CONDITIONING: FILTERS

A *filter* is used to remove undesirable frequency information from a dynamic signal. Filters can be broadly classified as being low pass, high pass, bandpass, and notch. The ideal gain characteristics of filters can be described by the magnitude ratio plots shown in Figure 6.27(a)–(d), which are described as follows. A *low-pass filter* permits frequencies

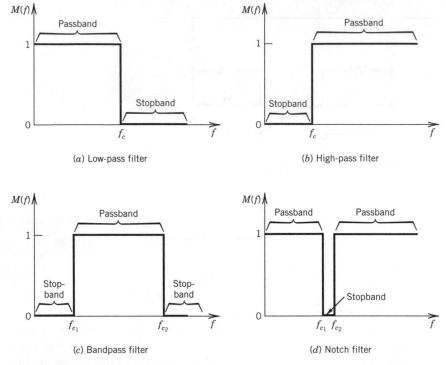

Figure 6.27 Ideal filter characteristics.

below a prescribed cut-off frequency to pass while blocking the passage of frequency information above the cutoff frequency, f_c. Similarly, a *high-pass filter* permits only frequencies above the cutoff frequency to pass. A *bandpass filter* combines features of both the low- and high-pass filters. It is described by a low cutoff frequency, f_{c_1}, and a high cutoff frequency, f_{c_2}, to define a band of frequencies that are permitted to pass through the filter. A *notch filter* permits the passage of all frequencies except those within a narrow frequency band. An intensive treatment of filters for analog and digital signals can be found in many specialized texts (e.g., [3–7]).

Filters perform a well-defined mathematical operation on the input signal as specified by their transfer function. Passive analog filter circuits consist of combinations of resistors, capacitors, and inductors. Active filters incorporate operational amplifiers into the circuit.

The sharp cutoff of the ideal filter cannot be realized in a practical filter. As an example, a plot of the magnitude ratio for a real low-pass filter is shown in Figure 6.28. All filter response curves will contain a transition band over which the magnitude ratio decreases relative to the frequency. This rate of transition is known as the filter *roll-off*, usually specified in units of dB/decade. In addition, the filter will introduce a phase shift between its input and output signal. Depending on the transfer function used, filters can be designed to achieve certain desirable response features. For example, a relatively flat magnitude ratio over its passband with a moderately steep initial roll-off and acceptable phase response is a characteristic of a *Butterworth filter* response. On the other hand, a very linear phase shift over its passband but with a relatively nonflat magnitude ratio and a fairly gradual initial roll-off is a characteristic of a *Bessel filter* response.

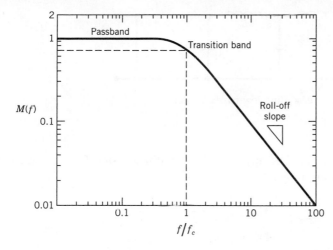

Figure 6.28 Magnitude ratio for a low-pass Butterworth filter.

The magnitude behavior of low-pass, bandpass, and high-pass filters to input frequency can be explored in the Labview® programs *Butterworth_filters_overall* and *Bessel_filters_overall*.

Butterworth Filter Design

A simple passive low-pass filter can be constructed using the resistor and capacitor (*RC*) circuit of Figure 6.29. By applying Kirchhoff's law about the input loop, we derive the model relating the input voltage, E_i, to the output voltage, E_o:

$$RC\dot{E}_o(t) + E_o(t) = E_i(t) \tag{6.57}$$

This real filter model is a first-order system. The magnitude ratio for this filter when subjected to a sinusoidal waveform input has already been given by equation (3.10) with $\tau = RC$ and $\omega = 2\pi f$ and is described by Figure 6.28. The roll-off slope is 20 dB/decade. The phase shift is given in equation (3.9).

A filter is designed around its cutoff frequency, f_c, defined as the frequency at which the signal power is reduced by one-half. This is equivalent to the magnitude ratio being reduced to 0.707. In terms of the decibel (dB),

$$dB = 20 \log M(f) \tag{3.11}$$

f_c occurs at -3 dB. For the filter of Figure 6.29, this requires that $\tau = RC = 1/2\pi f_c$.

The roll-off slope of a filter can be improved by staging filters in series, called cascading filters. This is done by adding additional reactive elements, such as inductors or resistors, to the circuit as shown in Figure 6.30. A *k*th-order or *k*-stage filter contains *k*

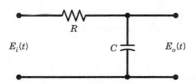

Figure 6.29 Low-pass *RC* Butterworth filter circuit.

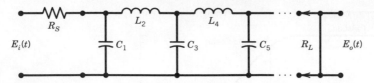

Figure 6.30 Multistage or cascading low-pass *LC* filters.

reactive elements in stages. The magnitude ratio and phase shift for a *k*-stage low-pass Butterworth filter are given by

$$M(f) = \frac{1}{\left[1 + (f/f_c)^{2k}\right]^{1/2}} \tag{6.58a}$$

$$\Phi(f) = \sum_{i=1}^{k} \Phi_i(k) \tag{6.58b}$$

The attenuation at any frequency can be estimated from the dynamic error, $\delta(f)$, or directly in decibels by

$$\text{Attenuation(dB)} = 10 \log\left[1 + (f/f_c)^{2k}\right] \tag{6.59}$$

The general magnitude response curve for a *k*-stage *RC* Butterworth filter is shown in Figure 6.31. The roll-off slope is $20 \times k$ [dB/decade]. Equation (6.59) is useful for specifying the required order of a filter based on magnitude response needs.

Table 6.1 lists the required values for elements C_i and L_i for a two- through five-stage Butterworth filter of the form of Figure 6.30 [3, 6, 7]. The values in Table 6.1 assume that

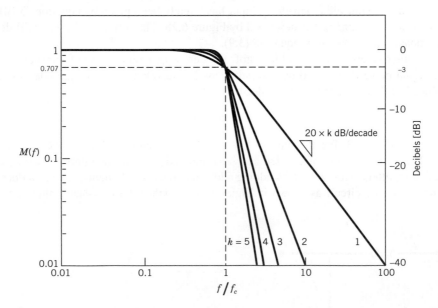

Figure 6.31 Magnitude characteristics for Butterworth low-pass filters of various stages.

Table 6.1 Values for Low-Pass LC Butterworth Filter[1]

k	C_1	L_2	C_3	L_4	C_5
2	1.414	1.414			
3	1.000	2.000	1.000		
4	0.765	1.848	1.848	0.765	
5	0.618	1.618	2.000	1.618	0.618

[1]Values for C_i in farads L_i in henrys are referenced to $R_s = R_L = 1\,\Omega$ and $\omega_c = 1$ rad/s. See discussion for proper scaling. For $k = 1$, use $C_1 = 1$ or $L_1 = 1$.

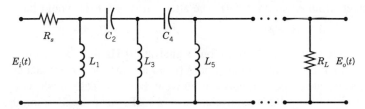

Figure 6.32 Multistage high-pass LC filter.

$R_s = R_L = 1\,\Omega$ and $\omega_c = 2\pi f_c = 1$ rad/s. For other values, L_i and C_i are scaled by

$$L = L_i R_s / 2\pi f_c \qquad (6.60a)$$
$$C = C_i / (R_s 2\pi f_c) \qquad (6.60b)$$

Butterworth filters are quite common and can be found in consumer products, such as audio equipment.

A Butterworth high-pass filter is shown in Figure 6.32. In comparing with Figure 6.30, we see that the capacitors and inductors have been swapped. To estimate the high-pass filter values for Figure 6.32, let $(L_i)_{\mathrm{HP}} = (1/C_i)_{\mathrm{LP}}$ and $(C_i)_{\mathrm{HP}} = (1/L_i)_{\mathrm{LP}}$ where subscript LP refers to the low-pass values from Table 6.1 and HP refers to the high-pass filter values to be used with Figure 6.32. The scaling equations of (6.60a) and (6.60b) and the phase shift properties described by equation (6.58b) apply. However, the magnitude ratio for a high-pass filter is given by

$$M(f) = \frac{f/f_c}{\left[1 + (f_c/f)^{2k}\right]^{1/2}} \qquad (6.61)$$

EXAMPLE 6.6

Design a one-stage Butterworth RC low-pass filter with a cutoff frequency of -3 dB set at 100 Hz. Calculate the dynamic error and attenuation at 60 and 200 Hz.

KNOWN

$f_c = 100$ Hz $k = 1$

$M(100\text{ Hz}) = 0.707$

FIND R, C, and δ

SOLUTION   A single-stage Butterworth RC filter circuit is shown in Figure 6.30. Using $\omega = 2\pi f$, the magnitude ratio for this circuit is given by

$$M(f) = \frac{1}{\left[1 + (2\pi f \tau)^2\right]^{1/2}}$$

Setting $M(f) = 0.707 = -3$ dB with $f = f_c = 100$ Hz yields, from equation (6.58a),

$$\tau = 1/2\pi f_c = RC = 0.0016 \, \text{s}$$

One possible combination might be $R = 800 \, \Omega$ and $C = 2 \, \mu\text{F}$.

The attenuation is found directly from the dynamic error, $\delta(f) = M(f) - 1$ or directly in decibels from equation (6.59). The attenuation at 60 Hz would be

$$\delta(60) = M(60 \, \text{Hz}) - 1 = 0.86 - 1 = -0.14$$

or -1.33 dB. So the amplitude of the input signal at 60 Hz will be down 14% after passing through this filter. Ideally, there would be no attenuation for frequencies below the cutoff frequency and complete attenuation above it. But practically this is not possible. At 200 Hz, $\delta(200) = -0.55$, equivalent to -46 dB. So over 55% of the signal amplitude at 200 Hz has been eliminated.

COMMENT   A two-stage Butterworth filter would decrease the undesirable signal attenuation seen at 60 Hz to $\delta(60) = -0.06$ or 6% while providing an even better signal attenuation at 200 Hz to $\delta(200) = -0.76$ or 76%. In general, higher order Butterworth filters give better response in the passband and sharper attenuation in the stopband, but with a penalty of increased phase shift.

Bessel Filter Design

A Bessel filter sacrifices a flat gain over its passband with a gradual initial rolloff in exchange for a very linear phase shift. A k-stage low-pass Bessel filter has the transfer function

$$G(s) = \frac{a_o}{a_o + a_1 s + \cdots + a_k s^k} \tag{6.62}$$

For design purposes, this can be rewritten as

$$G(s) = \frac{a_o}{D_k(s)} \tag{6.63}$$

where $D_k(s) = (2k - 1)D_{k-1}(s) + s^2 D_{k-2}(s)$, $D_o(s) = 1$ and $D_1(s) = s + 1$.

A k-stage LC low-pass Bessel filter can be based on the filter shown in Figure 6.30. Table 6.2 lists the required corresponding values for elements L_i and C_i for a two- through five-stage Bessel filter [6, 7]. The values in Table 6.2 assume that $R_s = R_L = 1 \, \Omega$ and

Table 6.2   Values for Low-Pass LC Bessel Filters[1]

k	C_1	L_2	C_3	L_4	C_5
2	1.577	0.423			
3	1.255	0.553	0.192		
4	1.060	0.512	0.318	0.110	
5	0.930	0.458	0.331	0.209	0.072

[1]Values for C_i in farads and L_i in henrys are referenced to $R_s = R_L = 1 \, \Omega$ and $\omega_c = 1$ rad/s. See discussion for proper scaling.

$\omega_c = 2\pi f_c = 1$ rad/s. For other values, L_i and C_i are found from equations (6.60a) and (6.60b).

Active Filters

An active filter uses the high-frequency gain characteristics of the operational amplifier to form an effective analog filter. A low-pass active filter is shown in Figure 6.33(a) using the type 741 operational amplifier. This is a first-order, single-stage, low-pass Butterworth filter. It has a low-pass cut-off frequency given by

$$f_c = \frac{1}{2\pi R_2 C_2} \tag{6.64}$$

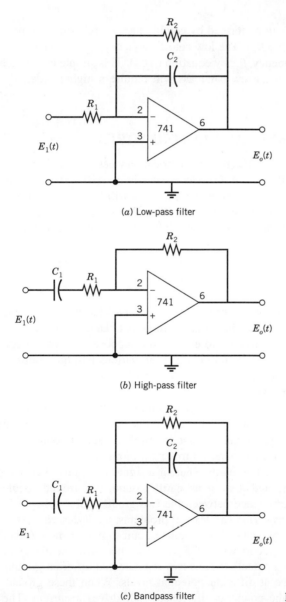

(a) Low-pass filter

(b) High-pass filter

(c) Bandpass filter

Figure 6.33 Basic active filters.

The static sensitivity (or gain) of the filter is given by $K = R_2/R_1$. The filter retains the Butterworth characteristics described by equation (6.58).

A first-order, single-stage, high-pass Butterworth active filter is shown in Figure 6.33(b), using the 741 operational amplifier. The filter has a high-pass cut-off frequency of

$$f_c = \frac{1}{2\pi R_1 C_1}$$ (6.65)

with static sensitivity (or gain) of $K = R_2/R_1$. The magnitude ratio is given by

$$M(f) = \frac{(f/f_c)}{\left[1 + (f/f_c)^2\right]^{1/2}}$$ (6.66)

An active bandpass filter can be formed by combining the high- and low-pass filters above and is shown in Figure 6.33(c). The low cutoff frequency, f_{c_1}, is given by equation (6.65) and the high cutoff frequency, f_{c_2}, by equation (6.64). This simple circuit is limited in the width of its bandpass. A narrow bandpass will require a higher order filter than shown.

6.9   GROUNDS, SHIELDING, AND CONNECTING WIRES

The type of connecting wires used between electrical devices can have a significant impact on the noise level of the signal. Low-level signals of <100 mV are particularly susceptible to errors induced by noise. Some simple rules will help to keep noise levels low: (1) keep the connecting wires as short as possible; (2) keep signal wires away from noise sources; (3) use a wire shield and proper ground; and (4) twist wire pairs along their lengths.

Ground and Ground Loops

The voltage at the end of a wire that is connected to a rod driven far into the ground would likely be at the same voltage level as the earth—a reference datum called zero or *earth ground*. A *ground* is simply a return path to earth. Now suppose that wire is connected from the rod through an electrical box and then through various building routings to the ground plug of an outlet. Would the ground potential at the outlet still be at zero? The answer is probably not. The network of wires that form the return path to earth would likely act as antennae and pick up some voltage potential relative to earth ground. Any instrument grounded at the outlet would be referenced back to this voltage potential, not to earth ground. Thus, an electrical ground is not an absolute value. Ground values vary between ground points because the ground returns pass through different equipment or building wiring on their return path to earth. Thus, if a signal is grounded at two points, say at a power source ground and then at an earth ground, the grounds could be at different voltage levels. The difference between the two ground point voltages is called the *common-mode voltage (cmv)*. This can lead to problems, as discussed next.

Ground loops are caused by connecting a signal circuit to two or more grounds that are at different potentials. A ground wire of finite resistance will usually carry some current, and so it will develop a potential. Thus two separate and different grounds, even two in close proximity, can be at different potential levels. When these ground points are connected into a circuit the potential difference itself drives a current. The result

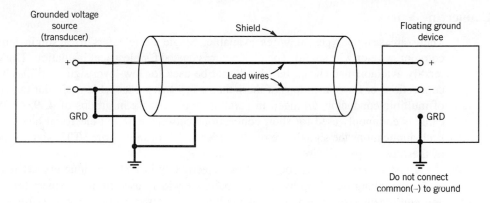

Figure 6.34 Signal grounding arrangements: grounded source signal with shield; floating signal at measuring device.

is a ground loop—an electrical interference superimposed onto the signal. A ground loop can manifest itself in various forms, such as a sinusoidal signal or simply a voltage offset.

Ensure that a system, including its sensors and all electrical components, has only one ground point. Figure 6.34 shows a proper connection between a grounded voltage source, such as a transducer, and a measuring device. Note that the common ($-$) line at the measuring device is not grounded. As such, it is referred to as being isolated or as a floating ground device. Incidentally, many devices are grounded through their AC power lines by means of a third prong. This can set up a ground loop relative to other circuit grounds. To create a floating ground, it is necessary to break this ground connection using a three- to two-prong adapter and ground the device separately.

Shields

Long wires act as antennas and will pick up stray signals from nearby electrical fields. The most common problem is 60 Hz AC line noise. Electrical shields are effective against such noise. A *shield* is a piece of metal foil or wire braid wrapped around the signal wires and connected to ground. The shield intercepts external electrical fields, returning them to ground. A shield ground loop is prevented by grounding the shield to only one point, usually the signal ground at the transducer. Figure 6.34 shows such a shield-to-ground arrangement for a grounded voltage source connected to a floating ground measuring device.

A common source of electrical fields is an ac power supply transformer. A capacitance coupling between the 60- or 50-Hz power supply wires and the signal wires is set up. For example, a 1-pF capacitance will superimpose a 40-mV interference on a 1-mV signal.

A different source of noise is from a magnetic field. When a lead wire moves within a magnetic field, a voltage is induced in the signal path. The same effect occurs with a lead wire in proximity to an operating motor. The best prevention is to separate the signal lead wires from such sources. Also, twisting the lead wires together tends to cancel any induced voltage, as the currents through the two wires are in opposite directions; the more twists per meter, the better. A final recourse is a magnetic shield made from a material having a high ferromagnetic permeability.

Connecting Wires

There are several types of wires available. Single cable refers to a single length of common wire or wire strand. The conductor is coated for electrical insulation. The wire is readily available and cheap, but should not be used for low-level signal (millivolt level) connections or for connections of more than a few wires. Flat cable is similar but consists of multiple conductors arranged in parallel strips, usually in groups of 4, 9, or 25. Flat cable is commonly used for short connections between adjacent electrical boards, but in such applications the signals are on the order of 1 V or more or are TTL signals. Neither of these two types of wires offers any shielding.

Twisted pairs of wires consist of two electrically insulated conductors twisted about each other along their lengths. Twisted pairs are widely used to interconnect transducers and equipment. The intertwining of the wires offers some immunity to noise with the effectiveness increasing with the number of twists per meter. Cables containing several twisted pairs are available; e.g., CAT5 network cable (such as internet cable) consists of four twisted pairs of wires. Shielded twisted pairs wrap the twisted pairs within a metallic foil shield. CAT6 network cable consists of four twisted pairs with a shield separating its wire pairs. Shielded twisted pairs are one of the best choices for applications requiring low-level signals or high data transfer rates.

Coaxial cable consists of an electrically insulated inner single conductor surrounded within an outer conductor made of stranded wire and a metal foil shield. In general, current flows in one direction along the inner wire and the opposite direction along the outer wire. Any electromagnetic fields generated will cancel. Coaxial cable is often the wire of choice for low-level high-frequency signals. Signals can be sent over very long distances with little loss. A variation of coaxial cable is triaxial cable, which contains two inner conductors. It is used in applications as with twisted pairs but offers superior noise immunity.

Optical cable is widely used to transmit low-level signals over long distances. This cable may contain one or more fiber-optic ribbons within a polystyrene shell. A transmitter converts the low-level voltage signal into infrared light. The light is transmitted through the cable to a receiver, which converts it back to a low-level voltage signal. The cable is virtually noise-free from magnetic fields and harsh environments and is incredibly light and compact compared to its metal wire counterpart.

6.10 SUMMARY

This chapter has focused on classic but basic analog electrical measurement devices. Devices that are sensitive to current, resistance, and voltage were presented. Signal conditioning devices that can modify or condition a signal include amplifiers, current loops, and filters, and these were also presented. Because all of these devices are widely used, often in combination and often masked within more complicated circuits, the reader should strive to become familiar with the workings of each.

Loading errors, which are due to the act of measurement or due to the presence of a sensor, can only be minimized by proper choice of sensor and technique. Loading errors that occur between the connecting stages of a measurement system can be minimized by proper impedance matching. A discussion of such impedance matching has been presented, including using the versatile voltage follower circuit for this purpose.

REFERENCES

1. Hickman, I., *Oscilloscopes*, 5th ed., Newnes, 2000.
2. Bleuler, E., and R. O. Haxby, *Methods of Experimental Physics*, 2d ed., Vol. 2, Academic, New York, 1975.
3. Lacanette, K., *A Basic Introduction to Filters: Active, Passive, and Switched-Capacitor Application Note 779*, National Semiconductor, Santa Clara, CA, 1991.
4. Stanley, W. D., G. R. Dougherty, and R. Dougherty, *Digital Signal Processing*, 2d ed., Reston (a Prentice Hall company), Reston, VA, 1984.
5. DeFatta, D. J., J. Lucas, and W. Hodgkiss, *Digital Signal Processing*, Wiley, New York, 1988.
6. Lam, H. Y., Analog and Digital Filters: Design and Realization, Prentice-Hall, Englewood Cliffs, 1979.
7. Niewizdomski, S., Filter Handbook – A Practical Design Guide, CRC Press, Boca Raton, 1989.

NOMENCLATURE

c	damping coefficient	G	amplifier closed-loop gain
e	error	$G(s)$	transfer function
f	frequency $[t^{-1}]$	I	electric current $[C\text{-}t^{-1}]$
f_c	filter cutoff frequency $[t^{-1}]$	I_e	effective current $[C\text{-}t^{-1}]$
f_m	maximum analog signal frequency $[t^{-1}]$	I_g	galvanometer or current
$h\{E_i(t)\}$	function		device current $[C\text{-}t^{-1}]$
k	cascaded filter stage order	K	static sensitivity
$\hat{k}$	unit vector aligned with current flow	L	inductance [H]
l	length $[l]$	$M(f)$	magnitude ratio at frequency, f
$\hat{n}$	unit vector normal to current loop	N	number of turns in a current-carrying
q	charge [C]		loop
r	radius, vector $[l]$	R	resistance $[\Omega]$
t	time $[t]$	R_B	effective bridge resistance $[\Omega]$
u_R	uncertainty in resistance R	R_g	galvanometer or detector resistance $[\Omega]$
A	cross-sectional area of a current carrying loop; operational amplifier open-loop gain	R_{eq}	equivalent resistance $[\Omega]$
		R_m	meter resistance $[\Omega]$
		δR	change in resistance $[\Omega]$
E	voltage [V]	$\mathbf{T}_\mu$	torque on a current carrying loop in a magnetic field $[m\text{-}l^2\text{-}t^{-2}]$
E_1	open circuit potential [V]		
E_i	input voltage [V]	α	angle
E_m	indicated output voltage [V]	τ	time constant $[t]$
E_o	output voltage [V]	$\Phi(f)$	phase shift at frequency, f
F	force $[m\text{-}l\text{-}t^{-2}]$	θ	angle

PROBLEMS

6.1 Determine the maximum torque on a current loop having 20 turns, a cross-sectional area of 1 in.^{2}, and experiencing a current of 20 mA. The magnetic field strength is 0.4 Wb/m^2.

6.2 A 5000-Ω voltage-dividing circuit is used in a particular application using the arrangement of Figure 6.15. What is the minimum allowable meter internal resistance such that the loading error in measuring the open circuit potential, E_o, will be less than 12% of the full-scale value? Note: a family of solutions will result.

6.3 Determine the loading error as a percentage of the output for the voltage dividing circuit of Figure 6.15, if $R_T = R_1 + R_2$ and $R_1 = kR_T$. The parameters of the circuit are

$$R_T = 500\,\Omega \quad E_i = 10\,\text{V} \quad R_m = 10\,000\,\Omega \quad k = 0.5$$

What would be the loading error if expressed as a percentage of the full-scale output? Show that the two answers for the loading error are equal when expressed in volts.

6.4 Consider the Wheatstone bridge shown in Figure 6.13. Suppose

$$R_3 = R_4 = 200\,\Omega$$
$$R_2 = \text{variable calibrated resistor}$$
$$R_1 = \text{transducer resistance} = 40x + 100$$

  a. When $x = 0$, what is the value of R_2 required to balance the bridge?

  b. If the bridge is operated in a balanced condition in order to measure x, determine the relationship between R_2 and x.

6.5 For the Wheatstone bridge shown in Figure 6.13, R_1 is a sensor whose resistance is related to a measured variable x by the equation $R_1 = 20x^2$. If $R_3 = R_4 = 100\,\Omega$ and the bridge is balanced when $R_2 = 46\,\Omega$, determine x.

6.6 A force sensor has as its output a change in resistance. The sensor forms one leg (R_1) of a basic Wheatstone bridge. The sensor resistance with no force load is 500 Ω, and its static sensitivity is 0.5 Ω/N. Each arm of the bridge is initially 500 Ω.

  a. Determine the bridge output for applied loads of 100, 200, and 350 N. The bridge is operated as a deflection bridge, with an input voltage of 10 V.

  b. Determine the current flow through the sensor.

  c. Repeat parts (a) and (b) with $R_m = 10\,k\Omega$ and $R_s = 600\,\Omega$.

6.7 A reactance bridge arrangement replaces the resistor in a Wheatstone bridge with a capacitor or inductor. Such a reactance bridge is then excited by an ac voltage. Consider the bridge arrangement shown in Figure 6.35. Show that the balance equations for this bridge are given by $C_2 = C_1 R_1/R_2$.

6.8 Circuits containing inductance- and capacitance-type elements exhibit varying impedance depending on the frequency of the input voltage. Consider the bridge circuit of Figure 6.36. For a capacitor and an inductor connected in a series arrangement, the impedance is a function of frequency, such that a minimum impedance occurs at the resonance frequency $f = 1/2\pi\sqrt{LC}$ where f is the frequency [Hz], L is the inductance [H], and C

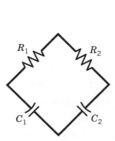

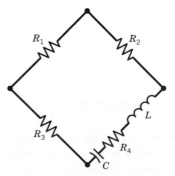

Figure 6.35 Bridge circuit for Problem 6.7.

Figure 6.36 Bridge circuit for Problem 6.8.

is the capacitance [F]. Design a bridge circuit that could be used to calibrate a frequency source at 500 Hz.

6.9 A Wheatstone bridge initially has resistances equal to $R_1 = 200\,\Omega$, $R_2 = 400\,\Omega$, $R_3 = 500\,\Omega$, and $R_4 = 600\,\Omega$. For an input voltage of 5 V, determine the output voltage at this condition. If R_1 changes to 250 Ω, what is the bridge output?

6.10 Construct a plot of the voltage output of a Wheatstone bridge having all resistances initially equal to 500 Ω, with a voltage input of 10 V for the following cases:

a. R_1 changes over the range 500–1000 Ω.

b. R_1 and R_2 change equally, but in opposite directions, over the range 500–600 Ω.

c. R_1 and R_3 change equally over the range 500–600 Ω.

Discuss the possible implications of these plots for using bridge circuits for measurements with single and multiple transducers connected as arms of the bridge.

6.11 Consider the simple potentiometer circuit shown in Figure 6.10 with reference to Figure 6.9. Perform a design-stage uncertainty analysis to determine the minimum uncertainty in measuring a voltage. The following information is available concerning the circuit (assume 95% confidence):

$$E_i = 10 \pm 0.1\,\text{V} \quad R_T = 100 \pm 1\,\Omega \quad R_g = 100\,\Omega \quad R_x = \text{reading} \pm 2\%$$

where R_g is the internal resistance of a galvanometer. The null condition of the galvanometer may be assumed to have negligible error. The uncertainty associated with R_x is associated with the reading obtained from the location of the sliding contact. Estimate the uncertainty in the measured value of voltage at nominal values of 2 and 8 V.

Problems 6.12 through 6.14 relate to the comparison of two sinusoidal input signals using a dual trace oscilloscope. A schematic diagram of the measurement system is shown in Figure 6.37, where one of the signals is assumed to be a reference standard signal, having a known frequency and amplitude. The oscilloscope trace for these inputs is called a Lissajous diagram. The problems can also be worked in the laboratory.

6.12 Develop the characteristic shape of the Lissajous diagrams for two sinusoidal inputs having the same amplitude, but with the following phase relationships.

a. in phase

b. $\pm 90°$ out of phase

c. 180° out of phase

6.13 Show that the phase angle for two sinusoidal signals can be determined from the relationship

$$\sin \Phi = y_i / y_a$$

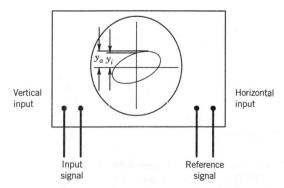

Vertical input

Horizontal input

Input signal

Reference signal

Figure 6.37 Dual-trace oscilloscope for measuring signal characteristics by using Lissajous diagrams.

where the values of y_i and y_a are illustrated in Figure 6.37, and represent the vertical distance to the y intercept and the maximum y value, respectively.

6.14 Draw a schematic diagram of a measurement system to determine the phase lag resulting from an electronic circuit. Your schematic diagram should include a reference signal, the electronic circuit, and the dual trace oscilloscope. Discuss the expected result and how a quantitative estimate of the phase lag can be determined.

6.15 Construct Lissajous diagrams for sinusoidal inputs to a dual trace oscilloscope having the following horizontal signal to vertical signal frequency ratios: (a) 1:1, (b) 1:2, (c) 2:1, (d) 1:3, (e) 2:3, (f) 5:2
These plots can be easily developed using spreadsheet software and plotting a sufficient number of cycles of each of the input signals.

6.16 A single-stage *RC* filter with $f_c = 100$ Hz is used to filter an analog signal. Determine the attenuation of the filtered analog signal at 10, 50, 75, and 200 Hz.

6.17 A three-stage *LC* Bessel filter with $f_c = 100$ Hz is used to filter an analog signal. Determine the attenuation of the filtered analog signal at 10, 50, 75, and 200 Hz.

6.18 Design a cascading *RC* low-pass filter that has a magnitude ratio flat to within 3 dB from 0 to 5 kHz but with an attenuation of at least 30 dB for all frequencies at and above 10 kHz.

6.19 Consider the circuit of Figure 6.38, which consists of an ideal current source I_n and a parallel resistor, R_n. Find its Thevenin equivalents, E_{th} and R_{th}.

6.20 An electrical displacement transducer has an output impedance of 500 Ω. Its voltage is input to a voltage measuring device that has an input impedance of 100 kΩ. Estimate the ratio of true voltage to the voltage measured by the measuring device.

6.21 Consider a transducer connected to a voltage measuring device. Plot the ratio E_m/E_1 as the ratio of output impedance of the transducer to input impedance of the measuring device varies from 1:1 to 1:1,000,000 (look at each order of magnitude). Comment on the effect that input impedance has on measuring a voltage.

6.22 The pH meter of Figure 6.39 consists of a transducer of glass containing paired electrodes. This transducer can produce voltage potentials up to 1 V with an internal

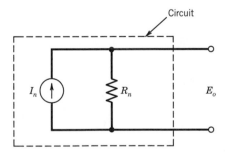

Figure 6.38 Circuit for Problem 6.19.

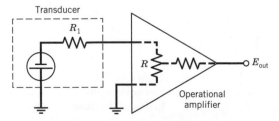

Figure 6.39 pH transducer circuit for Problem 6.22.

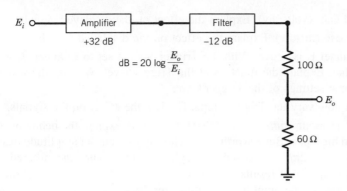

$$dB = 20 \log \frac{E_o}{E_i}$$

Figure 6.40 Multistage signal path for Problem 6.23.

output impedance of up to 10^9 Ω. If the signal is to be conditioned, estimate the minimum input impedance required to keep loading error below 0.1% for the op-amp shown.

6.23 Consider the circuit of Figure 6.40, which consists of an amplifier providing 32-dB gain, followed by a filter, which causes an attenuation of 12 dB at the frequency of interest, followed by a voltage divider. Find E_o/E_i across the 60 Ω resistor.

6.24 The temperature sensor and bridge circuit described in Example 6.2 is used to measure a temperature; the sensor output indicates a temperature of 30°C. It is desired to measure this temperature within ± 1°C. Assume that the errors in temperature measurement result from the two sources:

1. The galvanometer has limited sensitivity, allowing some current flow when indicating a balanced bridge condition. Take $R_g = 5\,\Omega$.

2. The ratio R_3/R_4, and the value of R_2 must be known. Uncertainties in these values contribute to the uncertainty in the value of R_1.

Specify a combination of galvanometer current at balanced conditions, and uncertainty in the bridge resistances to yield the required uncertainty in measured temperature.

6.25 The input to a subwoofer loudspeaker is to pass through a passive high-pass Butterworth filter having a cutoff frequency of 100 Hz. Specify a filter that meets the following specifications: At 50 Hz, the signal magnitude ratio should be at least at 0.95 and at 200 Hz, the magnitude ratio should be no more than 0.01. The sensor and load resistances are 10 Ω. You will need to specify the number of stages and the values for the components.

6.26 The input to a high-frequency loudspeaker is to pass through a low-pass Butterworth filter with cutoff frequency of 5000 Hz. Specify a filter that meets the following specifications: At 8000 Hz, the signal attenuation should be no more than -0.45 dB, and at 2500 Hz, the attenuation should be at least -20 dB. The source and load resistances are 10 Ω. You will need to specify the number of stages and the values for the components.

6.27 Design an active-RC low-pass first-order Butterworth filter for a cutoff frequency of 10 kHz, and a passband gain of 20. Use a 741 op amp and 0.1-µF capacitors for your design. Program Lowpass Butterworth Filter can be used.

6.28 Design an active-RC first-order high-pass filter for a cutoff frequency of 10 kHz and passband gain of 10. Use a 741 op-amp.

6.29 Use the Labview® program *Oscilloscope* to explore the workings of an actual oscilloscope. Vary between Channel A or B or both.

a. Characterize each signal by its waveform (i.e., triangle, square, sine). Determine the amplitude and the period of both signals.

b. Vary the signal gain (volts/div). Explain the effect on the displayed signals.

c. Vary the time base (ms/div). Explain the effect on the displayed signals.

d. Make sure Channel B is active. With the Trigger Source set to Channel B, vary the Trigger level dial. Explain the function of the Trigger level. Why do the waveforms disappear at some settings of the Trigger?

e. Repeat (c), but now vary the Trigger slope. Explain the effect on the signals.

6.30 Use the Labview$^{\circledR}$ program *Butterworth_Filters_Overall* to explore the behavior of low-pass, bandpass, and high-pass Butterworth filters chosen for their flat amplitude passband. The signal $y(t) = 2 \sin 2\pi ft$ is passed through the filter and the filtered signal $y^*(t) = B \sin[2\pi ft + \phi(f)]$ that results is shown. The single-tone results show the effect at the input frequency, f. The amplitude spectrum results show the effect over the full-frequency band of interest. Describe the amplitude behavior of the filtered signal as a single input frequency is slowly increased from 1 to 1000 Hz. Pay particular attention near the filter cutoff frequencies.

6.31 Use the Labview$^{\circledR}$ program *Bessel_Filters_Overall* to explore the behavior of low-pass, bandpass, and high-pass Bessel filters. Using the information from Problem 6.30, describe the amplitude behavior of the filtered signal as a single-input frequency is slowly increased from 1 to 1000 Hz.

6.32 The program *Filtering_Noise* demonstrates the effect of using a low-pass Butterworth filter to treat a signal containing high-frequency noise. Set the signal frequency at 10 Hz. Set the cutoff frequency at 25 Hz. Discuss the behavior of the filter as the number of stages is increased. Then vary cutoff frequency and discuss the behavior. You should be able to tie your discussion back to filter roll-off. What happens when the cutoff frequency is less than the signal frequency? Why?

6.33 Program *Monostable* provides a simulation of a monostable integrated circuit based on the type 555 op amp. The trigger controls the simulation. The user can vary the 'on' time and values for R and C. Discuss the output vs. time results for a simulation. Vary R and C to create a 1s square wave.

6.34 Program 741 *Op amp* simulates an op amp gain characteristics. Determine if it is noninverting or inverting and explain. For an input voltage of 1V, find a resistor combination for a gain of 5. What is this op amp's maximum output? Repeat for a gain of 0.5.

Chapter 7

Sampling, Digital Devices, and Data Acquisition

7.1 INTRODUCTION

Integrating analog electrical transducers with digital devices is cost effective and commonplace. Digital microprocessors are central to most controllers and data-acquisitions systems today. Plus, there are many advantages to a hybrid arrangement, including the efficient handling and rapid processing of large amounts of data and varying degrees of artificial intelligence. But there are fundamental differences between analog and digital systems, which impose some limitations and liabilities upon the engineer. As pointed out in Chapter 2, the most important difference is that analog signals are continuous in both amplitude and time whereas digital signals are discrete (noncontinuous) in both amplitude and time. It is not immediately obvious how a digital signal can be used to represent the continuous behavior of a process variable.

This chapter begins with an introduction to the fundamentals of sampling, the process by which continuous signals are made discrete. The major pitfalls are explored. Criteria are presented that circumvent the loss or the misinterpretation of signal information while undergoing the sampling process. We show how a discrete series of data can actually contain all of the information available in a continuous signal, or at least provide a very good approximation.

The discussion moves on to the devices most often involved in analog and digital systems. Analog devices interface with digital devices through an analog-to-digital converter. The reverse process of a digital device interfacing with an analog device occurs through a digital-to-analog converter. A digital device interfaces with another digital device through a digital I/O port. These interfaces are the major components of computer-based data-acquisition systems. Necessary components and the basic layout of these systems are introduced, and standard methods for communication between digital devices are presented.

7.2 SAMPLING CONCEPTS

Consider an analog signal and its discrete time series representation in Figure 7.1. The information contained in the analog and discrete representations may appear to be quite different. However, the important analog signal information concerning amplitude and frequency can be well represented by such a discrete series. Just how well represented will depend on the frequency content of the analog signal, the size of the time increment between each discrete number, and the total sample period of the measurement.

In Chapter 2, we discussed how a continuous dynamic signal could be represented by a Fourier series. The discrete Fourier transform (DFT) was also introduced as a method

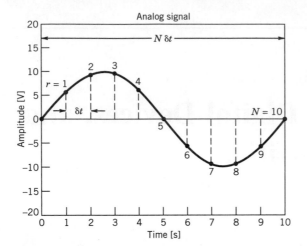

Figure 7.1 Analog and discrete representations of a time-varying signal.

for reconstructing a dynamic signal from a discrete set of data. The DFT, through equation (2.40), conveys all the information needed to reconstruct the Fourier series of a continuous dynamic signal from a representative discrete time series. Hence, Fourier analysis will provide certain guidelines for sampling continuous data. In this section, the concept of converting a continuous analog signal into an equivalent discrete time series is examined. Extended discussions on this subject of signal analysis can be found in many texts (e.g., [1–3]).

Sample Rate

Just how frequently must a time-dependent signal be measured to determine its frequency content? Consider Figure 7.2(a) in which the magnitude variation of a 10-Hz sine wave is plotted versus time over time period, t_f. Suppose this sine wave is measured repeatedly at successive sample time increments, δt. This corresponds to measuring the signal with a sample frequency or sample rate (in Hz) of

$$f_s = 1/\delta t \tag{7.1}$$

For this discussion, we will assume that the signal measurement occurs at a constant sample rate. For each measurement, the amplitude of the sine wave is converted into a number. For comparison, in Figures 7.2(b)–(d) the resulting series versus time plots are given when the signal is measured using sample time increments (or the equivalent sample rates) of (b) 0.010 s ($f_s = 100$ Hz), (c) 0.037 s ($f_s = 27$ Hz), and (d) 0.083 s ($f_s = 12$ Hz). We can see that the sample rate has a significant effect on our perception and reconstruction of the continuous analog signal in the time domain. As sample rate decreases, the amount of information per unit time describing the signal decreases. In Figures 7.2(b) and (c), we can still discern the 10-Hz frequency content of the original signal. But we see in Figure 7.2(d) that an interesting phenomenon occurs if the sample rate becomes too slow: The sine wave appears to be of a lower frequency.

We can conclude that the sample time increment or the corresponding sample rate plays a significant role in signal frequency representation. The *sampling theorem* states that to reconstruct the frequency content of a measured signal accurately, *the sample rate must be more than twice the highest frequency contained in the measured signal.*

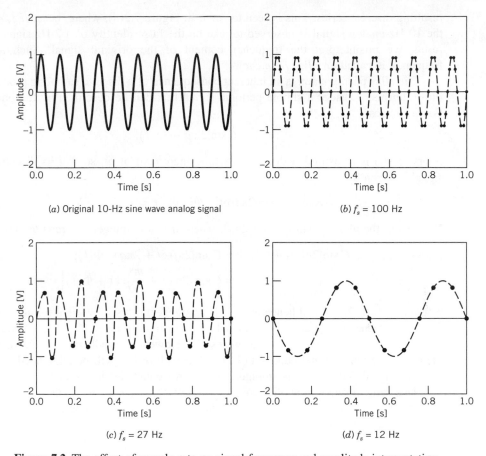

Figure 7.2 The effect of sample rate on signal frequency and amplitude interpretation.

Denoting the maximum frequency in the analog signal as f_m, the sampling theorem requires

$$f_s > 2f_m \tag{7.2}$$

or, equivalently, in terms of sample time increment,

$$\delta t < \frac{1}{2f_m} \tag{7.3}$$

When signal frequency content is important, equations (7.2) and (7.3) provide a criterion for the minimum sample rate or maximum sample time increment, respectively, to be used in converting data from a continuous to a discrete form. The frequencies that will be extracted from the DFT of the resulting discrete series will provide an accurate representation of the original signal frequencies regardless of the sample rate used, provided that the requirements of the sampling theorem are satisfied.

Alias Frequencies

When a signal is sampled at a sample rate that is less than $2f_m$, the higher frequency content of the analog signal will take on the false identity of a lower frequency in the resulting discrete series. This is seen to occur in Figure 7.2(d), where, because $f_s < 2f_m$, the 10-Hz analog signal is observed to take on the false identity of a 2-Hz signal. As a result, we misinterpret the frequency content of the original signal! Such a false frequency is called an *alias frequency*.

The alias phenomenon is an inherent consequence of a discrete sampling process. To illustrate this, consider a simple periodic signal that can be described by the one-term Fourier series:

$$y(t) = C\sin[2\pi ft + \Phi(f)]$$

Suppose $y(t)$ is measured with a sample time increment of δt, so that its discrete time signal is given by

$$y(r\delta t) = C\sin[2\pi fr\delta t + \Phi(f)] \quad r = 1, 2, \ldots, N \tag{7.4}$$

Now using the identity, $\sin x = \sin(2\pi q)$, where q is any integer, we rewrite $y(r\delta t)$ as

$$
\begin{aligned}
C\sin[2\pi fr\delta t + \Phi(f)] &= C\sin[2\pi fr\delta t + 2\pi q + \Phi(f)] \\
&= C\sin\left[2\pi\left(f + \frac{m}{\delta t}\right)r\delta t + \Phi(f)\right]
\end{aligned}
\tag{7.5}
$$

where $m = 0, 1, 2, \ldots$ (and hence, $mr = q$ is an integer). This shows that for any value of δt, the frequencies of f and $f + m/\delta t$ will be indistinguishable. Hence, all frequencies given by $f + m/\delta t$ are the alias frequencies of f. However, by adherence to the sampling theorem criterion of either equation (7.2) or (7.3), all $m \geq 1$ will be eliminated from the sampled signal and, thus, this ambiguity between frequencies is avoided.

This same discussion applies equally to complex periodic, aperiodic, and nondeterministic waveforms. This is shown by examining the general Fourier series used to represent such signals. Such a discrete series,

$$y(r\delta t) = \sum_{n=1}^{\infty} C_n \sin[2\pi nfr\delta t + \Phi_n(f)] \tag{7.6}$$

for $r = 1, 2, \ldots, N$, can be rewritten as

$$y(r\delta t) = \sum_{n=1}^{\infty} C_n \sin\left[2\pi\left(nf + \frac{nm}{\delta t}\right)r\delta t + \Phi_n(f)\right] \tag{7.7}$$

which displays the same aliasing phenomenon shown in equation (7.5).

In general, the Nyquist frequency defined by

$$f_N = \frac{f_s}{2} = \frac{1}{2\delta t} \tag{7.8}$$

represents a folding point for the aliasing phenomenon. All frequency content in the analog signal that is at frequencies above f_N will appear as alias frequencies of less than f_N in the sampled signal. That is, such frequencies will be folded back and superimposed on the signal as lower frequencies. The aliasing phenomenon occurs in spatial sampling [i.e., $y(x)$ at sampling intervals δx] as well, and the above discussion can be applied equally.

The alias frequency, f_a, can be predicted from the folding diagram of Figure 7.3. Here the original input frequency axis is folded back over itself at the folding point of f_N and again for each of its harmonics, mf_N, where $m = 1, 2, \ldots$. For example and as noted by the solid arrows in Figure 7.3, the frequencies $f = 0.5\,f_N,\ 1.5\,f_N,\ 2.5\,f_N, \ldots$ that may be

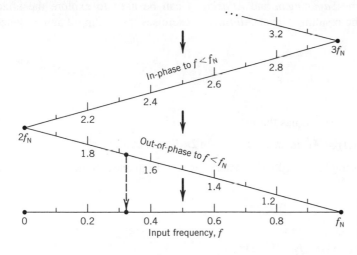

Figure 7.3 The folding diagram for alias frequencies.

present in the original input signal will all appear as the frequency $0.5\,f_N$ in the discrete series $y(r\delta t)$. Use of the folding diagram is illustrated further in Example 7.1.

How does one avoid this alias phenomenon when sampling a signal of unknown frequency content? Use a low-pass filter set at f_N to remove signal content at and above f_N prior to sampling and use an appropriate sample rate for the signal chosen based on equation (7.2). This type of filter is called an anti-aliasing filter. Or choose a sample rate that is so large that the measured signal will not have data frequency content above f_N.

EXAMPLE 7.1

A 10-Hz sine wave is sampled at 12 Hz. Compute the maximum frequency that can be represented in the resulting discrete signal. Compute the alias frequency.

KNOWN

$$f = 10\,\text{Hz}$$
$$f_s = 12\,\text{Hz}$$

ASSUMPTION   Constant sample rate

FIND f_N and f_a

SOLUTION   The Nyquist frequency, f_N, sets the maximum frequency that can be represented in a resulting data set. Using equation (7.8) with $f_s = 12$ Hz gives the Nyquist frequency, $f_N = 6$ Hz. This is the maximum frequency that can be sampled correctly. So the measured frequency will be in error.

All frequency content in the analog signal above f_N will appear as an alias frequency, f_a, between 0 and f_N. Because $f/f_N \approx 1.67$, f will be folded back and appear in the sampled signal as a frequency between 0 and f_N. Reading Figure 7.3, $f = 1.67 f_N$ is located directly above $0.33 f_N$. So f will be folded back to appear as $f_a = 0.33 f_N$ or 2 Hz. This folding is indicated by the dashed arrow in Figure 7.3. As a consequence, the 10-Hz sine wave sampled at 12 Hz will be completely indistinguishable from a 2-Hz sine wave sampled at 12 Hz in the discrete time series. The 10-Hz signal will take on identity of a 2-Hz signal—that's aliasing.

COMMENT   The programs *Sampling.m* and *Aliasing.vi* can be used to explore the effects of sample rate on the resulting signal. *Aliasing.vi* exercises the folding diagram concept.

EXAMPLE 7.2

A complex periodic signal has the form

$$y(t) = A_1 \sin 2\pi(25)t + A_2 \sin 2\pi(75)t + A_3 \sin 2\pi(125)t$$

If the signal is sampled at 100 Hz, determine the frequency content of the resulting discrete series.

KNOWN

$$f_s = 100\,\text{Hz}$$
$$f_1 = 25\,\text{Hz}   f_2 = 75\,\text{Hz}   f_3 = 125\,\text{Hz}$$

ASSUMPTION   Constant sample rate

FIND   The alias frequencies f_{a_1}, f_{a_2}, and f_{a_3}, and discrete set $\{y(r\delta t)\}$

SOLUTION   From equation (7.8) for a sample rate of 100 Hz, the Nyquist frequency is 50 Hz. All frequency content in the analog signal above f_N will take on an alias frequency between 0 and f_N in the sampled data series. With $f_1 = 0.5\,f_N$, $f_2 = 1.5\,f_N$, and $f_3 = 2.5\,f_N$, we can use Figure 7.3 to determine the respective alias frequencies.

i	f_i	f_{a_i}
1	25 Hz	25 Hz
2	75 Hz	25 Hz
3	125 Hz	25 Hz

So in the resulting series, $f_{a_1} = f_{a_2} = f_{a_3} = 25$ Hz. Because of the aliasing phenomenon, the 75- and 125-Hz components would be completely indistinguishable from the 25-Hz component. The discrete series would be described by

$$y(r\delta t) = (B_1 + B_2 + B_3)\sin 2\pi(25)r\delta t   r = 0, 1, 2, \ldots$$

where $B_1 = A_1$, $B_2 = -A_2$, and $B_3 = A_3$. Note from Figure 7.3 that when the original frequency component is out of phase with the alias frequency, the corresponding amplitude is negative. Clearly, this series misinterprets the original analog signal in both its frequency and amplitude content.

COMMENT   Example 7.2 illustrates the potential for misrepresenting a signal through improper sampling. For signals for which the frequency content is not known prior to their measurement, sample frequencies should be increased according to the following scheme:

- Sample the input signal at increasing sample rates using a fixed total sample time, and examine the time plots for each signal. Look for changes in the shape of the waveform (e.g., see Figure 7.2).

- Compute the amplitude spectrum for each signal and compare the resulting frequency content.

Always use (anti-aliasing) analog filters set at the Nyquist frequency.

Amplitude Ambiguity

For simple and complex periodic waveforms, the discrete Fourier transform of the sampled discrete time signal will remain unaltered by a change in the total sample period, $N\delta t$, provided that (1) the total sample period remains an integer multiple of the fundamental period, T_1, of the measured continuous waveform, i.e., $mT_1 = N\delta t$ where m is an integer, and (2) the sample time increment meets the sampling theorem criterion. As a consequence, the amplitudes associated with each frequency in the periodic signal, the spectral amplitudes, will be accurately represented by the DFT. This means that an original periodic waveform can be completely reconstructed from a discrete time series regardless of the sample time increment used. The total sample period will define the frequency resolution of the DFT:

$$\delta f = \frac{1}{N\,\delta t} = \frac{f_s}{N} \tag{7.9}$$

The frequency resolution plays a crucial role in the reconstruction of the signal amplitudes, as noted below.

An important difficulty arises when $N\delta t$ is not coincident with an integer multiple of the fundamental period of $y(t)$: The resulting DFT cannot *exactly* represent the spectral amplitudes of the sampled continuous waveform. The problem is brought on by the truncation of one complete cycle of the signal (see Figure 7.4), and from spectral resolution, because the associated fundamental frequency and its harmonics will not be coincident with a center frequency of the DFT. However, this error will decrease either as the value of $N\delta t$ more closely approximates an exact integer multiple of T_1 or as f_s becomes very large relative to f_m.

This situation is illustrated in Figure 7.4, which compares the amplitude spectrum resulting from sampling the signal, $y(t) = 10\cos 628t$ over different sample periods. Two sample periods of 0.0256 and 0.1024 s were used with $\delta t = 0.1$ ms, and a third sample period of 0.08 s with $\delta t = 0.3125$ ms.[1] These sample periods provide for a DFT frequency resolution, δf, of about 39, 9.8, and 12.5 Hz, respectively.

The two spectra shown in Figures 7.4(a)–(b) display a spike near the correct 100 Hz with surrounding noise spikes, known as leakage, at adjacent frequencies. Note that the original signal $y(t)$ cannot be exactly reconstructed from either of these spectra. The spectrum in Figure 7.4(c) has been constructed using a longer sample time increment (i.e., lower sample rate) than that of the spectra of Figures 7.4(a) and (b), and with fewer data points than the spectrum of Figure 7.4(b). Yet the spectrum of Figure 7.4(c) provides an exact representation of $y(t)$. The N and δt combination used has reduced leakage to zero, which maximizes the amplitude at 100 Hz. Its sample period corresponds to exactly (eight) periods of $y(t)$, and the frequency of 100 Hz corresponds to the center frequency of the eighth frequency interval of the DFT. As seen in Figure 7.4, the loss of accuracy in a DFT representation occurs in the form of amplitude "leakage" to adjacent frequencies. To the DFT, the truncated segment of the sampled signal appears as an aperiodic signal. The DFT returns the correct spectral amplitudes for both the periodic and aperiodic signal portions. But as a result, the spectral amplitudes for any truncated segment will be superimposed onto portions of the spectrum adjacent to the correct frequency. Recall how the amplitude varied in Figure 7.2(c). This is how leakage affects the time domain

---

[1]Most DFT (including FFT) algorithms require that $N = 2^M$, where M is an integer. This affects the selection of δt.

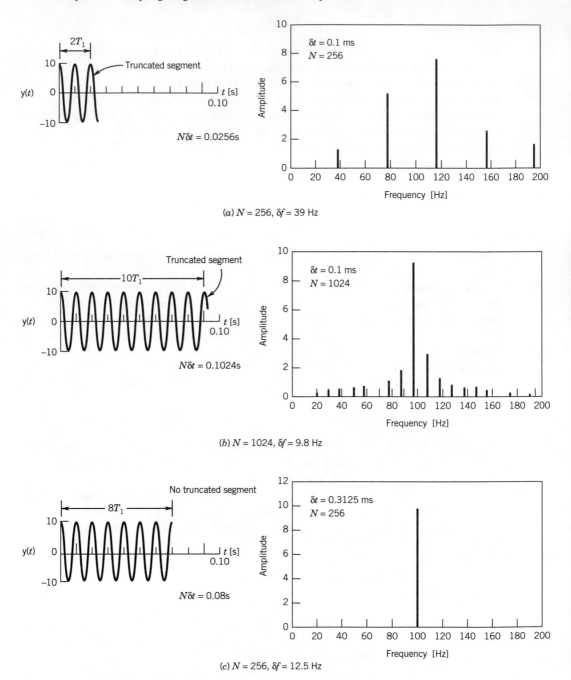

Figure 7.4 Amplitude spectra for $y(t)$. (a) $\delta f = 39$ Hz. (b) $\delta f = 9.8$ Hz. (c) $\delta f = 12.5$ Hz.

reconstruction. By varying the sample period or its equivalent, the DFT resolution, one can minimize leakage and control the accuracy of the spectral amplitudes.

If $y(t)$ is an aperiodic or nondeterministic waveform, there may not be a fundamental period. In such situations, one controls the accuracy of the spectral amplitudes by varying the DFT resolution, δf, to minimize leakage. As δf tends toward zero, leakage will decrease.

In summary, *the reconstruction of a measured waveform from a discrete signal is controlled by the sampling rate and the DFT resolution.* By adherence to the sampling theorem, one controls the frequency content of both the measured signal and the resulting spectrum. By variation of δf, one can control the accuracy of the spectral amplitude representation.

Programs *Sampling.m* and *Signal_C7.vi* explore the concept of sample rate and amplitude ambiguity. Program *Leakage.vi* allows the user to vary sample rate and number of points; it follows the discussion surrounding Figure 7.4. Program *DataSpec.m* explores sampling concepts on user-generated data series.

Selecting Sample Rate and Data Number

For an exact discrete representation in both frequency and amplitude of any periodic, analog waveform, both the number of data points and the sample rate should be chosen based on the preceding discussion using the criteria of equations (7.2) and (7.9). Equation (7.2) sets the maximum value for δt, or equivalently, the minimum sample rate f_s, and equation (7.9) sets the total sampling time, $N\delta t$, from which the data number, N is estimated.

In practice, exact discrete representations of the input analog signal frequency and amplitude content are rarely possible. As a general rule, setting, $f_s \geq 5\, f_m$ together with long sample periods, $N\delta t$, relative to the signal period will minimize spectral leakage and provide a good approximation of the original signal. An anti-alias filter should be used to ensure that no frequency above a desired maximum frequency is encountered. Really large data sets are cumbersome regardless of the computational power and storage available. So a little planning will pay off.

A good reconstruction of the original time-based signal without using DFT is possible through

$$y^*(t) = \frac{1}{\pi} \sum_{r=1}^{r=N} y(r\delta t) \frac{\sin[\pi(t/\delta t) - r]}{t/\delta t - r} \tag{7.10}$$

where $y^*(t)$ is the signal reconstructed from the data sequence $\{y(r\delta t)\}$. The equation improves as the value of N increases.

7.3   DIGITAL DEVICES: BITS AND WORDS

Digital signals are discrete in both time and amplitude. Digital systems use some variation of a binary numbering system both to represent and to transmit signal information. Binary systems use the binary digit or *bit* as the smallest unit of information. A bit is a single digit, either a 1 or a 0. Bits are like electrical switches and are used to convey both logical and numerical information. From a logic standpoint, the 1 or 0 are represented by the on or off switch settings. By appropriate action a bit can be reset to either on or off positions, thereby permitting control and logic actions. By combining bits it is possible to define integer numbers greater than a 1 or 0. A numerical *word* is an ordered sequence of bits with a *byte* being a specific sequence of 8 bits. Computer memory is usually byte addressed. The memory location where numerical information is stored is known as a *register* with each register assigned its own *address*.

A combination of M bits can be arranged to represent 2^M different words. For example, a combination of 2 bits can represent 2^2 or four possible combinations of bit arrangements: 00, 01, 10, or 11. We can alternatively reset this 2-bit word to produce these four different arrangements that represent the decimal, that is, base 10, integer numbers 0, 1, 2, or 3, respectively. So an 8-bit word can represent the numbers 0 through 255; a 16-bit word can represent 0 through 65,535.

The numerical value for a word is computed by moving by bit from right to left. From the right side, each successive bit will increase the value of the word by a unit, a 2, a 4, an 8, and so forth through the progression of 2^M, provided that the bit is in its on (value of 1) position; otherwise, the particular bit increases the value of the word by zero. A weighting scheme of an M-bit word is given as:

$$
\begin{array}{cccccc}
\text{Bit } M-1 & \ldots & \text{Bit 3} & \text{Bit 2} & \text{Bit 1} & \text{Bit 0} \\
2^{M-1} & \ldots & 2^3 & 2^2 & 2^1 & 2^0 \\
2^{M-1} & \ldots & 8 & 4 & 2 & 1
\end{array}
$$

Using this scheme, bit $M-1$ is known as the most significant bit (MSB) because its contribution to the numerical level of the word is the largest relative to the other bits. Bit 0 is known as the least significant bit (LSB). In hexadecimal (base 16), a sequence of 4 bits is used to represent a single digit, so that the 2^4 different digits (0 through 9 plus A through F) form the alphabet for creating a word.

Several binary codes are in common use. In 4-bit straight binary, 0000 is equivalent to analog zero and 1111 is equivalent to decimal 15 or hexadecimal F. Straight binary is considered to be a unipolar code because all numbers are of like sign. Bipolar codes allow for sign changes. *Offset binary* is a bipolar code with 0000 equal to -7 and 1111 equal to $+7$. The comparison between these two codes and several others is shown in Table 7.1. In offset binary there are two zero levels, true zero being centered between them. Note how the MSB is used as a sign bit in a bipolar code.

The *ones-* and *twos-complement binary* codes shown in Table 7.1 are both bipolar. They differ in the positioning of the zero with the twos-complement code containing one more negative level (-8) than positive ($+7$). Twos-complement code is widely used in

Table 7.1 Binary Codes (Example: 4-Bit Words)

Bits	(Hex) Straight	Offset	Twos Complement	Ones Complement	AVPS
0000	0	-7	$+0$	$+0$	-0
0001	1	-6	$+1$	$+1$	-1
0010	2	-5	$+2$	$+2$	-2
0011	3	-4	$+3$	$+3$	-3
0100	4	-3	$+4$	$+4$	-4
0101	5	-2	$+5$	$+5$	-5
0110	6	-1	$+6$	$+6$	-6
0111	7	-0	$+7$	$+7$	-7
1000	8	$+0$	-8	-7	$+0$
1001	9	$+1$	-7	-6	$+1$
1010	A	$+2$	-6	-5	$+2$
1011	B	$+3$	-5	-4	$+3$
1100	C	$+4$	-4	-3	$+4$
1101	D	$+5$	-3	-2	$+5$
1110	E	$+6$	-2	-1	$+6$
1111	F	$+7$	-1	-0	$+7$

digital computers. The bipolar absolute-value-plus-sign code (AVPS) uses the MSB as the sign bit but uses the lower bits in a straight binary representation; for example, 1100 and 0100 represent −4 and +4, respectively.

A code often used for digital decimal readouts and for communication between digital instruments is binary coded decimal or BCD. In this code, each individual digit of a decimal number is represented separately by its equivalent value coded in straight binary. For example, the three digits of the base 10 number 532_{10} are represented by the BCD number 0101 0011 0010 (i.e., binary 5, binary 3, binary 2). In this code each binary number need span but a single decade, that is, have values from 0 to 9. Hence, in digital readouts, when the limits of the lowest decade are exceeded, that is, 9 goes to 10, the code just carries into the next higher decade.

EXAMPLE 7.3

A 4-bit register contains the binary word 0101. Convert this to its decimal equivalent assuming a straight binary code.

KNOWN   4-bit register

FIND   Convert to base 10

ASSUMPTION   Straight binary code

SOLUTION   The content of the 4-bit register is the binary word 0101. This will represent

$$0 \times 2^3 + 1 \times 2^2 + 0 \times 2^1 + 1 \times 2^0 = 5$$

The equivalent of 0101 in decimal is 5.

7.4   TRANSMITTING DIGITAL NUMBERS: HIGH AND LOW SIGNALS

Electrical devices transmit binary code by using differing voltage levels. Because a bit can have a value of a 1 or 0, the presence of a particular voltage (call it HIGH) at a junction could be used to represent a 1-bit value whereas a different voltage (LOW) could represent a 0. Most simply, this voltage can be affected by use of an open or closed switch using a flip-flop (Chapter 6), such as depicted in Figure 7.5(*a*).

For example, a signal method common to many digital devices is a $+5$ V form of TTL (true transistor logic), which uses a nominal $+5$ V HIGH/0 V LOW scheme for representing the value of a bit. To avoid any ambiguity due to voltage fluctuations and provide for higher definition, a voltage level between $+2$ and $+5.5$ V is taken as HIGH and therefore a bit value of 1, whereas a voltage between -0.6 and $+0.8$ V is taken as LOW or a 0-bit value. But this scheme is not unique.

Binary numbers can be formed through a combination of HIGH and LOW voltages. Several switches can be grouped in parallel to form a register. Such an M-bit register forms a number based on the value of the voltages at its M output lines. This is illustrated in Figure 7.5(*b*) for a 4-bit register forming the number 1010. So the bit values can be changed simply by changing the opened or closed state of the switches that connect to the output lines. This defines a parallel form in which all of the bits needed to form a word are available simultaneously. But another form is serial, where the bits are separated in a time sequence of HIGH/LOW pulses, each pulse lasting for only one predefined time

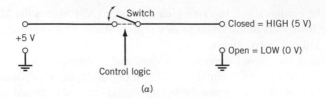

(a)

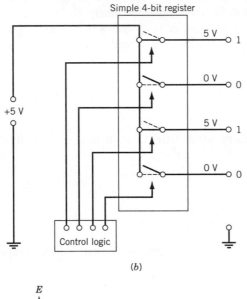

(b)

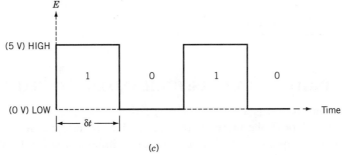

(c)

Figure 7.5 Methods for transmitting digital information. (a) Simple on/off (1 or 0) switch. (b) 4-bit register transmitting 1010 in parallel. (c) Serial transmission of 1010.

duration, δt. This is illustrated in the pulse sequence of Figure 7.5(*c*), which also forms the number 1010.

Communication between components internal to a digital device is usually in a bit parallel, but word serial form. That is, groups of bits are transmitted one group at a time. Another approach is to communicate in a fully serial manner, 1 bit at a time. Communication is discussed in detail later in this chapter.

7.5   VOLTAGE MEASUREMENTS

Digital measurement devices will consist of several components that interface the digital device with the analog world. In particular, the digital-to-analog converter and

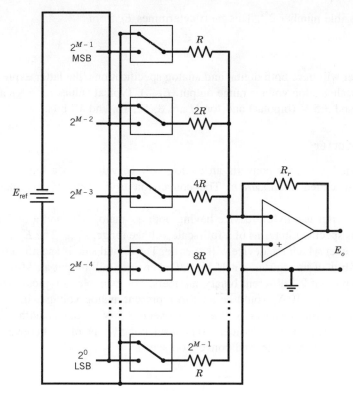

Figure 7.6 Digital-to-analog converter.

the analog-to-digital converter are discussed next. These form the major components of a digital voltmeter and a analog-to-digital/digital-to-analog data acquisition system.

Digital-to-Analog Converter

A digital-to-analog converter is an M-bit digital device that converts a digital binary word into an analog voltage (e.g., [4–7]). One possible scheme uses an M-bit register with a weighted resistor network and operational amplifier, as depicted in Figure 7.6. The network consists of M binary weighted resistors having a common summing point. The resistor associated with the register MSB will have a value of R. At each successive bit the resistor value is doubled, so that the resistor associated with the LSB will have a value of $(2^{M-1} R)$. The network output is a current, I, given by

$$I = E_{\text{ref}} \sum_{m=1}^{M} \frac{c_m}{2^{m-1}R} \tag{7.11}$$

The values for c_m are either 0 or 1 depending on the associated mth bit value of the register that controls the switch setting. The output voltage from the amplifier is

$$E_o = IR_r \tag{7.12}$$

Note that this is equivalent to an operation in which an M-bit D/A converter would compare the magnitude of the actual input binary number, X, contained within its register to the largest possible number 2^M. This ratio determines E_o from

$$E_o = \frac{X}{2^M}$$

(7.13)

The D/A converter will have both digital and analog specifications, the latter expressed in terms of its full-scale analog voltage range output (E_{FSR}). Typical values for E_{FSR} are 0 to 10 V (unipolar) and ± 5 V (bipolar) and for M are 8, 12, 16, and 18 bits.

Analog-to-Digital Converter

An analog-to-digital converter converts an analog voltage value into a binary number through a process called *quantization*. The conversion is discrete, taking place one number at a time.

The A/D converter is a hybrid device having both an analog side and a digital side. The analog side is specified in terms of a full-scale voltage range, E_{FSR}. The E_{FSR} defines the voltage range over which the device will operate. The digital side is specified in terms of the number of bits of its register. An M-bit A/D converter will output an M-bit binary number. It can represent 2^M different binary numbers. For example, a typical 8-bit A/D converter with an $E_{FSR} = 10$ V would be able to represent analog voltages in the range between 0 and 10 V (unipolar) or the range between ± 5 V (bipolar) with $2^8 = 256$ different binary values. Principal considerations in selecting a type of A/D converter will include: resolution, voltage range, and conversion speed.

Resolution

The A/D converter resolution is defined in terms of the smallest voltage increment that will cause a bit change. Resolution is specified in volts and determined by

$$Q = E_{FSR}/2^M$$

(7.14)

Primary sources of error intrinsic to any A/D converter are the resolution and associated quantization error, saturation error; and conversion error. Each will be discussed in order.

Quantization Error

An A/D converter will have a finite resolution, and any input voltage that falls between two adjacent output codes will result in an error. This error is referred to as the *quantization error*. It behaves as noise imposed on the digital signal.

The analog input to digital output relationship for a 0–4 V, 2-bit A/D converter is represented in Figure 7.7. From equation (7.14), the resolution of this device is $Q = 1$ V. For such a converter, an input voltage of 0.0 V would result in the same binary output of 00 as would an input of 0.9 V. Yet an input of 1.1 V results in an output of 01. Clearly, the output error is directly related to the resolution. In this encoding scheme, e_Q is bounded between 0 and 1 LSB above E_i, so we estimate $e_Q = Q$. This scheme is common in digital readout devices.

A second common scheme makes the quantization error symmetric about the input voltage. For this, the analog voltage is shifted internally within the A/D converter by a bias voltage, E_{bias}, of an amount equivalent to ½ LSB. This shift is transparent to the user. The effect of such a shift is shown on the second lower axis in Figure 7.7. This now makes e_Q bounded by $\pm 1/2$ LSB about E_i, that is, $e_Q = \pm 1/2Q$. This scheme is common in data-acquisition systems. Regardless of the scheme used, the span of the quantization error remains 1 LSB, and its effect is significant at small voltages.

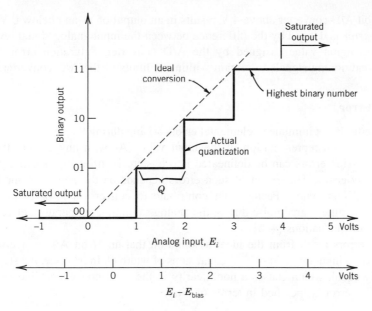

Figure 7.7 Binary quantization and saturation.

A/D converter resolution is sometimes specified in terms of *signal-to-noise ratio (SNR)*. The SNR relates the power of the signal, given by Ohm's law as E^2/R, to the power that can be resolved by quantization, given by $E^2/R2^M$. The SNR is just the ratio of these values. Defined in terms of the decibel [dB], this gives

$$SNR[dB] = 20 \log 2^M \qquad (7.15)$$

The effect of bit number on resolution and SNR is detailed in Table 7.2.

Saturation Error

The analog range of an A/D converter limits the minimum and maximum analog voltage that it can convert. If either limit is exceeded, the A/D converter output saturates and will not change with a subsequent increase in input level. As noted in Figure 7.7, an input to

Table 7.2   Conversion Resolution

Bits M	Q^a [V]	SNR [dB]
2	2.5	12
4	0.625	24
8	0.039	48
12	0.0024	72
16	$0.15 \ (10^{-3})$	96
18	$0.0381 \ (10^{-3})$	108

^{a}Assumes $E_{FSR} = 10$ V.

the 0–4-V, 2-bit A/D converter above 4 V results in an output of 11 and below 0 V of 00. A *saturation error* is defined by the difference between the input analog signal level and the equivalent digital value assigned by the A/D converter. Saturation error can be avoided by conditioning signals to remain within the limits of the A/D converter.

Conversion Error

An A/D converter is not immune to elemental errors arising during the conversion process that can lead to a misrepresentation of the input value. As with any device, the A/D converter conversion errors can be delineated into hysteresis, linearity, sensitivity, zero, and repeatability errors. The extent of such errors depends on the particular method of analog-to-digital conversion. Factors that contribute to conversion error include A/D converter settling time, signal noise during the analog sampling, temperature effects, and excitation power fluctuations [4–6].

Linearity errors result from the ideal assumption that an M-bit A/D converter will resolve the analog input range into 2^{M-1} equal steps of width Q. In practice, the steps may not be exactly equal, which causes a nonlinearity in the ideal conversion line drawn in Figure 7.7. This error is specified in terms of bits.

EXAMPLE 7.4

Compute the relative effect of quantization error (e_Q/E_i) in the quantization of a 100-mV and a 1-V analog signal using an 8-bit and a 12-bit A/D converter, both having a full-scale range of 0–10 V.

KNOWN

$E_i = 100\,\text{mV}$ and $1\,\text{V}$

$M = 8$ and 12

$E_{FSR} = 0–10\,\text{V}$

FIND e_Q/E_i, where e_Q is the quantization error.

SOLUTION   The resolutions for the 8-bit and 12-bit converters can be estimated from equation (7.14) as

$$Q_8 = \frac{E_{FSR}}{2^8} = \frac{10}{256} = 39\,\text{mV}$$

$$Q_{12} = \frac{E_{FSR}}{2^{12}} = \frac{10}{4096} = 2.4\,\text{mV}$$

Assume the A/D converter is designed so that the absolute quantization error is given by $\pm\,\tfrac{1}{2}Q$. The relative effect of the quantization error can be computed by e_Q/Ei. The results are tabulated as follows:

E_i	M	e_Q	$100 \times e_Q/E_i$
100 mV	8	±19.5 mV	19.5%
100 mV	12	±1.2 mV	1.2%
1 V	8	±19.5 mV	1.95%
1 V	12	±1.2 mV	0.12%

From a relative point of view, the error in converting a voltage is much greater at lower voltage levels.

EXAMPLE 7.5

The A/D converter with the specifications listed below is to be used in an environment in which the A/D converter temperature may change by $\pm 10°C$. Estimate the contributions of conversion and quantization errors to the uncertainty in the digital representation of an analog voltage by the converter.

Analog-to-Digital Converter

E_{FSR}	0–10 V
M	12 bits
Linearity	± 3 bits
Temperature drift	1 bit/5°C

KNOWN   12-bit resolution (see Example 7.4)

FIND $(u_c)_E$ measured

SOLUTION   We can estimate a design-stage uncertainty as a combination of uncertainty due to quantization errors, e_Q, and due to conversion errors, e_c:

$$(u_d) = \sqrt{u_o^2 + u_c^2}$$

The resolution of a 12-bit A/D converter with full-scale range of 0–10 V is (see Example 7.3) $e_Q = 2.4$ mV, so that the quantization error is estimated by the zero-order uncertainty

$$u_o = e_Q = \pm 1/2 Q = \pm 1.2 \text{ mV}$$

Now the conversion error is affected by two elements:

$$\text{linearity error} = e_2 = \pm 3 \, bits \times 2.4 \text{ mV}$$
$$= \pm 7.2 \text{ mV}$$

$$\text{temperature error} = e_3 = \frac{1 bit}{5°C} \times 10°C \times 2.4 \text{ mV}$$
$$= \pm 4.8 \text{ mV}$$

An estimate of the uncertainty due to conversion errors is found using the RSS method,

$$u_c = \sqrt{e_2^2 + e_3^2} = \sqrt{(7.2 \, \text{mV})^2 + (4.8 \, \text{mV})^2} = 8.6 \, \text{mV}$$

The combined uncertainty in the digital representation of the analog value due to these uncertainties is then

$$(u_d)_E = \pm \sqrt{(1.2 \, \text{mV})^2 + (8.6 \, \text{mV})^2}$$
$$= \pm 8.7 \, \text{mV} \quad (95\% \text{ assumed})$$

The effects of the conversion errors dominate the uncertainty.

Successive Approximation Converters

We next discuss several common methods for converting voltage signals into binary words. Additional methods for analog-to-digital conversion are discussed in specialized texts [4, 6].

The most common type of A/D converter uses the *successive approximation* technique. This technique uses a trial-and-error approach for estimating the input voltage to be converted. Basically, the successive approximation A/D converter guesses successive

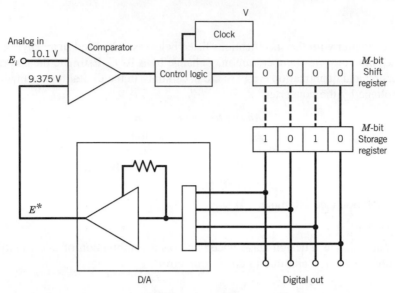

Figure 7.8 Successive approximation A/D converter. Four-bit scheme is shown with register $= 1010$ with $E_i = 10.1$ V.

binary values as it narrows in on the appropriate binary representation for the input voltage. As depicted in Figure 7.8, this A/D converter uses an M-bit register to generate a trial binary number, a D/A converter to convert the register contents into an analog voltage, and a voltage comparator (see Chapter 6.7) to compare the input voltage to the internally generated voltage. The conversion sequence is as follows:

1. The MSB is set to 1. All other bits are set to 0. This produces a value of E* at the D/A converter output equivalent to the register setting. If $E^* > E_i$, the comparator goes LOW causing the MSB to be reset to 0. If $E^* < E_i$, the MSB is kept HIGH at 1.

2. The second MSB is set to 1. Again, if $E^* > E_i$, it is reset to 0; otherwise its value remains 1.

3. The process continues through to the LSB. The final register value gives the quantization of E_i.

The process requires a time of one clock tick per bit.

An example of this sequence is shown in Figure 7.9 for an input voltage of $E_i = 10.1$ V and using the 0–15-V, 4-bit successive approximation A/D converter of

	Register	E^*	E_i	Comparator
Initial status	0000	0	10.1	
MSB set to 1	1000	7.5	10.1	High
Leave at 1	1000	7.5		
Second MSB set to 1	1100	11.25		Low
Reset to 0	1000	7.5	10.1	
Third MSB set to 1	1010	9.375		High
Leave at 1	1010	9.375		
LSB set to 1	1011	10.3125	10.1	Low
Reset to 0	1010	9.375		

Figure 7.9 Example of successive approximation conversion sequence for a 4-bit converter.

Figure 7.8. For this case, the converter has a resolution of 0.9375 V. The final register count of 1010 or its equivalent of 9.375 V is the output from the A/D converter. Its value differs from the input voltage of 10.1 V as a result of the quantization error. It should be clear that this error can be reduced by increasing the bit size of the register.

The successive approximation converter is typically used when conversion speed at a reasonable cost is important. The number of steps required to perform a conversion equals the number of bits in the A/D converter register. With one clock tick per step, a 12-bit A/D converter operating with a 1-MHz clock would require a maximum time of 12 µs per conversion. But this also reveals the trade-off between increasing the number of bits to lower quantization error and the resulting increase in conversion time. Characteristic maximum sample rates are on the order of 300 kHz at 8–10 bits and 100 kHz at 16 bits.

Sources of conversion error originate in the accuracies of the D/A converter and the comparator. Noise is the principle weakness of this type of converter, particularly at the decision points for the higher-order bits. The successive approximation process requires that the voltage remain constant during the conversion process. Because of this, a sample and hold circuit (SHC), discussed in Chapter 6, is used ahead of the converter input to measure and to hold the input voltage value constant throughout the duration of the conversion. The SHC also minimizes noise during the conversion.

Ramp (Integrating) Converters

Low-level ($<$1-mV) measurements usually rely on ramp converters for their low-noise features. Ramp converter analog-to-digital converters use the voltage level of a linear reference ramp signal to discern the voltage level of the analog input signal and convert it to its binary equivalent. Principal components, as shown in Figure 7.10, consist of an analog comparator, ramp function generator, and counter and M-bit register. The reference signal, initially at zero, is increased at set time steps, within which the ramp level is compared to the input voltage level. This comparison is continued until the two are equal.

The usual method for generating the ramp signal is to use a capacitor of known fixed capacitance, C, which is charged from a constant current source of amperage, I. Because the charge is related to current and time by

$$q = It \qquad (7.16)$$

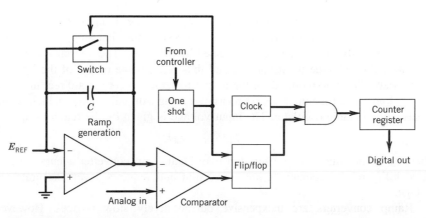

Figure 7.10 Ramp A/D converter.

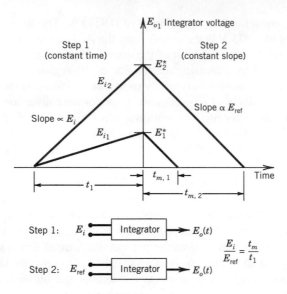

Figure 7.11 Dual-ramp analog voltage to digital conversion.

the reference ramp voltage will be linearly related to the elapsed time by

$$E_{\text{ref}} = \frac{q}{C} = \text{constant} \times t \qquad (7.17)$$

Time is integrated by a counter that increases the register value by 1 bit at each time step. Time step size depends on the value of 2^M. When the input voltage and ramp voltage magnitudes cross during a time step, the comparator output will go to zero which flips a flip-flop halting the process. The register count value will then indicate the digital binary equivalent of the input voltage.

The accuracy of this single ramp operation is limited by the accuracy of the timing clock and the known values and constancy of the capacitor and the charging current. Increased accuracy can be achieved using a dual ramp converter, sometimes referred to as a dual slope integrator, in which the measurement is accomplished in two steps, as illustrated in Figure 7.11. In the first step, the input voltage, E_i, is applied to an integrator for a fixed time period, t_1. The integrator output voltage increases in time with a slope proportional to the input voltage. At the end of the time interval, the output of the integrator has reached the level E^*. The second step in the process involves the application of a fixed reference voltage, E_{ref}, to the integrator. This step occurs immediately at the end of the fixed time interval in the first step, with the integrator voltage at exactly the same level established by the input. The reference voltage has the opposite polarity to the input voltage, and thus reduces the output of the integrator at a known rate. The time required for the output to return to zero is a direct measure of the input voltage. The time intervals are measured using a digital counter, which accumulates pulses from a stable oscillator. The input voltage is given by the relationship

$$E_i / E_{\text{ref}} = t_m / t_1 \qquad (7.18)$$

where t_m is the time required for step two. Dual ramp converter accuracy depends on the stability of the counter during conversion and requires a very accurate reference voltage.

Ramp converters are inexpensive but relatively slow devices. However, their integration process tends to average out any noise in the input signal. This is the feature that makes them so attractive for measuring low-level signals. The maximum conversion time using a ramp converter is estimated as

$$\text{maximum conversion time} = 2^M/\text{clock speed} \qquad (7.19)$$

For a dual ramp converter, this time is doubled. If one assumes that a normal distribution of input values are applied over the full range, the average conversion time can be taken to be one-half of the maximum value. For a 12-bit register with a 1-MHz clock and required to count the full 2^{12} pulses, a dual ramp converter would require $\sim$8 ms for a single conversion.

Parallel Converters

A very fast A/D converter is the parallel or flash converter depicted in Figure 7.12. These converters are common to high-end stand-alone digital oscilloscopes and spectral analyzers. An M-bit parallel converter uses $2^M - 1$ separate voltage comparators to compare a reference voltage to the applied input voltage. As indicated in Figure 7.12, the reference voltage applied to each successive comparator is increased by the equivalent of 1 LSB by using a voltage dividing resistor ladder. If the input voltage is less than its reference voltage, a comparator will go LOW; otherwise it will go HIGH.

Consider the 2-bit converter shown in Figure 7.12. If $E_{\text{in}} \geq \frac{1}{2} E_{\text{ref}}$ but $E_{\text{in}} < \frac{3}{4} E_{\text{ref}}$, then comparators 1 and 2 will be HIGH but comparator 3 will be LOW. In this manner, there can only be $2^M = 2^2$ different HIGH/LOW combinations from $2^2 - 1$ comparators, as noted in Table 7.3. The table shows how these combinations correspond to the 2^2

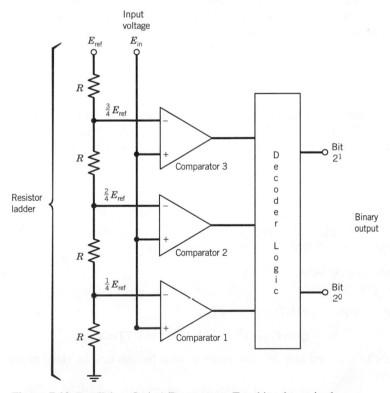

Figure 7.12 Parallel or flash A/D converter. Two-bit scheme is shown.

Table 7.3   Logic Scheme of a 2-Bit Parallel A/D Converter

Comparator States			Binary Output
HIGH	HIGH	HIGH	11
LOW	HIGH	HIGH	10
LOW	LOW	HIGH	01
LOW	LOW	LOW	00

possible values of a 2-bit binary output. Logic circuits transfer this information to a register.

Because all the comparators act simultaneously, the conversion occurs in a single clock count. So an attraction of this converter is its speed, with typical sample rates on the order of 150 MHz at 8 bits. Its disadvantage lies in the cost associated with its $2^M - 1$ comparators and the associated logic circuitry. For example, an 8-bit converter will require 255 comparators.

Digital Voltmeters

The digital voltmeter must convert an analog input signal to a digital output. This conversion can be done through several basic techniques [6]. The most common method used in digital meters is with dual ramp converters. The limits of performance for digital voltmeters can be significantly higher than for analog meters. Because of their very high input impedance, loading errors are very small at low-voltage levels, nearly as low as null balance instruments such as the potentiometer. The resolution (1 LSB) of these meters may be significantly better than their accuracy, so care must be taken in estimating uncertainty in their output. These devices are able to perform integration in both dc and true rms ac architecture.

EXAMPLE 7.6

A 0-to 10-V, three-digit digital voltmeter is built around a 10-bit, single-ramp A/D converter. A 100-kHz clock is used. An input voltage of 6.372 V is applied. What should be the DVM output value? How long will the conversion process take?

KNOWN

$E_{FSR} = 10\,V; M = 10$

Clock speed $= 100\,kHz$
Three significant digit readout (x.xx)
Input voltage, $E_i = 6.372\,V$

FIND   Digital output display value
Conversion time

SOLUTION   The resolution of the A/D converter is

$$Q = E_{FSR}/2^M = 10\,V/1024 = 9.77\,mV$$

Because each ramp step increases the counter value by one bit, the ramp converter will have a slope

$$slope = Q \times 1\,bit/step = 9.77\,mV/step$$

The number of steps required for conversion is

$$\text{steps required} = \frac{E_i}{\text{slope}} = \frac{6.372 \text{ V}}{0.00977 \text{ V/step}} = 652.29 \text{ or } 653 \text{ steps}$$

so the indicated output voltage is

$$E_0 = \text{slope} \times \text{number of steps}$$
$$= 9.77 \text{ mV/step} \times 653 \text{ steps} = 6.3769 \text{ V} \approx 6.38 \text{ V}$$

A three-digit display will round E_0 to 6.38 V. The difference between the input and output values is attributed to quantization error. With a clock speed of 100 kHz, each ramp step requires 10 μs/step, and

$$\text{conversion time} = 653 \text{ steps} \times 10 \text{ μs/step} = 6530 \text{ μs}$$

COMMENT   A similar dual ramp converter would require twice as many steps and twice the conversion time.

7.6   DATA-ACQUISITION SYSTEMS

A *data-acquisition system* is the portion of a measurement system that quantifies and stores data. There are many ways to do this. An engineer who reads a transducer dial associates a number with the dial position, and records the information in a log book performs all of the tasks germane to a data-acquisition system. In this section, we focus on microprocessor-based data-acquisition systems, which are used to perform automated data quantification and storage.

Figure 7.13 shows how a data-acquisition system (DAS) might fit into the general measurement scheme between the actual measurement and the subsequent data reduction. A typical signal flow scheme is shown in Figure 7.14 for multiple input signals to a single microprocessor-based/controller DAS.

Dedicated microprocessor systems can continuously perform their programming instructions to measure, store, interpret and provide process control without any intervention. Such microprocessors have input/output (I/O) ports to interface with other devices to measure and to output instructions. Programming allows for such operations as which sensors to measure, and when and how often, and for data reduction. Programming can allow for decision-making and feedback to control process variables.

Computer-based data-acquisition systems are hybrid systems combining a data-acquisition package with both the microprocessor and human interface capability of a personal computer (PC). The interface between external instruments and the PC is done using input/output (I/O) boards, which either mate to a board in an expansion slot in the computer or to an external communication port. These provide direct access to the computer's bus, the main path used for all computer operations.

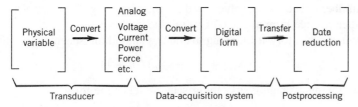

Figure 7.13 Typical signal and measurement scheme.

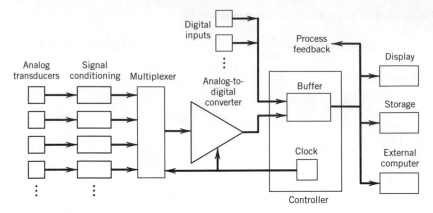

Figure 7.14 Signal flow scheme for an automated data-acquisition system.

7.7   DATA-ACQUISITION SYSTEM COMPONENTS

Signal Conditioning: Filters and Amplification

Analog signals will usually require some type of signal conditioning for the proper inter-face with a digital system. Filters and amplifiers are the most common components used.

Filters

Analog filters are used to control the frequency content of the signal being sampled. Anti-alias analog filters remove signal information above the Nyquist frequency prior to sampling. Not all data-acquisition boards contain analog filters, so these necessary components are often overlooked.

Digital filters, which are software-based algorithms, are effective for signal analysis after sampling. They cannot be used to prevent aliasing or to remove its effects. A typical digital filtering scheme involves taking the Fourier transform of the sampled signal, multiplying the signal amplitude in the targeted frequency domain to attain the desired frequency response (i.e., type of filter and desired filter settings), and transforming the signal back into time domain using the inverse Fourier transform [1, 2].

A simpler digital filter is the moving average or smoothing filter, which is used for removing noise. Essentially, this filter replaces a current data point value with an average based on a series of successive data point values. A center-weighted moving averaging scheme takes the form

$$y_i^* = (y_{i-n} + \cdots + y_{i-1} + y_i + y_{i+1} + \cdots + y_{i+n})/(2n + 1) \tag{7.20}$$

where y_i^* is the averaged value that is calculated and used in place of y_i and $(2n + 1)$ represents the number of successive values used for the averaging. For example, if $y_4 = 3$, $y_5 = 4$, $y_6 = 2$, then a three-term average of y_5 produces $y_5^* = 3$.

In a similar manner, a forward-moving averaging smoothing scheme takes the form

$$y_i^* = (y_i + y_{i+1} + \cdots + y_{i+n})/(n + 1) \tag{7.21}$$

and a backward moving averaging smoothing scheme takes the form

$$y_i^* = (y_{i-n} + \cdots + y_{i-1} + y_i)/(n + 1) \tag{7.22}$$

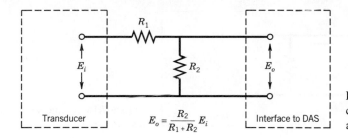

$$E_o = \frac{R_2}{R_1 + R_2} E_i$$

Figure 7.15 Voltage divider circuit for signal amplitude attenuation.

Light filtering might use three terms while heavy filtering might use 10 or more terms. This filtering scheme is easily accomplished within a spreadsheet program.

Amplifiers

All data-acquisition systems are input range limited; that is, there is a minimum value of signal that they can resolve, and a maximum value that initiates the onset of saturation. So some transducer signals will need amplification or attenuation prior to conversion. Most data-acquisition systems contain on-board instrumentation amplifiers, as depicted in Figure 7.14, with selectable gains ranging from less than to greater than unity. Gain is varied either by a resistor jumper or by logic switches set by software, which effectively reset resistor ratios across op-amplifiers. Chapter 6.6 discusses amplifiers.

Although instrument amplifiers offer good output impedance characteristics, voltages can also be attenuated using a voltage divider. The output voltage from the divider circuit of Figure 7.15 is determined by

$$E_o = E_i \frac{R_2}{R_1 + R_2} \tag{7.23}$$

For example, a 0–50-V signal can be measured by a 0–10-V A/D converter using $R_1 = 40\,k\Omega$ and $R_2 = 10\,k\Omega$.

When only the dynamic content of time-dependent signals is important, amplification may require a strategy. For example, suppose the mean value of a voltage signal is large but the dynamic content is small, such as with the 5-Hz signal

$$y(t) = 2 + 0.1 \sin 10\pi t \ [\text{V}] \tag{7.24}$$

Setting the amplifier gain at or more than $G = 2.5$ to improve the resolution of the dynamic content would saturate a ±5-V A/D converter. In such situations, you can remove the mean component from the signal prior to amplification by (1) adding a mean voltage of equal but opposite sign, such as −2 V here, or (2) passing the signal through a very low frequency high-pass filter (also known as *ac coupling*).

Shunt Circuits

An A/D converter requires a voltage signal at its input. It is straightforward to convert current signals into voltage signals using a shunt resistor. The circuit in Figure 7.16 provides a voltage

$$E_o = IR_{\text{shunt}} \tag{7.25}$$

for signal current I. For a transducer using a standard 4-to-20-mA current loop, a 250 Ω shunt would convert this current output to a 1-to-5-V signal.

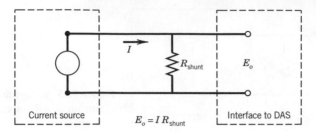

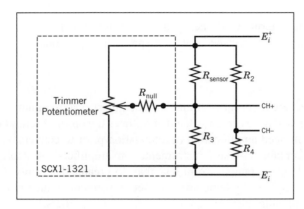

Figure 7.16 Simple shunt resistor circuit.

Figure 7.17 Circuit for applying a null offset voltage. (Courtesy of National Instruments, Inc.)

Offset Nulling Circuit

Offset nulling is used to subtract out a small voltage from a signal, such as to zero out a transducer output signal. The technique uses a trim potentiometer (R_{null}) in conjunction with a bridge circuit, such as shown in Figure 7.17, to produce the null voltage, E_{null}. A common use of this circuit is in conjunction with strain gauges. The available null voltage is

$$E_{null} = \pm \left[\frac{E_i}{2} - \frac{E_i R_3 (R_{null} + R_{sensor})}{R_{null} R_{sensor} + R_3 (R_{null} + R_{sensor})} \right] \tag{7.26}$$

Multiplexer

When multiple input signal lines are connected by a common throughput line to a single A/D converter, a multiplexer is used to switch between connections, one at a time. This is illustrated by the multiplexer depicted in Figure 7.18, which uses parallel flip-flops

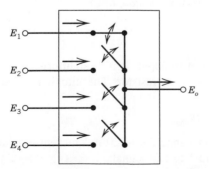

Figure 7.18 Multiplexer (four-channel shown).

(see Chapter 6) to open and close the connection paths sequentially. The switching rate is determined by the conversion timing control logic.

A/D Converters

High-speed data-acquisition boards employ successive approximation converters with conversion rates typically up to the 1-kHz to 10-MHz range or parallel converters for rates up to and over 150 MHz. Low-level voltage measurements require the high noise rejection of the dual-ramp converter. Here the trade-off is in speed with maximum conversion rates of 1–100 Hz more common.

D/A Converters

A digital-to-analog converter permits a DAS to convert digital numbers into analog voltages, which might be used for process control, such as to change a process variable, activate a device, or to drive a sensor positioning motor. The digital-to-analog signal is initiated by software or a controller.

Digital Input/Output

Digital signals represent a discrete state, either high or low. Digital input/output lines may be used to communicate between instruments (see Section 7.9), control relays (to power equipment on or off), or indicate the state or status of a device. A common way to transmit digital information is the use of a 5-V TTL signal (e.g., see Figure 7.5).

Digital I/O signals may be as a single state (HIGH or LOW; 5 V or 0 V) or as a series of pulses of HIGH/LOW states. The single-state signal might be used to operate a switch or relay or to signal an alarm. A series of pulses transmits a data series. Gate pulse counting entails counting pulses that occur over a specified period of time. This enables frequency determination (number of pulses/unit time) and counting/timing applications. Pulse stepping, sending a predetermined number of pulses in a series, is used to drive stepper motors and servos. Several I/O ports can be grouped to send parallel information.

Central Processing Unit: Microprocessor

Data-acquisition and control systems are usually built around a microprocessor. A PC is commonly used as it offers all the necessary components for effective measurement, data collection, program logic, and feedback control. This includes large amounts of memory for programming and storage, control circuits, input/output peripherals, and a clock built around a central processing unit (CPU).

The CPU controls operations, processes data, and sends and receives information to and from memory and peripherals via a bus. It consists of a control unit, arithmetic and logic unit (ALU), and registers. CPU timing is regulated about the regular pulses from its clock, and this is used to sequence actions. It performs arithmetic operations, comparisons, logical operations, and data management by combining the contents of registers. The CPU moves bits between registers in blocks. For example, a 32-bit CPU can access blocks of 32 bits representing 2^{32} different integer numbers.

Special-purpose, direct-application microprocessors are also common. These are found in many devices, including stand-alone data-acquisition systems and closed-loop controllers, and are quite flexible in their use. In addition to data acquisition from sensors,

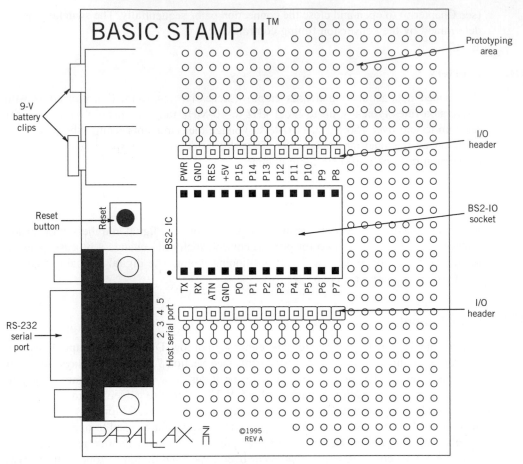

Figure 7.19 BASIC Stamp® typical of special-purpose, reprogrammable, multiport microprocessors. (Basic Stamp is a registered trademark of Parallax, Inc.)

they are used to drive devices, such as motors, and to operate relays and to send control signals. To illustrate this class of device, Figure 7.19 shows the BASIC Stamp2®. There are several versions of this 16-bit, 4-to 20-MHz battery-operated unit, but all offer fully programmable I/O ports that are easily interfaced with 5-V TTL-level devices and, with a control card, are interfaced with non-TTL devices. This permits use of sensors and control of devices ranging from data registers, A/D converters, and alarms to sensors, servos, relays, and serial lines. Communication with the microprocessor is also available through a synchronous three-wire interface that can couple with a serial communication port, such as on a PC. It can be programmed using the serial port to run interpreter-based software algorithms, which are stored in nonvolatile, reprogrammable memory called EEPROM (see Memory section). The utility and portability of such microprocessors is incredible, with applications ranging from multipurpose robots to remote sensing stations.

Program logic controllers (PLC) are widely used to measure and output signals and to perform logic, timing, sequencing, counting, and arithmetic operations in the control of machines and processes. In this respect, they are similar to PC-based data-acquisition and control systems but their simple programming is primarily dedicated to logic and switching operations. They are rugged and stand-alone units.

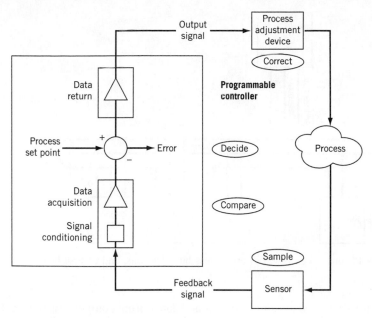

Figure 7.20 Closed-loop control concept built around a data-acquisition-based programmable controller.

Closed-Loop Controller

In closed-loop control, the controller is used to compare the state of a process, as determined through the value of a measured variable, with the value of a set condition and to take appropriate action to reduce the difference in value of the two. This difference in value is the error signal on which the controller acts. This error can change as part of the dynamics of the process, which causes the measured variable to change value, or because the set condition itself is changed. The controller action is to adjust the process so as keep the error to within a desired range.

The control process consists of this sequence: sample, compare, decide, and correct. An example of a digital control loop is depicted in Figure 7.20. Here the controller receives input about the process through a sensor that monitors the measured variable. The controller measures the sensor signal by sampling through an A/D converter or some digital input. The controller compares the measured value to the set value, computes the error signal, executes calculations through its control algorithm to decide on the correction needed, and acts to correct the process. In this example, we anticipate that the corrective elements are analog based and so the controller sends the necessary signal for appropriate corrective action at the output of a D/A converter. The controller repeats this procedure at each cycle. A more extensive treatment of feedback control is provided in Chapter 12.

Memory

Memory provides a means to store and retrieve information. Memory consists of registers. Registers are locations for the storage of numbers, the basic form of information storage on a computer. Computers contain three types of memory: *Random Access Memory (RAM)*, *Read-Only Memory (ROM)*, and *external storage device memory.*

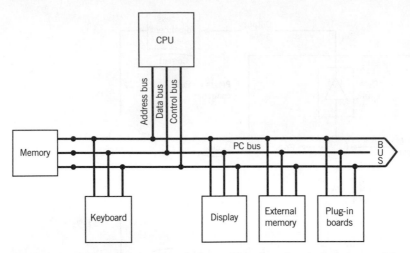

Figure 7.21 A PC bus is composed of an address bus, data bus, and control bus.

RAM is memory used to store programs and data during computer operation. RAM is temporary, reusable memory. ROM is dedicated memory typically used to store the *basic input/output system (BIOS)* instructions of a computer or a dedicated component algorithm. The BIOS contains tables and machine subroutines related to the most basic operations. Variations of these memory forms include *Electrically Erasable Programmable ROM (EEPROM)*, which is nonvolatile memory that may be repeatedly reprogrammed, a useful feature for controlling devices. External memory refers to peripheral storage devices, such as hard disk drives, and memory cards.

Central Bus

The path for communication between the CPU and memory and peripheral devices is the central bus. As depicted in Figure 7.21, the "central bus" is a collective term for three information paths: the address bus, the data bus, and the control bus. It can be considered as a three-lane highway, one lane for each bus, connecting all devices that reside along branch roads on the highway. When information or data are to be sent between devices, the address bus will contain the address of the device and the location within that device where the information is to be stored. The data bus will contain the execution instructions or the data being relayed. The control bus sends status information over the central bus. While information is being sent between any two devices connected to the central bus, a busy flag is sent out over the control bus to alert all other devices that the data bus is busy. This allows devices of different speeds to operate over the same lines.

7.8   ANALOG INPUT–OUTPUT COMMUNICATION

Data-Acquisition Boards

Analog interfacing with a computer is most often affected using a general purpose DAS I/O board. Typical units are available in the form of an expansion plug-in board or a PCMCIA card (Personal Computer Memory Card International Association). So the discussion narrows on these devices. A layout of a typical board is given in Figure 7.22. These boards use an expansion slot on the computer to interface with the computer bus. Field

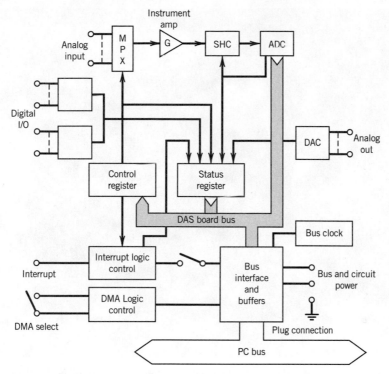

Figure 7.22 Typical layout for a data-acquisition plug-in board.
(Courtesy of National Instruments, Inc.)

wiring from transducers or other analog equipment is usually made to a screw terminal board with subsequent connection directly to the rear of the I/O board. A typical board allows for data transfer both to and from the computer memory using A/D conversion, D/A conversion, digital I/O, and counter/timer ports. A multipurpose, multi-channel high-speed data-acquisition board is shown in Figure 7.23. Hardware setup and amplifier gain are set either manually, by a resistor jumper such as in Figure 7.24, or by software controlled switches.

For example, the board in Figure 7.23 uses a 16-channel multiplexer and instrument amplifier. With its 12-bit successive approximation A/D converter with an 8–9-μs conversion rate and its 800-ns sample-hold time, sample rates of up to 100,000 Hz are possible. Input signals may be unipolar (e.g., 0–10 V) or bipolar (e.g., ±5 V). For the board shown, this allows input resolution for the 12-bit converter with an $E_{FSR} = 10$ V range of approximately

$$Q = \frac{E_{FSR}}{2^M} = \frac{10\,\text{V}}{2^{12}} = 2.44\,\text{mV}$$

The amplifier permits signal conditioning with gains from $G = 0.5$ to 1000. This improves the minimum detectable voltage when set at maximum gain to

$$Q = \frac{E_{FSR}}{(G)(2^M)} = \frac{10\,\text{V}}{(1000)(2^{12})} = 2.44\,\mu\text{V}$$

Digital I/O can usually be accomplished through these boards for instrument control applications and external triggering. For the board shown, 16-TTL compatible lines are available allowing for one 8-bit output and one 8-bit input line.

Figure 7.23 Photograph of a data-acquisition plug-in board. (Courtesy of National Instruments, Inc.)

Single- and Differential-Ended Connections

Analog signal input connections to a DAS board may be single or differential ended. *Single-ended connections* use only one signal line (+ or HIGH) that is measured relative to ground (GRD), as shown in Figure 7.25. The return line (− or LOW) and ground are connected together. There is a common external ground point, usually through the DAS board. Multiple single-ended connections to a DAS board are shown in Figure 7.26.

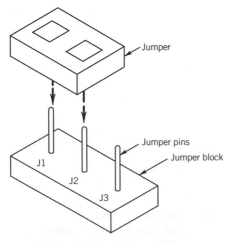

Pins J1 and J2 are to be connected

Figure 7.24 A resistor jumper block.

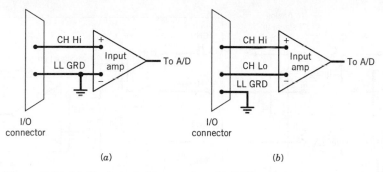

Figure 7.25 (a) Single-ended connection. (b) Differential-ended connection.

Single-ended connections are suitable only when all of the analog signals can be made relative to the common ground point. There should be no additional local ground at the signal source. Single-ended connecting wires should never be long as they are susceptible to EMI noise.

Why the concern over ground points? Electrical grounds are not all at the same voltage value (see "Grounds and Ground Loops," Chapter 6). So if a signal is grounded at two different points, such as at its source and at the DAS board, the grounds could be at different voltage levels. When a grounded source is wired as a single-ended connection, the difference between the source ground voltage and the board ground voltage gives rise to a *common-mode voltage* (*CMV*). The CMV will combine with the input signal superimposing interference and noise on the signal. This effect is referred to as a "ground loop." The measured signal will be unreliable.

Differential-ended connections allow the voltage difference between two distinct input signals to be measured. Here the signal input (+ or HIGH) line is paired with a signal return (− or LOW) line, which is isolated from ground (GRD), as shown in Figure 7.25. By using twisted pairs, the effects of noise are greatly reduced. A

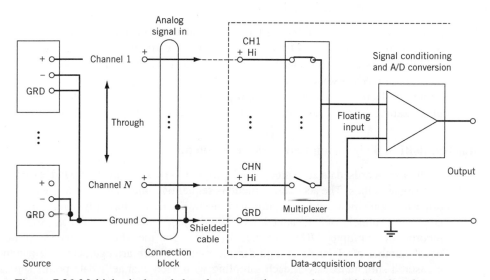

Figure 7.26 Multiple single-ended analog connections to a data-acquisition board.

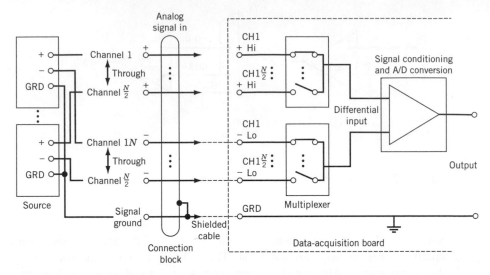

Figure 7.27 Multiple differential-ended analog connections to a data-acquisition board.

differential-ended connection to a DAS board is shown in Figure 7.27. The measured voltage is the voltage difference between these two (+ and −) lines for each channel.

When signals from various instruments are connected to a single DAS board, differential-ended connections are usually required. This is also the preferred way to measure low-level signals. However, for low-level measurements, a 10-k to 100-kΩ resistor should be connected between the signal return (− or LOW) line and ground (GRD) at the DAS board. The differential-ended connection is less prone to common-mode voltage errors. But be careful, if the measured voltage exceeds the board's common-mode (CMV) range limit specification, you can damage it!

Special Signal Conditioning Modules

Signal conditioning modules exist for different transducer applications. These connect between the transducer and the data-acquisition card. For example, resistance bridge modules allow the direct interfacing of strain gauges or other resistance sensors through an on-board Wheatstone bridge, such as depicted in Figure 7.28. Temperature modules allow for electronic thermocouple cold junction compensation and can provide for signal linearization for reasonably accurate (down to ±0.5°C) temperature measurements.

Data-Acquisition Triggering and Sample Sequence

With DAS boards, data acquisition can be triggered by software command, external pulse, or on-board clock. The acquisition mode is sequential, one channel at a time. The channel order of acquisition can be programmed on most boards. Normally, measured signals on adjacent channels are separated by the sample time increment, δt. Some systems incorporate multiple A/D converters to allow for the simultaneous sampling of an equal number of channels at high sample rates. In such arrangements, all channels are measured simultaneously at each sampling with each channel's data points separated in time by the sample time increment.

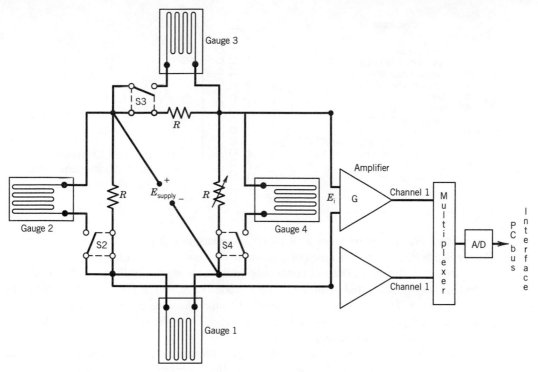

Figure 7.28 A strain-gauge interface. Gauges are connected by wires to the interface.

7.9   DIGITAL INPUT–OUTPUT COMMUNICATION

Certain standards exist for the manner in which digital information is communicated between digital devices [5, 6]. Serial communication methods transmit data bit by bit. Parallel communication methods transmit data in simultaneous groups of bits, for example, byte by byte. Both methods use a *handshake*, an interface procedure, which controls the data transfer between devices and TTL-level signals. Most lab equipment can be equipped to communicate by at least one of these means, and standards have been defined to codify communications between devices of different manufacturers.

Serial Communications

RS-232C

The RS-232C protocol, which was initially set up to translate signals between telephone lines and computers via a modem (*mo*dulator–*dem*odulator), is an interface for communication between a computer and any serial device. Basically, the protocol allows two-way communication using two single-ended signal (+) wires, noted as TRANSMIT and RECEIVE, between data-communications equipment (DCE), such as the modem or an instrument, and data terminal equipment (DTE), such as the computer. These two signals are analogous to a telephone's mouthpiece and earpiece signals. A signal GROUND wire allows signal return (−) paths. The remaining wires in the original standard are used to assess the state of the signal lines.

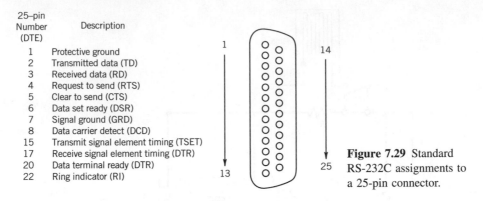

25-pin Number (DTE)	Description
1	Protective ground
2	Transmitted data (TD)
3	Received data (RD)
4	Request to send (RTS)
5	Clear to send (CTS)
6	Data set ready (DSR)
7	Signal ground (GRD)
8	Data carrier detect (DCD)
15	Transmit signal element timing (TSET)
17	Receive signal element timing (DTR)
20	Data terminal ready (DTR)
22	Ring indicator (RI)

Figure 7.29 Standard RS-232C assignments to a 25-pin connector.

Most PC computers have an RS-232C compatible I/O port. The popularity of this interface is due to the wide range of equipment that can utilize it. Either a 9-pin or 25-pin connector can be used. The full connection protocol is shown in Figure 7.29. Officially, devices may be separated by distances up to about 15 m, but in practice longer distances seem to work if using a well-shielded cable. Communications can be half-duplex or full-duplex. Half-duplex allows one device to transmit while the other receives. Full-duplex allows for both devices to transmit simultaneously.

The RS-232C is a rather loose standard, and there are no strict guidelines on how to implement the handshake. As such, compatibility problems will sometimes arise. The minimum number of wires required between DTE and DCE equipment is the three-wire connection shown in Figure 7.30. This connects only the TRANSMIT, RECEIVE, and GROUND lines while bypassing the handshake lines. The handshaking lines can be jumpered to fool either device into handshaking with itself thereby allowing the communication.

Communication between similar equipment, DTE to DTE or DCE to DCE, needs only the nine lines connected as shown in Figure 7.31, so a nine-pin connector can be used. The nine-pin connector wiring scheme is shown in Figure 7.32.

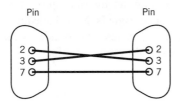

Figure 7.30 Minimum serial connections between DTE to DCE or DTE to DTE equipment (RS-232C).

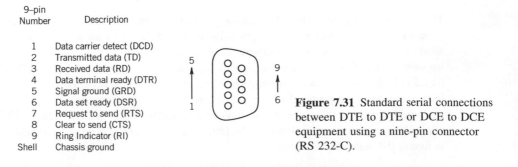

9-pin Number	Description
1	Data carrier detect (DCD)
2	Transmitted data (TD)
3	Received data (RD)
4	Data terminal ready (DTR)
5	Signal ground (GRD)
6	Data set ready (DSR)
7	Request to send (RTS)
8	Clear to send (CTS)
9	Ring Indicator (RI)
Shell	Chassis ground

Figure 7.31 Standard serial connections between DTE to DTE or DCE to DCE equipment using a nine-pin connector (RS 232-C).

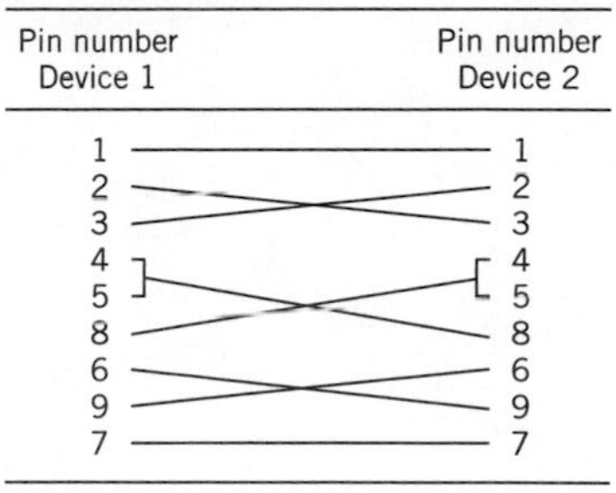

Pin number Device 1	Pin number Device 2

Figure 7.32 Nine-wire serial connection between DTE to DTE or DCE to DCE equipment.

Serial communication implies that data are sent in successive streams of information, one bit at a time. The value of each bit is represented by an analog voltage pulse with a 1 and 0 distinguished by two equal voltages of opposite polarity in the 3-to 25-V range. Communication rates are measured in baud, which refers to the number of signal pulses per second.

A typical asynchronous transmission is 10 serial bits comprised of a start bit followed by a 7-or 8-bit data stream, either one or no parity bit, and terminated by 1 or 2 stop bits. The start and stop bits form the serial "handshake" at the beginning and end of each data byte. *Asynchronous transmission* means that information may be sent at random intervals. So the start and stop bits are signals used to initiate and to end the transmission of each byte of data transmitted. The start bit allows for synchronization of the clocks of the two communicating devices. The parity bit allows for limited error checking. *Parity* involves counting the number of 1's in a byte. In 1 byte of data, there will be an even or odd number of bits with a value of 1. An additional bit added to each byte to make the number of 1 bits a predetermined even or odd number is called a parity bit. The receiving device will count the number of transmitted bits checking for the predetermined even (even parity) or odd (odd parity) number. In *synchronous transmission*, the two devices will initialize synchronize communication with each other. Even when there is no data to send, the devices will transmit characters just to maintain synchronization. Stop and start bits are then unnecessary allowing for higher data transfer rates.

Devices that employ data buffers, a region of preassigned RAM that serves as a data holding area, will use a software handshaking protocol such as XON/XOFF. With this, the receiving device will transmit an XOFF signal to halt transmission as the buffer nears full and an XON signal when it has emptied and is again ready to receive.

RS-422A/423A/449/485

RS-422A/423A provide recommendations for the electrical interface, and RS-449 specifies functional and mechanical characteristics of a serial communication standard. Equipment designed for RS-232C can be used with these standards with slight modification. Using a 37-pin connector and differential-ended connections to reduce noise, they allow for communication at rates up to 2-M baud and distances up to 1000 m. The standard uses a +5 V TTL (transistor–transistor logic) pulse signal to distinguish between a 1 bit (+2:+5.5 V) and a 0 bit (−0.6:+0.8 V).

The RS-485 protocol allows for multidrop (allows up to 32:255 devices on one line) operation that is well suited to local area networks (LAN). It communicates in half-duplex along a two-wire bus which makes it slower than RS-422.

Pin number	Description
Data lines	
1	Digital I/O data line 1
2	Digital I/O data line 2
3	Digital I/O data line 3
4	Digital I/O data line 4
13	Digital I/O data line 5
14	Digital I/O data line 6
15	Digital I/O data line 7
16	Digital I/O data line 8
Handshake lines	
6	Data valid (DAV)
7	Not ready for data (NRFD)
8	Not data accepted (NDAC)
Bus management lines	
5	End or identify (EOI)
9	Interface clear (IFC)
10	Service request (SRQ)
11	Attention (ATN)
17	Remote enable (REN)
Ground lines	
12	Shield
18–24	Ground

Figure 7.33 GPIB bus assignments to a 25-pin connector.

Universal Serial Bus and IEEE 1394

The *universal serial bus* (*USB*) permits peripheral expansion for up to 128 devices at low- to high-speed data transfer rates. The original USB (USB1.0/1.1) supports transfer rates from 1.5 Mbs up to 12 Mbs, and the high-speed USB (USB2.0) supports rates up to 480 Mbs. The USB supports a "hot swap" feature that allows the user to plug in a USB-compatible device and to use it without reboot. This bus connects USB devices to a single computer host through a USB root hub. The *IEEE 1394* standard defines a high-speed serial bus, which is also called *Firewire*, supporting transfer rates up to 400 Mbs between devices.

The USB physical interconnect is a tiered star topology. In this set-up, the root hub permits one to four attachments, which can be a combination of USB peripheral devices and additional USB hubs. Each successive hub can in turn support up to four devices or hubs. Cable length between a device and hub or between two hubs is limited to 5 m for USB and 4.5 m for IEEE 1394. The connecting cable (Figure 7.33) is a four-line wire consisting of a hub power line, two signal lines (+ and −), and a ground (GRD) line. IEEE 1394 uses six wires consisting of two pairs of signal lines and two power lines.

Parallel Communications

GPIB (IEEE-488)

The *general purpose interface bus (GPIB)* is a high-speed parallel interface. Originally developed by Hewlett-Packard, it is sometimes referred to as the HP-IB. The GPIB is usually operated under the IEEE-488 communication standard. The bus allows for the control of other devices through a central controller, and it allows devices to receive/transmit information from/to the controller. This standard is well defined and widely used to interface communication between computers and printers and scientific instrumentation.

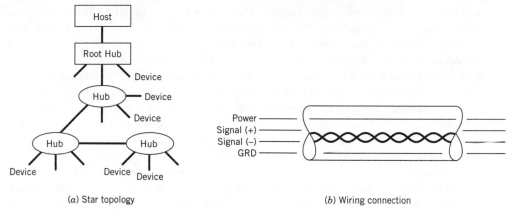

(a) Star topology   (b) Wiring connection

Figure 7.34 Cable configuration for the Universal Serial Bus (USB).

The IEEE-488 standard for the GPIB operates from a 16-wire bus with a 24-wire connector (Figure 7.34). A 25-pin connector is standard. The bus is formed by eight data lines plus eight lines for bus management and handshaking (two-way control communication). The additional eight lines are used for grounds and shield. Bit parallel, byte serial communication at data rates up to 1 Mbytes/s are possible, with 1:10 kbytes/s most common. Connector lines are limited to a length of roughly 4 m.

The standard requires a controller, a function usually served by the laboratory computer, and permits multidrop operation, allowing up to 15 devices to be attached to the bus at any time. Each device has its own bus address (addresses 1–14). The bus controller (address 0) controls all the bus activities and sequences all communications to and between devices, such as which bus device transmits or receives and when. This is done along the bus management and handshaking lines. The communication along the bus is bidirectional. Normally, ground true TTL logic (≤ 0.8 V HIGH, ≥ 2V LOW) is used.

The IEEE-488 standard specifies the following:

Data Bus: This is an 8-bit parallel bus formed by the eight digital input–output data lines (lines 1–4 and 13–16). Data are transmitted as one 8-bit byte at a time. The handshake and bus management lines communicate through the transmission of a HIGH or LOW signal along the line. A HIGH signal asserts a predetermined situation.

Handshake Bus: A three-wire handshake is specified. The *Data Valid* line (line 6) asserts that data are available on the data bus and are valid. The *Not Ready for Data* line (line 7) is asserted by a device until it is ready to receive data or instructions. The *Not Data Accepted* line (line 8) is asserted by a device until it has accepted a set of data. To illustrate the handshake concept, consider the handshake between a controller (computer) and a measuring instrument. The controller unasserts NRFD indicating that it is ready to accept data. When ready, the measuring instrument asserts DAV indicating to the controller that valid data are being sent over the data bus. When the controller has read and accepted the data, it unasserts NDAC.

Bus Management: There are five lines reserved to manage the flow of information on the data bus. The *Interface Clear* line (line 9) is used by the controller to clear devices, such as during the initial boot. The *Service Request* line (line 10) is used by a device to assert that it is ready to be serviced by the controller, such as when it is ready to transmit data. The *Attention* line (line 11) asserts that the data on the data bus are commands from the controller, devices are not to transmit. A low signal allows device messages onto the data line, such as the device data readings. The *End or Identify* line (line 5) is used by the transmitting device to indicate the end of a data transmission. With the *Attention* line asserted, it is used by the controller to poll its devices to determine which device asserted its Service Request. The *Remote Enable* line (line 17) is used by the controller to place a device in its remote mode. When asserted, the front panels of the device are deactivated and the controller has command over device programming.

The interplay between the controller and the devices on the bus is controlled by software programs. Most scientific devices rely on software drivers; many of these are built around menu-driven programs, which are designed to provide some flexibility in the specific set-up or operating parameters of the device.

Although a parallel interface is more costly than a serial interface, the GPIB allows multiple devices to be attached to a single computer, microprocessor, or network through a common interface, lowering overall cost. It is faster and more efficient at data transfer than a serial interface. The parallel interface and 8-bit data bus is similar to those used by microprocessor buses and more readily adapted.

EXAMPLE 7.7

A strain transducer has a static sensitivity of 2.5 V/unit strain (2.5 µV/µε) and requires a supply voltage of 5VDC. It is to be connected to a DAS having a ±5 V, 12-bit A/D converter and its signal measured at 1000 Hz. The transducer signal is to be amplified and filtered. For an expected measurement range of 1–500µε, specify appropriate values for amplifier gain, filter type and cut-off frequency, and show a signal flow diagram for the connections.

SOLUTION   The signal flow diagram is shown in Figure 7.35. Power is drawn off the data-acquisition board and routed to the transducer. Transducer signal wires are shielded and routed through the amplifier, filter, and connected to the data-acquisition board connector block using twisted pairs to channel 0, as shown. The differential-ended connection at the board will reduce noise.

The amplifier gain is determined by considering the minimum and maximum signal magnitudes expected, and the quantization error of the DAS. The nominal signal magnitude will range from 2.5 µV to 1.25 mV. An amplifier gain of $G = 1000$ will boost this from 2.5 mV to 1.25 V. This lower value is on the order of the quantization error of the 12-bit converter. A gain of $G = 3000$ will raise the low end of the signal out of quantization noise while keeping the high end out of saturation.

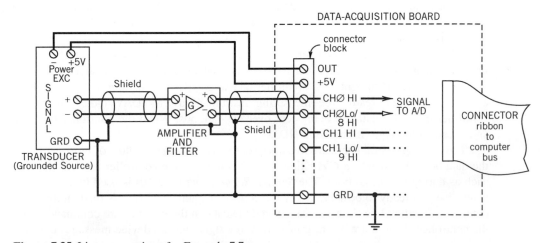

Figure 7.35 Line connections for Example 7.7.

With a sample rate of $f_s = 1000\,\text{Hz}$, an anti-alias, low-pass Butterworth filter with 12 dB/octave roll-off and a cutoff frequency set at $f_N = 500$ Hz would meet the task.

EXAMPLE 7.8

The output from an analog device (nominal output impedance of 600 W) is input to the 12-bit A/D converter (nominal input impedance of 1 MW) of a data-acquisition system. For a 2-V signal, will interstage loading be a problem?

KNOWN

$$Z_1 = 600\,\Omega \quad Z_m = 1\,\text{M}\,\Omega \quad E_1 = 2\,\text{V}$$

FIND

e_I

SOLUTION   The loading error is given by $e_I = E_\text{m} - E_\text{I}$, where E_1 is the true voltage and E_m is the measured voltage. From equation (6.39),

$$e_I = E_1 \left(\frac{1}{1 + Z_1/Z_m} - 1 \right) = -1.2\,\text{mV}$$

The interstage loading error at 2-V input actually will be less than the ± 2.4-mV quantization error of the 12-bit device.

EXAMPLE 7.9

An analog signal is to be sampled at a rate of 200 Hz with a 12-bit A/D converter that has an input range of -10 to 10-V input range. The signal contains a 200-Hz component, f_1, with an amplitude of 100 mV. Specify a suitable LC filter that will attenuate the 200-Hz component down to the A/D converter quantization resolution level and will act as an anti-alias filter for quantization.

KNOWN

$$f_s = 200 \text{ Hz}$$

$$f_1 = 200 \text{ Hz} \quad A_1 = 100 \text{ mV}$$

$$E_\text{FSR} = 10\,\text{V} \quad M = 12$$

FIND   Specify a suitable filter

SOLUTION   From equation (7.8), for $f_s = 200$ Hz, f_N is 100 Hz. So an appropriate design for an anti-alias filter would have the properties of $f_c = 100$ Hz and $M\,(100\,\text{Hz}) = -3$ dB. The A/D converter quantization resolution level is

$$Q = \frac{E_\text{FSR}}{2^M} = \frac{20\,\text{V}}{4096} = 4.88\,\text{mV}$$

For a 100-mV signal at 200 Hz, this requires a low-pass Butterworth filter with an attenuation of

$$M(200\,\text{Hz}) = \frac{4.88\,\text{mV}}{100\,\text{mV}} = 0.0488 \quad (\text{or} - 26\,\text{dB})$$

$$= \left[1 + (f/f_c)^{2k} \right]^{-1/2} = \left[1 + (200/100)^{2k} \right]^{-1/2}$$

or $k = 4.3 \approx 5$. Appropriate values for L and C can be set from Table 6.1 and Figure 6.30.

EXAMPLE 7.10

To consider how the GPIB lines might be used, suppose we wish to use a computer to instruct a voltmeter to take a reading and display it. In this case, the computer acts as the controller and transmits commands to the voltmeter, the receiver. With bus addresses set by software to: Transmit address = 0, Receive address = 3, the process begins with the Attention line 11 set HIGH (this halts activity while the bus devices listen to the computer commands):

(Below : C = Computer; DV = voltmeter)

C : Initialize all devices : Line 9 (IFC) HIGH then LOW.

DV : Voltmeter Ready (to receive instructions) : Line 7 (NRFD) LOW.

C : Computer Sends Instructions : Line 5 (EOI) HIGH.

The computer addresses the voltmeter. The eight parallel data lines become active as instructions are sent to the voltmeter. Instructions are 7- or 8-bit ASCII codes using DI01–8.

DV : Voltmeter Accepts : Line 8 (NDAC) LOW.

The voltmeter indicates it accepts the instructions.

DV : Voltmeter Active : Line 7 (NRFD) HIGH.

The voltmeter takes a reading. The line 7 status informs the computer that the voltmeter is busy so no new instructions should be sent.

DV : Voltmeter Ready : Line 7 (NRFD) LOW.

The voltmeter now displays its new reading.

This signals that the voltmeter awaits new instructions. The next instruction might be to send the new reading along the data lines to be stored in computer memory.

7.10 SUMMARY

This chapter has focused on sampling concepts, the interfacing of analog and digital devices, and data-acquisition systems. Despite the advantages of digital systems, they must interact with an analog world, and the back-and-forth exchange between an analog and digital signal has limitations. With this in mind, a thorough discussion has been provided on the selection of sample rate and the number of measurements required to reconstruct a continuous process variable from a discrete representation. Equations 7.1 and 7.9 explain the criteria by which a periodic waveform can be accurately represented by such a discrete time series. The limitations resulting from improper sampling and the discrete representation include frequency alias and leakage, and amplitude ambiguity.

The fact that practically any electrical instrument can be interfaced with a microcontroller or a data-acquisition system is significant. The working mechanics of analog-to-digital and digital-to-analog converters is basic to interfacing such analog and digital devices with data-acquisition systems. The limitations imposed by the resolution and range of these devices mandate certain signal conditioning requirements in terms of signal gain and filtering prior to acquisition. These require analog amplifiers and filters, devices discussed previously. Communication between digital devices surrounds us. But their proliferation means that standards for communication must be set. These standards are evaluated and replaced as newer devices demand faster data transfer.

REFERENCES

1. Bendat, J., and A. Piersol, *Random Data: Analysis and Measurement Procedures*, Wiley-Interscience, New York, 1971.
2. Stanley, W. D., G. R. Dougherty, and R. Dougherty, *Digital Signal Processing*, 2d ed., Reston (a Prentice Hall company), Reston, VA, 1984.
3. Lyons, R., *Understanding Digital Signal Processing*, 2d ed., Prentice-Hall, Englewood Cliffs, 2004.
4. Vandoren, A., *Data-Acquisition Systems*, Reston (a Prentice Hall company), Reston, VA, 1982.
5. Krutz, R., *Interfacing Techniques in Digital Design: Emphasis on Microprocessors*, Wiley, New York, 1988.
6. Hnatek, E., *A User's Handbook of D/A and A/D Converters*, Wiley-Interscience, New York, 1976.
7. Money, S. A., *Microprocessors in Instrumentation and Control*, McGraw-Hill, New York, 1985.

Suggested Reading

Ahmed, H., and P. J. Spreadbury, *Analogue and Digital Electronics for Engineers*, 2d ed., Cambridge University Press, Cambridge, UK, 1984.

Beauchamp, K.G., and C.K. Yuen., *Data Acquisition for Signal Analysis*, Allen & Unwin, London, 1980.

Evans, Alvis J., J. D. Mullen, and D. H. Smith, *Basic Electronics Technology*, Texas Instruments Inc., Dallas, TX, 1985.

Seitzer, D., G. Pretzl, and N. Hamdy., *Electronic Analog-to-Digital Converters: Principles, Circuits, Devices, Testing*, Wiley, New York, 1983.

Wobschall, D., *Circuit Design for Electronic Instrumentation: Analog and Digital Devices from Sensor to Display*, McGraw Hill, New York, 1979.

NOMENCLATURE

e	error	E_o	output voltage [V]
e_Q	quantization error	E_{FSR}	full-scale analog voltage range
f	frequency $[t^{-1}]$	G	amplifier gain
f_a	alias frequency $[t^{-1}]$	I	electric current [A]
f_c	filter cutoff frequency $[t^{-1}]$	M	digital component bit size
f_m	maximum analog signal frequency $[t^{-1}]$	$M(f)$	magnitude ratio at frequency f
f_N	Nyquist frequency $[t^{-1}]$	N	data set size
f_s	sample rate frequency $[t^{-1}]$	$N\,\delta t$	total digital sample period [t]
k	cascaded filter stage number	R	resistance [Ω]
t	time [t]	Q	A/D converter resolution [V]
$y(t)$	analog signal	δf	frequency resolution of DFT $[t^{-1}]$
y_i, $y(r\,\delta t)$	discrete values of a signal, $y(t)$	δR	change in resistance [Ω]
$\{y(r\,\delta t)\}$	complete discrete time signal of $y(t)$	δt	sample time increment [t]
A	amplifier open loop gain; also, constant	t	time constant [t]
E	voltage [V]	$\Phi(f)$	phase shift at frequency f
E_i	input voltage [V]		

PROBLEMS

For many of these problems, using spreadsheet software or the accompanying software will facilitate solution.

7.1 Convert the analog voltage, $E(t) = 5 \sin 2\pi t$ mV, into a discrete time signal. Specifically, using sample time increments of (a) 0.125 s, (b) 0.30 s, and (c) 0.75 s, plot each series as a function of time over at least one period. Discuss apparent differences between the discrete representations of the analog signal.

7.2 Compute the DFT for each of the three discrete signals in Problem 7.1. Discuss apparent differences. Use a data set of 128 points.

7.3 Compute the DFT for the discrete time signal that results from sampling the analog signal, $T(t) = 2 \sin 4\pi t °C$, at sample rates of 4 and 8 Hz. Use a data set of 128 points. Discuss and compare your results.

7.4 Determine the alias frequency that results from sampling f_1 at sample rate f_s:

    a. $f_1 = 60$ Hz; $f_s = 90$ Hz   **c.** $f_1 = 10$ Hz; $f_s = 6$ Hz

    b. $f_1 = 1.2$ kHz; $f_s = 2$ kHz   **d.** $f_1 = 16$ Hz; $f_s = 8$ Hz

7.5 A particular data-acquisition system is used to convert the analog signal, $E(t) = (\sin 2\pi t + 2 \sin 8\pi t)$ V, into a discrete time signal using a sample rate of 16 Hz. Build the discrete time signal and from that use the Fourier transform to reconstruct the Fourier series.

7.6 Consider the continuous signal found in Example 2.3. What would be an appropriate sample rate and sample period to use in sampling this signal if the resulting discrete series must have a size of 2^M where M is an integer and the signal is to be filtered at and above 2 Hz?

7.7 Convert the following straight binary numbers to positive integer base 10 numbers:

    a. 1010   **c.** 10111011

    b. 11111   **d.** 1100001

7.8 Convert: (a) 1100111.1101 (binary) into a base 10 number; (b) 4B2F into straight binary; (c) 278.632 (base 10) into straight binary.

7.9 Convert the following decimal (base 10) numbers into bipolar binary numbers using a two's complement code:

    a. 10   **c.** -247

    b. -10 **d.** 1013

7.10 A personal computer does integer arithmetic in two's complement binary code. How is the largest positive binary number represented in this code for an 8-bit byte? Add one to this number. What base 10 decimal numbers do these represent?

7.11 How is the largest negative binary number represented in twos complement code for an 8-bit byte. Subtract one from this number. What base 10 decimal numbers do these represent?

7.12 List some possible sources of uncertainty in the dual-slope procedure for A/D conversion. Derive a relationship between the uncertainty in the digital result and the slope of the integration process.

7.13 Compute the resolution and SNR for an M-bit A/D converter having a full-scale range of ± 5 V. Let M be 4, 8, 12, and 16.

7.14 A 12-bit A/D converter having an $E_{FSR} = 5$ V has a relative accuracy of 0.03% FS (full-scale). Estimate its quantization error in volts. What is the total possible error expected in volts? What value of relative uncertainty might be used for this device?

7.15 An 8-bit single ramp A/D converter with $E_{FSR} = 10$ V uses a 2.5-MHz clock and a comparator having a threshold voltage of 1 mV. Estimate:

   (i)   The binary output when the input voltage is $E = 6.000$ V. When $E = 6.035$ V.

   (ii) The actual conversion time for the 6.000-V input and average conversion times.

   (iii) The resolution of the converter.

7.16 Compare the maximum conversion times of a 10-bit successive approximation A/D converter to a dual-slope ramp converter if both use a 1-MHz clock and $E_{FSR} = 10$ V.

7.17 An 8-bit D/A converter shows an output of 3.58 V when straight binary 10110011 is applied. What is the output voltage when 01100100 is applied?

7.18 A 0–10 V, 4-bit successive approximation A/D converter is used to measure an input voltage of 4.9 V.

   (i)   Determine the binary representation and its analog approximation of the input signal. Explain your answer in terms of the quantization error of the A/D converter.

   (ii) If we wanted to ensure that the analog approximation be within 2.5 mV of the actual input voltage, estimate the number of bits required of an A/D converter.

7.19 Discuss the trade-offs between resolution and conversion rate for successive approximation, ramp, and parallel converters.

7.20 A 0–10-V, 10-bit A/D converter displays an output in straight binary code of 1010110111. Estimate the input voltage to within 1 LSB.

7.21 A ± 5-V, 8-bit A/D converter displays an output in twos-complement code of 10101011. Estimate the input voltage to within 1 LSB.

7.22 A dual-slope A/D converter has 12-bit resolution and uses a 10-kHz internal clock. Estimate the conversion time required for an output code equivalent to 2011_{10}.

7.23 How long does it take an 8-bit single ramp A/D converter using a 1-MHz clock to convert the number 173_{10}?

7.24 A successive approximation A/D converter has a full-scale output of 0–10 V and uses an 8-bit register. An input of 6.2 V is applied. Estimate the final register value.

7.25 The voltage from a 0-to 5-kg strain gauge balance scale is expected to vary from 0 to 3.50 mV. The signal is to be recorded using a 12-bit A/D converter having a ± 5-V range with the weight displayed on a computer monitor. Suggest an appropriate amplifier gain for this situation.

7.26 An aircraft wing will oscillate under wind gusts. The oscillations are expected to be at about 2 Hz. Wing mounted strain gauge sensors are connected to a ± 5-V, 12-bit A/D converter and data-acquisition system to measure this. For each test block, 10 s of data are sampled.

   (i)   Suggest an appropriate sample rate. Explain.

   (ii) If the signal oscillates with an amplitude of 2 V, express the signal as a Fourier series.

   (iii) Based on (i) and (ii), sketch a plot of the expected amplitude spectrum. What is the frequency spacing on the abscissa? What is its Nyquist frequency?

7.27 An electrical signal from a transducer is sampled at 20,000 Hz using a 12-bit successive approximation A/D converter. A total of 5 s of data are acquired and passed to computer memory. How much 16-bit computer memory (in kbytes) is required?

7.28 How many data points can be sampled by passing a signal through a 12-bit parallel A/D converter and stored in computer memory, if 8 MB of 32-bit computer memory is available? If the A/D converter uses a 100-MHz clock and acquisition is by DMA, estimate the duration of signal that can be measured.

7.29 Select an appropriate sample rate and data number to acquire with minimal leakage the first five terms of a square wave signal having a fundamental period of 1 s. Select an appropriate cutoff frequency for an anti-alias filter. Hint: approximate the square wave as a Fourier series.

7.30 A triangle wave with a period of 2 s can be expressed by the Fourier series,

$$y(t) = \Sigma[2D_1(1 - \cos n\pi)/n\pi]\cos \pi nt \qquad n = 1, 2, \ldots$$

Specify an appropriate sample rate and data number to acquire the first seven terms with minimal leakage. Select an appropriate cutoff frequency for an anti-alias filter.

7.31 Using Fourier transform software (or equivalent software), generate the amplitude spectrum for Problem 7.29.

7.32 Using Fourier transform software (or equivalent software), generate the amplitude spectrum for Problem 7.30. Use $D_1 = 1$ V.

7.33 A single-stage low-pass RC Butterworth filter with $f_c = 100$ Hz is used to filter an analog signal. Determine the attenuation of the filtered analog signal at 10, 50, 75, and 200 Hz.

7.34 A three-stage LC Bessel filter with $f_c = 100$ Hz is used to filter an analog signal. Determine the attenuation of the filtered analog signal at 10, 50, 75, and 200 Hz.

7.35 Design a cascading RC Butterworth low-pass filter that has a magnitude ratio flat to within 3 dB from 0 to 5 kHz but with an attenuation of at least 30 dB for all frequencies at and above 10 kHz.

7.36 A complex periodic analog signal consisting of frequency f_1 and its harmonics is to be sampled at 500 Hz. The signal is passed through a low-pass single-stage RC Butterworth filter rated at -3 dB at 250 Hz. What is the maximum frequency in the filtered signal for which the amplitude will be affected by a dynamic error of no more than 10%?

7.37 Choose an appropriate cascading low-pass filter to remove a 500-Hz component contained within an analog signal which is to be passed through an 8-bit A/D converter having 10-V range and 200-Hz sample rate. Attenuate the component to within the A/D converter quantization error.

7.38 The voltage output from a J-type thermocouple referenced to 0°C is to be used to measure temperatures from 50 to 70°C. The output voltages will vary linearly over this range from 2.585 to 3.649 mV.

   a. If the thermocouple voltage is input to a 12-bit A/D converter having a ±5-V range, estimate the percent quantization error in the digital value.

   b. If the analog signal can be first passed through an amplifier circuit, compute the amplifier gain required to reduce the quantization error to 5% or less.

   c. If the ratio of signal-to-noise level (SNR) in the analog signal is 40 dB, compute the magnitude of the noise after amplification. Discuss the results of (b) in light of this.

7.39 Specify an appropriate ±5-V M-bit A/D converter (8- or 12-bit), sample rate (up to 100 Hz) and signal conditioning to convert these analog signals into digital series. Estimate the quantization error and dynamic error resulting from the system specified:

   a. $E(t) = 2 \sin 20\pi t$ V

   b. $E(t) = 1.5 \sin \pi t + 20 \sin 32\pi t - 3 \sin (60\pi t + \pi/4)$ V

   c. $P(t) = -10 \sin 4\pi t + 5 \sin 8\pi t$ psi; $K = 0.4$ V/psi

7.40 The following signal is to be sampled using a 12-bit, ±5-V data-acquisition board

$$y(t) = 4 \sin 8\pi t + 2 \sin 20\pi t + 3 \sin 42\pi t$$

Select an appropriate sample rate and sample size that provide minimal spectral leakage.

7.41 Static pressures are to be measured at eight locations under the hood of a NASCAR race car. The pressure transducers to be used have an output span of ± 1 V for an input span of ± 25 cm H_2O. The signals are measured and recorded on a portable DAS which uses a 10-bit, ± 5-V A/D converter. Pressure needs to be resolved to within 0.25 cm H_2O. The dynamic content of the signals is important and has a fundamental period of about 0.5 s. The system has 4-MB memory, powers all instruments, and has 10 min of usable battery life. Suggest an appropriate sample rate, total sample time, and signal conditioning for this application. Sketch a signal flow diagram through the measurement system.

7.42 The transducer in Problem 7.41 has a rated accuracy of 0.25%. Estimate the design stage uncertainty of the transducer-A/D converter system based on known information. Would a 12-bit converter improve things notably? If following measurements the engineer notes a 0.10 cm H_2O uncertainty due to data scatter alone, what uncertainty dominates the measurement?

7.43 A strain gauge sensor is used with a bridge circuit and connected to a DAS as indicated in Figure 7.17. Estimate the range of offset nulling voltage available if: The bridge excitation is 3.333 V, sensor and bridge resistors are each at a nominal value of 120 Ω, and the adjustable trim potentiometer is rated at 39 kΩ.

7.44 Design a low-pass Butterworth filter around a 10 Hz cut-off (-3 dB) frequency. The filter is to pass 95% of signal magnitude at 5 Hz but no more than 10% at 20 Hz. Source and load impedances are 10 Ω.

7.45 A two-stage LC Butterworth filter with $f_c = 100$ Hz is used as an anti-alias filter for an analog signal. Determine the signal attenuation experienced at 10, 50, 75, and 200 Hz.

The following problems make use of the accompanying software.

7.46 In the discussion surrounding Figure 7.4, we point out the effects of sample rate and total sample period on the reconstructed time signal and its amplitude spectrum. Use program *Leakage* to duplicate Figure 7.4. Then develop a similar example (signal frequency, sample rate, and sample period) in which fewer points and a slower sample rate lead to a better reconstruction in both time and frequency domains. Incidentally, for a fixed sample rate, you can increment N and watch the acquired waveform develop such that the leakage decreases to zero as the acquired signal reaches an exact integer period of the waveform. For a fixed N, the same can be shown for changes in sample rate.

7.47 Using program *Aliasing* solve Example 7.1 to find the alias frequency. Observe the plot of the original and the acquired signal, as well as the amplitude spectrum. Decrement the signal frequency 0.1 Hz at a time until within the region where there no longer is aliasing. Based on these observations, discuss how the acquired time signal changes and how this is related to aliasing.

7.48 Use program *Aliasing* to understand the folding diagram of Figure 7.3. For a sample rate of 10 Hz, vary the signal frequency over its full range. Determine the frequencies corresponding to f_N, $2f_N$, $3f_N$, and $4f_N$ on Figure 7.3. Determine the alias frequencies corresponding to $1.6f_N$, $2.1f_N$, $2.6f_N$, and $3.2f_N$.

7.49 Using program *Signal_C7* examine the rule that when exact discrete representations are not possible, then a sample rate of at least 5 times the maximum signal frequency should be used. Select a sine wave of 2 Hz and discuss the acquired waveform as the sample rate is increased incrementally from a low to a high value. At what sample rate does the signal look like a sine wave? Compare with the corresponding frequency and amplitude content from the amplitude spectrum. Write up a short discussion of your observations and conclusions.

7.50 Program *Leakage* samples a single frequency signal at a user-defined sample rate and period. Describe how sample period corresponds to the length of the signal measured and how this affects leakage in the amplitude spectrum. Does frequency resolution matter?

Chapter 8

Temperature Measurements

8.1 INTRODUCTION

Temperature is one of the most commonly used and measured engineering variables. Much of our lives is affected by the diurnal and seasonal variations in ambient temperature, but the fundamental, scientific definition of temperature and a scale for the measurement of temperature are not commonly understood. Although temperature is one of the most familiar engineering variables, it is unfortunately not easy to define. This chapter explores the establishment of a practical temperature scale and common methods of temperature measurement. In addition, errors associated with the design and installation of a temperature sensor are discussed.

Historical Background

Early exploration of thermodynamic temperatures was accomplished by a French scientist, Guillaume Amontons (1663–1705). His efforts examined the behavior of a constant volume of air that was subject to temperature changes. The modern liquid-in-glass bulb thermometer traces its origin to Galileo (1565–1642), who attempted to use the volumetric expansion of liquids in tubes as a relative measure of temperature. Unfortunately, this open tube device was actually sensitive to both barometric pressure and temperature changes, and thus could be called a "barothermoscope." A major advance in temperature measurement occurred in 1630 as a result of a seemingly unrelated event: the development of the technology to manufacture capillary glass tubes. These tubes were then used with water and alcohol in a thermometric device resembling the bulb thermometer, and eventually led to the development of a practical temperature-measuring instrument.

A temperature scale proposed by Sir Isaac Newton (1642–1727) used the freezing point of water and the armpit temperature of a healthy man as extremes of temperature on a linear 0 to 12 scale. The scale proposed by Gabriel D. Fahrenheit, a German physicist (1686–1736), in 1715 attempted to incorporate body temperature as the median point on a scale having 180 divisions between the freezing point and the boiling point of water. Fahrenheit also successfully used mercury as the liquid in a bulb thermometer, making significant improvements over the attempts of Ismael Boulliau in 1659.

In 1742, the Swedish astronomer Anders Celsius[1] (1701–1744) described a temperature scale that divided the interval between the boiling and freezing points of water at 1 atm pressure into 100 equal parts. The boiling point of water was fixed as 0, and the freezing point of water as 100. Shortly after Celsius' death, Carolus Linnaeus (1707–1778) reversed the scale so that the 0 point corresponded to the freezing point of water

---

[1]It is interesting to note that in addition to his work in thermometry, Celsius published significant papers on the aurora borealis and the falling level of the Baltic Sea.

at 1 atm. Even though this scale may not have been originated by Celsius [1], in 1948 the change from degrees centigrade to degrees Celsius was officially adopted. It is also interesting to note that despite the many practical applications for temperature measurement, a practical temperature scale and measuring devices were initially developed as measurement tools for research. As stated by H. A. Klein in *The Science of Measurement: A Historical Survey,*[2]

> *From the original thermoscopes of Galileo and some of his contemporaries, the measurement of temperature has pursued paths of increasing ingenuity, sophistication and complexity. Yet temperature remains in its innermost essence the average molecular or atomic energy of the least bits making up matter, in their endless dance. Matter without motion is unthinkable. Temperature is the most meaningful physical variable for dealing with the effects of those infinitesimal, incessant internal motions of matter.*

We begin our discussion of temperature by examining a method for measuring temperature, before attempting a precise definition.

8.2   TEMPERATURE STANDARDS AND DEFINITION

Temperature can be loosely described as the property of an object that describes its hotness or coldness, concepts that are clearly relative. Our experiences indicate that heat transfer tends to equalize temperature, or more precisely, systems that are in thermal communication will eventually have equal temperatures. The zeroth law of thermodynamics states that two systems in thermal equilibrium with a third system are in thermal equilibrium with each other. Thermal equilibrium implies that no heat transfer occurs between the systems, which also indicates equality of temperature. *Although the zeroth law of thermodynamics essentially provides the definition of the equality of temperature, it provides no means for defining a temperature scale.*

A *temperature scale* provides for three essential aspects of temperature measurement: (1) the definition of the size of the degree (2) fixed reference points for establishing known temperatures, and (3) a means for interpolating between these fixed temperature points. These provisions are consistent with the requirements for any standard, as described in Chapter 1. To construct a temperature scale, the three aspects listed above for a temperature scale must be established.

Fixed Point Temperatures and Interpolation

To begin, consider the definition of the triple point of water as having a value of 0.01 for our temperature scale, as is done for the Celsius scale (0.01°C). This provides for an arbitrary starting point for a temperature scale; in fact, the number value assigned to this temperature could be anything. On the Fahrenheit temperature scale it has a value very close to 32. Consider another fixed point on our temperature scale. Fixed points are typically defined by phase-transition temperatures or the triple point of a pure substance. The point at which pure water boils at one standard atmosphere pressure is an easily reproducible fixed temperature. For our purposes let's assign this fixed point a numerical value of 100.

The next problem is to define the size of the degree. Since we have two fixed points on our temperature scale, we can see that the degree is 1/100th of the temperature difference between the ice point and the boiling point of water at atmospheric pressure.

---

[2]Dover, Mineola, NY, 1988.

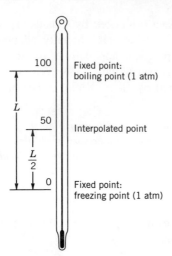

Figure 8.1 Calibration and interpolation for a liquid-in-glass thermometer.

Conceptually, this defines a workable scale for the measurement of temperature; however, as yet we have made no provision for interpolating between the two fixed-point temperatures.

The calibration of a temperature measurement device entails not only the establishment of fixed temperature points, but also the indication of any temperature between fixed points. The operation of a mercury-in-glass thermometer is based on the thermal expansion of mercury contained in a glass capillary, where the level of the mercury is read as an indication of the temperature. Imagine that we submerged the thermometer in water at the ice point, made a mark on the glass at the height of the column of mercury, and labeled it 0°C, as illustrated in Figure 8.1. Next we submerged the thermometer in boiling water, and again marked the level of the mercury, this time labeling it 100°C. Clearly we want to be able to measure temperatures other than these two fixed points. How can we determine the appropriate place on the thermometer to mark, say, 50°C?

The process of establishing 50°C without a fixed-point calibration is called interpolation. The simplest option would be to divide the distance on the thermometer between the marks representing 0 and 100 into equally spaced degree divisions. This places 50°C as shown in Figure 8.1. What assumption is implicit in this method of interpolation? It is obvious that we do not have enough information to appropriately divide the interval between 0 and 100 on the thermometer into degrees. Some theory of the behavior of the mercury in the thermometer, or many fixed points for calibration are necessary to resolve our dilemma. Even by the late eighteenth century, there was no standard for interpolating between fixed points on the temperature scale; the result was that different thermometers indicated different temperatures away from fixed points, sometimes with surprisingly large errors.

Temperature Scales and Standards

At this point, it is necessary to reconcile this arbitrary temperature scale with the idea of thermodynamic and absolute temperature. Thermodynamics defines a temperature scale that has an absolute reference, and defines an absolute zero for temperature. For example, this absolute temperature governs the energy behavior of an ideal gas, and is used in the ideal gas equation of state. The behavior of real gases at very low pressure may be used as a temperature standard to define a practical measure of temperature that approximates the thermodynamic temperature.

Table 8.1   Temperature Fixed Points as Defined by ITS-90

Defining State	Temperature	
	K	C
Triple point of hydrogen	13.8033	−259.3467
Liquid–vapor equilibrium for hydrogen at 25/76 atm	≈17	≈−256.15
Liquid–vapor equilibrium for hydrogen at 1 atm	≈20.3	≈−252.87
Triple point of neon	24.5561	−248.5939
Triple point of oxygen	54.3584	−218.7916
Triple point of argon	83.8058	−189.3442
Triple point of water	273.16	0.01
Solid–liquid equilibrium for gallium at 1 atm	302.9146	29.7646
Solid–liquid equilibrium for tin at 1 atm	505.078	231.928
Solid–liquid equilibrium for zinc at 1 atm	692.677	419.527
Solid–liquid equilibrium for silver at 1 atm	1234.93	961.78
Solid–liquid equilibrium for gold at 1 atm	1337.33	1064.18
Solid–liquid equilibrium for copper at 1 atm	1357.77	1084.62

The modern engineering definition of the temperature scale is provided by a standard called the International Temperature Scale of 1990 (ITS-90) [2]. This standard establishes fixed points for temperature, and provides standard procedures and devices for interpolating between fixed points. Temperatures established according to ITS-90 do not deviate from the thermodynamic temperature scale by more than the uncertainty in the thermodynamic temperature at the time of adoption of ITS-90. The primary fixed points from ITS-90 are shown in Table 8.1. In addition to these fixed points, other fixed points of secondary importance are available in ITS-90.[3]

Along with the fixed temperature points established by ITS-90, a standard for interpolation between these fixed points is necessary. Standards for acceptable thermometers and interpolating equations are provided in ITS-90. In the range of interest for most engineering applications, accurate interpolation between fixed points is provided by the variation of resistance with temperature for a platinum wire. For temperatures ranging from 13.8033 to 1234.93 K, ITS-90 establishes a platinum resistance thermometer as the standard interpolating instrument, and establishes interpolating equations that relate temperature to resistance. The practical application of these physical phenomena for temperature measurement is discussed in this chapter. Above 1234.93 K the temperature is defined in terms of blackbody radiation, without specifying an instrument for interpolation [2].

In summary, temperature measurement, and a practical temperature scale and standards for fixed points and interpolation have evolved over a period of about two centuries. Present standards for fixed-point temperatures and interpolation allow for practical and accurate measurements of temperature. In the United States, the National Institute of Standards and Technology provides for a means to obtain accurately calibrated platinum wire thermometers for use as secondary standards in the calibration of a temperature measuring system to any practical level of uncertainty.

---

[3]From 1968 through 1989 the International Practical Temperature Scale of 1968 (IPTS-68) (Metrologia 12:7.17) established fixed temperature points and interpolation standards. The differences between IPTS-68 and ITS-90 are documented in [3], which provides the necessary information to assess the importance of changes for instruments calibrated under IPTS-68.

8.3   THERMOMETRY BASED ON THERMAL EXPANSION

Most materials exhibit a change in size with changes in temperature. Since this physical phenomenon is well defined and repeatable, it is useful for temperature measurement. The liquid-in-glass thermometer and the bimetallic thermometer are based on this phenomenon.

Liquid-in-Glass Thermometers

A liquid-in-glass thermometer measures temperature by virtue of the thermal expansion of a liquid. The construction of a liquid-in-glass thermometer is shown in Figure 8.2. The liquid is contained in a glass structure that consists of a bulb and a stem. The bulb serves as a reservoir and provides sufficient fluid for the total volume change of the fluid to cause a detectable rise of the liquid in the stem of the thermometer. The stem contains a glass capillary tube, and the level of the liquid in the capillary is an indication of the temperature. The difference in thermal expansion between the liquid and the glass produces a practical change in the level of the liquid in the glass capillary. Principles and practices of temperature measurement using liquid-in-glass thermometers are described in [4].

During calibration, such a thermometer is subject to one of three measuring environments:

1. For a *complete immersion thermometer*, the entire thermometer is immersed in the calibrating temperature environment or fluid.

2. For a *total immersion thermometer*, the thermometer is immersed in the calibrating temperature environment up to the liquid level in the capillary.

3. For a *partial immersion thermometer*, the thermometer is immersed to a predetermined level in the calibrating environment.

For the most accurate temperature measurements, the thermometer should be immersed in the same manner in use as it was during calibration. In practice, it may not be possible

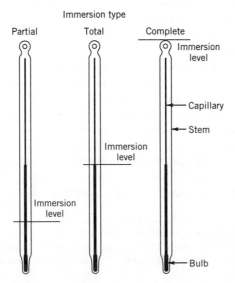

Figure 8.2 Liquid-in-glass thermometer.

to employ the thermometer in exactly the same way as when it was calibrated. In this case, stem corrections can be applied to the temperature reading [5].

Temperature measurements using liquid-in-glass thermometers can provide accuracies to $\pm 0.01°C$ under very carefully controlled conditions; however, such extraneous variables as pressure and changes in bulb volume over time can introduce significant errors in scale calibration. For example, pressure changes increase the indicated temperature by approximately $0.1°C$ per atmosphere [6]. Practical measurements using liquid-in-glass thermometers typically result in total uncertainties that range from ± 0.2 to $\pm 2°C$, depending on the specific instrument.

Bimetallic Thermometers

The physical phenomenon employed in a bimetallic temperature sensor is the *differential thermal expansion* of two metals. Figure 8.3 shows the construction and response of a bimetallic sensor to an input signal. The sensor is constructed by bonding two strips of different metals, A and B. The resulting bimetallic strip may be in a variety of shapes, depending on the particular application. Consider the simple linear construction shown in Figure 8.3. At the assembly temperature, T_1, the bimetallic strip will be straight; however, for temperatures other than T_1 the strip will have a curvature. The physical basis for the relationship between the radius of curvature and temperature is given as

$$r_c \propto \frac{d}{\left[(C_\alpha)_A - (C_\alpha)_B \right] (T_2 - T_1)} \tag{8.1}$$

where

r_c = radius of curvature

C_α = material thermal expansion coefficient

T = temperature

d = thickness

Bimetallic strips employ one metal having a high coefficient of thermal expansion with another having a low coefficient, providing increased sensitivity. Invar is often used as

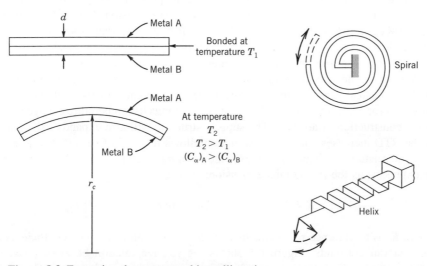

Figure 8.3 Expansion thermometry: bimetallic strip.

one of the metals, since for this material $C_\alpha = 1.7 \times 10^{-8}$ m/m°C, as compared to typical values for other metals, such as steels, which range from approximately 2×10^{-5} to 20×10^{-5} m/m°C.

The bimetallic sensor is used in many temperature control systems, and is the primary element in most dial thermometers. The geometries shown in Figure 8.3 serve to provide the desired deflection in the bimetallic strip for a given application. Dial thermometers using a bimetallic strip as their sensing element typically provide temperature measurements with uncertainties of ±1°C.

8.4   ELECTRICAL RESISTANCE THERMOMETRY

As a result of the physical nature of the conduction of electricity, electrical resistance of a conductor or semiconductor varies with temperature. Using this behavior as the basis for temperature measurement is extremely simple in principle, and leads to two basic classes of resistance thermometers: resistance temperature detectors (conductors) and thermistors (semiconductors). Resistance temperature detectors (RTD) may be formed from a solid metal wire, which exhibits an increase in electrical resistance with temperature. The physical basis for the relationship between resistance and temperature is the temperature dependence of the resistivity, ρ_e, of a material. The resistance of a conductor of length l and cross-sectional area A_c may be expressed in terms of the resistivity as

$$R = \frac{\rho_e l}{A_c} \tag{8.2}$$

Thermistors are semiconductor devices that display a very large decrease in resistance as temperature increases. Current manufacturing techniques provide thermistors that are stable and sufficiently accurate to function as temperature sensors.

Resistance Temperature Detectors

In the case of a resistance temperature detector, or RTD, the sensor is generally constructed by mounting a metal wire on an insulating support structure to eliminate mechanical strains, and by encasing the wire to prevent changes in resistance due to influences from the sensor's environment, such as corrosion. Figure 8.4 shows such a typical RTD construction. Mechanical strain changes a conductor's resistance and must be eliminated if accurate temperature measurements are to be made. This factor is essential because the resistance changes with mechanical strain are significant, as evidenced by the use of metal wire as sensors for the direct measurement of strain. Such mechanical stresses and resulting strains can be created by thermal expansion. Thus, provision for strain-free expansion of the conductor as its temperature changes is essential in the construction of an RTD. The support structure will also expand as the temperature of the RTD increases, and the construction allows for strain-free differential expansion.

The relationship between the resistance of a metal conductor and its temperature can be expressed as the polynomial expansion:

$$R = R_0 \left[1 + \alpha(T - T_0) + \beta(T - T_0)^2 + \cdots \right] \tag{8.3}$$

where R_0 is a reference resistance measured at temperature T_0. The coefficients $\alpha, \beta, \ldots$ are material constants. Figure 8.5 shows the relative relation between resistance and temperature for three common metals. This figure provides evidence that the relationship

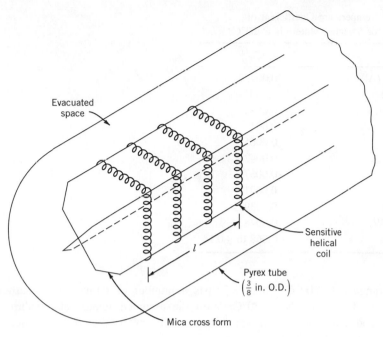

Figure 8.4 Construction of a platinum RTD. (From R. P. Benedict, *Fundamentals of Temperature, Pressure and Flow Measurements*, 3d ed., copyright © 1984 by John Wiley and Sons, New York).

between temperature and resistance over specific small temperature ranges is linear. This approximation can be expressed as

$$R = R_0[1 + \alpha(T - T_0)] \tag{8.4}$$

where α is the temperature coefficient of resistivity. For example, for platinum conductors the linear approximation is accurate to within $\pm0.3\%$ over the range 0–200°C and $\pm1.2\%$

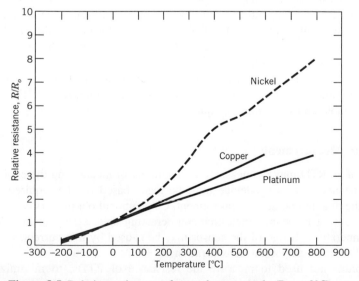

Figure 8.5 Relative resistance of several pure metals (R_o at 0°C).

Table 8.2   Temperature Coefficient of
Resistivity for Selected Materials at 20°C

Substance	$\alpha[°C^{-1}]$
Aluminum (Al)	0.00429
Carbon (C)	−0.0007
Copper (Cu)	0.0043
Gold (Au)	0.004
Iron (Fe)	0.00651
Lead (Pb)	0.0042
Nickel (Ni)	0.0067
Nichrome	0.00017
Platinum (Pt)	0.003927
Tungsten (W)	0.0048
Thermistors	−0.068 to +0.14

over the range 200–800°C. Table 8.2 lists a number of temperature coefficients of resistivity, α, for materials at 20°C. With the assumed linear relationship between resistance and temperature, an appropriate value of α should be chosen for the temperature range of interest.

The most common material chosen for the construction of RTDs is platinum. The RTD relies on the change in electrical resistance of a platinum wire to provide a precise measure of temperature, as expressed in equation (8.4). The principle of operation is quite simple: platinum exhibits a predictable and reproducible change in electrical resistance with temperature, which can be calibrated and interpolated to a high degree of accuracy. The linear approximation for the relationship between temperature and resistance is valid over a wide temperature range, and platinum is highly stable. To be suitable for use as a secondary temperature standard, a platinum resistance thermometer should have a value of α not less than $0.003925°C^{-1}$. This minimum value is an indication of the purity of the platinum. In general, RTDs may be used for the measurement of temperatures ranging from cryogenic to approximately 650°C. By properly constructing an RTD, and correctly measuring its resistance, an uncertainty in temperature measurement as low as ±0.005°C is possible. Because of this potential for low uncertainties and the predictable and stable behavior of platinum, the platinum RTD is widely used as a local standard.

Temperature measurements using an RTD are accomplished by measuring the resistance of the platinum wire and relating this resistance to temperature. For a NIST-certified RTD, a table and interpolating equation would be available.

RTD Resistance Measurement

The resistance of an RTD may be measured by a number of means, and the choice of an appropriate resistance measuring device must be made based on the required level of uncertainty in the final temperature measurement. Conventional ohmmeters cause a small current to flow during resistance measurements, creating self-heating in the RTD. An appreciable temperature change of the sensor may be caused by this current, in effect a loading error. This is an important consideration for RTDs.

Bridge circuits are used to measure the resistance of RTDs, to minimize loading errors and provide low uncertainties in measured resistance values. Wheatstone bridge

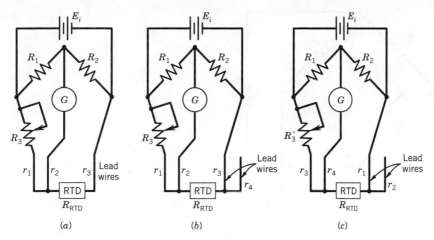

Figure 8.6 Bridge circuits. Average of the two readings (in b and c) eliminates the effect of lead wire resistances.

circuits are commonly used for these measurements. However, the basic Wheatstone bridge circuit does not compensate for the resistance of the leads in measuring the resistance of an RTD, which is a major source of error in electrical resistance thermometers. When greater accuracies are required, three-wire and four-wire bridge circuits are used. Figure 8.6(a) shows a three-wire Callendar-Griffiths bridge circuit. The lead wires numbered 1, 2, and 3 have resistances r_1, r_2, and r_3, respectively. At balanced conditions,

$$\frac{R_1}{R_2} = \frac{R_3}{R_{\text{RTD}}} \tag{8.5}$$

but with the lead wire resistances included in the circuit analysis,

$$\frac{R_1}{R_2} = \frac{R_3 + r_1}{R_{\text{RTD}} + r_3} \tag{8.6}$$

and with $R_1 = R_2$, the resistance of the RTD, R_{RTD}, can be found as

$$R_{\text{RTD}} = R_3 + r_1 - r_3 \tag{8.7}$$

If $r_1 = r_3$, the effect of these lead wires is eliminated from the determination of the RTD resistance by this bridge circuit. Note that the resistance of lead wire 2 does not contribute to any error in the measurement at balanced conditions, since no current flows through the galvanometer, G.

The four-wire Mueller bridge, as shown in Figure 8.6(b), provides increased compensation for lead-wire resistances compared to the Callendar-Griffiths bridge and is used with four-wire RTDs. The four-wire Mueller bridge is typically used when low uncertainties are desired, as in cases where the RTD is used as a laboratory standard. A circuit analysis of the bridge circuit in the first measurement configuration yields

$$R_{\text{RTD}} + r_3 = R_3 + r_1 \tag{8.8}$$

and in the second measurement configuration

$$R_{\text{RTD}} + r_1 = R_3' + r_3 \tag{8.9}$$

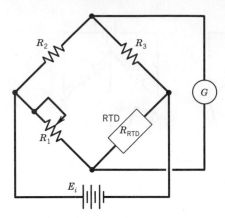

Figure 8.7 RTD Wheatstone bridge arrangement.

where R_3 and R'_3 represent the indicated values of resistance in the first and second configurations, respectively. Adding equations (8.8) and (8.9) results in an expression for the resistance of the RTD in terms of the indicated values for the two measurements:

$$R_{\text{RTD}} = \frac{R_3 + R'_3}{2} \tag{8.10}$$

With this approach, the effect of variations in lead wire resistances is minimized.

EXAMPLE 8.1

An RTD forms one arm of an equal-arm Wheatstone bridge, as shown in Figure 8.7. The fixed resistances, R_2 and R_3 are equal to 25 Ω. The RTD has a resistance of 25 Ω at a temperature of 0°C and is used to measure a temperature that is steady in time.

The resistance of the RTD over a small temperature range may be expressed, as in equation (8.4):

$$R_{\text{RTD}} = R_0[1 + \alpha(T - T_0)]$$

Suppose the coefficient of resistance for this RTD is $0.003925°\text{C}^{-1}$. A temperature measurement is made by placing the RTD in the measuring environment and balancing the bridge by adjusting R_1. The value of R_1 required to balance the bridge is 37.36 Ω. Determine the temperature of the RTD.

KNOWN   An RTD having a resistance of 25 Ω at 0°C with $\alpha = 0.003925°\text{C}^{-1}$ is used to measure a temperature. A value of $R_1 = 37.36$ Ω is required to balance the bridge circuit.

FIND   The temperature of the RTD.

SOLUTION   The resistance of the RTD is measured by balancing the bridge; recall that in a balanced condition,

$$R_{\text{RTD}} = R_1 \frac{R_3}{R_2}$$

The resistance of the RTD is found to be 37.36 Ω. With $R_0 = 25$ Ω at $T = 0°C$, and $\alpha = 0.003925°\text{C}^{-1}$, equation (8.4) becomes

$$37.36\,\Omega = 25(1 + \alpha T)\Omega$$

The temperature of the RTD is found to be 126°C.

EXAMPLE 8.2

Consider the bridge circuit and RTD of Example 8.1. To select or design a bridge circuit for measuring the resistance of the RTD in this example, the required uncertainty in temperature would be specified. If the required uncertainty in the measured temperature is $\leq 0.5°C$, would a 1% total uncertainty in each of the resistors that make up the bridge be acceptable? Neglect the effects of lead wire resistances for this example.

KNOWN   A required uncertainty in temperature of $\pm 0.5°C$, measured with the RTD and bridge circuit of Example 8.1.

FIND   The uncertainty in the measured temperature for a 1% total uncertainty in each of the resistors that make up the bridge circuit.

ASSUMPTION   All uncertainties are provided and evaluated at the 95% confidence level.

SOLUTION   Perform a design-stage uncertainty analysis. Assuming at the design stage that the total uncertainty in the resistances is 1%, then with initial values of the resistances in the bridge equal to 25 Ω, the design-stage uncertainties are set at

$$u_{R_1} = u_{R_2} = u_{R_3} = (0.01)(25) = 0.25\,\Omega$$

The RSS method is used to estimate the propagation of uncertainty in each resistor to the uncertainty in determining the RTD resistance by

$$u_{RTD} = \sqrt{\left[\frac{\partial R}{\partial R_1}u_{R_1}\right]^2 + \left[\frac{\partial R}{\partial R_2}u_{R_2}\right]^2\left[\frac{\partial R}{\partial R_3}u_{R_3}\right]^2}$$

where

$$R = R_{RTD} = \frac{R_1 R_3}{R_2}$$

Then, the design-stage uncertainty in the resistance of the RTD is

$$u_{RTD} = \sqrt{\left[\frac{R_3}{R_2}u_{R_1}\right]^2 + \left[\frac{-R_1 R_3}{R_2^2}u_{R_2}\right]^2 + \left[\frac{R_1}{R_2}u_{R_3}\right]^2}$$

$$u_{RTD} = \sqrt{(1 \times 0.25)^2 + (1 \times -0.25)^2 + (1 \times 0.25)^2} = 0.433\,\Omega$$

To determine the uncertainty in temperature, we know

$$R = R_{RTD} = R_0[1 + \alpha(T - T_0)]$$

and

$$u_T = \sqrt{\left(\frac{\partial T}{\partial R}u_{RTD}\right)^2}$$

Setting $T_0 = 0°C$ with $R_0 = 25\,\Omega$, and neglecting uncertainties in T_0, α, and R_0, we have

$$\frac{\partial T}{\partial R} = \frac{1}{\alpha R_0}$$

$$\frac{1}{\alpha R_0} = \frac{1}{(0.003925°C^{-1})(25\,\Omega)}$$

Then the design-stage uncertainty in temperature is

$$u_T = u_{\text{RTD}}\left(\frac{\partial T}{\partial R}\right) = \frac{0.433\,\Omega}{0.098\,\Omega/°C} = 4.4°C$$

The desired uncertainty in temperature is not achieved with the specified levels of uncertainty in the pertinent variables.

COMMENT   Uncertainty analysis, in this case, would have prevented performing a measurement that would not provide meaningful results.

EXAMPLE 8.3

Suppose the total uncertainty in the bridge resistances of Example 8.1 was reduced to 0.1%. Would the required level of uncertainty in temperature be achieved?

KNOWN   The uncertainty in each of the resistors in the bridge circuit for temperature measurement from Example 8.1 is $\pm 0.1\%$.

FIND   The resulting uncertainty in temperature.

SOLUTION   The uncertainty analysis from the previous example may be directly applied, with the uncertainty values for the resistances appropriately reduced. The uncertainties for the resistances are reduced from 0.25 to 0.025, yielding

$$u_{\text{RTD}} = \pm\sqrt{(1 \times 0.025)^2 + (1 \times -0.025)^2 + (1 \times 0.025)^2} = \pm 0.0433\,\Omega$$

and the resulting uncertainty in temperature is $\pm 0.44°C$, which satisfies the design constraint.

COMMENT   This result provides confidence that the effect of the resistors' uncertainties will not cause the uncertainty in temperature to exceed the target value. However, the uncertainty in temperature not only is a result of the uncertainty in the sensor, the RTD, but will depend on other aspects of the measurement system as well. The design-stage uncertainty analysis performed in this example may be viewed as ensuring that the factors considered do not produce a higher than acceptable uncertainty level. Additional sources of error will exist in any actual measurement using an RTD.

Practical Considerations

The transient thermal response of typical commercial RTDs is generally quite slow compared to other temperature sensors, and for transient measurements bridge circuits must be operated in a deflection mode. For these reasons, RTDs are not generally chosen for transient temperature measurements. A notable exception is the use of very small platinum wires for temperature measurements in noncorrosive flowing gases. In this application, wires having diameters on the order of 10 μm can have frequency responses higher than any other temperature sensor, because of their extremely low thermal capacitance. Obviously, the smallest impact would destroy this sensor. Other resistance sensors in the form of thin metallic films provide fast transient response temperature measurements, often in conjunction with anemometry or heat flux measurements. Such platinum films are constructed by depositing a platinum film onto a substrate, and coating the film with a ceramic glass for mechanical protection [7]. Typical film thickness ranges from 1 to 2 μm, with a 10-μm protective coating. Continuous exposure at temperatures of

600°C is possible with this construction. Some practical uses for film sensors include temperature control circuits for heating systems and cooking devices and surface temperature monitoring on electronic components subject to overheating. Uncertainty levels range from about ±0.1 to 2°C.

Thermistors

Thermistors (from *therm*ally sensitive re*sistors*) are ceramic-like semiconductor devices. The resistance of a typical thermistor decreases rapidly with temperature, which is in contrast to the small increases of resistance with temperature for RTDs. The functional relationship between resistance and temperature for a thermistor is generally assumed to be of the form

$$R = R_0 e^{\beta(1/T - 1/T_0)} \tag{8.11}$$

The parameter β ranges from 3500 to 4600 K, depending on the material, temperature, and individual construction for each sensor, and therefore must be determined for each thermistor. Figure 8.8 shows the variation of resistance with temperature for two common thermistor materials; the ordinate is the ratio of the resistance to the resistance at 25°C. Thermistors exhibit large resistance changes with temperature in comparison to typical RTD as indicated by comparison of Figures 8.5 and 8.8. Equation 8.11 is not accurate over a wide range of temperature, unless β is taken to be a function of temperature; typically the value of β specified by a manufacturer for a sensor is assumed to be constant over a limited temperature range. A simple calibration is possible for determining β as a function of temperature, as illustrated in the circuits shown in Figure 8.9. Other circuits and a more complete discussion of measuring β may be found in the Electronic Industries Association standard Thermistor Definitions and Test Methods [8].

Thermistors are generally used when high sensitivity, ruggedness, or fast response times are required [9]. Thermistors are often encapsulated in glass, and thus can be used in corrosive or abrasive environments. The resistance characteristics of the semiconductor material may change at elevated temperatures, and some aging of a thermistor will occur

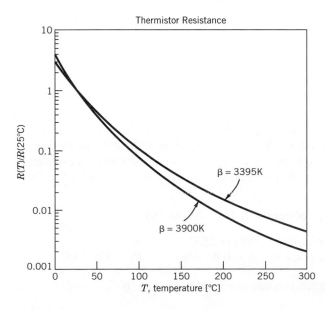

Figure 8.8 Representative thermistor resistance variations with temperature.

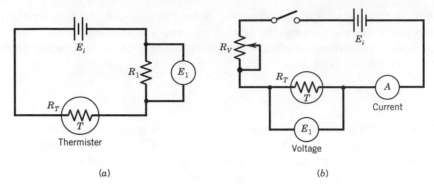

Figure 8.9 Circuits for determining β for thermistors. (a) Voltage divider method: $R_T = R_1(\frac{E_i}{E_1} - 1)$. Note: Both R_1 and E_i must be known values. The value of R_1 may be varied to achieve appropriate values of thermistor current. (b) Volt-ammeter method. Note: Both current and voltage are measured.

at temperatures above 200°C. The high resistance of a thermistor, compared to that of an RTD, eliminates the problems of lead wire resistance compensation. But thermistors are not interchangeable, and variations in room temperature resistances for ordinary thermistors having the same nominal characteristics may be as much as 20%.

The zero-power resistance of a thermistor is the resistance value of the thermistor with no flow of electric current. The dissipation constant for a thermistor is defined at a given ambient temperature as

$$\delta = \frac{P}{T - T_\infty} \tag{8.12}$$

where

δ = dissipation constant

P = power supplied to the thermistor

T, T_∞ = thermistor and ambient temperatures, respectively

The zero power resistance should be measured such that a decrease in the current flow to the thermistor will result in not more than a 0.1% change in resistance.

EXAMPLE 8.4

The material constant β is to be determined for a particular thermistor using the circuit shown in Figure 8.9(a). The thermistor has a resistance of 60 kΩ at 25°C. The reference resistor in the circuit, R_1, has a resistance of 130.5 kΩ. The dissipation constant, δ, is 0.09 mW/°C. The voltage source used for the measurement is constant at 1.564 V. The thermistor is to be used at temperatures ranging from 100 to 150°C. Determine the value of β.

KNOWN   The temperature range of interest is from 100 to 150°C.

$$R_0 = 60,000 \ \Omega, \quad T_0 = 25°C$$
$$E_i = 1.564 \ V, \quad \delta = 0.09 \ mW/°C, \quad R_1 = 130.5 \ k\Omega$$

FIND   The value of β over the temperature range from 100 to 150°C.

SOLUTION   The voltage drop across the fixed resistor is measured for three known values of thermistor temperature. The thermistor temperature is controlled and determined by placing the thermistor in a laboratory oven, and measuring the temperature of the oven. For each measured voltage across the reference resistor, the thermistor resistance, R_T, is determined from

$$R_T = R_1 \left(\frac{E_i}{E_1} - 1 \right) \tag{8.13}$$

The results of these measurements are as follows:

Temperature	R_1 Voltage	R_T
[°C]	[V]	[Ω]
100	1.501	5477.4
125	1.531	2812.9
150	1.545	1604.9

Equation 8.11 can be expressed in the form of a linear equation as

$$\ln \frac{R_T}{R_0} = \beta \left(\frac{1}{T} - \frac{1}{T_0} \right) \tag{8.14}$$

Applying this equation to the measured data, with $R_0 = 60,000\,\Omega$, the three data points above yield the following:

$\ln(R_T/R_0)$	$(1/T - 1/T_0)[\mathrm{K}^{-1}]$	β [K]
−2.394	-6.75×10^{-4}	3546.7
−3.060	-8.43×10^{-4}	3629.9
−3.621	-9.92×10^{-4}	3650.2

COMMENT   These results are for constant β and are based on the behavior described by equation (8.11), over the temperature range from T_0 to the temperature T. The significance of the measured differences in β will be examined further.

The measured values of β in Example 8.4 are different at each value of temperature. If β were truly a temperature independent constant, and these measurements had negligible uncertainty, all three measurements would yield the same value for β. The variation in β may be due to a physical effect of temperature, or may be attributable to the uncertainty in the measured values.

Are the measured differences significant, and, if so, what value of β best represents the behavior of the thermistor over this temperature range? To perform the necessary uncertainty analysis, additional information must be provided concerning the instruments and procedures used in the measurement.

EXAMPLE 8.5

Perform an uncertainty analysis to determine the uncertainty in each measured value of β in Example 8.4, and evaluate a single best estimate of β for this temperature range. The measurement of β involved the measurement of voltages, temperatures, and resistances. For temperature there is a random error associated with spatial and temporal variations in the oven temperature such that $S_{\bar{T}} = 0.19$°C for 20 measurements. In addition, based on a manufacturer's specification, there is a known measurement systematic uncertainty for temperature of ±0.36°C in the thermocouple.

The systematic errors in measuring resistance and voltage are negligible, and estimates of the instrument repeatability based on manufacturer's specifications in the measured values are ±1.5% for resistance and ±0.002 V for the voltage.

KNOWN   Standard deviation of the means for oven temperature, $S_{\bar{T}} = 0.19°C$, $N = 20$. The remaining errors are assigned systematic uncertainties:

$$B_T = \pm 0.36°C$$
$$B_R/R = \pm 1.5\%$$
$$B_E = \pm 0.002\,V$$

FIND   The uncertainty in β at each measured temperature, and a best estimate for β over the measured temperature range.

SOLUTION   Consider the problem of providing a single best estimate of β. One method of estimation might be to average the three measured values. This results in a value of 3609 K. However, since the relationship between (R_T/R_0) and $(1/T - 1/T_0)$ is expected to be linear, a least-squares fit can be performed on the three data points, and include the point $(0, 0)$. The resulting value of β is 3638 K. Is this difference significant, and which value best represents the behavior of the thermistor? To answer these questions, an uncertainty analysis must be performed for β.

For each measured value,

$$\beta = \frac{\ln(R_T/R_0)}{1/T - 1/T_0}$$

Errors in voltage, temperature, and resistance are propagated into the resulting value of β for each measurement.

Consider first the sensitivity indices, θ_i, for each of the variables R_T, R_0, T, and T_0. These may be tabulated by computing the appropriate partial derivatives of β, evaluated at each of the three temperatures, as follows:

T [°C]	θ_{R_T} [K/Ω]	θ_{R_0} [K/Ω]	θ_T	θ_{T_0}
100	−0.270	0.0247	−37.77	59.17
125	−0.422	0.0198	−27.18	48.48
150	−0.628	0.0168	−20.57	41.45

The determination of the uncertainty in β, u_β, requires the uncertainty in the measured value of resistance for the thermistor, u_{R_T}. But R_T is determined from the expression

$$R_T = R_1[(E_i/E_1) - 1]$$

and thus requires an analysis of the uncertainty in the resulting value of R_T, from measured values of R_1, E_i, and E_1. All errors in R_T are treated as systematic errors, yielding

$$B_{R_T} = \sqrt{\left[\frac{\partial R_T}{\partial R_1}B_{R_1}\right]^2 + \left[\frac{\partial R_T}{\partial E_i}B_{E_i}\right]^2 + \left[\frac{\partial R_T}{\partial E_1}B_{E_1}\right]^2}$$

To arrive at a representative value, we compute B_{R_T} at 125°C. The uncertainty in R_1 is $\pm 1.5\%$ of 130.5 kΩ, or ± 1.96 kΩ. The uncertainty in E_i and E_1 are each ± 0.002 V. Using the appropriate values to compute the sensitivity indices, the value of B_{R_T} is found as ± 247 Ω. (See Problem 8.25.)

An uncertainty for β will be determined for each of the measured temperatures. The propagation of the measurement systematic errors for temperature and resistance is found as

$$B_\beta = \sqrt{[\theta_T B_T]^2 + [\theta_{T_0}B_{T_0}]^2 + [\theta_{R_T}B_{R_T}]^2 + [\theta_{R_0}B_{R_0}]^2}$$

where

$$B_T = \pm 0.36°C \qquad B_{R_T} = \pm 247\,\Omega$$
$$B_{T_0} = \pm 0.36°C \qquad B_{R_0} = \pm 900\,\Omega$$

Table 8.3 Uncertainties in β

	Uncertainty (95%)		
	Random	Systematic	Total
T	P_β	B_β	u_β
[°C]	[K]	[K]	[K]
100	13.3	74.7	79.7
125	10.6	107.6	109.9
150	8.8	156.7	157.8

The standard random uncertainty for β contains contributions only from the statistically determined oven temperature characteristics and is found from

$$P_\beta = \sqrt{\left(\theta_T S_{\bar{T}}\right)^2 + \left(\theta_{T_0} S_{\bar{T}_0}\right)^2}$$

where both $S_{\bar{T}}$ and $S_{\bar{T}_0}$ are 0.19, as determined with $N = 20$.

The resulting values of uncertainty in β are found from

$$u_\beta = \sqrt{B_\beta^2 + \left(t_{19,95} P_\beta\right)^2}$$

where $t_{19,95}$ is 2.093. At each temperature the uncertainty in β is determined as shown in Table 8.3. The effect of increases in the sensitivity indices, θ_i, on the total uncertainty is to cause increased uncertainty in β as the temperature increases.

The original results of the measured values of β must now be reexamined. The results, from Table 8.3, are plotted as a function of temperature in Figure 8.10, with uncertainty limits on each data point. Clearly, there is no justification for assuming that the measured values indicate a trend of changes with temperature, and it would be appropriate to use either the average value of β or the value determined from the linear least-squares curve fit.

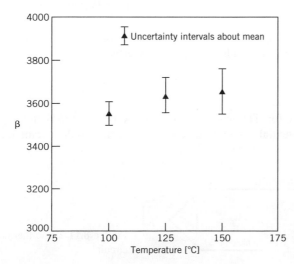

Figure 8.10 Measured values of β and associated uncertainties for three temperatures.

8.5   THERMOELECTRIC TEMPERATURE MEASUREMENT

The most common method of measuring and controlling temperature uses an electrical circuit called a thermocouple. A *thermocouple* consists of two electrical conductors that are made of dissimilar metals and have at least one electrical connection. This electrical connection is referred to as a *junction*. A thermocouple junction may be created by welding, soldering, or by any method that provides good electrical contact between the two conductors, such as twisting the wires around one another. The output of a thermocouple circuit is a voltage, and there is a definite relationship between this voltage and the temperatures of the junctions that make up the thermocouple circuit. We will examine the causes of this voltage, and develop the basis for using thermocouples to make engineering measurements of temperature.

Consider the thermocouple circuit shown in Figure 8.11. The junction labeled 1 is at a temperature T_1 and the junction labeled 2 is at a temperature T_2. If T_1 and T_2 are not equal, a finite open-circuit electric potential, emf_1, will be measured. The magnitude of the potential will depend on the difference in the temperatures and the particular metals the thermocouple circuit contains. A thermocouple junction is the source of an *electromotive force* (emf), which gives rise to the potential difference in a thermocouple circuit. It is the basis for temperature measurement using thermocouples. The circuit shown in Figure 8.11 is the most common form of a thermocouple circuit used for measuring temperature. This thermocouple circuit measures the difference between T_1 and T_2.

Thermoelectric phenomena result from the simultaneous flows of heat and electricity in an electrical conductor. More precisely, these phenomena result from coupled flows of entropy and electricity. The precise mathematical description of the source of these phenomena is provided by irreversible thermodynamics, but such a description of these phenomena is not necessary to use thermocouples to measure temperature. It is our goal to understand the origin of thermoelectric phenomena and the requirements for providing accurate temperature measurements using thermocouples.

In an electrical conductor that is subject to a temperature gradient, there will be both a flow of thermal energy and a flow of electricity. These phenomena are closely tied to the behavior of the free electrons in a metal; it is no coincidence that good electrical conductors are, in general, good thermal conductors. The characteristic behavior of these free electrons in an electrical circuit composed of dissimilar metals results in a useful relationship between temperature and emf. There are three basic phenomena that can occur in a thermocouple circuit: (1) the *Seebeck effect,* (2) the *Peltier effect*, and (3) the *Thomson effect.*

Under ideal measurement conditions, with no loading errors, the emf generated by a thermocouple circuit would be the result of the Seebeck effect only.

Seebeck Effect

The Seebeck effect, named for Thomas Johann Seebeck (1770–1831), refers to the generation of a voltage potential, or emf, in an open thermocouple circuit due to a

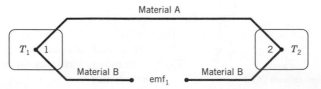

Figure 8.11 Basic thermocouple circuit.

difference in temperature between junctions in the circuit. The Seebeck effect refers to the case when there is no current flow in the circuit, as for an open circuit. There is a fixed, reproducible relationship between the emf and the junction temperatures T_1 and T_2 (Figure 8.11). This relationship is expressed by the Seebeck coefficient, α_{AB}, defined as

$$\alpha_{AB} = \left[\frac{\partial(\text{emf})}{\partial T}\right]_{\text{open circuit}} \tag{8.15}$$

where A and B refer to the two materials that comprise the thermocouple. Since the Seebeck coefficient specifies the rate of change of voltage with temperature for the materials A and B, it is equal to the static sensitivity of the open-circuit thermocouple.

Peltier Effect

A familiar concept is that of I^2R or Joule heating in a conductor through which an electrical current flows. Consider the two conductors having a common junction, shown in Figure 8.12, through which an electrical current, I, flows due to an externally applied emf. For any portion of either of the conductors, the energy removal rate required to maintain a constant temperature is I^2R, where R is the resistance to a current flow and is determined by the resistivity and size of the conductor. However, at the junction of the two dissimilar metals the removal of a quantity of energy different than I^2R is required to maintain a constant temperature. The difference in I^2R and the amount of energy generated by the current flowing through the junction is due to the Peltier effect. The Peltier effect is due to the thermodynamically reversible conversion of energy as a current flows across the junction, in contrast to the irreversible dissipation of energy associated with I^2R losses. *The Peltier heat is the quantity of heat in addition to the quantity I^2R that must be removed from the junction to maintain the junction at a constant temperature.* This amount of energy is proportional to the current flowing through the junction; the proportionality constant is the Peltier coefficient π_{AB}, and the heat transfer required to maintain a constant temperature is

$$Q_\pi = \pi_{AB}I \tag{8.16}$$

caused by the Peltier effect alone. This behavior was discovered by Jean Charles Athanase Peltier (1785–1845) during experiments with Seebeck's thermocouple. He observed that passing a current through a thermocouple circuit having two junctions, as in Figure 8.11, raised the temperature at one junction, while lowering the temperature at the other junction. This effect forms the basis of a device known as a Peltier refrigerator, which provides cooling without moving parts.

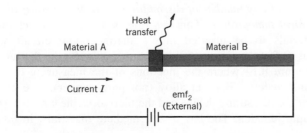

Figure 8.12 Peltier effect due to current flow across a junction of dissimilar metals.

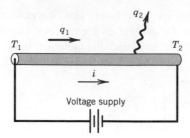

q_1 Energy flow as a result of a temperature gradient
q_2 Heat transfer to maintain constant temperature

Figure 8.13 Thomson effect due to simultaneous flows of current and heat.

Thomson Effect

In addition to the Seebeck effect and the Peltier effect, there is a third phenomenon that occurs in thermoelectric circuits. Consider the conductor shown in Figure 8.13, which is subject to a longitudinal temperature gradient and also subject to a potential difference, such that there is a flow of current and heat in the conductor. Again, to maintain a constant temperature in the conductor it is found that a quantity of energy different than the Joule heat, I^2R, must be removed from the conductor. First noted by William Thomson (1824–1907, Lord Kelvin from 1892) in 1851, this energy is expressed in terms of the Thomson coefficient, σ, as

$$Q_\sigma = \sigma I(T_1 - T_2) \tag{8.17}$$

For a thermocouple circuit, all three of these effects may be present and may contribute to the overall emf of the circuit.

Fundamental Thermocouple Laws

The use of thermocouple circuits to measure temperature is based upon observed behaviors of carefully controlled thermocouple materials and circuits. The following laws provide the basis necessary for temperature measurement with thermocouples:

1. **Law of Homogeneous Materials.** *A thermoelectric current cannot be sustained in a circuit of a single homogeneous material by the application of heat alone, regardless of how it might vary in cross section.* Simply stated, this law requires that at least two materials be used to construct a thermocouple circuit for the purpose of measuring temperature. It is interesting to note that a current may occur in an inhomogeneous wire that is nonuniformly heated; however, this is neither useful nor desirable in a thermocouple.

2. **Law of Intermediate Materials.** *The algebraic sum of the thermoelectric forces in a circuit composed of any number of dissimilar materials is zero if all of the circuit is at a uniform temperature.* This law allows a material other than the thermocouple materials to be inserted into a thermocouple circuit without changing the output emf of the circuit. As an example, consider the thermocouple circuit shown in Figure 8.14 where the junctions of the measuring device are made of copper and material B is an alloy (not pure copper). The electrical connection between the measuring device and the thermocouple circuit forms yet another thermocouple junction. The law of intermediate materials, in this case,

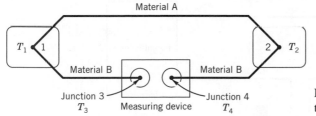

Figure 8.14 Typical thermocouple measuring circuit.

provides that the measured emf will be unchanged from the open-circuit emf, which corresponds to the temperature difference between T_1 and T_2, if $T_3 = T_4$. Another practical consequence of this law is that copper extension wires may be used to transmit thermocouple emfs to a measuring device. This is very beneficial because most materials used in thermocouple wire can be significantly more expensive than copper.

3. **Law of Successive or Intermediate Temperatures.** *If two dissimilar homogeneous materials that form a thermocouple circuit produce emf₁ when the junctions are at T_1 and T_2 and produce emf₂ when the junctions are at T_2 and T_3, the emf generated when the junctions are at T_1 and T_3 will be emf₁ + emf₂.* The law of intermediate temperatures is of great importance in practical temperature measurements. This law allows a thermocouple calibrated for one reference temperature to be used at another reference temperature.

Basic Temperature Measurement with Thermocouples

The basic thermocouple circuit shown in Figure 8.14 can be used to measure the difference between the two temperatures T_1 and T_2. For practical temperature measurements, one of these junctions becomes a reference junction, and is maintained at some known, constant reference temperature. The other junction then becomes the measuring junction, and the emf existing in the circuit for any temperature T_1 provides a direct indication of the temperature of the measuring junction.

Let's first examine a historically significant method of using a thermocouple circuit to measure temperature. Figure 8.15 shows two basic thermocouple circuits, using a chromel-constantan thermocouple and an ice bath to create a reference temperature. In Figure 8.15(*a*), the thermocouple wires are connected directly to a potentiometer to measure the emf. In Figure 8.15(*b*), copper extension wires are employed, creating two reference junctions. The law of intermediate materials ensures that neither the potentiometer nor the extension wires will change the emf of the circuit, as long as the two connecting junctions at the potentiometer and the two in the ice bath experience no temperature difference. All that is required to be able to measure temperature with this circuit is to know the relationship between the output emf and the temperature of the measuring junction, for the particular reference temperature. One method of determining this relationship is to calibrate the thermocouple. *However, we shall see that for reasonable levels of uncertainty for temperature measurement, standard materials and procedures allow thermocouples to be accurate temperature measuring devices without the necessity of calibration.*

The provisions for a reference junction should provide a temperature that is accurately known, stable, and reproducible. A very common reference junction temperature is provided by the ice point, 0°C, because of the ease with which it can be obtained.

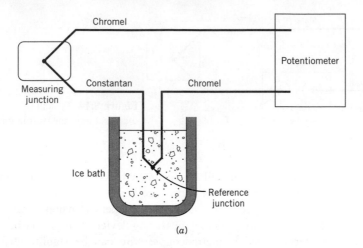

(a)

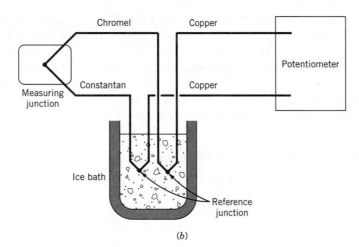

(b)

Figure 8.15 Thermocouple temperature measurement circuits.

The creation of a reference junction temperature of 0°C is accomplished in either of two basic ways. Prior to the development of an electronic means of creating a reference point in the electric circuit, an ice bath served to provide the reference junction temperature. An ice bath is typically made by filling a vacuum flask, or Dewar, with finely crushed ice, and adding just enough water to create a transparent slush. Surprising results are often obtained when the temperature of a mixture of ice and water is measured to verify the ice point is achieved. A few ice cubes floating in water does not create a 0°C environment! Ice baths can be constructed to provide a reference junction temperature to an uncertainty within ±0.01°C.

Electronic reference junctions provide a convenient means of the measurement of temperature without the necessity to construct an ice bath. Numerous manufacturers produce commercial temperature measuring devices with built-in reference junction compensation, and many digital data acquisition cards for personal computers include built-in reference junction compensation. The electronics generally rely on a thermistor to determine the local environment temperature, as shown in Figure 8.16. Uncertainties for the reference junction temperature in this case are on the order of ±0.1°C, with ±0.5°C as typical.

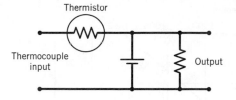

Figure 8.16 Basic thermistor circuit for thermocouple reference junction compensation.

Thermocouple Standards

NIST provides specifications for the materials and construction of standard thermocouple circuits for temperature measurement [10]. Many material combinations exist for thermocouples; these material combinations are identified by a thermocouple type and denoted by a letter. Table 8.4 shows the letter designations and the polarity of common thermocouples, along with some basic application information for each type. The choice of a type of thermocouple depends on the temperature range to be measured, the particular application, and the desired uncertainty level.

To determine the emf output of a particular material combination, a thermocouple is formed from a candidate material and a standard platinum alloy to form a thermocouple circuit having a 0°C reference temperature. Figure 8.17 shows the output of various materials in combination with platinum-67. The notation indicates the thermocouple type. The law of intermediate temperatures then allows the emf of any two materials whose emf relative to platinum is known to be determined. Figure 8.18 shows a plot of the emf as a function of temperature for some common thermocouple material combinations. The slope of the curves in this figure corresponds to the static sensitivity of the thermocouple measuring circuit.

Standard Thermocouple Voltage

Table 8.5 provides the standard composition of thermocouple materials, along with standard limits of error for the various material combinations. These limits specify the expected maximum errors resulting from the thermocouple materials. The NIST uses high-purity materials to establish the standard value of voltage output for a thermocouple composed of two specific materials. This results in standard tables or equations used to

Table 8.4   Thermocouple Designations

| Type | Material Combination | | Applications |
	Positive	Negative	
E	Chromel(+)	Constantan(−)	Highest sensitivity (<1000°C)
J	Iron(+)	Constantan(−)	Nonoxidizing environment (<760°C)
K	Chromel(+)	Alumel(−)	High temperature (<1372°C)
S	Platinum/ 10% rhodium	Platinum(−)	Long-term stability high temperature (<1768°C)
T	Copper(+)	Constantan(−)	Reducing or vacuum environments (<400°C)

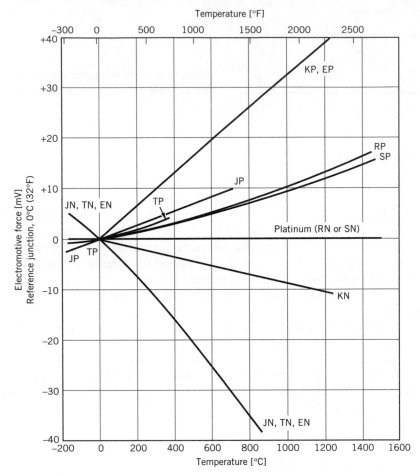

Figure 8.17 Thermal emf of thermocouple materials relative to platinum-67. Note: JP indicates the positive leg of a J thermocouple, or iron. (From R. P. Benedict, *Fundamentals of Temperature, Pressure and Flow Measurements*, 3d ed., copyright © 1984 by John Wiley and Sons, New York. Reprinted by permission.)

determine a measured temperature from a measured value of emf [10]. An example of such a table is provided in Table 8.6 for an iron/constantan thermocouple, usually referred to as a J-type thermocouple. Table 8.7 provides polynomial equations that relate emf and temperature for standard thermocouples. Because of the widespread need to measure temperature, an industry has grown up to supply high-grade thermocouple wire. Manufacturers can also provide thermocouples having special tolerance limits relative to the NIST standard voltages ranging from ±1.0°C to perhaps ±0.1°C. Thermocouples constructed of standard thermocouple wire do not require calibration to provide measurement of temperature within the tolerance limits given in Table 8.5.

Thermocouple Voltage Measurement

The Seebeck voltage for a thermocouple circuit is measured with no current flow in the circuit. From our discussion of the Thomson and Peltier effects, it is clear that the emf will be slightly changed from the open-circuit value when there is a current flow in the

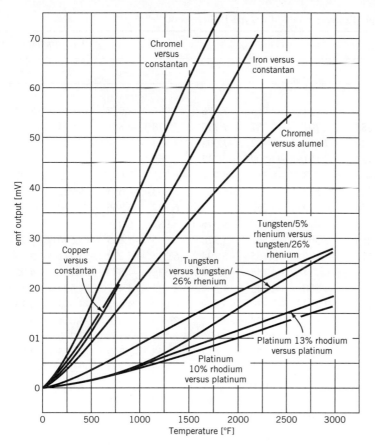

Figure 8.18 Thermocouple voltage output as a function of temperature for some common thermocouple materials. Reference junction is at 0°C. (From R. P. Benedict, *Fundamentals of Temperature, Pressure and Flow Measurements*, 3d ed., copyright © 1984 by John Wiley and Sons, New York. Reprinted by permission.)

Table 8.5   Standard Thermocouple Compositions[a]

| | Wire | | Expected |
Type	Positive	Negative	Systematic Error[b]
S	Platinum	Platinum/10% rhodium	±1.5°C or 0.25%
R	Platinum	Platinum/13% rhodium	±1.5°C
B	Platinum/30% rhodium	Platinum/6% rhodium	±0.5%
T	Copper	Constantan	±1.0°C or 0.75%
J	Iron	Constantan	±2.2°C or 0.75%
K	Chromel	Alumel	±2.2°C or 0.75%
E	Chromel	Constantan	±1.7°C or 0.5%

Alloy Designations

Constantan: 55% copper with 45% nickel

Chromel: 90% nickel with 10% chromium

Alumel: 94% nickel with 3% manganese, 2% aluminum, and 1% silicon

[a]From Temperature Measurements ANSI PTC 19.3-1974.
[b]Use greater value; these limits of error do not include installation errors.

Table 8.6   Thermocouple Reference Table for Type-J Thermocouple[a]

Temperature (°C)									Thermocouple emf (mV)	
	0	−1	−2	−3	−4	−5	−6	−7	−8	−9
−210	−8.095									
−200	−7.890	−7.912	−7.934	−7.955	−7.976	−7.996	−8.017	−8.037	−8.057	−8.076
−190	−7.659	−7.683	−7.707	−7.731	−7.755	−7.778	−7.801	−7.824	−7.846	−7.868
−180	−7.403	−7.429	−7.456	−7.482	−7.508	−7.534	−7.559	−7.585	−7.610	−7.634
−170	−7.123	−7.152	−7.181	−7.209	−7.237	−7.265	−7.293	−7.321	−7.348	−7.376
−160	−6.821	−6.853	−6.883	−6.914	−6.944	−6.975	−7.005	−7.035	−7.064	−7.094
−150	−6.500	−6.533	−6.566	−6.598	−6.631	−6.663	−6.695	−6.727	−6.759	−6.790
−140	−6.159	−6.194	−6.229	−6.263	−6.298	−6.332	−6.366	−6.400	−6.433	−6.467
−130	−5.801	−5.838	−5.874	−5.910	−5.946	−5.982	−6.018	−6.054	−6.089	−6.124
−120	−5.426	−5.465	−5.503	−5.541	−5.578	−5.616	−5.653	−5.690	−5.727	−5.764
−110	−5.037	−5.076	−5.116	−5.155	−5.194	−5.233	−5.272	−5.311	−5.350	−5.388
−100	−4.633	−4.674	−4.714	−4.755	−4.796	−4.836	−4.877	−4.917	−4.957	−4.997
−90	−4.215	−4.257	−4.300	−4.342	−4.384	−4.425	−4.467	−4.509	−4.550	−4.591
−80	−3.786	−3.829	−3.872	−3.916	−3.959	−4.002	−4.045	−4.088	−4.130	−4.173
−70	−3.344	−3.389	−3.434	−3.478	−3.522	−3.566	−3.610	−3.654	−3.698	−3.742
−60	−2.893	−2.938	−2.984	−3.029	−3.075	−3.120	−3.165	−3.210	−3.255	−3.300
−50	−2.431	−2.478	−2.524	−2.571	−2.617	−2.663	−2.709	−2.755	−2.801	−2.847
−40	−1.961	−2.008	−2.055	−2.103	−2.150	−2.197	−2.244	−2.291	−2.338	−2.385
−30	−1.482	−1.530	−1.578	−1.626	−1.674	−1.722	−1.770	−1.818	−1.865	−1.913
−20	−0.995	−1.044	−1.093	−1.142	−1.190	−1.239	−1.288	−1.336	−1.385	−1.433
−10	−0.501	−0.550	−0.600	−0.650	−0.699	−0.749	−0.798	−0.847	−0.896	−0.946
0	0.000	−0.050	−0.101	−0.151	−0.201	−0.251	−0.301	−0.351	−0.401	−0.451
	0	+1	+2	+3	+4	+5	+6	+7	+8	+9
0	0.000	0.050	0.101	0.151	0.202	0.253	0.303	0.354	0.405	0.451
10	0.507	0.558	0.609	0.660	0.711	0.762	0.814	0.865	0.916	0.968
20	1.019	1.071	1.122	1.174	1.226	1.277	1.329	1.381	1.433	1.485
30	1.537	1.589	1.641	1.693	1.745	1.797	1.849	1.902	1.954	2.006
40	2.059	2.111	2.164	2.216	2.269	2.322	2.374	2.427	2.480	2.532
50	2.585	2.638	2.691	2.744	2.797	2.850	2.903	2.956	3.009	3.062
60	3.116	3.169	3.222	3.275	3.329	3.382	3.436	3.489	3.543	3.596
70	3.650	3.703	3.757	3.810	3.864	3.918	3.971	4.025	4.079	4.133
80	4.187	4.240	4.294	4.348	4.402	4.456	4.510	4.564	4.618	4.672
90	4.726	4.781	4.835	4.889	4.943	4.997	5.052	5.106	5.160	5.215
100	5.269	5.323	5.378	5.432	5.487	5.541	5.595	5.650	5.705	5.759
110	5.814	5.868	5.923	5.977	6.032	6.087	6.141	6.196	6.251	6.306
120	6.360	6.415	6.470	6.525	6.579	6.634	6.689	6.744	6.799	6.854
130	6.909	6.964	7.019	7.074	7.129	7.184	7.239	7.294	7.349	7.404
140	7.459	7.514	7.569	7.624	7.679	7.734	7.789	7.844	7.900	7.955
150	8.010	8.065	8.120	8.175	8.231	8.286	8.341	8.396	8.452	8.507
160	8.562	8.618	8.673	8.728	8.783	8.839	8.894	8.949	9.005	9.060
170	9.115	9.171	9.226	9.282	9.337	9.392	9.448	9.503	9.559	9.614
180	9.669	9.725	9.780	9.836	9.891	9.947	10.002	10.057	10.113	10.168
190	10.224	10.279	10.335	10.390	10.446	10.501	10.557	10.612	10.668	10.723

(*Continued*)

Table 8.6 (*Continued*)

Temperature (°C)									Thermocouple emf (mV)	
	0	+1	+2	+3	+4	+5	+6	+7	+8	+9
200	10.779	10.834	10.890	10.945	11.001	11.056	11.112	11.167	11.223	11.278
210	11.334	11.389	11.445	11.501	11.556	11.612	11.667	11.723	11.778	11.834
220	11.889	11.945	12.000	12.056	12.111	12.167	12.222	12.278	12.334	12.389
230	12.445	12.500	12.556	12.611	12.667	12.722	12.778	12.833	12.889	12.944
240	13.000	13.056	13.111	13.167	13.222	13.278	13.333	13.389	13.444	13.500
250	13.555	13.611	13.666	13.722	13.777	13.833	13.888	13.944	13.999	14.055
260	14.110	14.166	14.221	14.277	14.332	14.388	14.443	14.499	14.554	14.609
270	14.665	14.720	14.776	14.831	14.887	14.942	14.998	15.053	15.109	15.164
280	15.219	15.275	15.330	15.386	15.441	15.496	15.552	15.607	15.663	15.718
290	15.773	15.829	15.884	15.940	15.995	16.050	16.106	16.161	16.216	16.272
300	16.327	16.383	16.438	16.493	16.549	16.604	16.659	16.715	16.770	16.825
310	16.881	16.936	16.991	17.046	17.102	17.157	17.212	17.268	17.323	17.378
320	17.434	17.489	17.544	17.599	17.655	17.710	17.765	17.820	17.876	17.931
330	17.986	18.041	18.097	18.152	18.207	18.262	18.318	18.373	18.428	18.483
340	18.538	18.594	18.649	18.704	18.759	18.814	18.870	18.925	18.980	19.035
350	19.090	19.146	19.201	19.256	19.311	19.366	19.422	19.477	19.532	19.587
360	19.642	19.697	19.753	19.808	19.863	19.918	19.973	20.028	20.083	20.139
370	20.194	20.249	20.304	20.359	20.414	20.469	20.525	20.580	20.635	20.690
380	20.745	20.800	20.855	20.911	20.966	21.021	21.076	21.131	21.186	21.241
390	21.297	21.352	21.407	21.462	21.517	21.572	21.627	21.683	21.738	21.793
400	21.848	21.903	21.958	22.014	22.069	22.124	22.179	22.234	22.289	22.345
410	22.400	22.455	22.510	22.565	22.620	22.676	22.731	22.786	22.841	22.896
420	22.952	23.007	23.062	23.117	23.172	23.228	23.283	23.338	23.393	23.449
430	23.504	23.559	23.614	23.670	23.725	23.780	23.835	23.891	23.946	24.001
440	24.057	24.112	24.167	24.223	24.278	24.333	24.389	24.444	24.499	24.555
450	24.610	24.665	24.721	24.776	24.832	24.887	24.943	24.998	25.053	25.109
460	25.164	25.220	25.275	25.331	25.386	25.442	25.497	25.553	25.608	25.664
470	25.720	25.775	25.831	25.886	25.942	25.998	26.053	26.109	26.165	26.220
480	26.276	26.332	26.387	26.443	26.499	26.555	26.610	26.666	26.722	26.778
490	26.834	26.889	26.945	27.001	27.057	27.113	27.169	27.225	27.281	27.337
500	27.393	27.449	27.505	27.561	27.617	27.673	27.729	27.785	27.841	27.897
510	27.953	28.010	28.066	28.122	28.178	28.234	28.291	28.347	28.403	28.460
520	28.516	28.572	28.629	28.685	28.741	28.798	28.854	28.911	28.967	29.024
530	29.080	29.137	29.194	29.250	29.307	29.363	29.420	29.477	29.534	29.590
540	29.647	29.704	29.761	29.818	29.874	29.931	29.988	30.045	30.102	30.159
550	30.216	30.273	30.330	30.387	30.444	30.502	30.559	30.616	30.673	30.730
560	30.788	30.845	30.902	30.960	31.017	31.074	31.132	31.189	31.247	31.304
570	31.362	31.419	31.477	31.535	31.592	31.650	31.708	31.766	31.823	31.881
580	31.939	31.997	32.055	32.113	32.171	32.229	32.287	32.345	32.403	32.461
590	32.519	32.577	32.636	32.694	32.752	32.810	32.869	32.927	32.985	33.044
600	33.102	33.161	33.219	33.278	33.337	33.395	33.454	33.513	33.571	33.630
610	33.689	33.748	33.807	33.866	33.925	33.984	34.043	34.102	34.161	34.220
620	34.279	34.338	34.397	34.457	34.516	34.575	34.635	34.694	34.754	34.813
630	34.873	34.932	34.992	35.051	35.111	35.171	35.230	35.290	35.350	35.410

(*Continued*)

Table 8.6 *(Continued)*

Temperature (°C)										Thermocouple emf (mV)
	0	+1	+2	+3	+4	+5	+6	+7	+8	+9
640	35.470	35.530	35.590	35.650	35.710	35.770	35.830	35.890	35.950	36.010
650	36.071	36.131	36.191	36.252	36.312	36.373	36.433	36.494	36.554	36.615
660	36.675	36.736	36.797	36.858	36.918	36.979	37.040	37.101	37.162	37.223
670	37.284	37.345	37.406	37.467	37.528	37.590	37.651	37.712	37.773	37.835
680	37.896	37.958	38.019	38.081	38.142	38.204	38.265	38.327	38.389	38.450
690	38.512	38.574	38.636	38.698	38.760	38.822	38.884	38.946	39.008	39.070
700	39.132	39.194	39.256	39.318	39.381	39.443	39.505	39.568	39.630	39.693
710	39.755	39.818	39.880	39.943	40.005	40.068	40.131	40.193	40.256	40.319
720	40.382	40.445	40.508	40.570	40.633	40.696	40.759	40.822	40.886	40.949
730	41.012	41.075	41.138	41.201	41.265	41.328	41.391	41.455	41.518	41.581
740	41.645	41.708	41.772	41.835	41.899	41.962	42.026	42.090	42.153	42.217
750	42.281	42.344	42.408	42.472	42.536	42.599	42.663	42.727	42.791	42.855
760	42.919	42.983	43.047	43.110	43.174	43.238	43.303	43.367	43.431	43.495

[a]Reference junction at 0°C.

thermocouple circuit. As such, the best method for the measurement of thermocouple voltages is a device that minimizes current flow. For many years, potentiometer was the laboratory standard for voltage measurement in thermocouple circuits. A potentiometer, as described in Chapter 6, has nearly zero loading error at a balanced condition. However, modern voltage-measuring devices, such as digital voltmeters or data-acquisition cards, have such a high input impedance that they can be used with minimal loading error. These devices can also be used in either static or dynamic measuring situations where the loading error created by the measurement device is acceptable for the particular application. For such needs, high-impedance voltmeters have been incorporated into commercially available temperature indicators, temperature controllers, and digital data-acquisition systems.

EXAMPLE 8.6

The thermocouple circuit shown in Figure 8.19 is used to measure the temperature T_1. The thermocouple junction labeled 2 is at a temperature of 0°C, maintained by an ice-point bath. The voltage output is measured using a potentiometer, and found to be 9.669 mV. What is T_1?

KNOWN A thermocouple circuit having one junction at 0°C and a second junction at an unknown temperature. The circuit produces an emf of 9.669 mV.

FIND The temperature T_1.

ASSUMPTION Thermocouple follows NIST standard.

SOLUTION Standard thermocouple tables such as Table 8.6 are referenced to 0°C. The temperature of the reference junction for this case is 0°C. Therefore, the temperature corresponding to an output voltage may simply be determined from Table 8.6, in this case as 180°C.

Table 8.7   Reference Functions for Selected Letter Designated Thermocouples

The relationship between emf and temperature is provided in the form of a polynomial in temperature [10]

$$E = \sum_{i=0}^{n} c_i T^i$$

where E is in mV and T is in °C. Constants are provided below.

Thermocouple Type	Temperature Range	Constants
J-type	−210–760°C	$c_0 = 0.000\,000\,000\,0$
		$c_1 = 5.038\,118\,781\,5 \times 10^1$
		$c_2 = 3.047\,583\,693\,0 \times 10^{-2}$
		$c_3 = -8.568\,106\,572\,0 \times 10^{-5}$
		$c_4 = 1.322\,819\,529\,5 \times 10^{-7}$
		$c_5 = -1.705\,295\,833\,7 \times 10^{-10}$
		$c_6 = 2.094\,809\,069\,7 \times 10^{-13}$
		$c_7 = -1.253\,839\,533\,6 \times 10^{-16}$
		$c_8 = 1.563\,172\,569\,7 \times 10^{-20}$
T-type	−270–0°C	$c_0 = 0.000\,000\,000\,0$
		$c_1 = 3.874\,810\,636\,4 \times 10^1$
		$c_2 = 4.419\,443\,434\,7 \times 10^{-2}$
		$c_3 = 1.184\,432\,310\,5 \times 10^{-4}$
		$c_4 = 2.003\,297\,355\,4 \times 10^{-5}$
		$c_5 = 9.013\,801\,955\,9 \times 10^{-7}$
		$c_6 = 2.265\,115\,659\,3 \times 10^{-8}$
		$c_7 = 3.607\,115\,420\,5 \times 10^{-10}$
		$c_8 = 3.849\,393\,988\,3 \times 10^{-12}$
		$c_9 = 2.821\,352\,192\,5 \times 10^{-14}$
		$c_{10} = 1.425\,159\,477\,9 \times 10^{-16}$
		$c_{11} = 4.876\,866\,228\,6 \times 10^{-19}$
		$c_{12} = 1.079\,553\,927\,0 \times 10^{-21}$
		$c_{13} = 1.394\,502\,706\,2 \times 10^{-24}$
		$c_{14} = 7.979\,515\,392\,7 \times 10^{-28}$
T-type	0–400°C	$c_0 = 0.000\,000\,000\,0$
		$c_1 = 3.874\,810\,636\,4 \times 10^1$
		$c_2 = 3.329\,222\,788\,0 \times 10^{-2}$
		$c_3 = 2.061\,824\,340\,4 \times 10^{-4}$
		$c_4 = -2.188\,225\,684\,6 \times 10^{-6}$
		$c_5 = 1.099\,688\,092\,8 \times 10^{-8}$
		$c_6 = -3.081\,575\,877\,2 \times 10^{-11}$
		$c_7 = 4.547\,913\,529\,0 \times 10^{-14}$
		$c_8 = -2.751\,290\,167\,3 \times 10^{-17}$

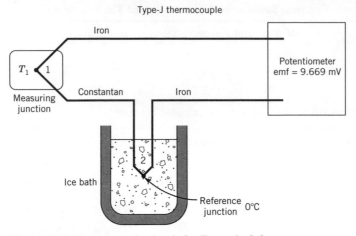

Figure 8.19 Thermocouple circuit for Example 8.6.

COMMENT Because of the law of intermediate metals, the junctions formed at the potentiometer do not affect the voltage measured for the thermocouple circuit, and the voltage output reflects accurately the temperature difference between junctions 1 and 2.

EXAMPLE 8.7

Suppose the thermocouple circuit in the previous example (Example 8.6) now has junction 2 maintained at a temperature of 30°C, and produces an output voltage of 8.132 mV. What temperature is sensed by the measuring junction?

KNOWN T_2 is 30°C, and the output emf is 8.132 mV.

ASSUMPTION Thermocouple follows NIST standard emf behavior.

FIND The temperature of the measuring junction.

SOLUTION By the law of intermediate temperatures the output emf for a thermocouple circuit having two junctions, one at 0°C and the other at T_1, would be the sum of the emfs for a thermocouple circuit between 0 and 30°C and between 30°C and T_1. Thus,

$$\text{emf}_{0-30} + \text{emf}_{30-T_1} = \text{emf}_{0-T_1}$$

This relationship allows the voltage reading from the nonstandard reference temperature to be converted to a 0°C reference temperature by adding $\text{emf}_{0-30} = 1.537$ to the existing reading. This results in an equivalent output voltage, referenced to 0°C as

$$1.537 + 8.132 = 9.669 \text{ mV}$$

Clearly, this thermocouple is sensing the same temperature as in the previous example, 180°C. This value is determined from Table 8.6.

COMMENT Note that the effect of raising the reference junction temperature is to lower the output voltage of the thermocouple circuit. Negative values of voltage, as compared with the polarity listed in Table 8.4, indicate that the measured temperature is less than the reference junction temperature.

> **EXAMPLE 8.8**
>
> A J-type thermocouple measures a temperature of 100°C and is referenced to 0°C. The thermocouple is AWG 30 (30-gauge or 0.010-in. wire diameter) and is arranged in a circuit as shown in Figure 8.15(*a*). The length of the thermocouple wire is 10 ft, in order to run from the measurement point to the ice bath and to a potentiometer. The resolution of the potentiometer is 0.005 mV. If the thermocouple wire has a resistance per unit length, as specified by the manufacturer, of 5.6 Ω/ft, estimate the residual current in the thermocouple when the circuit is balanced within the resolution of the potentiometer.

KNOWN   A potentiometer having a resolution of 0.005 mV is used to measure the emf of a J-type thermocouple that is 10 ft long.

FIND   The residual current in the thermocouple circuit.

SOLUTION   The total resistance of the thermocouple circuit is 56 Ω for 10 ft of thermocouple wire. The residual current is then found from Ohm's law as

$$I = \frac{E}{R} = \frac{0.005\,\mathrm{mV}}{56\,\Omega} = 8.9 \times 10^{-8}\,\mathrm{A}$$

COMMENT   The loading error due to this current flow is $\sim$0.005 mV/54.3 μV/°C $\approx$ 0.1°C.

> **EXAMPLE 8.9**
>
> Suppose a high-impedance voltmeter is used in place of the potentiometer in Example 8.8. Determine the minimum input impedance required for the voltmeter that will limit the loading error to the same level as the potentiometer.

KNOWN   Loading error should be less than 8.9×10^{-8} A.

FIND   Input impedance for a voltmeter that would produce the same current flow or loading error.

SOLUTION   At 100°C a J-type thermocouple referenced to 0°C will have a Seebeck voltage of $E_s = 5.269$ mV. At this temperature, the required voltmeter impedance to limit the current flow to 8.9×10^{-8} A is found from Ohm's law:

$$\frac{E_s}{I} = 5.269 \times 10^{-3}\,\mathrm{V}/8.9 \times 10^{-8}\,\mathrm{A} = 59.2\,\mathrm{k}\Omega$$

COMMENT   This input impedance is not high for microvoltmeters and indicates that such a voltmeter would be a reasonable choice in this situation. As always, the allowable loading error should be determined based on the required uncertainty in the measured temperature.

Multiple-Junction Thermocouple Circuits

A thermocouple circuit composed of two junctions of dissimilar metals produces an open-circuit emf that is related to the temperature difference between the two junctions. More than two junctions can be employed in a thermocouple circuit, and thermocouple circuits can be devised to measure temperature differences, or average temperature, or to amplify the output voltage of a thermocouple circuit.

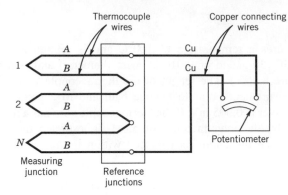

Figure 8.20 Thermopile arrangement. (From R. P. Benedict, *Fundamentals of Temperature, Pressure and Flow Measurements*, 3d ed., copyright © 1984 by John Wiley and Sons, New York. Reprinted by permission.)

Thermopiles

Thermopile is the term used to describe a multiple-junction thermocouple circuit that is designed to amplify the output of the circuit. Because thermocouple voltage outputs are typically in the millivolt range, increasing the voltage output may be a key element in reducing the uncertainty in the temperature measurement, or may be necessary to allow transmission of the thermocouple signal to the recording device. Figure 8.20 shows a thermopile for providing an amplified output signal; in this case the output voltage would be N times the single thermocouple output, where N is the number of junctions in the circuit. The average output voltage corresponds to the average temperature level sensed by the N junctions. This thermopile arrangement can be used to measure a spatially averaged temperature, or to measure a single value of temperature. The measurement of a single value of temperature entails considerations of the physical size of a thermopile, as compared to a single thermocouple. In transient measurements, a thermopile may have a more limited frequency range than a single thermocouple, due to its increased thermal capacitance. Thermopiles are particularly useful for reducing the uncertainty in measuring small temperature differences between the measuring and reference junctions. The principle has also been used to generate small amounts of power in spacecraft and to provide thermoelectric cooling.

Figure 8.21 shows a series arrangement of thermocouple junctions designed to measure the average temperature difference between junctions. This thermocouple circuit could be used in an environment where a uniform temperature was desired. In that case, a voltage output would indicate that a temperature difference existed between two of the thermocouple junctions. It should be noted, however, that a zero voltage output could also occur if the temperature differences that occurred in the circuit summed to a zero emf. Alternatively, junctions $1, 2, \ldots, N$ could be located at one physical location, while junctions $1', 2', \ldots, N'$ could be located at another physical location. Applications for such a circuit might be the measurement of heat flux through a solid.

Thermocouples in Parallel

When a spatially averaged temperature is desired, multiple thermocouple junctions can be arranged as shown in Figure 8.22. In such an arrangement of N junctions, a mean emf is produced, given by

$$\overline{\text{emf}} = \frac{1}{N} \sum_{i=1}^{N} (\text{emf})_i \tag{8.18}$$

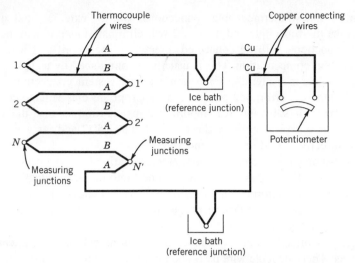

Figure 8.21 Thermocouples arranged to sense temperature differences. (From R. P. Benedict, *Fundamentals of Temperature, Pressure and Flow Measurements*, 3d ed., copyright © 1984 by John Wiley and Sons, New York. Reprinted by permission.)

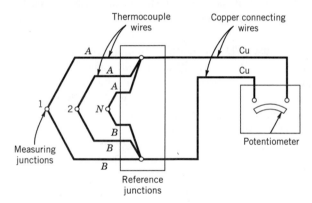

Figure 8.22 Parallel arrangement of thermocouples for sensing the average temperature of the measuring junctions. (From R. P. Benedict, *Fundamentals of Temperature, Pressure and Flow Measurements*, 3d ed., copyright © 1984 by John Wiley and Sons, New York. Reprinted by permission.)

The mean emf is indicative of a mean temperature,

$$\bar{T} = \frac{1}{N} \sum_{i=1}^{N} T_i \qquad (8.19)$$

Measuring temperatures by using thermocouples connected to a data-acquisition systems is common practice. However, the characteristics of the thermocouple, including the need for a reference or cold junction and the low signal voltages produced, complicate its use. Nevertheless, with a little attention and realistic expectation of achievable accuracy, the systems are quite acceptable for most monitoring and moderate accuracy measurements.

Data-Acquisition Considerations

Once the appropriate thermocouple type is selected, attention must be given to the cold junction compensation method. The two connection points of the thermocouple to the

DAS board form two new thermocouple connections. Use of external cold junction methods between the thermocouple and the board will eliminate this problem, but more frequently the thermocouple will be connected directly to the board and use built-in electronic cold junction compensation. This is usually accomplished by using a separate thermistor sensor, which measures the temperature at the system connection point to determine the cold junction error, and providing an appropriate bias voltage correction either directly or through software. An important consideration is that the internal correction method has a typical error of the order of 0.5–1.5°C and, as a systematic error, this error is directly passed on to the measurement.

These boards also may use internal polynomial interpolation for converting measured voltage into temperature. If not, this can be programmed into the data-reduction scheme by using, for example, the information of Table 8.7. Nonetheless, this introduces a "linearization" error, which is a function of thermocouple material and temperature range and typically specified with the DAS board.

Thermocouples are often used in harsh, industrial environments with significant EMI and rf noise sources. Thermocouple wire pairs should be twisted to reduce noise. Also, a differential-ended connection is preferred between the thermocouple and the DAS board. In this arrangement, though, the thermocouple becomes an isolated voltage source, meaning there is no longer a direct ground path keeping the input within its common mode range. As a consequence, a common complaint is that the measured signal may drift or suddenly jump in the level of its output. Usually, this interference behavior is eliminated by placing a 10-kΩ to 100-kΩ resistor between the low terminal of the input and low-level ground.

Because most DAS boards use A/D converters having a ±5 V full scale, the signal must be conditioned using an amplifier. Usually a gain of 100–500 is sufficient. High gain, very low noise amplifiers are important for accurate measurements. Consider a J-type thermocouple using a 12-bit A/D converter with a signal conditioning gain of 100. This allows for a full-scale input range of 100 m V, which is a suitable range for most measurements. Then, the A/D conversion resolution is

$$\frac{E_{\text{FSR}}}{(G)(2^M)} = \frac{10\,\text{V}}{(100)(2^{12})} = 24.4\,\mu\text{V/bit}$$

A J-type thermocouple has a sensitivity of ∼55 μV/°C. Thus, the measurement resolution becomes

$$(24.4\,\mu\text{V/bit})/(55\,\mu\text{V/}^\circ\text{C}) = 0.44^\circ\text{C/bit}.$$

Thermocouples tend to have long-time constants relative to the typical sample rate capabilities of general purpose DAS boards. If temperature measurements show greater than expected fluctuations, high-frequency sampling noise is a likely cause. Reducing the sample rate or using a smoothing filter are simple solutions. The period of averaging should be on the scale of the time constant of the thermocouple.

EXAMPLE 8.10

It is desired to create an off-the-shelf temperature measuring system for a personal computer (PC)-based control application. The proposed temperature measuring system is illustrated schematically in Figure 8.23. The temperature measurement system consists of:

- A PC-based data-acquisition system composed of a data-acquisition board, a computer, and appropriate software to allow measurement of analog input voltage signals.

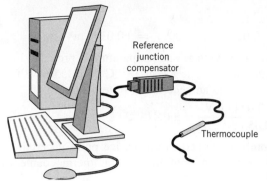

Figure 8.23 PC-based temperature measurement system.

- A J-type thermocouple and reference junction compensator. The reference junction compensator serves to provide an output emf from the thermocouple equivalent to the value that would exist between the measuring junction and a reference temperature of 0°C.
- A J-type thermocouple, which is uncalibrated but meets NIST standard limits of error.

The system is designed to measure and control a process temperature that varies slowly in time compared to sampling rate of the data-acquisition system. The process nominally operates at 185°C.

The following specifications are applicable to the measurement system components:

Component	Characteristics	Accuracy Specifications
Data-acquisition board	Analog voltage input range: 0–0.1 V	12-bit A/D converter accuracy: ±0.01% of reading
Reference junction compensator	J-type compensation range from 0 to 50°C	±0.5°C over the range 20–36°C
Thermocouple (J-type)	Stainless steel sheathed ungrounded junction	Accuracy: ±1.0°C based on NIST standard limits of error

The purpose of the measurement system requires that the temperature measurement have a total uncertainty of less than 1.5°C. Based on a design stage uncertainty analysis, does this measurement system meet the overall accuracy requirement?

SOLUTION   The design stage uncertainty for this measurement system will be determined by expressing the uncertainty of each system component as an equivalent uncertainty in temperature, and then combining these design stage uncertainties.

The 12-bit A/D converter divides the full-scale voltage range into 2^{12} or 4096 equal-sized intervals. Thus, the resolution (quantization error) of the A/D in measuring voltage is

$$\frac{0.1 \text{ V}}{4096 \text{ intervals}} = 0.0244 \text{ mV}$$

The uncertainty of the DAS is specified as 0.01% of the reading. A nominal value for the thermocouple voltage must be known or established. In the present case, the nominal process temperature is 185°C, which corresponds to a thermocouple voltage of approximately 10 mV. Thus the calibration uncertainty of the DAS is 0.001 mV.

The contribution to the total uncertainty of the temperature measurement system from the DAS can now be determined. First combine the resolution and calibration uncertainties as

$$u_{DAS} = \pm\sqrt{(0.0244)^2 + (0.001)^2} = \pm 0.0244 \text{ mV}$$

The relationship between uncertainty in voltage and temperature is provided by the static sensitivity, which can be estimated from Table 8.6 at 185°C as 0.055 mV/°C. Thus an uncertainty of 0.0244 mV corresponds to an uncertainty in temperature of

$$\frac{0.0244 \text{ mV}}{0.055 \text{ mV/°C}} = \pm 0.444°\text{C}$$

This uncertainty can now be combined directly with the ice point uncertainty and the uncertainty associated with the standard limits of error for the thermocouple, as

$$u_T = \pm\sqrt{(0.44)^2 + (1.0)^2 + (0.5)^2} = \pm 1.2°\text{C}$$

COMMENT   It would be appropriate in many cases to calibrate the thermocouple against a laboratory standard, such as an RTD, which has a calibration traceable to NIST standards. With reasonable expense and care, an uncertainty level in the thermocouple of ±0.1°C can be achieved, compared to the uncalibrated value of ±1°C. If the thermocouple were calibrated, the resulting uncertainty in the overall measurement of temperature is reduced to ±0.676°C, so that by reducing the uncertainty contribution of the thermocouple by a factor of 10, the system uncertainty would be reduced by a factor of 2.

EXAMPLE 8.11

An effective method of evaluating data acquisition and reduction errors associated with the use of multiple temperature sensors within a test facility is to provide a known temperature point at which the sensor outputs can be compared. Suppose the outputs from M similar thermocouple sensors (e.g., all T-type) are to be measured and stored on an M-channel data-acquisition system. Each sensor is referenced to the same reference junction temperature (e.g., ice point) and operated in the normal manner. The sensors are exposed to a known and uniform temperature. N (say 30) readings for each of the M thermocouples are recorded. What information can be obtained from the data?

KNOWN
$$M(j = 1, 2, \ldots, M) \text{ thermocouples}$$
$$N(i = 1, 2, \ldots, N) \text{ readings measured for each thermocouple}$$

SOLUTION   The mean value for all readings of the ith thermocouple is given as

$$\bar{T}_j = \frac{1}{N}\sum_{i=1}^{N} T_{ij}$$

The pooled mean for all the thermocouples is given as

$$\langle \bar{T} \rangle = \frac{1}{M}\sum_{j=1}^{M} \bar{T}_j$$

The difference between the pooled mean temperature and the known temperature would provide an estimate of the systematic uncertainty that can be expected from any channel during data acquisition. On the other hand, the differences between each $\bar{T}_j$ and $\langle \bar{T} \rangle$ must reflect the precision among the M channels. The standard random uncertainty for the data acquisition and reduction instrumentation system is then

$$S_{\bar{T}} = \frac{\langle S_T \rangle}{\sqrt{M}} \quad \text{where} \quad \langle S_T \rangle = \sqrt{\frac{\sum_{j=1}^{M}\sum_{i=1}^{N}\left(T_{ij} - \bar{T}_j\right)^2}{M(N-1)}}$$

with degrees of freedom, $v = M(N-1)$

COMMENT   Elemental errors accounted for in these estimates include:

- Reference junction random errors
- Random errors in the known temperature
- Data-acquisition system random errors
- Extension cable and connecting plug systematic errors
- Thermocouple emf-*T* correlation systematic errors

These estimates would not include instrument calibration errors or probe insertion errors.

8.6   RADIATIVE TEMPERATURE MEASUREMENTS

There is a distinct advantage to measuring temperature by detecting thermal radiation. The sensor for thermal radiation need not be in contact with the surface to be measured, making this method attractive for a wide variety of applications. The basic operation of a radiation thermometer is predicated upon some knowledge of the radiation characteristics of the surface whose temperature is being measured, relative to the calibration of the thermometer. The spectral characteristics of radiative measurements of temperature is beyond the scope of the present discussion; an excellent source for further information is found in [11].

Radiation Fundamentals

Radiation refers to the emission of electromagnetic waves from the surface of an object. This radiation has characteristics of both waves and particles, which leads to a description of the radiation as being composed of photons. The photons generally travel in straight lines from points of emission to another surface, where they are absorbed, reflected, or transmitted. This electromagnetic radiation exists over a large range of wavelengths that includes x-rays, ultraviolet radiation, visible light, and thermal radiation, as shown in Figure 8.24. The thermal radiation emitted from an object is related to its temperature, and has wavelengths ranging from approximately 10^{-7} to 10^{-3} m. It is necessary to

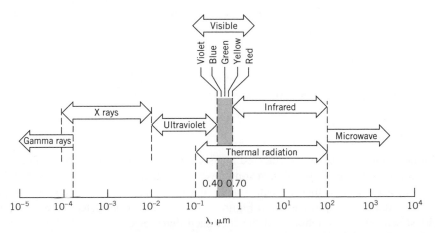

Figure 8.24 The electromagnetic spectrum. (From F. P. Incropera and D. P. DeWitt, *Fundamentals of Heat and Mass Transfer*, 2d ed., copyright © 1985 by John Wiley & Sons, New York. Reprinted by permission.)

understand two key aspects of radiative heat transfer in relation to temperature measurements. First, the radiation emitted by an object is proportional to the fourth power of its temperature. In the ideal case, this may be expressed as

$$E_b = \sigma T^4 \tag{8.20}$$

where E_b is the flux of energy radiating from an ideal surface, or the blackbody emissive power. The emissive power of a body is the energy emitted per unit area and per unit time. The term "blackbody" implies a surface that absorbs all incident radiation, and as a result emits radiation in an "ideal" manner.

Second, the emissive power is a direct measure of the total radiation emitted by an object. However, energy is emitted by an ideal radiator over a range of wavelengths, and at any given temperature the distribution of the energy emitted as a function of wavelength is unique. Max Planck (1858–1947) developed the basis for the theory of quantum mechanics in 1900 as a result of examining the wavelength distribution of radiation. He proposed the following equation to describe the wavelength distribution of thermal radiation for an ideal or blackbody radiator:

$$E_{b\lambda} = \frac{2\pi h_p c^2}{\lambda^5 \left[\exp\left(h_p c / k_B \lambda T\right) - 1\right]} \tag{8.21}$$

where

$E_{b\lambda}$ = total emissive power at the wavelength λ

λ = wavelength

c = speed of light

h_p = Planck's constant = 6.6256×10^{-34} J s

k_B = Boltzmann's constant = 1.3805×10^{-23} J/K

Figure 8.25 is a plot of this wavelength distribution for various temperatures. For the purposes of radiative temperature measurements, it is crucial to note that the maximum energy emission shifts to shorter wavelengths at higher temperatures. Our experiences confirm this behavior through observation of color changes as a surface is heated.

Consider an electrical heating element, as can be found in an electric oven. With no electric current flow through the element, it appears almost black, its room temperature color. With a current flow, the element temperature rises and it appears to change color to a dull red, and perhaps to a reddish orange. If its temperature continued to increase, eventually it would appear white. This change in color signifies a shift in the maximum intensity of the emitted radiation to shorter wavelengths, out of the infrared and into the visible. There is also an increase in total emitted energy. The Planck distribution provides a basis for the measurement of temperature through color comparison.

Radiation Detectors

Radiative energy flux can be detected in a sensor by two basic techniques. The detector is subject to radiant energy from the source whose temperature is to be measured. The first technique involves a thermal detector in which absorbed radiative energy elevates the detector temperature, as shown in Figure 8.26. These thermal detectors are certainly the oldest sensors for radiation, and the first such detector can probably be credited to Sir William Herschel, who verified the presence of infrared radiation using a thermometer and a prism. The equilibrium temperature of the detector is a direct measure of the

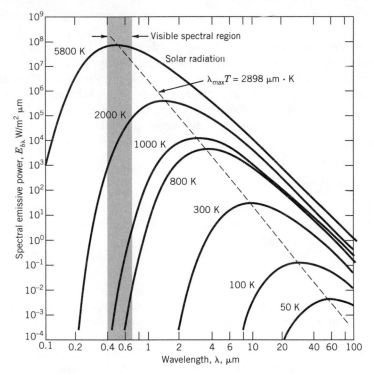

Figure 8.25 Planck distribution of blackbody emissive power as a function of wavelength. (From F. P. Incropera and D. P. DeWitt, *Fundamentals of Heat and Mass Transfer*, 2d ed., copyright © 1985 by John Wiley & Sons, New York. Reprinted by permission.)

amount of radiation absorbed. The resulting rise in temperature must then be measured. Thermopile detectors provide a thermoelectric power resulting from a change in temperature. A thermistor can also be used as the detector, and results in a change in resistance with temperature.

A second basic type of detector relies on the interaction of a photon with an electron, resulting in an electric current. In a photomultiplier tube, the emitted electrons are accelerated and used to create an amplified current, which is measured. Photovoltaic cells may be employed as radiation detectors. The photovoltaic effect results from the generation of a potential across a p-n junction in a semiconductor when it is subject to a flux of photons. Electron-hole pairs are formed if the incident photon has an energy level of sufficient magnitude. This process will result in the direct conversion of radiation into electrical energy, and results in high sensitivity and a fast response time when used as a detector. In general, the photon detectors tend to be spectrally selective, so that the

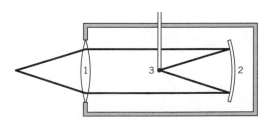

Figure 8.26 Schematic of a basic radiometer: (1) lens; (2) focusing mirror; (3) detector (thermopile or thermistor).

relative sensitivity of the detector tends to change with the wavelength of the measured radiation.

Many considerations enter into the choice of a detector for radiative measurements. If time response is important, photon detectors are significantly faster than thermopile or thermistor detectors, and therefore have a much wider frequency response. Photodetectors saturate, while thermopile sensors may slowly change their characteristics over time. Some instruments have variations of sensitivity with the incident angle of the incoming radiation; this factor may be important for solar insolation measurements. Other considerations include wavelength sensitivity, cost, and allowable operating temperatures.

Radiative Temperature Measurements

Commercially applicable radiation thermometers vary widely in their complexity and the accuracy of the resultant measurements. We will consider only the basic techniques that allow measurement of temperature.

Radiometer

Perhaps the simplest form, a radiometer measures a source temperature by measuring the voltage output from a thermopile detector. A schematic of such a device is shown in Figure 8.26. The increase in temperature of the thermopile is a direct indication of the temperature of the radiation source. One application of this principle is in the measurement of total solar radiation incident upon a surface. Figure 8.27 shows a schematic of a pyranometer, used to measure global solar irradiance. It would have a hemispherical field of view, and measures both the direct or beam radiation, and diffuse radiation. The diffuse and beam components of radiation can be separated by shading the pyranometer from the direct solar radiation, thereby measuring the diffuse component.

Pyrometry

Optical pyrometry identifies the temperature of a surface by its color, or more precisely the color of the radiation it emits. A schematic of an optical pyrometer is shown in Figure 8.28. A standard lamp is calibrated so that the current flow through its filament is controlled and calibrated in terms of the filament temperature. Comparison is made optically between the color of this filament and the surface of the object whose temperature is being measured. The comparator can be the human eye. Uncertainties in the measurement may be reduced by appropriately filtering the incoming light. Corrections must be applied for surface emissivity associated with the measured radiation; uncertainties vary with the skill of the user, and generally are on the order

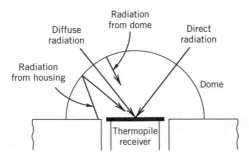

Figure 8.27 Pyranometer construction.

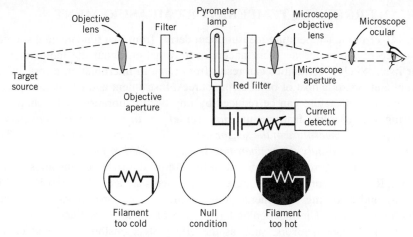

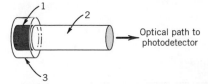

Appearance of lamp filament in eyepiece of optical pyrometer.

Figure 8.28 Schematic diagram of a disappearing filament optical pyrometer.

of 5°C. Replacing the human eye with a different detector extends the range of useful temperature measurement and reduces the random uncertainty.

The major advantage of an optical pyrometer lies in its ability to measure high temperatures remotely. For example, it could be used to measure the temperature of a furnace without having any sensor in the furnace itself. For many applications this provides a safe and economical means of measuring high temperatures.

Optical Fiber Thermometers

The optical fiber thermometer is based on the creation of an ideal radiator that is optically coupled to a fiber-optic transmission system [12, 13], as shown in Figure 8.29. The temperature sensor in this system is a thin, single-crystal aluminum oxide (sapphire) fiber; a metallic coating on the tip of the fiber forms a blackbody radiating cavity, which radiates directly along the sapphire crystal fiber. The single-crystal sapphire fiber is necessary because of the high-temperature operation of the thermometer. The operating range of this thermometer is 300–1900°C. Signal transmission is accomplished using standard, low-temperature fiber optics. A specific wavelength band of the transmitted radiation is detected and measured, and these raw data are reduced to yield the temperature of the blackbody sensor.

The absence of electrical signals associated with the sensor signal provides excellent immunity from electromagnetic and radio-frequency interference. The measurement system has superior frequency response and sensitivity. The system has been employed for measurement in combustion applications. Temperature resolution of 0.0001°C or better is possible.

Figure 8.29 Optical fiber thermometer: (1) blackbody cavity (iridium film); (2) sapphire fiber (single crystal); (3) protective coating (Al_2O_3).

8.7 PHYSICAL ERRORS IN TEMPERATURE MEASUREMENT

In general, errors in temperature measurement derive from two fundamental sources. The first source of errors derives from uncertain information about the temperature of the sensor itself. Such uncertainties can result from random interpolation errors, calibration systematic errors, or a host of other error sources. Instrument and procedural uncertainty in the sensor temperature can be reduced by improved calibration or by changes in the measuring and recording instruments. *However, errors in temperature measurement can occur even if the temperature of the sensor was measured exactly. In such cases, the probe does not sense accurately the temperature it was intended to measure.*

A list of typical errors associated with the use of temperature sensors is provided in Table 8.8. Random errors in temperature measurements are a result of resolution limits of measuring and recording equipment, time variations in extraneous variables, and other sources of variation. Thermocouples have some characteristics that can lead to both systematic and random errors, such as the effect of extension wires and connectors. Another major source of error for thermocouples involves the accuracy of the reference junction. Ground loops can lead to spurious readings, especially when thermocouple outputs are amplified for control or data-acquisition purposes. As with any measurement system, calibration of the entire measurement system, in place if possible, is the best means of identifying error sources and reducing the resulting measurement uncertainty to acceptable limits.

Insertion Errors

This discussion will focus on ensuring that a sensor output accurately represents the temperature it is intended to measure. For example, suppose it is desired to measure the outdoor temperature. This measurement could employ a large dial thermometer, which

Table 8.8   Measuring Errors Associated with Temperature Sensors

Random Errors
1. Imprecision of readings
2. Time and spatial variations

Systematic Errors
1. Insertion errors, heating or cooling of junctions
   a. Conduction errors
   b. Radiation errors
   c. Recovery errors
2. Effects of plugs and extension wires
   a. Nonisothermal connections
   b. Loading errors
3. Ignorance of materials or material changes during measurements
   a. Aging following calibration
   b. Annealing effects
   c. Cold work hardening
4. Ground loops
5. Magnetic field effects
6. Galvanic error
7. Reference junction inaccuracies

might be placed on a football field or a tennis court in the direct sunlight, and assumed to represent "the temperature," perhaps as high as 50°C (120°F). But what temperature is being indicated by this thermometer? Certainly the thermometer is not measuring the air temperature, nor is it measuring the temperature of the field or the court. The thermometer is subject to the very sources of error we wish to describe and analyze. *The thermometer, very simply, indicates its own temperature!* The temperature of the thermometer is the thermodynamic equilibrium temperature that results from the radiant energy from the sun, convective exchange with the air, and conduction heat transfer with the surface on which it is resting. Considering the fact that these thermometers typically have a glass cover, which acts as a solar collector, it is very likely that the thermometer temperature is significantly higher than the air temperature.

The physical mechanisms that may cause a temperature probe to indicate a temperature different than that intended include conduction, radiation, and recovery errors. In any real measurement system, their effects could be coupled, and therefore should not be considered independently. However, for simplicity, each error source will be considered separately. Our purpose is to provide only approximate analyses of the errors, and not to provide predictive techniques for correcting measured temperatures. These analyses should be used for selection and design of temperature measuring systems. The goal of the measurement engineer should be to minimize these errors, as far as possible, through the careful installation and design of temperature probes.

Conduction Errors

Errors that result from conduction heat transfer between the measuring environment and the ambient are often called *immersion errors*. Consider the temperature probe shown in Figure 8.30. In many circumstances, a temperature probe extends from the measuring environment through a wall into the ambient environment, where indicating or recording systems are located. The probe and the electrical leads form a path for the conduction of energy from the measuring environment to the ambient. The fundamental nature of the error created by conduction in measured temperatures can be illustrated by the model of a temperature probe shown in Figure 8.31. The essential physics of immersion errors associated with conduction can be discerned by modeling the temperature probe as a fin. Suppose we assume that the measured temperature is higher than the ambient temperature. If we consider a differential element of the fin, as shown in Figure 8.31(*b*), at steady state there is energy conducted along the fin, and transferred by convection from the surface. The surface area for convection is $P\,dx$, where P is the perimeter or circumference. Applying the first law of thermodynamics to this differential element yields

$$q_{x+dx} - q_x = hP\,dx[T(x) - T_\infty] \tag{8.22}$$

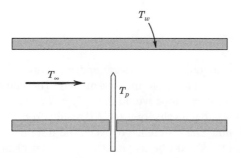

Figure 8.30 Temperature probe inserted into a measuring environment.

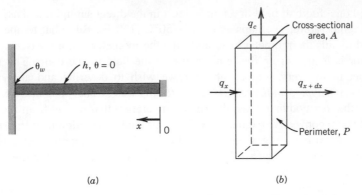

(a)   (b)

Figure 8.31 Model of a temperature probe as a one-dimensional fin.

where h is the convection coefficient. If q is expanded in a Taylor series about the point x, and the substitutions

$$\theta = T - T_\infty \quad q = -kA\frac{dT}{dx} \quad m = \sqrt{\frac{hP}{kA}} \tag{8.23}$$

are made, then the governing differential equation becomes

$$\frac{d^2\theta}{dx^2} - m^2\theta = 0 \tag{8.24}$$

Here k is the effective thermal conductivity of the temperature probe. The solution to this differential equation for the boundary conditions that the wall has a temperature T_w, or a normalized value $\theta_w = T_w - T_\infty$, and the end of the fin is small in surface area, is

$$\frac{\theta(x)}{\theta_w} = \frac{\cosh mx}{\cosh mL} \tag{8.25}$$

The point $x = 0$ is the location where the temperature is being assumed to be measured, and therefore the solution is evaluated at $x = 0$ as

$$\frac{\theta(0)}{\theta_w} = \frac{T(0) - T_\infty}{T_w - T_\infty} = \frac{1}{\cosh mL} \tag{8.26}$$

From this analysis, the error due to conduction, e_c, can be estimated. An ideal sensor would indicate the fluid temperature, T_∞; therefore, if the sensor temperature is $T_p = T(0)$, then the conduction error is

$$e_c = T_p - T_\infty = \frac{T_w - T_\infty}{\cosh mL} \tag{8.27}$$

Probe Design

The purpose of the preceding analysis is to gain some physical understanding of ways to minimize conduction errors (not to correct inaccurate measurements). The behavior of this solution is such that the ideal temperature probe would have $T_p = T_\infty$, or $\theta(0) = 0$, implying that $e_c = 0$. Equation 8.26 shows that a value of $\theta(0) \neq 0$ results from a nonzero value of θ_w, and a finite value of cosh mL. The difference between the fluid temperature being measured and the wall temperature should be as small as possible; clearly, this

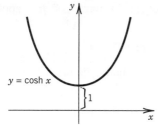

Figure 8.32 Behavior of the hyperbolic cosine.

implies that the wall should be insulated to minimize this temperature difference, and the resulting conduction error.

The term cosh mL should be as large as possible. The behavior of the cosh function is shown in Figure 8.32. Since the hyperbolic cosine monotonically increases for increasing values of the argument, the goal of a probe design should be to maximize the value of the product mL, or $(hP/kA)^{1/2} L$. In general, the thermal conductivity of a temperature probe and the convection coefficient are not design parameters. Thus, two important conclusions are that the probe should be as small in diameter as possible, and should be inserted as far as possible into the measuring environment away from the bounding surface, i.e., make L large. A small diameter increases the ratio of the perimeter, P, to the cross-sectional area, A. For a circular cross section, this ratio is $4/D$, where D is the diameter. A good rule of thumb based on equation (8.27) is to have an $L/D > 50$ for negligible conduction error.

Although this analysis clearly indicates the fundamental aspects of conduction errors in temperature measurements, it does not provide the capability to correct measured temperatures for conduction errors. Conduction errors should be minimized through appropriate design and installation of temperature probes. Usually, the physical situation is sufficiently complex to preclude accurate mathematical description of the measurement errors. Additional information on modeling conduction errors may be found in [14].

Radiation Errors

Consider a temperature probe used to measure a gas temperature. In the presence of significant radiation heat transfer, the equilibrium temperature of a temperature probe may be different than the fluid temperature being measured. Because radiation heat transfer is proportional to the fourth power of temperature, the importance of radiation effects increases as the absolute temperature of the measuring environment increases. The error due to radiation can be estimated by considering steady-state thermodynamic equilibrium conditions for a temperature sensor. Consider the case where energy is transferred to/from a sensor by convection from the environment and from/to the sensor by radiation to a body at a different temperature, such as a pipe or furnace wall. For this analysis conduction will be neglected. A first law analysis of a system containing the probe, at steady-state conditions, yields

$$q_c + q_r = 0 \tag{8.28}$$

where

q_c = convective heat transfer

q_r = radiative heat transfer

The heat-transfer components can then be expressed in terms of the appropriate fundamental relations as

$$q_c = hA_s(T_\infty - T) \qquad q_r = FA_s\varepsilon\sigma(T_w^4 - T^4) \tag{8.29}$$

Assuming that the surroundings may be treated as a blackbody, the first law for a system consisting of the temperature probe is

$$hA_s(T_\infty - T_p) = FA_s\varepsilon\sigma(T_p^4 - T_w^4) \tag{8.30}$$

A temperature probe is generally small compared to its surroundings, which justifies the assumption that the surroundings may be treated as black. The radiation error, e_r, is estimated by

$$e_r = (T_p - T_\infty) = \frac{F\varepsilon\sigma}{h}(T_w^4 - T_p^4) \tag{8.31}$$

where

σ = the Stefan-Boltzmann constant $(\sigma = 5.669 \times 10^{-8}\,\mathrm{W/m^2K^4})$

ε = emissivity of the sensor

F = radiation view factor

T_p = probe temperature

T_w = temperature of the surrounding walls

T_∞ = fluid temperature being measured

Again, if the sensor is small compared to the scale of the surroundings, the view factor from the sensor to the surroundings may be taken as 1.

EXAMPLE 8.12

A typical situation where radiation would be important occurs in measuring the temperature of a furnace. Figure 8.33 shows a small temperature probe for which conduction errors are negligible, which is placed in a high-temperature enclosure, where the fluid temperature is T_∞ and the walls of the enclosure are at T_w. Convection and radiation are assumed to be the only contributing heat transfer modes at steady state. Develop an expression for the equilibrium temperature of the probe. Also determine the equilibrium temperature of the probe and the radiation error in the case where $T_\infty = 800°\mathrm{C}$, $T_w = 500°\mathrm{C}$, and the emissivity of the probe is 0.8. The convective heat transfer coefficient is 100 $\mathrm{W/m^2\,°C}$.

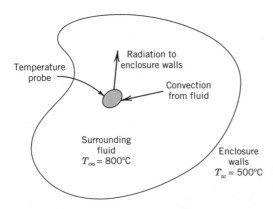

Figure 8.33 Analysis of a temperature probe in a radiative and convective environment.

KNOWN   Temperatures $T_\infty = 800°C$ and $T_w = 500°C$, with $h = 100$ W/m^2°C. The emissivity of the probe is 0.8.

FIND   An expression for the equilibrium temperature of the probe, and the resulting probe temperature for the stated conditions.

ASSUMPTIONS   The surroundings may be treated as black, and conduction heat transfer is neglected.

SOLUTION   The probe is modeled as a small, spherical body within the enclosed furnace; the radiation view factor from the probe to the furnace is 1.0. At steady state, an equilibrium temperature may be found from an energy balance. The first law for a system consisting of the temperature probe, from equation 8.30 is

$$h\left(T_\infty - T_p\right) + \sigma\varepsilon\left(T_w^4 - T_p^4\right) = 0 \qquad (8.32)$$

The probe is attempting to measure T_∞. Equation 8.32 yields an equilibrium temperature for the probe of $T_p = 642.5°C$. The radiation error in this case is (from 8.31) $-157.5°C$. This result indicates that radiative heat transfer can create significant measurement errors at elevated temperatures.

COMMENT   Equation (8.31) (and equation (8.32) above) does not allow direct solution for the radiation error because it contains the probe temperature. Any calculator or computer-based method for the solution of nonlinear equations may be employed to affect the solution, although a trial and error approach converges rapidly to the correct temperature.

Radiation Shielding

Radiation shielding is a key concept in controlling radiative heat transfer; shielding for radiation is analogous to insulation to reduce conduction heat transfer. A radiation shield is an opaque surface interposed between a temperature sensor and its radiative surroundings so as to reduce electromagnetic wave interchange. In principle, the shield attains an equilibrium temperature closer to the fluid temperature than the surroundings. Because the probe can no longer "see" the surroundings, with the radiation shield in place, the probe temperature is closer to the fluid temperature. Additional information on radiation error in temperature measurements may be found in [14]. The following example serves to demonstrate radiation errors and the effect of shielding.

EXAMPLE 8.13

Consider again Example 8.12, where the oven is maintained at a temperature of 800°C. Because of energy losses, the walls of the oven are cooler, having a temperature of 500°C. For the present case, consider the temperature probe as a small, spherical object located in the oven, having no thermal conduction path to the ambient. (All energy exchange is through convection and radiation). Under these conditions, the probe temperature is 642.5°C, as found in Example 8.12.

Suppose a radiation shield is placed between the temperature probe and the walls of the furnace, which blocks the path for radiative energy transfer. Examine the effect of adding a radiation shield on the probe temperature.

KNOWN   A radiation shield is added to a temperature probe in an environment with $T_\infty = 800°C$ and $T_w = 500°C$.

FIND   The radiation error in the presence of the shield.

ASSUMPTIONS   The radiation shield completely surrounds the probe, and the surroundings may be treated as a blackbody.

SOLUTION   The shield equilibrium temperature is higher than the wall temperature by virtue of convection with the fluid. As a result, the probe "sees" a higher temperature surface, and the probe temperature is closer to the fluid temperature, resulting in less measurement error.

For a single radiation shield placed so that it completely surrounds the probe, which is small compared to the size of the enclosure and has an emissivity of 1, the equilibrium temperature of the shield can be determined from equation (8.30). The temperature of the shield is found to be 628°C. Because the sensor now "sees" the shield, rather than the wall, the temperature measured by the probe will be 697°C, which is also determined from equation (8.30).

One shield with an emissivity of 1 provides for an improvement over the case of no shields, but a better choice of the surface characteristics of the shield material can result in much better performance. If the shield has an emissivity of 0.1, the shield temperature rises to 756°C, and the probe temperature to 771°C.

COMMENT   Shielding provides improved temperature measurements by reducing radiative heat transfer. Another area for improvement in this temperature measurement could be the elevation of the wall temperature through insulation.

This discussion of radiation shielding serves to demonstrate the usefulness of shielding as a means of improving temperature measurements in radiative environments. As with conduction errors, the development should be used to guide the design and installation of temperature sensors, not to correct measured temperatures. Further information on radiation errors may be found in [5].

Recovery Errors in Temperature Measurement

The kinetic energy of a gas moving at high velocity can be converted to sensible energy by reversibly and adiabatically bringing the flow to rest at a point. The temperature resulting from this process is called the *stagnation or total temperature*, T_t. On the other hand, the *static temperature* of the gas, T_∞, is the temperature that would be measured by an instrument moving at the local fluid velocity. From a molecular point of view, the static temperature measures the magnitude of the random kinetic energy of the molecules that comprise the gas, while the stagnation temperature includes both the directed and random components of kinetic energy. Generally, the engineer would be content with knowledge of either temperature, but in high-speed gas flows the sensor indicates neither temperature.

For negligible changes in potential energy, and in the absence of heat transfer or work, the energy equation for a flow may be written

$$h_1 + \frac{U^2}{2} = h_2 \tag{8.33}$$

where state 2 refers to the stagnation condition, and state 1 to a condition where the gas is flowing with the velocity U. Assuming ideal gas behavior, the enthalpy difference $h_2 - h_1$ may be expressed as $c_p(T_2 - T_1)$, or in terms of static and stagnation temperatures

$$\frac{U^2}{2} = c_p(T_t - T_\infty) \tag{8.34}$$

The term $U^2/2c_p$ is called the *dynamic temperature*.

What implication does this have for the measurement of temperature in a flowing gas stream? The physical nature of gases at normal pressures and temperatures is such that the velocity of the gas on a solid surface is zero, because of the effects of viscosity. Thus, when a temperature probe is placed in a moving fluid, the fluid is brought to rest on the surface of the probe. However, this process may not be thermodynamically reversible, and the gas in question may not behave as an ideal gas. *Deceleration of the flow by the probe converts some portion of the directed kinetic energy of the flow to thermal energy, and elevates the temperature of the probe above the static temperature of the gas.* The fraction of the kinetic energy recovered as thermal energy is called the recovery factor, r, defined as

$$r \equiv \frac{T_p - T_\infty}{U^2/2c_p} \tag{8.35}$$

where T_p represents the equilibrium temperature of the stationary (with respect to the flow) real temperature probe. In general, r may be a function of the velocity of the flow, or more precisely, the Mach number and Reynolds number of the flow, and the shape and orientation of the temperature probe. For thermocouple junctions of round wire, Moffat [15] reports values of

$$r = 0.68 \pm 0.07 \quad (95\%) \quad \text{for wires normal to the flow}$$
$$r = 0.86 \pm 0.09 \quad (95\%) \quad \text{for wires parallel to the flow}$$

These recovery factor values tend to be constant at velocities for which temperature errors are significant, usually flows where the Mach number is greater than 0.1. For thermo-couples having a welded junction, a spherical weld bead significantly larger than the wire diameter tends to a value of the recovery factor of 0.75, for the wires parallel or normal to the flow. The relationships between temperature and velocity for temperature probes with known recovery factors are

$$T_p = T_\infty + \frac{rU^2}{2c_p} \tag{8.36}$$

or in terms of the *recovery error*, e_U,

$$e_U = T_p - T_\infty = \frac{rU^2}{2c_p} \tag{8.37}$$

The probe temperature is related to the stagnation temperature by

$$T_p = T_t - \frac{(1-r)U^2}{2c_p} \tag{8.38}$$

Fundamentally, in liquids the stagnation and static temperatures are essentially equal [5], and the recovery error may generally be taken as zero for liquid flows. In any case, high-velocity flows are rarely encountered in liquids.

EXAMPLE 8.14

A temperature probe having a recovery factor of 0.86 is to be used to measure a flow of air at velocities up to the sonic velocity, at a pressure of 1 atm and a static temperature of 30°C. Calculate the value of the recovery error in the temperature measurement as the velocity of the air flow increases, from 0 to the speed of sound, using equation (8.37).

KNOWN

$r = 0.86$ $p_\infty = 1$ atm abs $= 101$ kPa abs
$M \leq 1$ $T_\infty = 30°C = 303$ K

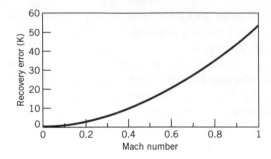

Figure 8.34 Behavior of recovery error as a function of Mach number.

FIND   The recovery error as a function of air velocity.

ASSUMPTION   Air behaves as an ideal gas.

SOLUTION   Assuming that air behaves as an ideal gas, the speed of sound is expressed as

$$a = \sqrt{kRT} \qquad (8.39)$$

For air at 101 kPa and 303 K, with $R = 0.287$ kJ/kg K, the speed of sound is approximately 349 m/s.

Figure 8.34 shows the error in temperature measurement as a function of Mach number for this temperature probe. The *Mach number* is defined as the ratio of the flow velocity to the speed of sound.

COMMENT   Typically, the static and total temperatures of the flowing fluid stream are to be determined from the measured probe temperature. In this case a second independent measurement of the velocity is necessary.

8.8   SUMMARY

Temperature is a fundamentally important quantity in science and engineering, both in concept and practice. As such, temperature is one of the most widely measured engineering variables, providing the basis for a variety of control and safety systems. This chapter provides the basis for the selection and installation of temperature sensors.

Temperature is defined for practical purposes through the establishment of a temperature scale, such as the Kelvin scale, that encompasses fixed reference points and interpolation standards. The International Temperature Scale 1990 is the accepted standard for temperature measurement.

The two most common methods of temperature measurement employ thermocouples and resistance temperature detectors. Standards for the construction and use of these temperature measuring devices have been established and provide the basis for selection and installation of commercially available sensors and measuring systems.

Installation effects on the accuracy of temperature measurements are a direct result of the influence of radiation, conduction, and convection heat transfer on the equilibrium temperature of a temperature sensor. The installation of a temperature probe into a measuring environment can be accomplished in such a way as to minimize the uncertainty in the resulting temperature measurement.

REFERENCES

1. Patterson, E. C., Eponyms: Why Celsius? *American Scientist* 77(4):413, 1989.
2. Committee Report, The International Temperature Scale of 1990, *Metrologia* 27(3), 1990. (The text of the superseded International Practical Temperature Scale of 1968

appears as an appendix in the National Bureau of Standards monograph 124. An amended version was adopted in 1975, and the English text published: *Metrologia* 12:7–17, 1976.)

3. Quinn, T. J., News from BIPM, *Metrologia* 26:69–74, 1989.
4. *Temperature Measurement*, Supplement to American Society of Mechanical Engineers PTC 19.3, 1974.
5. Benedict, R. P., *Fundamentals of Temperature, Pressure and Flow Measurements*, 3d ed., Wiley, New York, 1984.
6. McGee, T. D., *Principles and Methods of Temperature Measurement*, Wiley-Interscience, New York, 1988.
7. Diehl, W., Thin-film PRTD, *Measurements and Control*, 155–159, December 1982.
8. Thermistor Definitions and Test Methods, Electronic Industries Standard RS-275-A (ANSI Standard C83.68-1972), June 1971.
9. Thermometrics, Inc., Vol. 1. NTC Thermistors, 1993.
10. Burns, G. W., Scroger, M. G., and G. F. Strouse, Temperature-Electromotive Force Reference Functions and Tables for the Letter-Designated Thermocouple Types Based on the ITS-90, *NIST Monograph* 175, April 1993 (supersedes NBS Monograph 125).
11. Dewitt, D. P., and G. D. Nutter, *Theory and Practice of Radiation Thermometry*, Wiley-Interscience, New York, 1988.
12. Dils, R. R., High-temperature optical fiber thermometer, *Journal of Applied Physics*, 54(3):1198–1201, 1983.
13. Optical fiber thermometer, *Measurements and Control*, April 1987.
14. Sparrow, E. M., Error estimates in temperature measurement, in E. R. G. Eckert and R. J. Goldstein (Eds.), *Measurements in Heat Transfer*, 2d ed., Hemisphere, Washington, DC, 1976.
15. Moffat, R. J., Gas temperature measurements, in *Temperature—Its Measurement and Control in Science and Industry*, Vol. 3, Part 2, Reinhold, New York, 1962.

NOMENCLATURE

a	speed of sound $[l\,t^{-1}]$		A_c	cross-sectional area $[l^2]$
c	speed of light in a vacuum $[l\,t^{-1}]$		A_s	surface area $[l^2]$
c_p	specific heat $[l^2 t^{-2}/°]$		A_N	surface area $[l^2]$
d	thickness of bimetallic strip $[l]$		B	systematic uncertainty
e_c	conduction temperature error $[°]$		C_α	coefficient of thermal expansion $[l\,l^{-1}/°]$
e_U	recovery temperature error $[°]$		D	diameter $[l]$
e_r	radiation temperature error $[°]$		E_b	blackbody emissive power $[m\,t^{-3}]$
emf	electromotive force		$E_{b\lambda}$	spectral emissive power $[m\,t^{-3}]$
h	convective heat transfer coefficient $[m\,t^{-3}/°]$; enthalpy $[l^2 t^{-2}]$		E_i	input voltage [V]
			E_1	voltage drop across R_1 [V]
k	thermal conductivity $[m\,l\,t^{-3}/°]$		F	radiation view factor
k	ratio of specific heats (equation for the speed of sound)		I	current [A]
			L	length of temperature probe $[l]$
l	length $[l]$		P	standard random uncertainty; perimeter $[l]$
m	$\sqrt{hP/kA}$ (equation 7.15) $[l^{-2}]$		Q	heat transfer $[m\,l^2\,t^{-3}]$
q	heat flux $[m\,t^3]$		R	resistance $[\Omega]$; gas constant $[l^2\,t^{-2}/°]$
r	recovery factor		R_0	reference resistance $[\Omega]$
r_c	radius of curvature $[l]$		R_T	thermistor resistance $[\Omega]$
u	uncertainty		$S_{\bar{x}}$	standard deviation of the means for the variable x
u_d	design-stage uncertainty			

T	temperature $[°]$	π_{AB}	Peltier coefficient $[m\,l^2\,t^{-3}\,A^{-1}]$
T_0	reference temperature $[°]$	σ	Thomson coefficient $[m\,l^2\,t^{-3}\,A^{-1}]$;
T_p	probe temperature $[°]$		Stefan-Boltzmann constant for radiation
T_w	wall or boundary temperature $[°]$		$m\,t^{-3}/(°)^4]$
T_∞	fluid temperature $[°]$	θ	nondimensional temperature
U	fluid velocity $[l\,t^{-1}]$	θ_x	sensitivity index for variable x
α	temperature coefficient of resistivity		(uncertainty analysis)
	$[\Omega/°]$	ρ_e	resistivity $[\Omega\,l]$
α_{AB}	Seebeck coefficient	δ	thermistor dissipation constant
β	material constant for thermistor		$[m\,l^2\,t^{-3}/°]$
	resistance $[°]$; constant in	ε	emissivity
	polynomial expansion		

PROBLEMS

8.1 Define and discuss the significance of the following terms, as they apply to temperature and temperature measurements:

a. temperature scale

b. temperature standards

c. fixed points

d. interpolation

8.2 Fixed temperature points in the International Temperature Scale are phase equilibrium states for a variety of pure substances. Discuss the conditions necessary within an experimental apparatus to accurately reproduce these fixed temperature points. How would elevation, weather, and material purity affect the uncertainty in these fixed points?

8.3 Calculate the resistance of a platinum wire that is 2 m in length and has a diameter of 0.1 cm. The resistivity of platinum at 25°C is 9.83×10^{-6} Ω-cm. What implications does this result have for the construction of a resistance thermometer using platinum?

8.4 An RTD forms one arm of a Wheatstone bridge, as shown in Figure 8.35. The RTD is used to measure a constant temperature, with the bridge operated in a balanced mode. The RTD has a resistance of 25 Ω at a temperature of 0°C, and a thermal coefficient of resistance, $\alpha = 0.003925°C^{-1}$. The value of the variable resistance R_1 must be set to

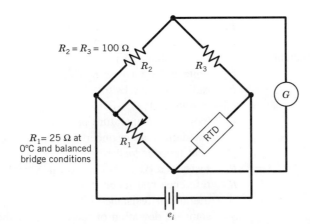

Figure 8.35 Wheatstone bridge circuit for Problem 8.4.

11.485 Ω to balance the bridge circuit, with the RTD in thermal equilibrium with the measuring environment.

    a. Determine the temperature of the RTD.

    b. Compare this circuit to the equal-arm bridge in Example 8.2. Which circuit provides the greater static sensitivity?

8.5 A thermistor is placed in a 100°C environment, and its resistance measured as 20,000 Ω. The material constant, β, for this thermistor is 3650°C. If the thermistor is then used to measure a particular temperature, and its resistance is measured as 500 Ω, determine the thermistor temperature.

8.6 Estimate the required level of uncertainty in the measurement of resistance for a platinum RTD if the RTD is to serve as a local standard for the calibration of a temperature measurement system for an uncertainty of ±0.005°C. Assume $R(0°C) = 100$ Ω.

8.7 Define and discuss the following terms related to thermocouple circuits:

    a. thermocouple junction   **d.** Peltier effect

    b. thermocouple laws   **e.** Seebeck coefficient

    c. reference junction

8.8 The thermocouple circuit in Figure 8.36 represents a J-type thermocouple with the reference junction having $T_2 = 0°C$. The output emf is 13.777 mV. What is the temperature of the measuring junction, T_1?

8.9 The thermocouple circuit in Figure 8.36 represents a J-type thermocouple. The circuit produces an emf of 15 mV for $T_1 = 750°C$. What is T_2?

8.10 The thermocouple circuit in Figure 8.36 is composed of copper and constantan and has an output voltage of 6 mV for $T_1 = 200°C$. What is T_2?

8.11 **a.** The thermocouple shown in Figure 8.37(*a*) yields an output voltage of 7.947 mV. What is the temperature of the measuring junction?

    b. The ice bath that maintains the reference junction temperature melts, allowing the reference junction to reach a temperature of 25°C. If the measuring junction of part (a) remains at the same temperature, what voltage would be measured by the potentiometer?

    c. Copper extension leads are installed as shown in Figure 8.37(*b*). For an output voltage of 7.947 mV, what is the temperature of the measuring junction?

8.12 A J-type thermocouple referenced to 70°F has a measured output emf of 2.878 mV. What is the temperature of the measuring junction?

8.13 A J-type thermocouple referenced to 0°C indicates 4.115 mV. What is the temperature of the measuring junction?

8.14 A temperature measurement requires an uncertainty of ±2°C at a temperature of 200°C. A standard T-type thermocouple is to be used with a readout device that provides electronic ice-point reference junction compensation, and has a stated instrument error of ±0.5°C with 0.1°C resolutions. Determine whether the uncertainty constraint is met at the design stage.

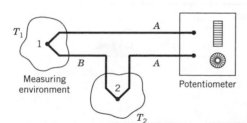

Figure 8.36 Thermocouple circuit for Problem 8.8; 1, measuring junction; 2, reference junction.

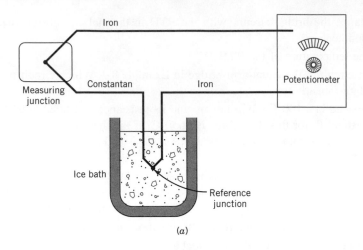

(a)

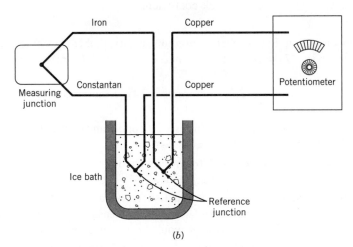

(b)

Figure 8.37 Schematic
diagram for Problem 8.9.

8.15 A temperature difference of 3.0°C is measured using a thermopile having three pairs of measuring junctions, arranged as shown in Figure 8.21.

   a. Determine the output of the thermopile for J-type thermocouple wire, if all pairs of junctions sense the 3.0°C temperature difference. The average temperature of the junctions is 80°C.

   b. If the thermopile is constructed of wire that has a maximum emf variation from the NIST standard values of ±0.8%, and the voltage measuring capabilities in the system are such that the uncertainty is ±0.0005 V, perform an uncertainty analysis to estimate the uncertainty in the measured temperature difference at the design stage.

8.16 Complete the following table for a J-type thermocouple:

Temperature [°C]		emf [mV]
Measured	Reference	
100	0	
	0	−0.5
100	50	
	50	2.5

8.17 A J-type thermopile is constructed as shown in Figure 8.20, to measure a single temperature. For a four-junction thermopile, referenced to 0°C, what would be the emf produced at a temperature of 125°C? If a voltage measuring device was available that had a total uncertainty of ±0.0001 V, how many junctions would be required in the thermopile to reduce the uncertainty in the measured temperature to 0.1°C?

8.18 You are employed as a heating, ventilating, and air conditioning engineer. Your task is to decide where in a residence to place a thermostat, and how it is to be mounted on the wall. A thermostat contains a bimetallic temperature measuring device that serves as the sensor for the control logic of the heating and air conditioning system for the house. Consider the heating season. When the temperature of the sensor falls 1°C below the set-point temperature of the thermostat, the furnace is activated; when the temperature rises 1°C above the set point, the furnace is turned off. Discuss where the thermostat should be placed in the house, what factors could cause the temperature of the sensor in the thermostat to be different from the air temperature, and possible causes of discomfort for the occupants of the house. How does the thermal capacitance of the temperature sensor affect the operation of the thermostat? Why are thermostats typically set 5°C higher in the air conditioning season?

8.19 A J-type thermocouple for use at temperatures between 0 and 100°C was calibrated at the steam point in a device called a hypsometer. A hypsometer creates a constant temperature environment at the saturation temperature of water, at the local barometric pressure. The steam-point temperature is strongly affected by barometric pressure variations. Atmospheric pressure on the day of this calibration was 30.1 in Hg. The steam-point temperature as a function of barometric pressure may be expressed as

$$T_{st} = 212 + 50.422 \left(\frac{p}{p_0} - 1 \right) - 20.95 \left(\frac{p}{p_0} - 1 \right)^2 \, [^\circ F]$$

where $p_0 = 29.921$ in Hg. At the steam point, the emf produced by the thermocouple, referenced to 0°C, is measured as 5.310 mV. Construct a calibration curve for this thermocouple by plotting the difference between the thermocouple reference table value and the measured value $(emf_{ref} - emf_{meas})$ versus temperature. What is this difference at 0°C? Suggest a means for measuring temperatures between 0 and 100°C using this calibration, and estimate the contribution to the total uncertainty.

8.20 A J-type thermocouple is calibrated against an RTD standard within ±0.01°C between 0 and 200°C. The emf is measured with a potentiometer having 0.001 mV resolution and less than 0.015 mV bias. The reference junction temperature is provided by an ice bath. The calibration procedure yields the following results:

$T_{RTD} [^\circ C]$	0.00	20.50	40.00	60.43	80.25	100.65
emf [mV]	0.010	1.038	2.096	3.207	4.231	5.336

a. Determine a polynomial to describe the relation between the temperature and thermocouple emf.

b. Estimate the uncertainty in temperature using this thermocouple and potentiometer.

c. Suppose the thermocouple is connected to a digital temperature indicator having a resolution of 0.1°C and better than 0.3°C accuracy. Estimate the uncertainty in indicated temperature.

8.21 A beaded thermocouple is placed in a duct in a moving gas stream having a velocity of 200 ft/s. The thermocouple indicates a temperature of 1400 R.

a. Determine the true static temperature of the fluid, based on correcting the reading for velocity errors. Take the specific heat of the fluid to be 0.6 Btu/lbm°R, and the recovery factor to be 0.22.

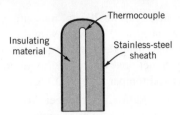

Figure 8.38 Typical construction of a sheathed thermocouple.

b. Estimate the error in the thermocouple reading due to radiation if the walls of the duct are at 1200°R. The view factor from the probe to the duct walls is 1, the convective heat-transfer coefficient, h, is 30 Btu/hr-ft^2-°R, and the emissivity of the temperature probe is 1.

8.22 It is desired to measure the static temperature of the air outside of an aircraft flying at 20,000 ft with a speed of 300 miles per hour, or 438.3 ft/s. A temperature probe is used that has a recovery factor, r, of 0.75. If the static temperature of the air is 413 R, and the specific heat is 0.24 Btu/lbm R, what is the temperature indicated by the probe? Local atmospheric pressure at 20,000 ft is approximately 970 lb/ft^2, which results in an air density of 0.0442 lbm/ft^3. Discuss additional factors that might affect the accuracy of the static temperature reading.

8.23 Consider the typical construction of a sheathed thermocouple, as shown in Figure 8.38. Analysis of this geometry to determine conduction errors in temperature measurement is difficult. Suggest a method for placing a realistic upper limit on the conduction error for such a probe, for a specified immersion depth into a convective environment.

8.24 An iron-constantan thermocouple is placed in a moving air stream in a duct, as shown in Figure 8.39. The thermocouple reference junction is maintained at 212°F. The emf output from the thermocouple is 14.143 mV.

a. Determine the thermocouple junction temperature.

b. By considering recovery and radiation errors, estimate the total error in the indicated temperature. Discuss whether this estimate of the measurement error is conservative, and why or why not. The heat-transfer coefficient may be taken as 70 Btu/hr-ft^2-°F.

Air Properties	**Thermocouple Properties**
$c_p = 0.24$ Btu/lbm°F	$r = 0.7$
$u = 200$ ft/s	$e = 0.25$

8.25 In Example 8.5, an uncertainty value for R_T was determined at 125°C as $B_{R_T} = \pm 247\,\Omega$. Show that this value is correct by performing an uncertainty analysis on R_T. In addition, determine the value of B_{R_T} at the temperatures 150 and 100°C. What error is introduced into the uncertainty analysis for β by using the value of B_{R_T} at 125°C?

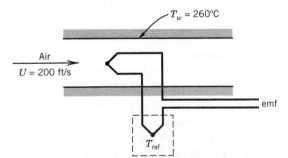

Figure 8.39 Schematic diagram for Problem 8.22.

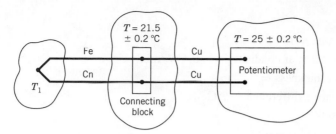

Figure 8.40 Thermocouple circuit for Problem 8.24.

8.26 The thermocouple circuit shown in Figure 8.40 measures the temperature T_1. The potentiometer limits of error are given as:

  Limits of error: $\pm 0.05\%$ of reading $+ 15$ μV at 25°C

  Resolution: 5 μV

  Give a best estimate for the temperature T_1, if the output emf is 9 mV.

8.27 A concentration of salt of 600 ppm in tap water will cause a 0.05°C change in the freezing point of water. For an ice bath prepared using tap water and ice cubes from tap water at a local laboratory, and having 1500 ppm of salts, determine the error in the ice-point reference. Upon repeated measurements, is this error manifested as a systematic or random error? Explain.

8.28 A platinum RTD ($\alpha = 0.00392°C^{-1}$) is to be calibrated in a fixed-point environment. The probe itself is used in a balanced mode with a Wheatstone bridge, as shown in Figure 8.41. The bridge resistances are known to an uncertainty of ± 0.001 Ω (95%). At 0°C the bridge balances when $R_c = 100.000$ Ω. At 100°C the bridge balances when $R_c = 139.200$ W.

a. Find the RTD resistance corresponding to 0 and 100°C, and the uncertainty in each value.

b. Calculate the uncertainty in determining a temperature using this RTD-bridge system for a measured temperature that results in $R_c = 300$ Ω. Assume $u_\alpha = \pm 1 \times 10^{-5}$ (95%).

8.29 A T-type thermopile is used to measure temperature difference across insulation in the ceiling of a residence in an energy monitoring program. The temperature difference across the insulation is used to calculate energy loss through the ceiling from the relationship

$$Q = kA_c(\Delta T / L)$$

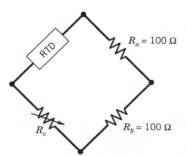

Figure 8.41 Bridge circuit for Problem 8.26.

where

$$A_c = \text{ceiling area} = 15\,\text{m}^2$$
$$k = \text{insulation thermal conductivity} -0.4\,\text{W/m°C}$$
$$L = \text{insulation thickness} = 0.25\,\text{m}$$
$$\Delta T = \text{temperature difference} = 5\text{°C}$$
$$Q = \text{heat loss}(W)$$

The value of the temperature difference is expected to be 5°C, and the thermocouple emf is measured with an uncertainty of ±0.04 mV. Determine the required number of thermopile junctions to yield an uncertainty in Q of ±5%, assuming the uncertainty in all other variables may be neglected.

8.30 A T-type thermocouple referenced to 0°C is used to measure the temperature of boiling water. What is the emf of this circuit at 100°C?

8.31 A T-type thermocouple referenced to 0°C develops an output emf of 1.2 mV. What is the temperature sensed by the thermocouple?

8.32 A temperature measurement system consists of a digital voltmeter and a T-type thermocouple. The thermocouple leads are connected directly to the voltmeter, which is placed in an air conditioned space at 25°C. The output emf from the thermocouple is 10 mV. What is the measuring junction temperature?

Chapter 9

Pressure and Velocity Measurements

9.1 INTRODUCTION

In this chapter, we introduce methods to measure the pressure and velocity within fluids. Instruments and procedures for establishing known values of pressure for calibration purposes, as well as various types of transducers for pressure measurement, are discussed. We also discuss well-established methods measuring the local and full-field velocity within a moving fluid. Finally, we present practical considerations, including common error sources, for pressure and velocity measurements.

9.2 PRESSURE CONCEPTS

Pressure represents a contact force per unit area. It acts inward and normal to the surface of any physical boundary that a fluid contacts. A basic understanding of the origin of a pressure involves the consideration of the forces acting between the fluid molecules and the solid boundaries containing the fluid. For example, consider the measurement of pressure at the wall of a vessel containing a perfect gas. As a molecule with some amount of momentum collides with the solid boundary, it will rebound off in a different direction. From Newton's second law, we know that the change in linear momentum of the molecule produces an equal but opposite force on the boundary. It is the net effect of these collisions that yields the pressure sensed at the boundary surface. Factors that affect the magnitude or frequency of the collisions, such as fluid temperature and fluid density, will affect the pressure. In fact, this reasoning is the basis of the kinetic theory from which the ideal gas equation of state is derived.

A pressure scale can be related to molecular activity, as well, since a lack of any molecular activity must form the limit of absolute zero pressure. A pure vacuum, which contains no molecules, would form the primary standard for absolute zero pressure. As shown in Figure 9.1, the absolute pressure scale is quantified relative to this absolute zero pressure. The pressure under standard atmospheric conditions is defined [1] as 1.01320×10^5 Pa absolute (where 1 Pa = 1 N/m^2). This is equivalent to

101.32 kPa absolute

1 atm absolute

14.696 lb/in.^{2} absolute (written as, psia)

1.013 bar absolute

Also indicated in Figure 9.1 is a gauge pressure scale. The gauge pressure scale is measured relative to some absolute reference pressure, which is defined in a manner

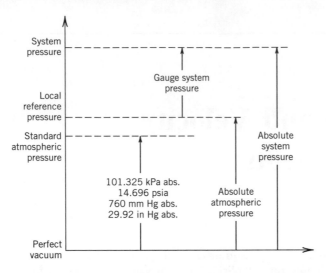

Figure 9.1 Relative pressure scales.

convenient to the measurement. The relation between an absolute pressure, p_{abs}, and its corresponding gauge pressure, p_{gauge}, is given by

$$p_{gauge} = p_{abs} - p_0 \qquad (9.1)$$

where p_0 is a reference pressure. A commonly used reference pressure is the local absolute atmospheric pressure existing at the place of the measurement. Absolute pressure will be a positive number. Gauge pressure can be positive or negative depending on the value of measured pressure relative to the reference pressure. A differential pressure, such as $p_1 - p_2$, is a relative measure and cannot be written as an absolute pressure.

Pressure can also be described in terms of the pressure exerted on a surface submerged in a column of fluid at a depth, h, as depicted in Figure 9.2. From hydrostatics, the pressure at any depth within a fluid of specific weight γ can be written as

$$p_{abs}(h) = p_0(h_0) + \gamma h \qquad (9.2)$$

In equation (9.2), p_0 is the pressure at an arbitrary datum line at h_0, and h is measured relative to h_0. The fluid specific weight is given by $\gamma = \rho g$. When equation (9.2) is rearranged, the equivalent head of fluid of depth, h, becomes

$$h = (p_{abs} - p_0)/\gamma \qquad (9.3)$$

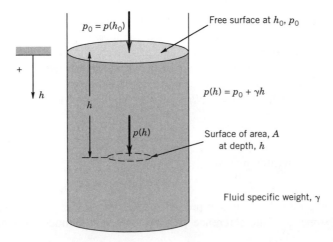

Figure 9.2 Hydrostatic equivalent pressure head and pressure.

The equivalent pressure head at one standard atmosphere is defined to be

$$760 \text{ mm Hg abs} - 760 \text{ torr abs} = 1 \text{ atm abs}$$
$$= 10,350.8 \text{ mm } H_2O \text{ abs} = 29.92 \text{ in. Hg abs}$$
$$= 407.513 \text{ in. } H_2O \text{ abs}$$

The standard is based on mercury with a density of $0.0135951 \text{ kg/cm}^3$ at $0°C$ and water at $0.000998207 \text{ kg/cm}^3$ at $20°C$ [1].

EXAMPLE 9.1

Determine the absolute and gauge pressures and the equivalent pressure head at a depth of 10 m below the free surface of a pool of water at $20°C$.

KNOWN

$h = 10 \text{ m}$, where $h = h_0 = 0$ is the free surface

$T = 20°C$

$\rho_{H_2O} = 998.207 \text{ kg/m}^3$

Specific gravity of mercury, $S_{Hg} = 13.57$.

ASSUMPTIONS   Water density constant

$p_0(h_0) = 1.0132 \times 10^5$ Pa abs

FIND p_{abs}, p_{gauge}, and h

SOLUTION   The absolute pressure can be determined directly from equation (9.2). Using the pressure at the free surface as the reference pressure and the datum line for h_0, the absolute pressure must be

$$p_{abs} = 1.0132 \times 10^5 N/m^2 + \frac{(997.4 \text{ kg/m}^3)(9.8 \text{ m/s}^2)(10 \text{ m})}{1 \text{ kg} - \text{m/N} - s^2}$$
$$= 1.9906 \times 10^5 N/m^2 \text{ abs}$$

This is equivalent to 199.06 kPa abs or 1.96 atm abs or 28.80 lb/in.^{2} abs or 1.99 bar abs. The gauge pressure is found from equation (9.1) to be

$$p_g = p_{abs} - p_0 = \gamma h$$
$$= 9.7745 \times 10^4 N/m^2$$

which is equivalent to 97.7 kPa or 0.96 atm or 14.1 lb/in.^{2} or 0.98 bar.
    In terms of equivalent head, the pressure is stated from equation (9.3):

$$h_{abs} = \frac{p_{abs}}{\rho g} = \frac{1.9906 \times 10^5 N/m^2 \text{abs}}{(997.4 \text{ kg/m}^3)(9.8 \text{ m/s}^2)}$$
$$= 20.36 \text{ m } H_2O \text{ abs} = 1.50 \text{ m Hg abs}$$

or in terms of gauge pressure relative to 760 mm Hg abs:

$$h_g = \frac{p_{abs} - p_g}{\rho g} = \frac{(1.9906 \times 10^5) - (1.0132 \times 10^5)N/m^2}{(998.2 \text{ kg/m}^3)(9.8 \text{ m/s}^2)(1N\text{-}s^2/\text{kg-m})}$$
$$= 10 \text{ m } H_2O = 0.73 \text{ m Hg}$$

9.3   PRESSURE REFERENCE INSTRUMENTS

The units of pressure are defined through the standards of the fundamental dimensions of mass, length, and time. In practice, pressure transducers are calibrated by comparison against certain reference instruments. In this section, several basic reference instruments are discussed.

McLeod Gauge

The McLeod gauge [2] is a pressure-measuring instrument and laboratory reference standard used to establish gas pressures in the subatmospheric range of 1 mm Hg abs down to 0.1 mm Hg abs. One variation of this instrument is sketched in Figure 9.3(*a*) in which the gauge is connected directly to the low-pressure source. The glass tubing is arranged so that a sample of the gas at an unkown low pressure can be trapped by inverting the gauge from the sensing position, depicted as Figure 9.3(*a*), to that of the measuring position, depicted as Figure 9.3(*b*). In this way, the gas trapped within the capillary is isothermally compressed by a rising column of mercury. Boyle's law is then used to relate the two pressures on either side of the mercury to the distance of travel of the mercury within the capillary. Mercury is the preferred working fluid because of its high density and very low vapor pressure.

At the equilibrium and measuring position, the capillary pressure, p_2, is related to the unknown gas pressure to be determined, p_1, by $p_2 = p_1(\forall_1/\forall_2)$ where $\forall_1$ is the gas volume of the gauge in Figure 9.3(*a*) (a constant for a gauge at any pressure) and $\forall_2$ is the capillary volume in Figure 9.3(*b*). But $\forall_2 = Ay$, where A is the known cross-sectional area of the capillary and y is the vertical length of the capillary occupied by the gas. With γ as the specific weight of the mercury, the difference in pressures is related by $p_2 - p_1 = \gamma y$ such that the unknown gas pressure is just a function of y:

$$p_1 = \gamma A y^2/(\forall_1 - Ay) \tag{9.4}$$

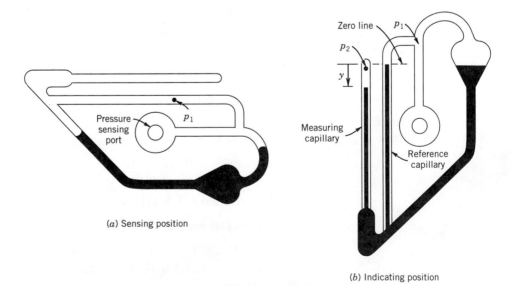

(*a*) Sensing position

(*b*) Indicating position

Figure 9.3 McLeod gauge.

In practice, a commercial McLeod gauge will have the capillary etched and calibrated to indicate either pressure, p_1, or its equivalent head, p_1/γ, directly.

The McLeod gauge generally does not require correction. The reference stem offsets capillary forces acting in the measuring capillary. Instrument systematic uncertainty will be on the order of 0.5% (95%) at 1 mm Hg abs and increasing to 3% (95%) at 0.1 mm Hg abs.

Barometer

A barometer consists of an inverted tube containing a fluid and is used to measure atmospheric pressure. To create the barometer, the tube, which is sealed at only one end, is evacuated to zero absolute pressure. With open end down, the tube is immersed within a liquid-filled reservoir as shown in the illustration of the Fortin barometer in Figure 9.4. The reservoir is open to atmospheric pressure, which forces the liquid to rise up the tube.

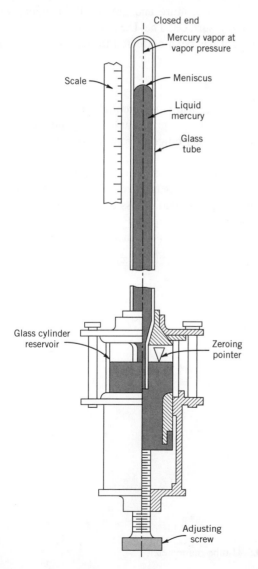

Figure 9.4 Fortin barometer.

From equations 9.2 and 9.3, the resulting height of the liquid column above the reservoir free surface is a measure of the absolute atmospheric pressure in equivalent head (equation 9.3). Evangelista Torricelli (1608–1647), a colleague of Galileo, can be credited with developing and interpreting the working principles of the barometer in 1644.

As Figure 9.4 shows, the closed end of the tube will be at the vapor pressure of the barometric liquid at room temperature. So the indicated pressure will be the atmospheric pressure minus the liquid vapor pressure. Mercury is the most common liquid used because it has a very low vapor pressure, and so, for practical use, the indicated pressure can be taken as the local barometric pressure. However, for accurate work the barometer will need to be corrected for temperature effects, which change the vapor pressure, and for temperature and altitude effects on the weight of mercury, and for deviations from standard gravity (9.80665 m/s^2 or 32.17405 ft/s^2). Correction curves are provided by instrument manufacturers.

Barometers are used as local standards for the measurement of atmospheric pressure. Under standard conditions for pressure temperature and gravity, the mercury will rise 760 mm (29.92 in.) above the reservoir surface. The U.S. National Weather Service always reports a barometric pressure that has been corrected to sea-level elevation.

Manometers

A manometer is an instrument used to measure differential pressure based on the relationship between pressure and the hydrostatic equivalent head of fluid. Several design variations are available allowing measurements ranging from the order of 0.001 mm of manometer fluid to several meters.

The U-tube manometer in Figure 9.5 consists of a transparent tube filled with an indicating liquid of specific weight, γ_m. This forms two free surfaces of the manometer liquid. The difference in pressures p_1 and p_2 applied across the two free surfaces brings about a deflection, H, in the level of the manometer liquid. For a measured fluid of specific weight γ, the hydrostatic equation can be applied to the manometer of Figure 9.5 as

$$p_1 = p_2 + \gamma x + \gamma_m H - \gamma(H + x)$$

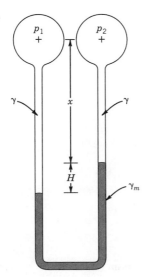

Figure 9.5 U-tube manometer.

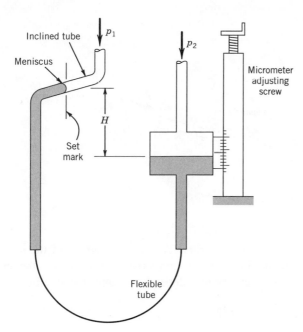

Figure 9.6 Micromanometer.

which yields the relation between the manometer deflection and applied differential pressure,

$$p_1 - p_2 = (\gamma_m - \gamma)H \qquad (9.5)$$

From equation (9.5), the static sensitivity of the U-tube manometer is given by $K = 1/(\gamma_m - \gamma)$. To maximize manometer sensitivity, we want to choose manometer liquids that minimize the value of $(\gamma_m - \gamma)$. From a practical standpoint, however, the manometer fluid should have a greater specific weight than and must not be soluble with the working fluid. The manometer fluid should be selected to provide a deflection that is measurable yet not so great that it becomes awkward to observe.

A variation in the U-tube manometer is the micromanometer shown in Figure 9.6. These special purpose instruments are used to measure very small differential pressures, down to 0.005 mm H_2O (0.0002 in. H_2O). In the micromanometer, the manometer reservoir is moved up or down until the level of the manometer fluid within the reservoir is at the same level as a set mark within a magnifying sight glass. At that point the manometer meniscus will be at the set mark, and this serves as a reference position. Changes in pressure bring about fluid displacement so that the reservoir must be moved up or down to bring the meniscus back to the set mark. The amount of this repositioning is equal to the change in equivalent pressure head. The position of the reservoir is controlled by a micrometer or other calibrated displacement measuring device so that relative changes in pressure can be measured with a high resolution.

The inclined tube manometer is also used to measure small changes in pressure. It is essentially a U-tube manometer with one leg inclined at an angle θ, typically from 10 to 30° relative to the horizontal. As indicated in Figure 9.7, a change in pressure equivalent to a deflection of height H in a U-tube manometer would bring about a change in position of the meniscus in the inclined leg of $L = H/\sin \theta$. This provides an increased sensitivity over the conventional U-tube by the factor of $1/\sin \theta$.

A number of elemental errors affect the instrument uncertainty of all types of manometers. These include scale and alignment errors, zero error, temperature error,

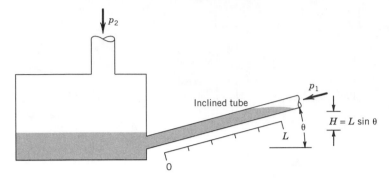

Figure 9.7 Inclined tube manometer.

gravity error, and capillary and meniscus errors. The specific weight of the manometer fluid will vary with temperature but can be corrected. For example, the common manometer fluid of mercury has a temperature dependence approximated by

$$\gamma_{Hg} = \frac{133.084}{1 + 0.00006T} \, [\text{N/m}^3] = \frac{848.707}{1 + 0.000101(T - 32)} \, [\text{lb/ft}^3]$$

with T in °C or °F, respectively. A gravity correction for elevation, z, and latitude, ϕ, corrects for gravity error effects using the dimensionless correction,

$$e_1 = -(2.637 \times 10^{-3} \cos 2\phi + 9.6 \times 10^{-8}z + 5 \times 10^{-5}) \qquad (9.6a)$$
$$= -(2.637 \times 10^{-3} \cos 2\phi + 2.9 \times 10^{-8}z + 5 \times 10^{-5}) \qquad (9.6b)$$

where ϕ is in degrees and z is in feet for equation (9.6a) and meters in equation (9.6b). Tube-to-liquid capillary forces lead to the development of a meniscus. Although the actual effect varies with purity of the manometer liquid, these effects can be minimized by using manometer tube bores of greater than about 6 mm (0.25 in.). In general, the instrument uncertainty in measuring pressure can be as low as 0.02–0.2% of the reading.

EXAMPLE 9.2

An inclined manometer with indicating leg at 30° is to be used at 20°C to measure an air pressure of nominal magnitude of 100 N/m² relative to ambient. "Unity" oil ($S = 1$) is to be used. The specific weight of the oil is 9770 ± 0.5% N/m² (95%) at 20°C, the angle of inclination can be set to within 1° using a bubble level, and the manometer resolution is 1 mm with a manometer zero error equal to its interpolation error. Estimate the uncertainty in indicated differential pressure at the design stage.

KNOWN

$$p = 100 \, \text{N/m}^2 \, (\text{nominal})$$

Manometer

   Resolution : 1 mm

   Zero error : 0.5 mm

$$\theta = 30 \pm 1° \quad (95\% \text{ assumed})$$
$$\gamma_m = 9770 \pm 0.5\% \, \text{N/m}^3 \quad (95\%)$$

ASSUMPTIONS   Temperature and capillary effects in manometer and gravity error in the specific weights of the fluids are negligible.

FIND

u_d

SOLUTION   The relation between pressure and manometer deflection is given by equation (9.5) with $H = L \sin \theta$:

$$\Delta p = p_1 - p_2 = L(\gamma_m - \gamma) \sin \theta$$

where p_2 is ambient pressure so that Δp is the nominal pressure relative to ambient. For a nominal $\Delta p = 100 \, \text{N/m}^2$, the nominal manometer rise L would be

$$L = \frac{\Delta p}{(\gamma_m - \gamma) \sin \theta} \approx \frac{\Delta p}{\gamma_m \sin \theta} = 21 \, \text{mm}$$

where $\gamma_m \gg \gamma$ and the value for γ and its uncertainty are neglected. For the design stage analysis, $p = f(\gamma_m, L, \theta)$, so that the uncertainty in pressure, Δp, is estimated by

$$(u_d)_p = \pm \sqrt{\left[\frac{\partial \Delta p}{\partial \gamma_m} (u_d)_{\gamma_m} \right]^2 + \left[\frac{\partial \Delta p}{\partial L} (u_d)_L \right]^2 + \left[\frac{\partial \Delta p}{\partial \theta} (u_d)_\theta \right]^2}$$

At assumed 95% confidence levels, the manometer specific weight uncertainty and angle uncertainty are estimated from the problem as

$$(u_d)_{\gamma_m} = (9770 \, \text{N/m}^3)(0.005) \approx 49 \, \text{N/m}^3$$

$$(u_d)_\theta = 1° = 0.0175 \, \text{rad}$$

The uncertainty in estimating the pressure from the indicated deflection is due both to the manometer resolution, u_o, and the zero point offset error, which we take as its instrument error, u_c. Using the uncertainties associated with these errors,

$$(u_d)_L = \sqrt{u_o^2 + u_c^2} = \sqrt{(0.5 \, \text{mm})^2 + (0.5 \, \text{mm})^2} = 0.7 \, \text{mm}$$

Evaluating the derivatives and substituting values gives a design-stage uncertainty in Δp of

$$(u_d)_{\Delta p} = \pm \sqrt{(0.26)^2 + (3.42)^2 + (3.10)^2} = \pm 4.6 \, \text{N/m}^2 \quad (95\%)$$

COMMENT   At a 30° inclination and for this pressure, the uncertainty in pressure is affected almost equally by the instrument inclination and the deflection uncertainties. As the manometer inclination is increased to a more vertical orientation, that is, toward the U-tube mano-meter, inclination uncertainty becomes less important and is negligible near 90°. However, for a U-tube manometer, the deflection is reduced to less than 11 mm, a 50% reduction in manometer sensitivity, with an associated design-stage uncertainty of 6.8 N/m^2.

Deadweight Testers

The deadweight tester makes use of the fundamental definition of pressure as a force per unit area to create and to determine the pressure within a sealed chamber. These are used as a laboratory standard for the calibration of pressure-measuring devices over the pressure range from 70 to 7×10^7 N/m^2 (0.01 to 10,000 psi). A deadweight tester, such as that shown in Figure 9.8, consists of an internal chamber filled with a liquid, and a close-fitting piston and cylinder. Chamber pressure is produced by the compression of the

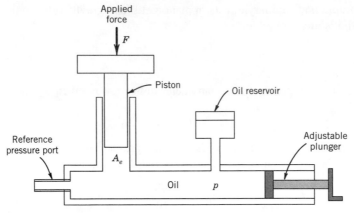

Figure 9.8 Deadweight tester.

liquid, usually oil, by the adjustable plunger. This pressure acts on the end of the carefully machined piston. A static equilibrium will exist when the external pressure exerted by the piston on the fluid balances with the chamber pressure. This external piston pressure is created by a downwards force acting over the equivalent area, A_e, of the piston. The weight of the piston plus the additional weight of calibrated masses are used to provide this external force, F. At static equilibrium the piston will float and the chamber pressure can be deduced as

$$p = \frac{F}{A_e} + \sum \text{errors} \tag{9.7}$$

A pressure-measuring device, such as a pressure transducer, can be connected to the tester's reference port and calibrated by comparison to the chamber pressure. For an approximate calibration, the errors can be ignored.

When errors are accounted, the instrument uncertainty in the chamber pressure using a deadweight tester can be as low as 0.05–0.01% of the reading. A number of elemental errors contribute to equation (9.7), including air buoyancy effects, variations in local gravity, uncertainty in the known mass of the piston and added masses, shear effects, thermal expansion of the piston area, and elastic deformation of the piston.

An indicated pressure, p_i, can be corrected for gravity effects, e_1 from equation (9.6a) or (9.6b), and for air buoyancy effects, e_2, by

$$p = p_i(1 + e_1 + e_2) \tag{9.8}$$

where
$$e_2 = -\gamma_{\text{air}}/\gamma_{\text{masses}} \tag{9.9}$$

The tester fluid lubricates the piston so that the piston will be partially supported by the shear forces in the oil in the gap separating the piston and the cylinder. This error varies inversely with the tester fluid viscosity, so high viscosity fluids are preferred. In a typical tester, this error is less than 0.01% of the reading. At high pressures, elastic deformation of the piston will affect the actual piston area. For this reason, the effective area is based on the average of the piston and cylinder diameters, another source of uncertainty.

EXAMPLE 9.3

A deadweight tester indicates 100.00 lb/in.^{2} (i.e., 100.00 psi), at 70°F in Clemson, SC ($\phi = 34°$, $z = 841$ ft). Manufacturer specifications for the effective piston area were stated at 72°F so that thermal expansion effects remain negligible. Take $\gamma_{air} = 0.076$ lb/ft^3 and $\gamma_{mass} = 496$ lb/ft^3. Correct the indicated reading for known errors.

KNOWN

$p_i = 100.00$ psi

$z = 841$ ft

$\phi = 34°$

ASSUMPTION   Systematic error corrections for altitude and latitude apply.

FIND

p

SOLUTION   The corrected pressure is found by equation (9.8). From equation (9.9), the correction for buoyancy effects is

$$e_2 = -\gamma_{air}/\gamma_{masses} = -0.076/496 = -0.000154$$

The correction for gravity effects is from equation (9.6a)

$$e_1 = -(2.637 \times 10^{-3} \cos 2\phi + 9.6 \times 10^{-8} z + 5 \times 10^{-5})$$
$$= -(0.0010 + 8 \times 10^{-5} + 5 \times 10^{-5}) = 0.001119$$

From equation (9.8), the corrected pressure becomes

$$p = 100.00 \times (1 - 0.000154 - 0.001119) \, \text{lb/in}^2 = 99.87 \, \text{lb/in}^2$$

COMMENT   This amounts to correcting an indicated signal for known systematic errors. Here that correction is $\approx 0.13\%$.

9.4   PRESSURE TRANSDUCERS

A pressure transducer converts a measured pressure into a mechanical or electrical signal. The transducer is actually a hybrid sensor-transducer. The primary sensor is usually an elastic element that deforms or deflects under pressure. Several common elastic elements used, shown in Figure 9.9, include the Bourdon tube, bellows, capsule, and diaphragm. A secondary transducer element converts the elastic element deflection into a readily measurable signal such as an electrical voltage or mechanical rotation of a pointer. There are many methods available to perform this secondary function, but electrical transducers will require additional external signal conditioning equipment and external power supplies to drive their electrical output signals.

Pressure transducers are subject to some or all of the following elemental errors: resolution, zero shift error, linearity error, sensitivity error, hysteresis, and drift due to environmental temperature changes. Electrical transducers are also subject to loading error between the transducer output and its indicating device (Chapter 6). This error increases transducer nonlinearity over its operating range. A voltage follower (Chapter 6) can be inserted at the output of the transducer to isolate transducer load.

Bourdon Tube

The Bourdon tube is a curved metal tube having an elliptical cross section that mechanically deforms under pressure. In practice, one end of the tube is held fixed and

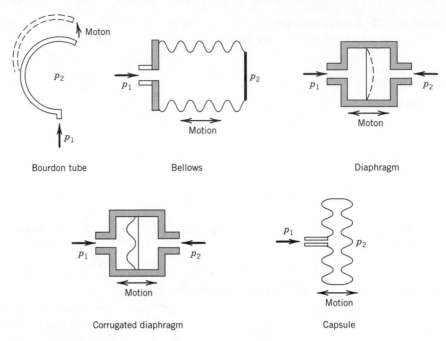

Figure 9.9 Elastic elements used as pressure sensors.

the input pressure applied internally. A pressure difference between the outside of the tube and the inside of the tube will bring about tube deformation and a deflection of the tube free end. This action of the tube under pressure can be likened to the action of a deflated balloon when it is inflated slightly. The magnitude of the deflection is proportional to the magnitude of the pressure difference. Several variations exist, such as the C shape (Figure 9.9), the spiral, and the twisted tube. The exterior of the tube is usually open to atmosphere (hence, the origin of the term "gauge" pressure referring to pressure referenced to atmospheric pressure), but in some variations the tube may be placed within a sealed housing and the tube exterior exposed to some other reference pressure, allowing for absolute and for differential designs.

The Bourdon tube mechanical dial gauge is perhaps the most commonly used pressure transducer. A typical design is shown in Figure 9.10, in which the secondary element is a mechanical linkage that converts the tube displacement into a rotation of a pointer. The instrument has a range in which pressure will be linearly related to the pointer rotation range of the instrument, and this is usually the range specified by its manufacturer. Designs exist that can be used for low or high pressures. Various ranges for gauges exist of from 10^4 to 10^9 Pa (0.1–100 000 psi). The best Bourdon tube gauges have instrument uncertainties as low as 0.1% of the full-scale deflection of the gauge, but values of 0.5 to 2% are more common.

Bellows and Capsule

A bellows sensing element is a thin-walled, flexible metal tube formed into deep convolutions and sealed at one end (Figure 9.9). One end is held fixed and pressure is applied internally. A difference between the internal and external pressures will cause the bellows to change in length. The bellows is housed within a chamber that can be sealed

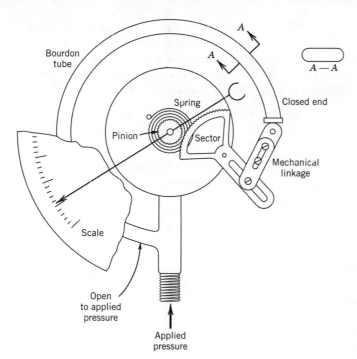

Figure 9.10 Bourdon tube pressure gauge.

and evacuated for absolute measurements, vented through a reference pressure port for differential measurements, or opened to atmosphere for gauge pressure measurements. A similar design, the capsule sensing element, is also a thin-walled, flexible metal tube whose length changes with pressure but its shape tends to be wider in diameter and shorter in length (Figure 9.9).

A mechanical linkage is used to convert the translational displacement of the bellows or capsule sensors into a measurable form. A common transducer is the sliding arm potentiometer (voltage-divider, Chapter 6) found in the potentiometric pressure transducer shown in Figure 9.11. Another type uses a linear variable displacement transducer (see LVDT, Chapter 12) to measure the bellows or capsule displacement. The LVDT design has a high sensitivity and is commonly found in pressure transducers rated for low pressures and for small pressure ranges, such as zero to several hundred mm Hg absolute, gauge, or differential.

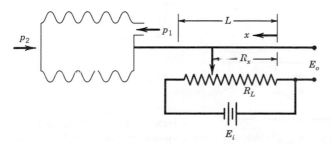

Figure 9.11 Potentiometer pressure transducer.

EXAMPLE 9.4

An engineer finds a desirable potentiometric bellows gauge with a stated range of 0 to 700 kPa in a catalog. It uses a sliding contact potentiometer having a terminal resistance of $50\,\Omega - 10\,k\Omega \pm 10\%$ over full scale. The following manufacturer information is available:

Calibration linearity: <0.5% full scale
Repeatability: $\pm0.1\%$ full scale
Excitation voltage: 5–10 V dc

A voltmeter having an input impedance of $100\,k\Omega$ is available to measure output. For an applied transducer excitation voltage of 5 V, estimate the expected instrument uncertainty in measuring a pressure of 350 kPa.

KNOWN

$$E_i = 5\,V$$
$$R_L = 50\,\Omega - 10\,k\Omega \pm 10\% \quad (95\% \text{ assumed})$$

FIND

u_c

SOLUTION   The electrical arrangement of a potentiometric gauge is the voltage divider circuit (Chapter 6). The circuit is illustrated in Figure 9.12. For an infinite measuring instrument input impedance and assuming that the terminal resistance is at its minimum at $p = 0$ kPa and at its maximum at $p = 700$ kPa, then

$$E_o(p) = \frac{R_L(p)E_i}{10\,k\,\Omega}$$

so that $E_o(0\,kPa) = 0.0250$ V, $E_o\ (350\,kPa) = 2.4875$ V, and $E_o\ (700\,kPa) = 5$ V. The output is nearly linear, so the static sensitivity will be approximately

$$K = \frac{E_o(700) - E_o(0)}{700 - 0\,kPa} = 7.11\,\mu V/Pa$$

The transducer instrument uncertainty will be affected by the elemental errors due to linearity, e_L, sensitivity, e_k, and loading, e_l. For a full-scale voltage of 5 V, the specifications indicate

$$e_L = 5\,V \times 0.005 = 0.0250\,V \qquad e_k = 5\,V \times 0.001 = 0.0050\,V$$

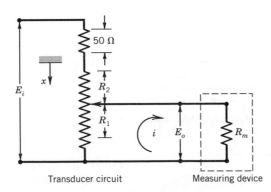

Transducer circuit   Measuring device

Figure 9.12 Potentiometer transducer circuit for Example 9.4.

Loading error will depend on the ratio of terminal resistance to measuring instrument input impedance, R_m. Using equation (6.35) with Figure 6.16, at $p = 350\,\text{kPa}$, $R_1 = 4975\,\Omega$, $R_2 = 5025\,\Omega$, and $R_m = 100{,}000\,\Omega$. So with interstage loading, the output voltage becomes

$$E_o(350) = 2.4268\,\text{V}$$

The error due to loading is $e_I = 2.4268 - 2.4875 = -0.0607\,\text{V}$. Unless we correct the reading for the loading error, we include it in the uncertainty estimate. At 350 kPa, the instrument uncertainty, u_c, is estimated from the three elemental errors to be

$$u_c = \left[e_L^2 + e_R^2 + e_I^2\right]^{1/2} = \left[0.0250^2 + 0.0050^2 + 0.0607^2\right] = 0.0652\,\text{V}$$

For a static sensitivity of 7.11 μV/Pa, this equates to

$$u_c = 9.17\,\text{kPa} \quad (95\%)$$

Diaphragms

An effective primary pressure element is a diaphragm, which is a thin elastic circular plate supported about its circumference. The action of a diaphragm within a pressure transducer is similar to the action of a trampoline and a pressure differential on the top and bottom diaphragm faces acts to deform it. The magnitude of the deformation is proportional to the pressure difference. Both membrane and corrugated designs are used. Membranes are made of metal or nonmetallic material, such as plastic or neoprene. The material chosen depends on the pressure range anticipated and the fluid in contact with it. Corrugated diaphragms contain a number of corrugations that serve to increase diaphragm stiffness and to increase the diaphragm effective surface area.

Pressure transducers that use a diaphragm sensor are well suited for either static or dynamic pressure measurements. They have good linearity and resolution over their useful range. An advantage of the diaphragm sensor is that the very low mass and relative stiffness of the thin diaphragm give the sensor a very high natural frequency with a small damping ratio. Hence, these transducers can have a very wide frequency response and very short 90% rise and settling times. The natural frequency of a circular diaphragm can be estimated by [4]

$$\omega_n = 64.15\sqrt{\frac{E_m t^2 g_c}{12(1 - v_p^2)\rho r^4}} \tag{9.10}$$

where E_m is the bulk modulus [psi or N/m²], t the thickness [in. or m], r the radius [in. or m], ρ the material density [lb/in.³ or kg/m³], and v_p the Poisson's ratio for the diaphragm material with $g_c = 386\,\text{lb}_m\text{-in./lb-s}^2 = 1\,\text{kg-m/N-s}^2$. The maximum elastic deflection of a uniformly loaded, circular diaphragm supported about its circumference occurs at its center and can be estimated by

$$y_{\text{max}} = \frac{3(p_1 - p_2)(1 - v_p^2)r^4}{16E_m t^3} \tag{9.11}$$

provided that the deflection does not exceed one-third the diaphragm thickness. Diaphragms should be selected so as to not exceed this maximum deflection over the anticipated operating range.

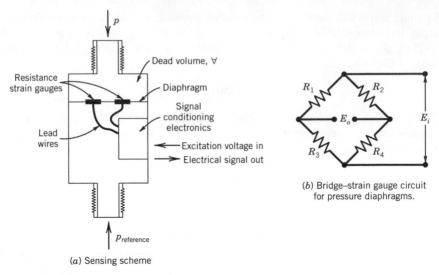

(a) Sensing scheme

(b) Bridge–strain gauge circuit for pressure diaphragms.

Figure 9.13 Diaphragm pressure transducer.

Various secondary elements are available to translate this displacement of the diaphragm into a measurable signal. Several methods are discussed below.

Strain Gauge Elements

The most common method for converting diaphragm displacement into a measurable signal is to sense the strain induced on the diaphragm surface as it is displaced. Strain gauges, devices whose measurable resistance is proportional to their sensed strain (Chapter 11), can be bonded directly onto the diaphragm or onto a deforming element (such as a thin beam) attached to the diaphragm so as to deform with the diaphragm and sense strain. Metal strain gauges can be used with liquids. Strain gauge resistance is reasonably linear over a wide range of strain and can be directly related to the diaphragm sensed pressure [5]. A diaphragm transducer using strain gauge detection is depicted in Figure 9.13.

The use of semiconductor technology in pressure transducer construction has allowed the development of a variety of very fast, very small, highly sensitive strain gauge diaphragm transducers. Silicone piezoresistive strain gauges can be diffused into a single crystal of silicone wafer, which forms the diaphragm. Semiconductor strain gauges have a static sensitivity that is 50 times greater than conventional metallic strain gauges. Because the piezoresistive gauges are integral to the diaphragm, they are relatively immune to the thermoelastic strains prevalent in conventional metallic strain gauge–diaphragm constructions. Furthermore, a silicone diaphragm will not creep with age (as will a metallic gauge), thus minimizing calibration drift over time. However, gauge failure is catastrophic and uncoated silicone does not tolerate liquids.

Capacitance Elements

When one or more fixed metal plates are placed directly above or below a metallic diaphragm, a capacitor is created that forms an effective secondary element. Such a transducer using this method is depicted in Figure 9.14. It is known as a capacitance

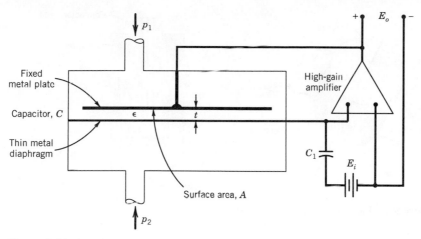

Figure 9.14 Capacitance pressure transducer.

transducer. The capacitance, C, developed between two parallel plates separated by a distance, t, is determined by

$$C = c\varepsilon A/t \qquad (9.12)$$

where ε is the dielectric constant of the material between the plates (for air, $\varepsilon = 1$), A is the overlapping area of the two plates, and c is the proportionality constant given by 0.0885 (when A is measured in [cm^2] and t in [cm]) or 0.225 (when A is measured in [in.^{2}] and t in [in.]). Displacement of the diaphragm changes the average gap separation. In the circuit shown, the measured voltage will be essentially linear with developed capacitance

$$E_o = \frac{C_1}{C} E_i \qquad (9.13)$$

and pressure can be inferred. The capacitance pressure transducer possesses the attractive features of other diaphragm transducers, including small size and a very wide operating range. However, it is sensitive to temperature changes and has a relatively high impedance output.

Piezoelectric Crystal Elements

Piezoelectric crystals form effective secondary elements for dynamic (transient) pressure measurements. Under the action of compression, tension, or shear, a piezoelectric crystal will deform and develop a surface charge, q, which is proportional to the force acting to bring about the deformation. In a piezoelectric pressure transducer, a preloaded crystal is mounted to the diaphragm sensor as indicated in Figure 9.15. Pressure acts normal to the crystal axis and changes the crystal thickness, t, by a small amount Δt. This sets up a charge

$$q = K_q pA$$

where p is the pressure acting over the electrode area A and K_q is the crystal charge sensitivity, a material property. The voltage developed across the electrodes is given by

$$E_o = q/C$$

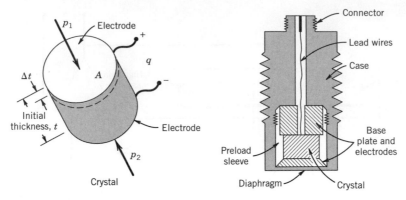

Figure 9.15 Piezoelectric pressure transducer.

where C is the capacitance of the crystal-electrode combination. The capacitance can be obtained from equation (9.12) to yield

$$E_o = K_q pt / c\varepsilon = K_E pt \qquad (9.14)$$

where K_E is the voltage sensitivity of the transducer. The crystal sensitivities for quartz, the most common material used, are $K_q = 2.2 \times 10^{-9}$ coulombs/N and $K_E = 0.055$ V-m/N. A charge amplifier (Chapter 6) is used to convert charge to voltage.

9.5   PRESSURE TRANSDUCER CALIBRATION

Static Calibration

The static calibration of a pressure transducer can be accomplished by direct comparison against any of the pressure reference instruments discussed (Section 9.3) or a certified laboratory standard transducer. For the low-pressure range, the manometric instruments along with the laboratory barometer serve as convenient and easy working standards. The approach is to pressurize a chamber and expose both the reference instrument, which serves as the standard, and the candidate transducer to that pressure for a side-by-side measurement. For the high-pressure range, the deadweight tester is a desirable pressure reference standard.

Dynamic Calibration

The rise time and frequency response of a pressure transducer are found by dynamic calibration. As discussed in Chapter 3, the rise time of an instrument is found through a step change in input. The frequency response is found through the application of periodic input signals.

An electrical switching valve is useful for creating a step change in pressure. But the mechanical lag of the valve limits its use to transducers having an expected rise time of 100 ms or more. Faster applications might use a shock tube calibration. As shown in Figure 9.16, the shock tube consists of a long pipe separated into two chambers by a thin diaphragm. The transducer is mounted into the pipe wall of one chamber at pressure p_1. The pressure in the other chamber is raised from p_1 to p_2. Some mechanism, such as a mechanically controlled needle, is used to burst the diaphragm on command. Upon bursting, the pressure differential causes a pressure shock wave to move down the low

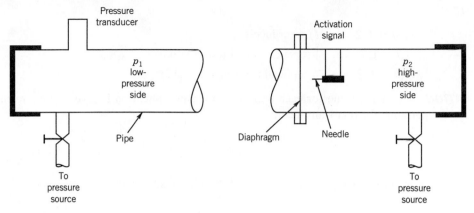

Figure 9.16 Shock tube facility.

pressure chamber. A shock wave has a thickness on the order of 1 μm and moves at the speed of sound, a. So as the shock passes the transducer, the transducer experiences a change in pressure from p_1 to p over a time $t = d/a$, where d is the diameter of the transducer pressure port and pressure p is found from

$$p = p_1[1 + (2k/k + 1)(M_1^2 - 1)] \tag{9.15}$$

where k is the gas-specific heat ratio and M_1 is the Mach number calculated at pressure p_1. The velocity of the shock wave can also be deduced from the output of fast-acting pressure sensors mounted in the shock tube wall. Typical values of t are on the order of 1–10 μs; so this method is at least 4 orders faster than a switching valve. The transducer rise time is calculated from the output record.

A reciprocating piston within a cylinder is a simple means to generate a sinusoidal variation in pressure for frequency response calibration. The piston can be driven by a variable speed motor and its displacement measured by a fast responding transducer, such as an LVDT (Chapter 12). Under properly controlled conditions (Example 1.2), the actual pressure variation can be estimated from the piston displacement. Other techniques include an encased loudspeaker or an acoustically resonant enclosure, which serves as a frequency driver instead of a piston, or using an oscillating flow control valve to vary pressure with time.

EXAMPLE 9.5

A common method to estimate the data acquisition and reduction errors present in the pressure-measuring instrumentation of a test rig is to apply a series of replication tests on the M calibrated pressure transducers used on that rig (when the number of transducers installed is large, a few transducers can be selected at random and tested). In such a test each transducer is interconnected through a common manifold to the output of a single deadweight tester (or suitable standard) so that each transducer is exposed to exactly the same static pressure. Each transducer is operated at its normal excitation voltage. The test proceeds as follows: record applied test pressure (from the deadweight tester), record each transducer output N (e.g., $N \geq 10$) times, vent manifold to local atmosphere, close manifold to reapply test pressure, repeat procedure at least K (e.g., $K \geq 5$) times. What information is available in such test data?

KNOWN

$M(j = 1, 2, \ldots, M)$ transducers

$N(i = 1, 2, \ldots, N)$ repetitions at a test pressure

$K(k = 1, 2, \ldots, K)$ replications of test pressure

SOLUTION   The mean value for each transducer for any replication is given by the mean of the repetitions for that transducer,

$$\bar{p}_{jk} = \frac{1}{N} \sum_{i=1}^{N} p_{ijk}$$

The mean value for the replications of the jth transducer is given by

$$\langle \bar{p}_j \rangle = \frac{1}{K} \sum_{k=1}^{K} \bar{p}_{jk}$$

The difference between the pooled mean pressure and the known applied pressure would provide an estimate of the systematic error to be expected from any transducer during data acquisition. On the other hand, differences between $\bar{p}_{jk}$ and $\langle \bar{p}_k \rangle$ must be due to the random error in the data acquisition and reduction procedure. This random error is estimated by considering the variation of the test pressure mean, $\bar{p}_{jk}$, about the pooled mean, $\langle \bar{p}_k \rangle$, for each transducer:

$$\langle S_j \rangle = \sqrt{\frac{\sum \left(\bar{p}_{jk} - \langle \bar{p}_j \rangle \right)^2}{K - 1}}$$

The pooled standard deviation of the mean for the M transducers provides the estimate of the random uncertainty of the data acquisition and reduction procedure:

$$S_{\bar{p}} = \frac{\langle S_p \rangle}{\sqrt{MK}} \qquad \text{where} \qquad \langle S_p \rangle = \sqrt{\frac{\sum_{j=1}^{M} \sum_{k=1}^{K} \left(\bar{p}_{jk} - \langle \bar{p}_j \rangle \right)^2}{M(K - 1)}}$$

with degrees of freedom, $\nu = M(K - 1)$.

COMMENT   The statistical estimate contains the effects of random error due to pressure standard (applied pressure repeatability), pressure transducer repeatability (repetition and replication), excitation voltages, and the recording system. It does not contain the effects of instrument calibration errors, large deviations in environmental conditions, pressure tap design errors, or dynamic pressure effects. The systematic errors for these must be set by the engineer based on other information.

Note that if the above procedure were to be repeated over a range of different applied known pressures, then this would be a calibration. This would allow instrument calibration errors and data reduction curve fit errors over the range to be entered into the analysis.

9.6   PRESSURE MEASUREMENTS IN MOVING FLUIDS

Pressure measurements in moving fluids deserve special consideration. Consider the flow over the bluff body shown in Figure 9.17. Assume that the upstream flow is uniform and steady with negligible losses. Along streamline A, the upstream flow moves with a velocity, U_1, such as at point 1. As the flow approaches point 2 it must slow down and

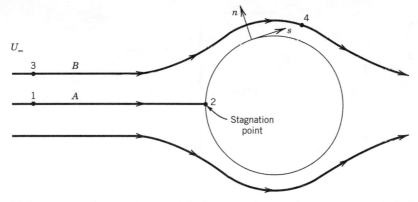

Figure 9.17 Streamline flow over a bluff body.

finally stop at the front end of the body. Above streamline A flow moves over the top of the bluff body, and below streamline A flow moves under the body. Point 2 is known as the stagnation point and streamline A the stagnation streamline for this flow. Along streamline B, the velocity at point 3 will be U_3 and because the upstream flow is considered to be uniform it follows that $U_1 = U_3$. As the flow along B approaches the body, it is deflected around the body. From conservation of mass principles, $U_4 > U_3$. Application of conservation of energy between points 1 and 2 and between 3 and 4 yields

$$p_1 + \rho U_1^2/2 = p_2 + \rho U_2^2$$
$$p_3 + \rho U_3^2/2 = p_4 + \rho U_4^2 \tag{9.16}$$

However, because point 2 is the stagnation point, $U_2 = 0$, and

$$p_2 = p_t = p_1 + \rho U_1^2/2 \tag{9.17}$$

Hence, it follows that $p_2 > p_1$ by an amount equal to $\rho U_1^2/2$, called the *dynamic pressure*, an amount equivalent to the kinetic energy per unit mass of the flow as it moves along the streamline. If no energy is lost through irreversible processes, such as through a transfer of heat,[1] this translational kinetic energy will be transferred completely into p_2. The value of p_2 is known as the *stagnation* or the *total pressure* and will be noted as p_t. The total pressure can be determined by bringing the flow to rest at a point in an isentropic manner.

The pressures at 1, 3, and 4 are known as static pressures[2] of the flow. The *static pressure* is that pressure sensed by a fluid particle as it moves with the same velocity as the local flow. The static pressure and velocity at points 1 and 3 are given the special names of the freestream static pressure and freestream velocity. Since $U_4 > U_3$, equation (9.16) shows that $p_4 < p_3$. It follows from equation (9.17) that the total pressure is the sum of the static and dynamic pressures anywhere in the flow.

---

[1]This is a realistic assumption for subsonic flows. In supersonic flows, the assumption will not be valid across a shock wave.

[2]The term "static pressure" is a misnomer in moving fluids, but its use here conforms to common expression. "Stream pressure" is more appropriate and is sometimes used.

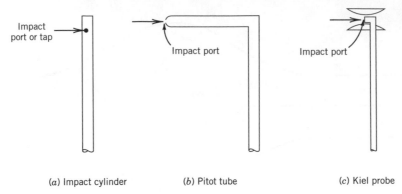

(a) Impact cylinder   (b) Pitot tube   (c) Kiel probe

Figure 9.18 Total pressure measurement devices. (a) Impact cylinder. (b) Pitot tube. (c) Kiel probe.

Total Pressure Measurement

In practice, the total pressure is measured using an impact probe, such as those depicted in Figure 9.18. A small hole in the impact probe is aligned with the flow so as to cause the flow to come to rest at the hole. The sensed pressure is transferred through the impact probe to a pressure transducer or other pressure sensing device such as a manometer. Alignment with the flow is somewhat critical, although the probes in Figure 9.18(a) and (b) are relatively insensitive (within ~1% error in indicated reading) to misalignment within a ±7°angle. A special type of impact probe shown in Figure 9.18(c), known as a Kiel probe, uses a shroud around the impact port. The effect of the shroud is to force the local flow to align itself with the shroud axis so as to impact directly onto the impact port. This effectively eliminates total pressure sensitivity to misalignment up to ±40°.

Static Pressure Measurement

Consider the coordinate system, s-n, fixed to the body as shown in Figure 9.17. When a flow passes over any real object, a boundary layer is formed. Within a boundary layer, the pressure gradient in the direction normal to the streamwise direction s will be $\partial p/\partial n \approx 0$ [6]. This means that the local value of static pressure can be measured by sensing the pressure in the direction normal to the flow streamline.

Within ducted flows, static pressure is sensed by wall taps, small, burr-free, holes drilled into the duct wall perpendicular to the flow direction at the measurement point. The tap is fitted with a hose or tube, which is connected to a pressure gauge or transducer. A recommended design for a wall tap is shown in Figure 9.19. The tap hole diameter is typically between 1 and 10% of the pipe diameter, with the smaller size preferred [5].

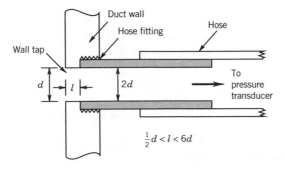

Figure 9.19 Anatomy of a static pressure wall tap.

Curvature of the flow streamlines introduces acceleration forces that change the local velocity. From equation (9.16) and Figure 9.17, we see that curvature of the flow streamlines must affect the local static pressure, since $p_1 = p_3$ but $p_3 \neq p_4$. The free-stream static pressure given by p_1 and p_3 can be measured only where there is no curvature of the streamlines. So if pressure at the wall is to be measured, the wall taps should not disturb the flow in any way, for a disturbance would cause streamline curvature. The tap should be perpendicular with the wall with no drilling burrs [8].

A static pressure probe can be inserted into the flow to measure local pressure. It should be a streamlined design to minimize the disturbance of the flow. It should be physically small so as not to cause more than a negligible increase in velocity in the vicinity of measurement. As a rule, the frontal area of the probe should not exceed 5% of the pipe flow area. The static pressure sensing port should be located well downstream of the leading edge of the probe so as to allow the streamlines to realign themselves parallel with the probe. Such a concept is built into the improved Prandtl tube design shown in Figure 9.20(a).

A Prandtl tube probe consists of eight holes arranged about the probe circumference and positioned 8 to 16 probe diameters downstream of the probe leading edge and 16 probe diameters upstream of its support stem. A pressure transducer or manometer is connected to the probe stem to measure the sensed pressure. The hole positions are chosen to minimize static pressure error caused by the disturbance to the flow streamlines due to the probe's leading edge and stem. This is illustrated in Figure 9.20(b) where the relative static error, $p_e/p_v = (p_i - p)/(\frac{1}{2}\rho U^2)$, as a function of tap location along the

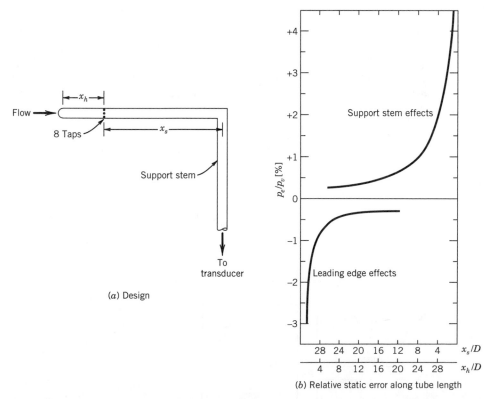

Figure 9.20 Improved Prandtl tube for static pressure. (a) Design. (b) Relative static error along tube length.

probe body is plotted and with p_i as the indicated (measured) pressure. Real viscous effects around the static probe cause a slight discrepancy between the actual static pressure and the indicated static pressure. To account for this a correction factor, C_0, is used with $p = C_0 p_i$ where $0.99 < C_0 < 0.995$.

9.7   DESIGN AND INSTALLATION: TRANSMISSION EFFECTS

The size of the pressure tap diameter and the length of tubing between a pressure tap and a pressure transducer form a pressure-measuring system that can have dynamic response characteristics very different from the pressure transducer itself. In the discussions to follow it is assumed that the transducer has frequency response and rise time characteristics that exceed those of the entire pressure measuring system (tubing plus transducer). Consider the configuration depicted in Figure 9.21 in which a rigid tube of length L and diameter d is used to connect a pressure tap to a pressure transducer of internal dead volume $\forall$ (e.g., Figure 9.13). We will assume that under static conditions the input pressure at the tap will be indicated by the pressure transducer. But if the pressure tap is exposed to a time-dependent pressure, $p_a(t)$, the response behavior of the tubing will dominate the indicated system output from the transducer, $p(t)$. By considering the one-dimensional pressure forces acting on a lumped mass of fluid within the connecting tube, a model for the pressure system can be developed. Be aware that this model is based on several very simplifying assumptions and should be used only as a guide in the design of a time-dependent pressure measurement system, not as a correction method.

Gases

Compressibility of a gas can be described through the fluid bulk modulus of elasticity, E_m. Pressure changes will act on the fluid in an effort to move it back and forth by a distance x within the tube. Consider a free body of a volume of fluid within the tube (Figure 9.22). At any instant, we can expect the fluid to be acted upon by the driving pressure force, $p_a \pi d^2/4$, a damping force due to fluid shear forces, $8\pi\mu L\dot{x}$, and a compression-restoring force, $\pi^2 E_m d^2 x/16\forall$. Summing the forces in Newton's second law yields the response equation

$$\frac{4L\rho\forall}{\pi E_m d^2}\ddot{p}_m + \frac{128\mu L\forall}{\pi E_m d^4}\dot{p}_m + p_m = p_a(t) \tag{9.18}$$

in which p_m is the measured pressure and p_a the applied pressure. Using equation (3.13), this gives

$$\omega_n = \frac{d\sqrt{\pi E_m/\rho L\forall}}{2} \tag{9.19}$$

$$\zeta = \frac{32\mu\sqrt{\forall L/\pi\rho E_m}}{d^3} \tag{9.20}$$

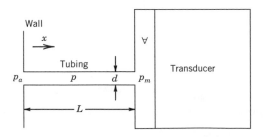

Figure 9.21 Wall tap to pressure transducer connection.

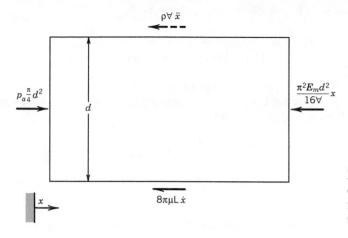

Figure 9.22 Free-body diagram of forces acting on a fluid volume in a pressure transmission line.

For a perfect gas with speed of sound, $a = \sqrt{kRT}$, where T is the absolute temperature of the gas, the natural frequency and damping ratio of the system become

$$\omega_n = \frac{d\sqrt{\pi a^2/L\forall}}{2} \tag{9.21}$$

$$\zeta = \frac{32\mu\sqrt{\forall L/\pi}}{a\rho d^3} \tag{9.22}$$

When the tube volume, $\forall_t = \pi d^2 L/4 \gg \forall$, then a series of standing pressure waves develop and a better predictor becomes [9]

$$\omega_n = \frac{a}{L(0.5 + 4\forall/\forall_t)} \tag{9.23}$$

$$\zeta = \frac{16\mu L\sqrt{0.5 + 4\forall/\forall_t}}{\rho a d^2} \tag{9.24}$$

Note that in either case, *larger diameter and shorter length tubes improve pressure system response*.

EXAMPLE 9.6

A pressure transducer with a natural frequency of 100 kHz is connected to a 0.10-in. static wall pressure tap using a 0.10-in. i.d. rigid tube that is 5 in. long. The transducer has a dead volume of 1 in.^{3}. Determine the pressure transmission system magnitude ratio response to fluctuating pressures of air at 72°F if the fluctuations are about 1 atm abs mean pressure. $R_{air} = 53.3$ ft-lb/lb$_m$-°R, and $\mu = 4 \times 10^{-7}$ lb-s/ft^2.

KNOWN

L = 5 in.   k = 1.4
d = 0.1 in.   T = 72°F = 532°R
∀ = 1 in.^{3} $\rho = p/RT = 0.075$ lb$_m$/ft^3

ASSUMPTION   Air behaves as a perfect gas.

FIND   Find $M(\omega)$

SOLUTION   The magnitude ratio is given by equation (3.19) as

$$M(\omega) = \frac{1}{\sqrt{[1 - (\omega/\omega_n)^2]^2 + [2\zeta(\omega/\omega_n)]^2}}$$

We need ω_n and ζ to solve $M(\omega)$ for various frequencies.

For this geometry, $\forall_t = \pi d^2 L/4 = 0.4\,\text{in.}^3$, meaning that $\forall_t \ll \forall$, so we use equations (9.21) and (9.22) With $a = \sqrt{kRT} = 1130\,\text{ft/s}$,

$$\omega_n = \frac{d\sqrt{\pi a^2/L\forall}}{2} = 537.4\,\text{rad/s}$$

$$\zeta = \frac{32\mu\sqrt{\forall L/\pi}}{a\rho d^3} = 0.08$$

This system is lightly damped. Solving for $M(\omega)$ yields the following representative values:

$\omega[\text{rad/s}]$	$M(\omega)$
63	1.01
315	1.50
535	6.27
3150	0.03

Liquids

In liquid flows, pressure changes are transported through the pressure system more readily (due to the higher speed of sound). Still the analysis leading to equation (9.18) holds. However, a momentum corection factor or equivalent mass is introduced to account for the inertial effects not included in a one-dimensional analysis (e.g., [10]). This in effect increases the inertial force by 1.33 to yield

$$\omega_n = \frac{d\sqrt{3\pi E_m/\rho L\forall}}{4} \tag{9.25}$$

$$\zeta = \frac{16\mu\sqrt{3\forall L/\pi\rho E_m}}{d^3} \tag{9.26}$$

Heavily Damped Systems

In systems in which we estimate a damping ratio greater than 1.5, the heavily-damped system can be simplified further. The behavior of the pressure-measuring system will closely follow that of a first-order system. A typical pressure transducer will have a rated compliance, C_{vp}, which is a measure of the transducer volume change relative to an applied pressure change. The response of the first-order system is indicated through its time constant which can be approximated by [11].

$$\tau = \frac{128\mu L C_{vp}}{\pi d^4} \tag{9.27}$$

An important aspect of equation (9.27) is that the time constant is proportional to L/d^4 ($\propto (L/d)^2/\forall_t$). Long and small diameter connecting tubes will result in relatively sluggish measurement system response to changes in pressure.

EXAMPLE 9.7: A Test Case

An engineer wishes to measure aerodynamic downforce on a race stock car as it moves on a track. The pressure difference between the top and bottom surfaces of the car is responsible for this downward force, which allows the tires to adhere better at high speeds. Pressure is measured using surface (wall) taps connected by 5-mm i.d. tubing to ± 25.4 cm H_2O, 0–5 V capacitance pressure transducers (accuracy: 0.25%), such as in Figures 9.19 and 9.21. Transducer output is measured by a portable data-acquisition system using a 12-bit, 5-V A/D converter (accuracy: 2 LSB). In testing at 180 kph, the maximum average pressure found is on the rear deck (rear window/trunk lid) and is about 8 cm $H_2O \pm 0.10$ cm H_2O (95%)—the random uncertainty is due to pressure fluctuations. The engineer is asked two questions: (1) Will increasing the resolution of the A/D conversion system reduce uncertainty (i.e., improve accuracy)? (2) How well can rear downforce be measured on the track?

SOLUTION   Force is pressure acting over an area. Aerodynamic downforce is estimated by integrating the pressure over the surface area of the car. An approximation is to measure the average pressure acting on discrete effective areas on the car surface.

If we ignore installation errors and ambient influences and just look at direct measurement errors, the measurement system errors and data random errors are important. The resolution of the transducer and A/D system is limited to

$$Q_{transducer} = 50.8 \, cm \; H_2O/5V = 10.16 \, cm \; H_2O/V$$
$$Q_{A/D} = 5 \, V/2^{12} = 0.00122 \, V/digit$$

Or, the total measuring system resolution is $Q = 0.00122 \times 10.16 = 0.0124 \, cm \, H_2O$. This yields a $u_o = 0.0124$ cm H_2O. However, improving the DAS system to a 16-bit system improves Q and u_o to 0.008 cm H_2O.

Pressure fluctuations occur as the flow fluctuates around the car. We can measure pressure at a moderate sample rate (25 Hz), use a smoothing filter, and average the readings. Statistical variations from these average values yield a random uncertainty, u_{pavr}. The random errors vary with position on the car as a result of local flow variations.

If we look at the rear deck of the car and use the instrument "accuracy" specifications as instrument systematic uncertainty,

$$u_{c \, transducer} \approx (0.0025) \, (8 \, cm \, H_2O) = 0.02 \, cm \, H_2O$$
$$u_{c \, A/D} = (2 \, bits)(0.0124 \, cm \, H_2O/bit) = 0.025 \, cm \, H_2O$$

which, combined with u_o for the 12-bit system gives the design-stage estimate

$$(u_d)_p = \pm \left[(0.02)^2 + (0.025)^2 + (0.0124)^2 \right]^{1/2} = \pm 0.034 \, cm \, H_2O \quad (95\%)$$

Using the 16-bit system will not change this estimate. The instrument errors dominate the design-stage uncertainty. Hence from design-stage analysis, it is clear that the higher resolution cannot improve the measurement.

The effective area on the rear deck can be estimated from carefully controlled calibration measurements within a wind tunnel with the car at various angles to the wind. A highly accurate balance scale is used to measure downforce. From this, the rear effective area is found to be

$$A_{eff} = 10832 \pm 147 \, cm^2 \quad (95\%)$$

The downforce is simply $F_D = pA_{eff}$, so the percent uncertainty is found by

$$u_F/F_D = \pm[(u_p/p)^2 + (u_A/A_{eff})^2] = 0.014 \text{ or } 1.4\% \quad (95\%)$$

On-track effects contribute to raise this number to about 2%.

COMMENT:

Here we see that the dominant uncertainty is due to the flow process and not the instrumentation. To improve the uncertainty requires alternate methods to estimate the downforce.

___

9.8   FLUID VELOCITY MEASURING SYSTEMS

Velocity measuring systems are used when information about a moving fluid within a localized portion of the flow is needed. Desirable information can consist of the mean velocity, as well as any of the dynamic components of the velocity. Dynamic components are found in pulsating or oscillating flows, or in turbulent flows. For most general engineering applications, information about the mean flow velocity is usually sufficient. The dynamic velocity information is often sought during applied and basic fluid mechanics research and development, such as in attempting to study airplane wing response to air turbulence, a complex periodic waveform as the wing sees it.

In general, the instantaneous velocity can be written as

$$U(t) = \overline{U} + u \quad (9.28)$$

where $\overline{U}$ is the mean velocity and u is the time-dependent dynamic (fluctuating) component of the velocity. The instantaneous velocity can also be expressed in terms of a Fourier series

$$U(t) = \overline{U} + \sum C_i \sin(\omega_i t + \phi_i') \quad (9.29)$$

so that the mean velocity and the amplitude and frequency information concerning the dynamic velocity component can be found through a Fourier analysis of the time-dependent velocity signal.

Pitot-Static Pressure Probe

For a steady, incompressible, isentropic flow, equation (9.16) can be written at any arbitrary point x in the flow field as

$$p_t = p_x + \frac{1}{2}\rho U_x^2 \quad (9.30)$$

or rearranging

$$p_v = p_t - p_x = \frac{1}{2}\rho U_x^2 \quad (9.31)$$

Here p_v, the difference between the total and static pressures at any point in the flow, is the *dynamic pressure*. Measuring the dynamic pressure of a moving fluid at point x provides a method for estimating the local velocity existing at point x. From equation (9.30),

$$U_x = \sqrt{\frac{2p_v}{\rho}} = \sqrt{\frac{2(p_t - p_x)}{\rho}} \quad (9.32)$$

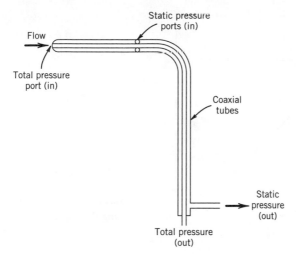

Static pressure
ports (in)

Flow

Total pressure
port (in)

Coaxial
tubes

Static
pressure
(out)

Total pressure
(out)

Figure 9.23 Pitot-static pressure probe.

In practice, equation (9.32) is utilized through a device known as a *pitot-static pressure probe*. Such an instrument has an outward appearance similar to that of an improved Prandtl static pressure probe [Figure 9.20(*a*)], except that the pitot–static probe contains an interior pressure tube attached to an impact port at the leading edge of the probe as shown in Figure 9.23. This creates two coaxial internal cavities within the probe, one exposed to the total pressure and the second exposed to the static pressure. The two pressures are typically measured using a differential pressure transducer so as to indicate p_v directly.

The pitot–static pressure probe is relatively insensitive to misalignment over the yaw angle range of $\pm 15°$. When possible, the probe can be rotated until a maximum signal is measured, a condition that indicates that it is aligned with the mean flow direction. However, the probes have a lower velocity limit of use that is brought about by strong viscous effects in the entry regions of the pressure ports. In general, viscous effects should not be a concern, provided that the Reynolds number based on the probe radius, $\mathrm{Re}_r = \overline{U}r/v \geq 500$ where v is the kinematic viscosity of the fluid. For $10 < \mathrm{Re}_r < 500$, a correction to the dynamic pressure should be applied, $p_v = C_v p_i$, where

$$C_v = 1 + (4/\mathrm{Re}_r) \tag{9.33}$$

and p_i is the indicated dynamic pressure from the probe. However, even with this correction, the measured dynamic pressure will have a systematic uncertainty on the order of 40% at $\mathrm{Re}_r \approx 10$ decreasing to 1% for $\mathrm{Re}_r \geq 500$.

In high-speed gas flows, compressibility effects near the probe leading edge require a closer inspection of the governing equation for a pitot-static pressure probe. Recall equation (8.34), which states the energy balance for a perfect gas between the freestream and a stagnation point. We can rewrite this in terms of the values at any point x and the stagnation value as

$$\frac{U^2}{2} = c_p(T_t - T_\infty) \tag{8.34}$$

For an isentropic process, the relationship between temperature and pressure can be stated as

$$\frac{T_x}{T_t} = \left(\frac{p_x}{p_t}\right)^{(k-1)/k} \tag{9.34}$$

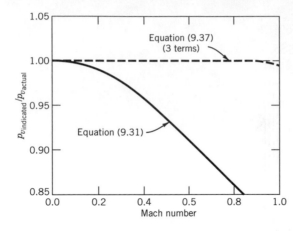

Figure 9.24 Relative error in the dynamic pressure between using equations (9.31) and (9.37) at increasing flow speeds.

where k is the ratio of specific heats for the gas, $k = c_p/c_v$. The Mach number of a moving fluid relates its local velocity to the local speed of sound,

$$M = U/a \tag{9.35}$$

where the acoustic wave speed, or speed of sound, is defined for a perfect gas as

$$a = \sqrt{kRT_x} \tag{9.36}$$

and where T_x is the absolute temperature of the gas at the point of interest. Combining equation (8.34) with equations (9.34)–(9.36) and using a binomial expansion yields the relationship between total pressure and static pressure at any point x in a moving compressible flow,

$$p_v = p_t - p_x = \frac{1}{2}\rho U_x^2[1 + M^2/4 + (2 - k)M^4/24 + \cdots] \tag{9.37}$$

Equation (9.37) reduces to equation (9.31) when $M \ll 1$. The error in the estimate of p_v based on use of equation (9.31) relative to the true dynamic pressure becomes significant for $M > 0.3$ as shown in Figure 9.24. Thus, $M \sim 0.3$ is used as the incompressible limit for perfect gas flows.

For $M > 1$, the local velocity can be estimated by the Rayleigh relation

$$U = \sqrt{2[k/(k - 1)][(p/\rho)^{(k-1)/k} - 1]} \tag{9.38}$$

where both p and p_t must be measured by independent means.

Thermal Anemometry

The rate at which energy, $\dot{Q}$, is transferred between a warm body at T_s and a cooler moving fluid at T_f is proportional both to the temperature difference between them and to the thermal conductance of the heat transfer path, hA. This thermal conductance increases with fluid velocity, thereby increasing the rate of heat transfer at any given temperature difference. Hence, a relationship between the rate of heat transfer and velocity will exist, thus forming the working basis of a thermal anemometer.

A thermal anemometer utilizes a sensor, a metallic resistance temperature device (RTD) element, which makes up one active leg of a Wheatstone bridge circuit, as

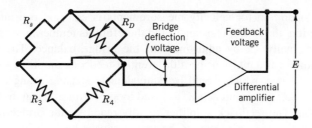

Figure 9.25 Thermal anemometer circuit. Shown in constant resistance mode.

indicated in Figure 9.25. The resistance-temperature relation for such a sensor was shown in Chapter 8 to be well represented by

$$R_s = R_0[1 + \alpha(T_s - T_0)] \tag{9.39}$$

so that sensor temperature can be inferred through a resistance measurement. A current is passed through the sensor to heat it to some desired temperature above that of the host fluid. The relationship between the rate of heat transfer from the sensor and the cooling fluid velocity is given by King's law [12] as

$$\dot{Q} = I^2R = A + BU^n \tag{9.40}$$

where A and B are constants that depend on the fluid and sensor physical properties and operating temperatures, and n is a constant that depends on sensor dimensions [13]. Typically, $0.45 \le n \le 0.52$ [14]. A, B, and n are found through calibration.

Two types of sensors are common: the hot-wire and the hot-film. As shown in Figure 9.26, the hot-wire sensor consists of a tungsten or platinum wire ranging from 1 to 4 mm in length and from 1.5 to 15 μm in diameter. The wire is supported between two rigid needles that protrude from a ceramic tube that houses the lead wires. A hot-film sensor usually consists of a thin (2 μm) platinum or gold film deposited onto a glass substrate and covered with a high thermal conductivity coating. The coating acts to electrically insulate the film and offers some mechanical protection. Hot wires are generally used in electrically nonconducting fluids, while hot films can be used in conducting fluids or in nonconducting fluids when a more rugged sensor is needed.

Two anemometer bridge operating modes are possible: (1) constant current and (2) constant resistance. In constant current operation, a fixed current is passed through the sensor to heat it. The sensor resistance, and therefore its temperature as in equation (9.39), are permitted to vary with the rate of heat transfer between the sensor and its environment. Bridge deflection voltage provides a measure of the cooling velocity. The

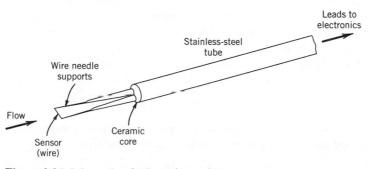

Figure 9.26 Schematic of a hot-wire probe.

more common mode of operation for velocity measurements is constant resistance. In constant resistance operation, the sensor resistance, and therefore its temperature again as in equation (9.39), is originally set by adjustment of the bridge balance. The sensor resistance is then maintained constant by using a differential feedback amplifier to sense small changes in bridge balance, which would be equivalent to sensing changes in the sensor set point resistance; i.e., the circuit acts as a closed loop controller using the bridge balance as the error signal. The feedback amplifier rapidly readjusts the bridge applied voltage, thereby adjusting the sensor current to bring the sensor back to its set point resistance and corresponding temperature. Because the current through the sensor will vary with changes in the velocity, the instantaneous power ($I^2 R_s$) required to maintain this constant temperature is equivalent to the instantaneous rate of heat transfer from the sensor ($\dot{Q}$). In terms of the instantaneous applied bridge voltage, E, required to maintain a constant sensor resistance, the velocity is found by the correlation

$$E^2 = C + DU^n \tag{9.41}$$

where constants, C, D, and n are found by calibration under a fixed sensor and fluid temperature condition. An electronic or digital linearizing scheme is usually employed to condition the signal by performing the transformation

$$E_1 = K\left(\frac{E^2 - C}{D}\right)^{1/n} \tag{9.42}$$

such that the measured output from the linearizer, E_1, is

$$E_1 = KU \tag{9.43}$$

where K is found through a static calibration.

For mean velocity measurements, the thermal anemometer is a straightforward device to use. It has a better usable sensitivity than the pitot-static tube at lower velocities. Multiple velocity components can be measured by using multiple sensors, each sensor aligned differently to the mean flow direction and operated by independent anemometer circuits [13, 15]. Because it has a high-frequency response, fluctuating (dynamic) velocities can be measured. In highly turbulent flows with rms fluctuations of $\sqrt{\overline{u^2}} \geq 0.1\overline{U}$, signal interpretation can become complicated but it has been well investigated [e.g., 15]. Low-frequency fluid temperature fluctuations can be compensated for by placing resistor R_3 directly adjacent to the sensor and exposed to the flow. An extensive bibliography of thermal anemometry theory and signal interpretation exists [16].

In constant temperature mode using a fast responding differential feedback amplifier, a hot-wire system can attain a frequency response that is flat up to 100,000 Hz, which makes it particularly useful in fluid mechanics turbulence research. However, less-expensive and more rugged systems are commonly used for industrial flow monitoring where a fast dynamic response is desirable. An upper frequency limit on a cylindrical sensor of diameter d is brought about by the natural oscillation in the flow immediately downstream of a body which vibrates the sensor. The frequency of this oscillation, known as the Strouhal frequency (and explained further in Section 10.6), occurs at approximately

$$f \approx 0.22[\overline{U}/d] \quad (10^2 < \text{Re}_d < 10^7) \tag{9.44}$$

The heated sensor will warm the fluid within its proximity. Under flowing conditions this will not cause any measurable problems so long as the condition

$$\text{Re}_d \geq Gr^{1/3} \tag{9.45}$$

is met where $\mathrm{Re}_d = \overline{U}d/\nu$, $Gr = d^3 g\beta(T_s - T_{\text{fluid}})/\nu^2$, and β is the coefficient of thermal expansion of the fluid. Equation (9.45) ensures that the inertial forces of the moving fluid dominate over the buoyant forces brought on by the heated sensor. For air, this forms a lower velocity limit on the order of 0.6 m/s for hotwire sensors.

Doppler Anemometry

The Doppler effect describes the phenomenon experienced by an observer whereby the frequency of light or sound waves emitted from a source that is traveling away from or toward the observer is shifted from its original value and by an amount proportional to its speed. Most readers are familiar with the change in pitch of a train heard by an observer as the train changes from approaching to receding. Any radiant energy wave, such as a sound or light wave, will experience a Doppler effect. The effect was recognized and modeled by Johann Doppler (1803–1853). The observed shift in frequency, called the Doppler shift, is directly related to the speed of the emitter relative to the observer. To an independent observer, the frequency of emission is perceived to be higher than actual if the emitter is moving toward the observer and lower if moving away, because the arrival of the emission at the observer location will be affected by the relative velocity of the emission source. The Doppler effect is used in astrophysics to measure the velocity of distant objects by monitoring the frequency of light emitted from a particular gas, usually hydrogen. Since in the visible light spectrum frequency is related to color, the common terms of red shift or blue shift refer to frequency shifts toward the red side of the spectrum or toward the blue side.

Doppler anemometry refers to a class of techniques that utilize the Doppler effect to measure the local velocity in a moving fluid. In these techniques, the emission source and the observer remain stationary. However, small scattering particles suspended in and moving with the fluid can be used to generate the Doppler effect. The emission source is a coherent narrow incident wave. Either acoustic waves or light waves are used.

When a laser beam is used as the incident wave source, the velocity measuring device is called a laser Doppler anemometer (*LDA*). Yeh and Cummins in 1964 [17] discussed the first practical laser Doppler anemometer system. A laser beam provides a ready emission source that is monochromatic and remains coherent over long distances. As a moving particle suspended in the fluid passes through the laser beam, it scatters light in all directions. An observer viewing this encounter between the particle and the beam will perceive the scattered light at a frequency, f_s:

$$f_s = f_i \pm f_D \qquad (9.46)$$

where f_i is the frequency of the incident laser beam and f_D is the Doppler shift. Using visible light, an incident laser beam frequency will be on the order of 10^{14} Hz. For most engineering applications, the velocities are such that the Doppler shift frequency, f_D, will be on the order of 10^3–10^7 Hz. Such a small shift in the incident frequency can be difficult to detect in a practical instrument. An operating mode that overcomes this difficulty is the dual-beam mode shown in Figure 9.27. In this mode, a single laser beam is divided into two coherent beams of equal intensity using an optical beam splitter. These incident beams are passed through a focusing lens which focuses the beams to a point in the flow. The focal point forms the effective measuring volume (sensor) of the instrument. Particles suspended in and moving with the fluid will scatter light as they pass through the beams. The frequency of the scattered light will be that given by equation (9.46) everywhere but at the measuring volume. There, the two beams cross and the incident information from the two beams mix, a process known as optical heterodyne. The outcome of this mixing is

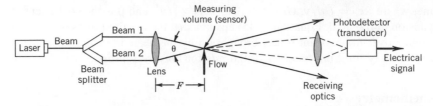

Figure 9.27 Laser Doppler anemometer. Shown in the dual-beam mode of operation.

a separation of the incident frequency from the Doppler frequency. A stationary observer, such as an optical photodiode, focused on the measuring volume will see two distinct frequencies, the Doppler shift frequency and the unshifted incident frequency, instead of seeing their sum. It is a simple matter to separate the much smaller Doppler frequency from the incident frequency by filtering.

For the setup shown in Figure 9.27, the velocity is related directly to the Doppler shift by

$$U = \frac{\lambda}{2\sin\theta/2} f_D = d_f/f_D \qquad (9.47)$$

where the component of the velocity measured is that which is in the plane of and bisector to the crossing beams. In theory, by using beams of different color or polarization, different velocity components can be measured simultaneously. However, the dependence of the lens focal length on color will cause a small displacement between the different measuring volumes formed by the different colors. For most applications this can be corrected. The LDA technique requires no direct calibration beyond determination of the parameters in d_f and the ability to measure f_D.

In dual-beam mode, the output from the photodiode transducer is a current of a magnitude proportional to the square of the amplitude of the scattered light seen and of a frequency equal to f_D. This effect is seen as a Doppler "burst" shown in the typical oscilloscope trace of Figure 9.28. The Doppler burst is the frequency signal created by a particle moving through the measuring volume. If the instantaneous velocity of a dynamic flow varies with time, the Doppler shift from successive scatters will vary with time. This time-dependent frequency information can be extracted by any variety of processing equipment that can interpret the signal current. The most common is the burst analyzer.

Burst analyzers extract Doppler frequency information by performing a Fourier analysis (see Chapter 2) on the input signal. This is done by first discretizing the photodetector analog signal at a high sample rate and then analyzing the signal.

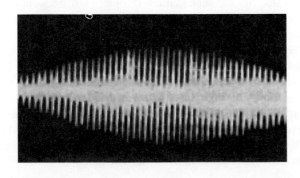

Figure 9.28 Oscilloscope trace of a photodiode output showing the Doppler frequency from a single particle moving through the measuring volume.

The analysis can work in one of two ways. In the first approach, the analyzer performs the FFT continuously on small blocks of data from which the Doppler frequency is directly determined. In the second mode, correlation mode, the sampled signal is correlated with itself through a mathematical transformation of the form

$$R_j = \sum_{i=1}^{n} x(i)x(i+j) \tag{9.48}$$

where i refers to the sample value at time t and j to the sample value at time delay Δt. This operation improves the signal-to-noise ratio (SNR) of the signal; the frequency is determined from the correlation function and the sample rate. From equation (9.47), the Doppler frequency can be converted to a velocity and output. The acquisition and analysis occur rapidly so that the signal appears nearly continuous in time and with only a short time lag.

All methods output a voltage that is proportional to the instantaneous velocity, which makes their signal easy to process, or a digital output path to a digital computer for signal analysis. At very low light levels and very few scattering particles, the signal level to noise level can be very low. In such cases, photon correlation techniques are successful [18, 19].

The LDA technique measures velocity at a point. Thus, its probe volume needs to be moved around to map out the flow field. If the SNR and seeding is good, the LDA can measure time-dependent velocities well. The LDA technique is particularly useful where probe blockage effects render other methods unsuitable, where fluid density and temperature fluctuations occur, or where environments hostile to physical sensors exist. An extended discussion of LDA techniques can be found in references [19, 20].

EXAMPLE 9.8

A laser Doppler anemometer made up of a He-Ne laser ($\lambda = 632.8\,\text{nm}$) is used to measure the velocity of water at a point in a flow. A 150-mm lens having a $\theta = 11°$ is used to operate the LDA in a dual-beam mode. If an average Doppler frequency of 1.41 MHz is measured, estimate the velocity of water.

KNOWN

$$\bar{f}_D = 1.41\,\text{MHz} \qquad F = 150\,\text{mm}$$
$$\lambda = 632.8\,\text{nm} \qquad \theta = 11°$$

ASSUMPTION   Scattering particles follow the water exactly.

FIND

$\overline{U}$

SOLUTION   Using equation (9.47),

$$\overline{U} = \frac{\lambda}{2 \sin \theta/2} \bar{f}_D = \left(\frac{632.8 \times 10^{-9}\text{m}}{2 \sin(11°/2)} \right)(1.41 \times 10^6\,\text{Hz}) = 4.655\,\text{m/s}$$

Particle Image Velocimetry

Particle image velocimetry measures the full-field instantaneous velocities in a planar cross section of a flow. The technique tracks the time displacement of particles, which are

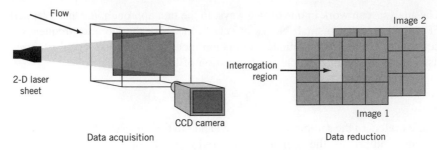

Figure 9.29 Digital particle image velocimeter.

assumed to follow the flow. The technique uses a coherent light source (laser beam), optics, CCD-camera, and dedicated signal interrogation software.

In a simple overview, the image of particles suspended in the flow are illuminated and recorded during repetitive flashes of a laser beam. These images are recorded and compared. The distance traveled by any particle during the period between flashes is a measure of its velocity. By repeatedly flashing the laser, in the manner of a strobe light, the particle positions can be tracked and velocity as a function of time can be obtained.

In a typical layout, such as shown in Figure 9.29, a pulsed flash laser beam is passed through a cylindrical lens, which converts the beam into a two-dimensional sheet of light. This laser sheet illuminates an appropriate cross section of the flow field. The camera is positioned and focused to record the view of the illuminated field. The laser flash and camera shutter are synchronized to capture the flow image. The acquired digital image is stored and processed by interrogation software resulting in a full-field instantaneous velocity mapping of the flow.

The operating principle is based on particle displacement with time

$$\vec{U} = \Delta\vec{x}/\Delta t \tag{9.49}$$

where $\vec{U}$ is the instantaneous particle velocity vector based on its spatial position $\vec{x}(x, y, z, t)$. The camera records particle position at each flash into separate image frames. To obtain velocity data in a rapid manner, each image is divided into small areas, called interrogation areas. The corresponding interrogation areas between two images, I_1 and I_2, are cross-correlated with each other, on a pixel by pixel basis. A particular particle movement from position $\vec{x}_1$ to $\vec{x}_2$ will show up as a signal peak in the correlation, $R_{12}(x)$, where

$$R_{12}(x) = \iint\limits_A I_1(x)I_2(x + \Delta x)dx \tag{9.50}$$

so identifying the common particle and allowing the estimate of the particle displacement, $\Delta\vec{x}$. By repeating the cross-correlation between images for each interrogation area, a velocity vector map of the full area results.

A number of variations of this process have been developed but the concept remains the same. The technique works in gases or liquids. Three-dimensional information can be obtained by using two cameras. As with LDA (see discussion above), particle size and properties should be chosen relative to the fluid and flow velocities expected so that the particles move with the fluid. The maximum flow speed measurable is limited by the interrogation area size. The resolution depends on laser flash width and separation time, flow velocity, camera recording time, and image magnification.

Selection of Velocity Measuring Methods

Selecting the best velocity measuring system for a particular application involves a number of factors that an engineer needs to weight accordingly. These factors include:

1. Required spatial resolution
2. Required velocity range
3. Sensitivity to velocity changes only
4. Required need to quantify dynamic velocity
5. Acceptable probe blockage of flow
6. Ability to be used in hostile environments
7. Calibration requirements
8. Low cost and ease of use

When used under appropriate conditions, the uncertainty in velocity determined by any of the discussed methods can be as low as 1% of the measured velocity, although under special conditions LDA methods can have an uncertainty one order of magnitude lower [21].

Pitot-Static Pressure Methods

The pressure probe methods are best suited for finding the mean velocity in fluids of constant density. Relative to other methods, they are the simplest and cheapest method available to measure velocity at a point. Probe blockage of the flow is not a problem in large ducts and away from walls. Fluid particulate will block the impact ports, but aspirating models are available for such situations. They are subject to mean flow misalignment errors. They require no calibration and are frequently used in the field and laboratory alike.

Thermal Anemometer

Thermal anemometers are best suited for use in clean fluids of constant temperature and density. They are well suited for measuring dynamic velocities with very high resolution. However, signal interpretation in strongly dynamic flows can be complicated [15, 22]. Hot-film sensors are less fragile and less susceptible to contamination than hot-wire sensors. Probe blockage is not significant in large ducts and away from walls. Thermal anemometers are 180° directionally ambiguous (i.e., flows from the left or right give the same output signal), an important factor in flows that may contain flow reversal regions. An industrial grade system can be built rather inexpensively. The thermal anemometer is usually calibrated against either pressure probes or an LDA.

Laser Doppler Anemometer

The laser Doppler anemometer (LDA) is a relatively expensive and technically advanced point velocity measuring technique that can be used for most types of flows but is also well suited to hostile, combusting, or highly dynamic flow environments. It offers good frequency response, small spatial resolution, no probe blockage, and simple signal interpretation, but requires optical access and the presence of scattering particles. This method provides very good temporal resolution for time-accurate measurements in turbulent flows. The method measures the velocity of particles suspended in the moving fluid, not the fluid velocity. So careful planning is required in particle selection to ensure that the particle velocities represent the fluid velocity exactly. The size and concentration of the particles govern the system frequency response [23, 24].

Particle Image Velocimetry

Particle image velocimetry (PIV) is a relatively expensive and technically advanced full-field velocity measuring technique that can be used for most types of flows, including hostile and combusting flows. There is no probe blockage of the flow but it requires optical access and the presence of scattering particles. The method provides an instantaneous snapshot of the flow providing excellent views of flow structures, but time-dependent quantification of such dynamic flows, while possible, pushes the state-of-the-art. The method measures the velocity of particles suspended in the moving fluid, not the fluid velocity. So careful planning is required in particle selection to ensure that the particle velocities represent the fluid velocity exactly.

9.9 SUMMARY

Several reference pressure instruments have been presented that form the working standards for pressure transducer calibration. Pressure transducers convert sensed pressure into an output form that is readily quantifiable. These transducers come in many forms but tend to operate on either hydrostatic principles, expansion techniques, or force-displacement methods.

In moving fluids, special care must be taken in the measurement of the pressure to delineate between static and total pressure. Methods for the separate measurement of static and total pressure or for the measurement of the dynamic pressure are readily available and well documented.

The measurement of the local velocity within a moving fluid can be accomplished in a number of ways. Selecting the proper tool requires assessment of the need: mean and/or fluctuating velocity, point or full field measurement, optical assess or opaque boundaries or fluid. Specifically, dynamic pressure, thermal anemometry, Doppler anemometry and particle velocimetry methods have been presented. As discussed, each method offers advantages over the other and the best technique must be carefully weighed against the needs and constraints of a particular application.

REFERENCES

1. Brombacher, W. G., D. P. Johnson, and J. L. Cross, *NBS Monograph* 8, 1960.
2. McLeod, H., *Philos. Mag.* 48, 1874.
3. Sweeney, R. J., *Measurement Techniques in Mechanical Engineering*, Wiley, New York, 1953.
4. M. Hetenyi, ed., *Handbook of Experimental Stress Analysis*, Wiley, New York, 1950.
5. Way, S., Bending of circular plates with large deflection, *Transactions of the ASME* 56, 1934.
6. Schlicting, H., *Boundary Layer Theory*, McGraw-Hill, New York, 1969.
7. Franklin, R. E., and J. M. Wallace, Absolute measurements of static-hole error using flush transducers, *Journal of Fluid Mechanics* 42, 1970.
8. Rayle, R. E., Influence of orifice geometry on static pressure measurements, ASME Paper No. 59-A-234, 1959.
9. Iberall, A. S., Attenuation of oscillatory pressures in instrument lines, *Transactions of the ASME* 72, 1950.

10. Streeter, V. L., and E. B. Wylie, *Fluid Mechanics*, 8th ed., McGraw-Hill, New York, 1985.

11. Doebelin, E. O., *Measurement Systems: Application and Design*, 5th ed., McGraw-Hill Science/Engineering/Math, New York, 2003.

12. King, L. V., On the convection from small cylinders in a stream of fluid: Determination of the convection constants of small platinum wires with application to hot-wire anemometry, *Proceedings of the Royal Society, London* 90, 1914.

13. Hinze, J. O., *Turbulence*, McGraw-Hill, New York, 1959.

14. Collis, D. C., and M. J. Williams, Two-dimensional convection from heated wires at low Reynolds numbers, *Journal of Fluid Mechanics* 6, 1959.

15. Rodi, W., A new method for analyzing hot-wire signals in a highly turbulent flow and its evaluation in a round jet, *DISA Information*, 17, Dantek Electronics, Denmark, 1975. See also, Bruun, H. H., Interpretation of X-wire signals, *DISA Information*, 18, Dantek Electronics, Denmark, 1975.

16. Freymuth, P., A bibliography of thermal anemometry, *TSI Quarterly* 4, 1978.

17. Yeh, Y., and H. Cummins, Localized fluid flow measurement with a He-Ne laser spectrometer, *Applied Physics Letters* 4, 1964.

18. Photon Correlation and Light Beating Spectroscopy, H. Z. Cummins and E. R. Pike, eds., *Proc. NATO ASI*, Plenum, New York, 1973.

19. Durst, F., A. Melling, and J. H. Whitelaw, *Principles and Practice of Laser Doppler Anemometry*, Academic, New York, 1976.

20. R. J. Goldstein, ed., Fluid Mechanics Measurements, *Hemisphere*, New York, 1983.

21. Goldstein, R. J., and D. K. Kried, Measurement of laminar flow development in a square duct using a laser doppler flowmeter, *Journal of Applied Mechanics* 34, 1967.

22. Yavuzkurt, S., A guide to uncertainty analysis of hot-wire data, *Transactions of the ASME, Journal of Fluids Engineering* 106, 1984.

23. Maxwell, B. R., and R. G. Seaholtz, Velocity lag of solid particles in oscillating gases and in gases passing through normal shock waves, NASA-TN-D-7490.

24. Dring, R. P., and Suo, M. Particle trajectories in swirling flows, *Transactions of the ASME, Journal of Fluids Engineering* 104, 1982.

NOMENCLATURE

d	diameter $[l]$		z	altitude $[l]$
e_i	elemental errors		C	capacitance [F]
h	depth $[l]$		E	voltage [V]
h_0	reference depth $(h = 0)$ $[l]$		E_m	bulk modulus of elasticity $[m^{-1}lt^{-2}]$
k	ratio of specific heats		Gr	Grashof number
p	pressure $[m^{-1}lt^{-2}]$		H	manometer deflection height $[l]$
p_a	applied pressure $[m^{-1}lt^{-2}]$		K	static sensitivity
p_{abs}	absolute pressure $[m^{-1}lt^{-2}]$		K_q	charge sensitivity $[Cm^{-1}l^{-1}t^2]$
p_e	relative static pressure error $[m^{-1}lt^{-2}]$		K_E	voltage sensitivity $[Vm^{-1}t^2]$
p_i	indicated pressure $[m^{-1}lt^{-2}]$		M	Mach number
p_m	measured pressure $[m^{-1}lt^{-2}]$		Re_d	Reynolds number, $\mathrm{Re} = Vd/\nu$
p_t	total or stagnation pressure $[m^{-1}lt^{-2}]$		S	specific gravity
p_v	dynamic pressure $[m^{-1}lt^{-2}]$		U	velocity $[lt^{-1}]$
q	charge [C]		$\forall$	volume $[l^3]$
r	radius $[l]$		γ	specific weight $[ml^{-2}t^{-2}]$
t	thickness $[l]$		ε	dielectric constant
y	displacement $[l]$		λ	wavelength $[l]$

ρ density $[ml^{-3}]$ μ absolute viscosity $[mt^{-1}l^{-1}]$
τ time constant $[t]$ ν kinematic viscosity $[l^2/t]$
ϕ latitude ν_p Poisson ratio
ω_n natural frequency $[t^{-1}]$

PROBLEMS

9.1 Convert the following absolute pressures to gauge pressure units of N/m^2:
   a. 10.8 psia
   b. 1.75 bars abs
   c. 30.36 in. H_2O absolute
   d. 791 mm Hg abs

9.2 Convert the following gauge pressures into absolute pressure relative to one standard atmosphere:
   a. -0.55 psi
   b. 100 mm Hg
   c. 98.6 kPa
   d. 7.62 cm H_2O

9.3 A water-filled manometer is used to measure the pressure in an air-filled tank. One leg of the manometer is open to atmosphere. For a measured manometer deflection of 250 cm water, determine the tank static pressure. Barometric pressure is 101.3 kPa abs.

9.4 A deadweight tester is used to provide a standard reference pressure for the calibration of a pressure transducer. A combination of 25.3 kg_f of 7.62-cm.-diameter stainless steel disks is found to be necessary to balance the tester piston against its internal pressure. For an effective piston area of 5.065 cm.^{2} and a piston weight of 5.35 kg_f, determine the standard reference pressure in bars, N/m^2, and Pa abs. Barometric pressure is 770 mm Hg abs, elevation is 20 m, and latitude is $42°$.

9.5 An inclined tube manometer indicates a change in pressure of 5.6 cm H_2O when switched from a null mode (both legs at atmospheric pressure) to deflection mode (one leg measuring, one leg at atmospheric pressure). For an inclination of $30°$ relative to horizontal, determine the pressure change indicated.

9.6 Show that the static sensitivity of an inclined tube manometer is a factor of $1/\sin\theta$ higher than for a U-tube manometer.

9.7 Determine the static sensitivity of an inclined tube manometer set at an angle of $30°$. The manometer tube measures the pressure difference of air and uses mercury as its fluid.

9.8 Show that the instrument (systematic) uncertainty in the inclined tube manometer of Example 9.2 increases to 6.8 N/m^2 as θ goes to $90°$.

9.9 Determine the maximum deflection and the natural frequency of a 0.1-in.-thick diaphragm made of steel ($E_m = 30$ Mpsi, $\nu_p = 0.32$, $\rho = 0.28$ $lb_m/in.^3$) if the diaphragm must be 0.75 in. in diameter. Determine its differential pressure limit.

9.10 A strain gauge, diaphragm pressure transducer (accuracy: $<0.1\%$ reading) is subjected to a pressure differential of 10 kPa. If the output is measured using a voltmeter having a resolution of 10 mV and accuracy of better than 0.1% of the reading, estimate the uncertainty in pressure at the design stage. How does this change at 100 and 1000 kPa?

9.11 Select a practical manometeric fluid to measure pressures up to 10 psi of an inert gas ($\gamma = 0.066$ lb/ft^3), if water ($\gamma = 62.4$ lb/ft^3), oil ($S = 0.82$), and mercury ($S = 13.57$) are available.

9.12 An air pressure over the 200–400-N/m^2 range is to be measured relative to atmosphere using a U-tube manometer with mercury ($S = 13.57$). Manometer resolution will be 1 mm with a zero error uncertainty of 0.5 mm. Estimate the design-stage uncertainty in gauge pressure based on the manometer indication at 20°C. Would an inclined manometer ($\theta = 30°$) be a better choice if the inclination could be set to within 0.5°?

9.13 Calculate the design-stage uncertainty in estimating a nominal pressure of 10,000 N/m^2 using an inclined manometer (resolution: 1 mm; zero error: 0.5 mm) with water at 20°C for inclination angles of 10–90° (using 10° increments). The inclination angle can be set to within 1°.

9.14 A capacitance pressure transducer, such as shown in Figure 9.14, uses a C_1 of 0.01 ± 0.005 μF and an excitation voltage of $5 \pm 1\%$ V. The plates have an overlap area of 8 ± 0.01 mm^2 and are separated by an air gap of 1.5 ± 0.1 mm. If the plates move apart by 0.2 mm, estimate the change in capacitance and the output voltage.

9.15 A diaphragm pressure transducer is calibrated against a pressure standard that has been certified by NIST [accuracy: within ± 0.5 psi (95%)]. Both the standard and pressure transducer output a voltage signal, which is to be measured by a voltmeter (accuracy: ± 10 mV; resolution: 1mV). A calibration curve fit yields: $p = 0.564 + 24.0E \pm 1$ psi (95%) based on 6 points over the 0–100-psi range. When the transducer is installed for its intended purpose, installation effects are estimated to affect pressure up to ± 0.5 psi. Estimate the uncertainty associated with a pressure measurement using the installed transducer–voltmeter system.

9.16 A diaphragm pressure transducer has a water-cooled sensor for high-temperature environments. Its manufacturer claims that it has a rise time of 10 ms, a ringing frequency of 200 Hz, and damping ratio of 0.8.

(i) Describe a test plan to verify the manufacturer's specifications.

(ii) Would this transducer have a suitable frequency response to measure the pressure variations in a typical four-cylinder engine? Show your reasoning.

9.17 Find the natural frequency of a 1-mm-thick, 6-mm-diameter steel diaphragm to be used for high-frequency pressure measurements. What would be the maximum operating pressure difference that could be applied? What is the effect of a larger diameter for this application?

9.18 The pressure fluctuations in a pipe filled with air at 20°C is to be measured using a static wall tap, rigid connecting tubing, and a diaphragm pressure transducer. The transducer has a natural frequency of 100,000 Hz. For a tap and tubing diameter of 3.5 mm, a tube length of 0.25 m, and a transducer dead volume of 1600 mm^3, estimate the resonance frequency of the system. What is the maximum frequency that this system can measure with no more than a 10% dynamic error? Plot the frequency response of the system.

9.19 Estimate the sensitivity of a pitot-static tube pressure signal to the velocity that it senses.

9.20 A pitot-static pressure probe inserted within a large duct indicates a differential pressure of 20.3 cm H$_2$O. Determine the velocity measured.

9.21 A pitot-static tube is placed in a flow of 20°C air at the centerline of a round duct. The pressure difference is sensed by a differential piezoelectric pressure transducer-charge amplifier system whose voltage is noted by a voltmeter (accuracy: ± 10 mV; resolution: 1 mV). The transducer calibration (N = 30) is given by

$$p = 0.205 + 0.950 \, E[V] \pm 0.002 \, N/m^2 \quad (95\%)$$

Three replications for a desired operating condition give the following data:

Run	N	E [V]	S_E [V]
1	21	2.439	0.010
2	21	2.354	0.009
3	21	2.473	0.012

Estimate the flow velocity and its uncertainty.

9.22 A tall pitot-static tube is mounted through and 1-m above the roof of a performance car such that it senses the freestream. Estimate the static, stagnation, and dynamic pressure sensed at 325 kph, if: (a) The car is moving along a long, straight section of road, and (b) The car is stationary within a wind tunnel where the flow is blown over the car.

9.23 Wall pressure taps (e.g., Figures 9.19 and 9.21) are often used to sense surface pressure and are connected to transducers by connecting tubing. Two race engineers discuss the preferred diameter of the tubing to measure pressure changes on a car as it moves along a track. The tubing length may be up to 2 m. Engineer A suggests very small 2-mm-diameter tubing to reduce air volume so to increase response time. Engineer B disagrees and suggests 5-mm tubing to balance air friction with air volume to increase response time. Offer your opinion and its basis. (Hint: Look at length-to-diameter effects.)

9.24 A system similar to that described in Example 9.7 is used to measure surface pressures on a car during a wind tunnel test. Large pressure data sets are taken. Estimate the overall uncertainty of the measurements using the 12-bit A/D converter.

Typical Wind Tunnel Measurements Stock Race Car at 180 km/hr

position	Pressure [cm H_2O]	
	$|p_{avr}|$	S_p
hood	0.8	0.025
roof	3.3	0.0025
rear deck	8.0	0.05

9.25 Pressure is measured 20 times at random time intervals over the course of a test run on a gas turbine compressor section under fixed operating conditions. This procedure is duplicated at each of four measuring positions separated by 90° in the compressor's cross-plane. The results are

Station	1	2	3	4
p (MN/m^2)	153	142	161	157
S_p (MN/m^2)	7	9	9	7

What would be the significance of pooling the data in determining the mean pressure here? What information could be found by comparing the pooled mean value to the local mean values? Do these. What new information would replications provide?

9.26 Determine the resolution of a manometer required to measure the velocity of air from 5 to 50 m/s using a pitot-static tube and a manometric fluid of mercury ($S = 13.57$) to a zero-order uncertainty of 5% and 1%.

9.27 A long cylinder is placed into a wind tunnel and aligned perpendicular to an oncoming freestream. Static wall pressure taps are located circumferentially about the centerline of the cylinder at 45° increments with 0° at the impact (stagnation) position. Each tap is connected to a separate manometer referenced to atmosphere. A pitot-static tube indicates an upstream dynamic pressure of 20.3 cm H_2O which is used to determine the freestream velocity. The following static pressures are measured:

Tap	p (cm H_2O)	Tap (deg.)	p (cm H_2O)
0°	0.0	135°	23.1
45°	41.4	180°	23.9
90°	81.3		

Compute the local velocities around the cylinder if the total pressure in the flow remains constant. $p_{atm} = 101.3$ kPa abs, $T_{atm} = 16°C$

9.28 A 6-mm diameter pitot-static tube is used as a working standard to calibrate a hot-wire anemometer in 20°C air. If dynamic pressure is measured using a water-filled micro-manometer, determine the smallest manometer deflection for which the pitot-static tube can be considered as accurate without correction for viscous effects.

9.29 For the thermal anemometer in Figures 9.25 and 9.26, determine the decade resistance setting required to set a platinum sensor at 40°C above ambient if the sensor ambient resistance is 110 Ω and $R_3 = 500$ Ω and $R_4 = 500$ Ω. $\alpha = 0.00395°C^{-1}$

9.30 Determine the static sensitivity of the output from a constant resistance anemometer as a function of velocity. Is it more sensitive at high or at low velocities?

9.31 A laser Doppler anemometer setup in a dual-beam mode uses a 600-mm focal length lens ($\theta = 5.5°$) and an argon-ion laser ($\lambda = 514.4$ nm). Compute the Doppler shift frequency expected at 1, 10, and 100 m/s. Repeat for a 300-mm lens ($\theta = 7.3°$).

9.32 A set of 5000 measurements of velocity at a point in a flow using a dual-beam LDA gives the following results:

$$U = 21.37\,\text{m/s} \qquad S_U = 0.43\,\text{m/s}$$

If the Doppler shift can be measured with an uncertainty of better than 0.9%, the optical angle of $\theta = 6°$ can be measured to within 0.25°, and the laser can be tuned to $\lambda = 623.8 \pm 0.5\%$ nm, determine the best estimate of the velocity.

9.33 In order to measure the flow rate in a 2m × 2m air conditioning duct, an engineer uses a pitot-static probe to measure dynamic head. The duct is divided into nine equal rectangular areas with the pressure measured at the center of each. Based on the results below, estimate the flow rate for air at 15°C and 1 atm.

Position	1	2	3	4	5	6	7	8	9
H (mm H_2O)	5.0	6.0	6.5	6.0	5.0	6.5	7.5	7.0	5.0

9.34 In Problem 9.33, estimate the uncertainty in the average flow rate computed using available information. Expect that duct dimensional values contain only systematic uncertainties at no more than 10 mm (95%) and that random errors in dynamic head are limited to the data set variation. Neglect fluid property errors and probe misalignment errors.

9.35 The pressure drop across a valve through which air flows is expected to be 10 kPa. If this differential were applied to the two legs of a U-tube manometer filled with mercury, estimate the manometer deflection. What is the deflection if a 30° inclined tube manometer were used? $S_{Hg} = 13.6$.

9.36 Estimate the differential pressure limit for a 0.5-mm-thick, 25-mm-diameter steel diaphragm pressure transducer. $v_p = 0.32$, $E_m = 200$ GPa.

Chapter 10

Flow Measurements

10.1 INTRODUCTION

Flow rate can be expressed in terms of a volume per unit time, known as the *volume flow rate*, or as a mass per unit time, known as the *mass flow rate*. Not only is this quantity useful in flow metering, but many engineering systems require flow measurement information for proper process control. For example, in heat transfer processes the energy exchange rate between a solid and a fluid is directly proportional to the mass flow rate of the fluid. Many manufactured products require accurate blending of different fluids flowing at different rates. Properly selected and designed flow measurement equipment command a high engineering priority because whether it be the flow of water into our homes or the flow of petroleum at the oil well, flow measurements are vitally linked to the economy.

This chapter discusses some of the most common and accepted methods for flow quantification. Size, accuracy, cost, pressure drop, pressure losses, and compatibility with the fluid are important engineering design considerations for flow metering devices. All methods have both desirable and undesirable features that necessitate some compromise in the selection of the best method for the particular application, and many important considerations are included in this chapter. Inherent uncertainties in fluid properties, such as density, viscosity, or specific heat, can affect the accuracy of a flow measurement made using some metering methods. Techniques that preclude knowledge of fluid properties are being introduced for use in the more demanding of these engineering applications. The chapter objective is to present both an overview of basic flow metering techniques for proper meter selection, as well as those design considerations important in the integration of a flow metering system with the process system it will meter.

10.2 HISTORICAL COMMENTS

Their importance in engineering systems gives flow measurement methods a rich history. The earliest available accounts of flow metering were recorded by Hero of Alexandria (ca. 150 B.C.) who proposed a scheme to regulate water flow using a siphon pipe attached to a constant head reservoir. The early Romans developed elaborate water systems to supply public baths and private homes. In fact, Sextus Frontinius (A.D. 40–103), Commissioner of Water Works for Rome, authored a treatise on design methods for water distribution. Evidence suggests that Roman designers understood the correlation between volume flow rate and pipe flow area. Weirs were used to regulate bulk flow through aqueducts, and the cross-sectional area of terra-cotta pipe was used to regulate supplies to individual buildings.

Following a number of experiments conducted using olive oil and water, Leonardo da Vinci (1452–1519) first formally proposed the continuity principle: that area, velocity, and flow rate were related. However, most of his writings were lost until centuries later, and Benedetto Castelli (ca. 1577–1644), a student of Galileo, has been credited in some texts with developing the same steady, incompressible continuity concepts in his day. Isaac Newton (1642–1727), Daniel Bernoulli (1700–1782), and Leonhard Euler (1707–1783) built the mathematical and physical bases on which modern flow meters would later be developed. By the nineteenth century, the concepts of continuity, energy, and momentum were sufficiently understood for practical exploitation. Relations between flow rate and pressure losses were developed that would permit the tabulation of the hydraulic coefficients necessary for the quantitative engineering design of many modern flow meters.

10.3   FLOW RATE CONCEPTS

The flow rate through a pipeline, duct, or other flow system can be described by use of a control volume, a judiciously selected volume in space through which a fluid flows. The amount of fluid that passes through this volume in a given period of time will determine the flow rate. A geometrical boundary of a control volume is called a control surface. Such a control volume is shown in Figure 10.1, which consists here of a defined volume within a pipe.

The velocity of a fluid at a point can be described by the use of a three-dimensional velocity vectors given here in cylindrical coordinates by

$$\vec{U} = \vec{U}(x, r, \theta) = u\,\widehat{e}_x + v\,\widehat{e}_r + w\,\widehat{e}_\theta$$

where u, v, and w are the scalar velocity magnitudes and $\widehat{e}_x$, $\widehat{e}_r$, and $\widehat{e}_\theta$ are unit vectors in each of the component directions, x, r, and θ, respectively.

The amount of fluid of density ρ that passes through the control volume of volume $\forall$ at any instant in time depends on the amount of fluid that crosses the control surfaces. This can be expressed by examination of mass flow into, out of, and remaining within the control volume (CV) at any instant. Conservation of mass demands that the rate at which mass accumulates within the control volume plus the net flow of mass that physically crosses any of its control surfaces (CS) be zero. This is expressed by

$$\frac{\partial}{\partial t} \iiint_{CV} \rho\,d\forall + \iint_{CS} \rho\vec{U} \cdot \widehat{n}\,dA = 0 \tag{10.1}$$

where $\widehat{n}$ is the outward normal from a control surface of area A.

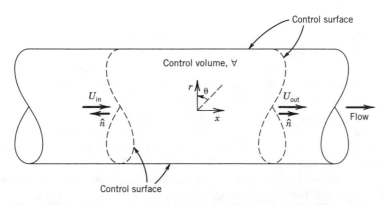

Figure 10.1 Control volume concept as applied to flow through a pipe.

A steady flow situation exists when the sum of the mass flow into the control volume equals the sum of the mass flow out of the control volume. For steady flows, equation (10.1) can be simplified to

$$\dot{m}_{\text{in}} = \dot{m}_{\text{out}} \tag{10.2}$$

where $\dot{m}$ is defined as the mass flow rate through any area A

$$\dot{m} = \iint_A \rho U dA$$

If the average mass flux, $\overline{\rho U}$, across a control surface is known, then equation (10.2) becomes

$$\dot{m} = \overline{\rho U} A \tag{10.3}$$

Mass flow rate has the dimensions of mass per unit of time [e.g., units of kg/s, lb_{m}/s, etc.]. For constant density flows, equation (10.3) reduces to

$$Q_{\text{in}} = Q_{\text{out}} \tag{10.4}$$

where Q is defined as the volume flow rate

$$Q = \iint_A U dA$$

For example, in a pipe of circular cross section, the volume flow rate at axial position x is found by

$$Q = \int_0^{2\pi} \int_0^{r_1} U(r, \theta) r dr d\theta \tag{10.5}$$

where r_1 is the pipe radius. If the average velocity, $\overline{U} = \frac{1}{A} \iint_A U dA$, is known, then the volume flow rate is found by

$$Q = \overline{U} A \tag{10.6}$$

From this analysis and equations (10.3) and (10.6), it is clear that to estimate the steady mass flow rate we need techniques that are sensitive to the average mass flux, $\overline{\rho U}$, and to estimate the steady volume flow rate we need techniques that are sensitive to the average velocity, $\overline{U}$. There are many direct or indirect methods to do either of these with volume flow rate methods being far more common.

The flow through a pipe or duct can be characterized as being laminar, turbulent, or a transition between the two. Flow character is determined through the nondimensional Reynolds number, defined by

$$\text{Re}_{d_1} = \frac{\overline{U} d_1}{\nu} = \frac{4Q}{\pi d_1 \nu} \tag{10.7}$$

where ν is the fluid kinematic viscosity and d_1 is the diameter for circular pipes. For noncircular ducts the hydraulic diameter, $4r_H$, is used in place of d_1, where r_H is the wetted area divided by the wetted perimeter. In pipes, the flow is laminar when $\text{Re}_{d_1} <$ 2000.

10.4   VOLUME FLOW RATE THROUGH VELOCITY DETERMINATION

The direct implementation of equation (10.5) for estimating the volume flow rate through a duct requires measuring the velocity at points along several cross sections of a flow control surface. Methods for determining the velocity at a point include any of those previously discussed in Chapter 9. This procedure is most often used for the one-time

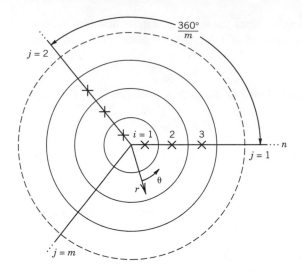

Figure 10.2 Location of n measurements along m radial lines in a pipe.

verification and/or calibration of system flow rates. For example, the procedure has been used in ventilation system setup and problem diagnosis where the installation of an in-line flow meter is uncommon because operation does not require continuous monitoring.

In using this technique in circular pipes, a number of discrete measuring positions are chosen along m flow cross sections (radii) spaced at $360/m$ degrees apart, such as shown in Figure 10.2. A velocity probe is traversed along each flow cross section with readings taken at each measurement position. There are several options as to the selection of measuring positions, and the details are specified in the references for circular and rectangular areas [2, 4, 13]. The simplest method is to divide the flow area into smaller equal areas with measurements made at the centroid of each of the smaller areas. Regardless of the option selected, the average flow rate is estimated along each cross section traversed using equation (10.5) and the pooled mean of the flow rates for the m cross sections determined to yield the best estimate of the duct flow rate. Example 10.1 illustrates this method for estimating volume flow rate.

EXAMPLE 10.1

A steady flow of air at 20°C passes through a 25.4-cm-i.d. circular pipe. A velocity-measuring probe is traversed along three cross-sectional lines ($j = 1, 2, 3$) of the pipe and measurements are made at four-radial positions ($i = 1, 2, 3, 4$) along each traverse line. The locations for each measurement are selected at the centroids of equally spaced areal increments as indicated below [2, 13]. Determine the volume flow rate in the pipe.

Radial Location, i	r/r_1	U_{ij} [m/s]		
		Line 1	Line 2	Line 3
1	0.3536	8.71	8.62	8.78
2	0.6124	6.26	6.31	6.20
3	0.7906	3.69	3.74	3.79
4	0.9354	1.24	1.20	1.28

KNOWN

$U_{ij}(r/r_1)$ for $i = 1,\ 2,\ 3,\ 4;\ j = 1,\ 2,\ 3$

$d_1 = 25.4\,\text{cm}\,(A = \pi d_1^2/4 = 0.051\,\text{m}^2)$

ASSUMPTIONS   Constant and steady pipe flow during all measurements
Incompressible flow

FIND

$\langle \bar{Q} \rangle$

SOLUTION   The flow rate is found by integrating the velocity profile across the duct along each line and subsequent averaging of the three values. For discrete velocity data, equation (10.5) is written along each line, $j = 1,\ 2,\ 3$, as

$$Q_j = 2\pi \int_0^{r_1} U r\,dr \approx 2\pi \sum_{i=1}^{4} U_{ij} r \Delta r$$

where Δr is the radial distance separating each position of measurement. This can be further simplified since the velocities are located at positions that make up centroids of equal areas:

$$Q_j = \frac{A}{4} \sum_{i=1}^{4} U_{ij}$$

Then, the mean flow rate along each line of traverse is

$$Q_1 = 0.252\,\text{m}^3/\text{s} \qquad Q_2 = 0.252\,\text{m}^3/\text{s} \qquad Q_3 = 0.254\,\text{m}^3/\text{s}$$

The average pipe flow rate is found from the pooled mean of the individual flow rates as

$$\langle \bar{Q} \rangle = \frac{1}{3} \sum_{j=1}^{3} Q_j = 0.252\,\text{m}^3/\text{s}$$

10.5   PRESSURE DIFFERENTIAL METERS

The operating principle of a pressure differential meter is based on the relationship between volume flow rate and the pressure drop $\Delta p = p_1 - p_2$, along the flow path,

$$Q \propto (p_1 - p_2)^n$$

where the value of n equals one for laminar flow occurring between the pressure measurement locations and n equals one-half in fully turbulent flow.

An intentional reduction in flow area will cause a measurable local pressure drop across the flow path. The reduced flow area causes a local increase in velocity. So the pressure drop is in part due to the so-called Bernoulli effect, the inverse relationship between local velocity and pressure, but also due to flow energy losses. Pressure differential meters that use area reduction methods are commonly called *obstruction meters*.

Obstruction Meters

Three common obstruction meters are the *orifice plate*, the *venturi*, and the *flow nozzle*. Flow area profiles of each are shown in Figure 10.3. These meters are usually inserted in-line with a pipe. This class of meters operates using similar physical reasoning to relate

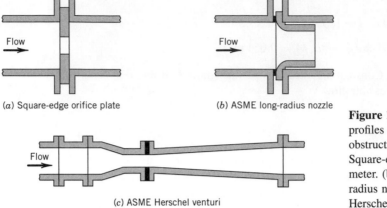

(a) Square-edge orifice plate   (b) ASME long-radius nozzle

(c) ASME Herschel venturi

Figure 10.3 Flow area profiles of common obstruction meters. (a) Square-edged orifice plate meter. (b) ASME long radius nozzle. (c) ASME Herschel venturi meter.

volume flow rate to pressure drop. Referring to Figure 10.4, consider the energy equation written between two control surfaces for an incompressible fluid flow through the arbitrary control volume shown. It can be assumed that (1) no external energy in the form of heat is added to the flow, (2) there is no shaft work done within the control volume, and that the flow is (3) steady and (4) one-dimensional. This yields

$$\frac{p_1}{\gamma} + \frac{\overline{U}_1^2}{2g} = \frac{p_2}{\gamma} + \frac{\overline{U}_2^2}{\gamma} + h_{L_{1-2}} \tag{10.8}$$

where $h_{L_{1-2}}$ denotes the head losses occurring due to frictional effects between control surfaces 1 and 2. For incompressible flows, equations (10.4) and (10.6) yield,

$$\overline{U}_1 = \overline{U}_2 \frac{A_2}{A_1}$$

Substituting $\overline{U}_1$ into equation (10.8) and rearranging yields the incompressible volume flow rate,

$$Q_I = \overline{U}_2 A_2 = \frac{A_2}{\sqrt{1 - (A_2/A_1)^2}} \sqrt{\frac{2(p_1 - p_2)}{\rho} + 2gh_{L_{1-2}}} \tag{10.9}$$

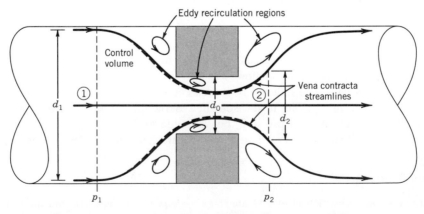

Figure 10.4 Control volume concept as applied between two streamlines for flow through an obstruction meter.

where the subscript I emphasizes that equation (10.9) yields an incompressible flow rate. Later we drop the subscript.

When the flow area changes abruptly, the effective flow area immediately downstream of the alteration will not necessarily be the same as the pipe flow area. This is the vena contracta effect, originally investigated by Jean Borda (1733–1799) and illustrated in Figure 10.4. When a fluid cannot exactly follow a sudden area expansion due to its own inertia, a central core flow, the vena contracta, forms and is bounded by regions of slower moving recirculating eddies. As a consequence, the pressure sensed with pipe wall taps will correspond to the higher moving velocity within the vena contracta with its unknown flow area, A_2. To account for this, we introduce the contraction coefficient C_c, where $C_c = A_2/A_0$ with A_0 based on the meter throat diameter, into equation (10.9). This yields

$$Q_I = \frac{C_c A_0}{\sqrt{1 - (C_c A_0/A_1)^2}} \sqrt{\frac{2(p_1 - p_2)}{\rho} + 2gh_{L_{1-2}}} \tag{10.10}$$

The frictional head losses can be incorporated into a friction coefficient, C_f, such that equation (10.10) becomes

$$Q_I = \frac{C_f C_c A_0}{\sqrt{1 - (C_c A_0/A_1)^2}} \sqrt{\frac{2(p_1 - p_2)}{\rho}} \tag{10.11}$$

For convenience, the coefficients are factored out of equation (10.11) and replaced by a single coefficient known as the *discharge coefficient*, C. The discharge coefficient represents the ratio of the actual flow rate through a meter to the ideal flow rate possible for the pressure drop measured, i.e., $C = Q_{I_{actual}}/Q_{I_{ideal}}$. Reworking equation (10.11) leads to the incompressible operating equation

$$Q_I = CEA_0\sqrt{\frac{2\Delta p}{\rho}} = K_0 A_0 \sqrt{\frac{2\Delta p}{\rho}} \tag{10.12}$$

where E, known as the velocity of approach factor, is defined by

$$E = \frac{1}{\sqrt{1 - (A_0/A_1)^2}} = \frac{1}{\sqrt{1 - \beta^4}} \tag{10.13}$$

with the beta ratio defined by $\beta = d_0/d_1$, and where $K_0 = CE$ is called the *flow coefficient*.

The discharge coefficient and the flow coefficient are tabulated quantities. Each is a function of the flow Reynolds number and the β ratio for each particular obstruction flow meter design, $C = f(\text{Re}_{d_1}, \beta)$ and $K_0 = f(\text{Re}_{d_1}, \beta)$. Because the magnitude of the vena contracta and head loss effects must vary along the length of a meter, equation (10.12) is very sensitive to pressure tap location, and consistency in tap placement is imperative for correct operation [2].

Compressibility Effects

In compressible gas flows, compressibility effects are accounted for by introducing the compressible adiabatic expansion factor, Y. Here Y is defined as the ratio of the actual compressible volume flow rate, Q, divided by the assumed incompressible flow rate Q_I. Combining with equation (10.12) yields

$$Q = YQ_I = CEA_0 Y \sqrt{2\Delta p/\rho_1} \tag{10.14}$$

where ρ_1 is the upstream fluid density. When $Y = 1$, the flow is incompressible and equation (10.14) reduces to equation (10.12). Equation (10.14) represents a general form of the working equation for obstruction meter volume flow rate determination.

The expansion factor, Y, depends on several values: the β ratio, the gas specific heat ratio, k, and the relative pressure drop across the meter, $(p_1 - p_2)/p_1$, for a particular meter type, i.e., $Y = f[\beta, k, (p_1 - p_2)/p_1]$. As a general rule, compressibility effects should be considered when $(p_1 - p_2)/p_1 \geq 0.1$.

Standards

The flow behaviors of the orifice plate, venturi, and flow nozzle have been studied to such an extent that these meters are used extensively without calibration. Values for the discharge coefficients, flow coefficients, and expansion factors are tabulated and available in standard flow handbooks along with standardized construction, installation, and operation techniques [2–4, 12]. Nonstandard installation requires an in-line calibration.

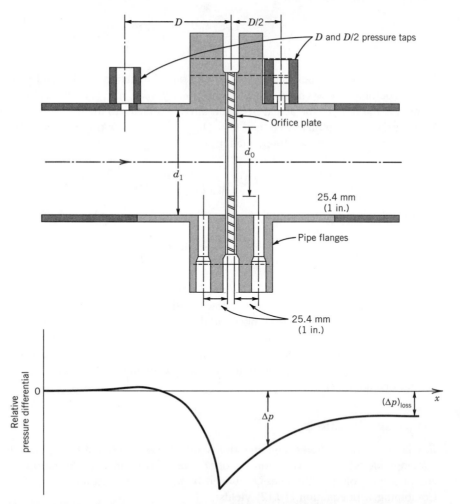

Figure 10.5 Square-edged orifice meter installed in a pipeline with optional 1 D and ½ D, and flange pressure taps shown. Relative flow pressure drop along pipe axis is shown.

Orifice Meter

An orifice meter consists of a circular plate, containing a hole (orifice), which is inserted into a pipe such that the orifice is concentric with the pipe inside diameter (i.d.). Several variations in the orifice design exist but the square-edged orifice, shown in Figure 10.5, is common. Installation is simplified by housing the orifice plate between two pipe flanges. With this technique an orifice plate is interchangeable with others of different β value. The simplicity of the installation and orifice design allows for a range of β values to be maintained on hand at modest expense.

Rudimentary versions of the orifice plate have existed for several centuries. Both Torricelli and Newton used orifice plates to study the relation between pressure head and efflux from reservoirs, although neither ever got the discharge coefficients quite right [5].

For an orifice plate, equation (10.14) is used with values of A and β being based on the orifice hole diameter. The exact location of the pressure taps is crucial when tabulated values for flow coefficient and expansion factor are used. Standard pressure tap locations include (1) flange taps where pressure tap centers are located 25.4 mm (1 in.) upstream and 25.4 mm (1 in.) downstream of the nearest orifice face, and (2) taps located one pipe diameter upstream and one-half diameter downstream of the upstream orifice face. Nonstandard tap locations require on-site meter calibration.

Values for the flow coefficient, $K_0 = f(\text{Re}_{d_1}, \beta)$ and for the expansion factor, $Y = f[\beta, k, (p_1 - p_2)/p_1]$ for a square-edged orifice plate are given in Figures 10.6 and 10.7 based on the use of flange taps. The relative instrument systematic uncertainty in the discharge coefficient [12] is $\sim \pm 0.6\%$ of C for $0.2 \le \beta \le 0.6$ and $\pm \beta\%$ of C for all $\beta > 0.6$. The relative instrument systematic uncertainty for the expansion factor is about

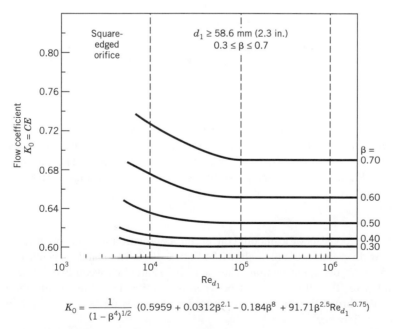

$$K_0 = \frac{1}{(1 - \beta^4)^{1/2}} (0.5959 + 0.0312\beta^{2.1} - 0.184\beta^8 + 91.71\beta^{2.5}\text{Re}_{d_1}^{-0.75})$$

Figure 10.6 Flow coefficients for a square-edged orifice meter having flange pressure taps. (Compiled from data in [2]).

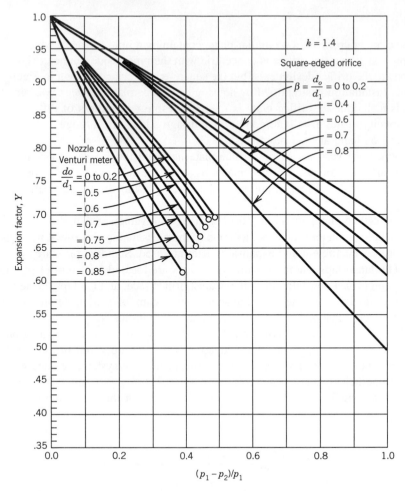

Figure 10.7 Expansion factors for common obstruction meters with $k = c_p/c_v = 1.4$. (Courtesy of American Society of Mechanical Engineers, New York; compiled and reprinted from [2].)

$\pm[4(p_1 - p_2)/p_1]\%$ of Y. Although the orifice plate represents a relatively inexpensive flow meter solution with an easily measurable pressure drop, it introduces a large permanent pressure loss, $(\Delta p)_{\text{loss}} = \rho g h_L$, into the flow system. The pressure drop is illustrated in Figure 10.5 with the pressure loss estimated from Figure 10.8.

Venturi Meter

A venturi meter consists of a smooth converging contraction to a narrow throat followed by a shallow diverging section, as shown in Figure 10.9. The standard venturi can utilize either a 15° or 7° divergent section. The meter is installed between two flanges intended for this purpose. Pressure is sensed between a location upstream of the throat and a location at the throat. Equation (10.14) is used with values for both A and β being based on the throat diameter.

The quality of a venturi meter ranges from cast to precision-machined units. The discharge coefficient varies little for pipe diameters above 7.6 cm (3 in.). In the operating

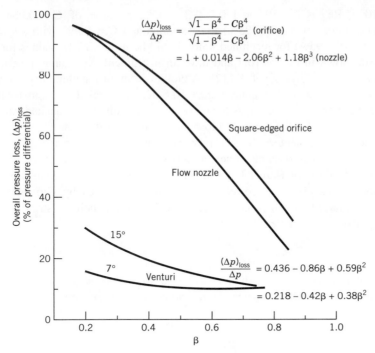

Figure 10.8 The permanent pressure loss associated with flow through common obstruction meters. (Courtesy of American Society of Mechanical Engineers, New York, NY; compiled and reprinted from [2].)

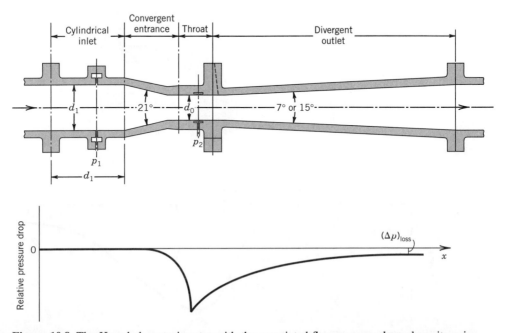

Figure 10.9 The Herschel venturi meter with the associated flow pressure drop along its axis.

range $2 \times 10^5 \leq \mathrm{Re}_{d_1} \leq 2 \times 10^6$ and $0.4 \leq \beta \leq 0.75$, a value of $C = 0.984$ with a systematic uncertainty of $\pm 0.7\%$ (95%) for cast units and $C = 0.995$ with a systematic uncertainty of $\pm 1\%$ (95%) for machined units should be used [2, 12]. Values for expansion factor are shown in Figure 10.7 and have an instrument systematic uncertainty of $\pm[(4 + 100\beta^2)\ (p_1 - p_2)/p_1]\%$ of Y [12]. Although a venturi meter presents a much higher initial cost over an orifice plate, Figure 10.8 demonstrates that the meter shows a much smaller permanent pressure loss for a given installation. This translates into lower system operating costs for the pump or blower used to move the flow.

The modern venturi meter was first proposed by Clemens Herschel (1842–1930). Herschel's design was based on his understanding of the principles developed by several men, most notably those of Daniel Bernoulli. However, he cited the studies of contraction/expansion angles and their corresponding resistance losses by Giovanni Venturi (1746–1822) and later those by James Francis (1815–1892) as being instrumental to his design of a practical flow meter.

Flow Nozzles

A flow nozzle consists of a gradual contraction to a narrow throat. It needs less installation space than a venturi meter and has about 80% of the initial cost. A common

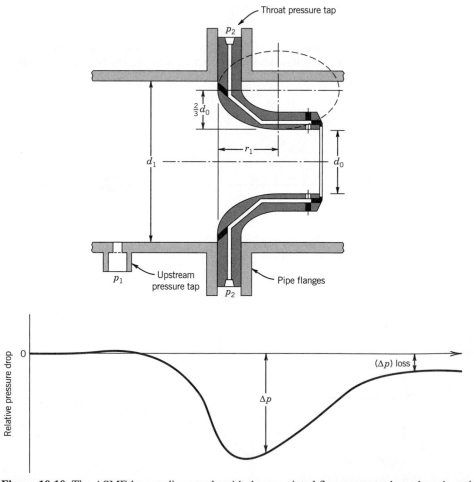

Figure 10.10 The ASME long-radius nozzle with the associated flow pressure drop along its axis.

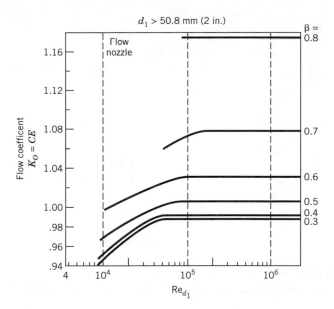

Figure 10.11 Flow coefficients for an ASME long-radius nozzle with a throat pressure tap. (Compiled from [2].)

form for the nozzle is the ASME long radius nozzle in which the nozzle contraction is that of the quadrant of an ellipse with major axis aligned with the flow axis, as shown in Figure 10.10. The nozzle is typically installed inline, but can also be used at the inlet to and the outlet from a plenum or reservoir or at the outlet of a pipe. Pressure taps are usually located at one pipe diameter upstream of the nozzle inlet and at the nozzle throat using either wall or throat taps. The flow rate is determined from equation (10.14) with values for A_o and β being based on the throat diameter. Typical values for the flow coefficient and expansion factor are given in Figures 10.11 and 10.7. The relative instrument systematic uncertainty for discharge coefficient is about $\pm 2\%$ of C and for expansion factor is about $\pm[2(p_1 - p_2)/p_1]\%$ of Y [12]. The permanent loss associated with a flow nozzle is larger than for a comparable venturi (see Figure 10.8) for the same pressure drop.

The idea of using a nozzle as a flow meter was first proposed in 1891 by John Ripley Freeman (1855–1932), an inspector and engineer employed by a factory fire insurance firm. His work required tedious tests to quantify pressure losses in pipes, hoses, and fittings. He noted a consistent relationship between pressure drop and flow rate through fire nozzles.

EXAMPLE 10.2

A U-tube manometer filled with manometer fluid (of specific gravity S_m) is used to measure the pressure drop across an obstruction meter. A fluid of specific gravity, S, flows through the meter. Determine a relationship between the meter flow rate and the measured manometer deflection, H.

KNOWN   Fluid (of specific gravity S and specific weight γ)
Manometer fluid (of specific gravity S_m and specific weight γ_m)

ASSUMPTION   Density of fluids remain constant.

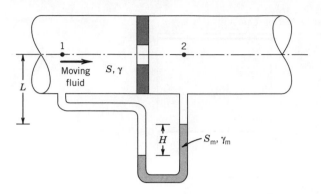

Figure 10.12 Manometer of Example 10.2.

FIND

$$Q = f(H)$$

SOLUTION   Equation (10.14) provides a relationship between flow rate and flow pressure drop. From Figure 10.12 and hydrostatic principles, the pressure drop measured by the manometer is

$$\Delta p = p_1 - p_2 = \gamma_m H - \gamma H = (\gamma_m - \gamma)H = \gamma H[(S_m/S) - 1]$$

Substituting this relation into equation (10.14) yields the working equation based on the equivalent pressure head

$$Q = CEA_o Y \sqrt{2g\,H[(S_m/S) - 1}$$ (10.15)

EXAMPLE 10.3

A 10-cm-diameter, square-edged orifice plate is used to meter the steady flow of 16°C water through an 20-cm pipe. Flange taps are used and the pressure drop measured is 50 cm Hg. Determine the pipe flow rate. The specific gravity of mercury is 13.5.

KNOWN

$$d_1 = 20\,\text{cm}$$

$$H = 50\,\text{cm Hg}$$

$$d_0 = 10\,\text{cm}$$

Properties (properties are from the Appendix)

$$\mu = 1.08 \times 10^{-3} \text{N-s/m}^2$$

$$\rho = 999\,\text{kg/m}^3$$

ASSUMPTIONS   Steady flow
Incompressible flow (Y = 1)

FIND   Q

SOLUTION   Equation (10.14) is used with an orifice plate, and it requires knowledge of E and C. The beta ratio is $\beta = d_0/d_1 = 0.5$, so the velocity of approach factor is calculated from equation (10.13)

$$E = \frac{1}{\sqrt{1 - \beta^4}} = \frac{1}{\sqrt{1 - 0.5^4}} = 1.0328$$

We know that $C = f(Re_{d_1}, \beta)$. The flow Reynolds number is estimated using equation (10.7) together with the relation $\nu = \mu/\rho$,

$$Re_{d_1} = \frac{4\rho Q}{\pi d_1 \mu} = \frac{4Q}{\pi d_1 \nu}$$

We see that without information concerning Q, we cannot estimate the Reynolds number, and so C cannot be determined explicitly.

Instead, a trial and error approach is undertaken: Guess a value for C (or for K_0) and iterate. A good start is to guess a value for either at a high value of Re_{d_1}. This is the flat region of Figure 10.6. From Figure 10.6, we choose a value of $K_0 = CE = 0.625$.

Based on the manometer deflection and equation (10.15) (see Example 10.2),

$$Q = CEA_0 Y \sqrt{2gH[(S_m/S) - 1]} = K_0 A_0 Y \sqrt{2gH[(S_m/S) - 1]}$$
$$= (0.625)(\pi/4)(0.10\,\text{m})^2(1)\sqrt{2(9.8\,\text{m/s}^2)(0.50\,\text{m})[(13.5/1) - 1]}$$
$$= 0.054\,\text{m}^3/\text{s}$$

Next we have to test the guessed value for K_0 to determine if it was correct. For this value of Q,

$$Re_{d_1} = \frac{4\rho Q}{\pi d_1 \mu} = \frac{4(999\,\text{kg/m}^3)(0.054\,\text{m}^3/\text{s})}{\pi(0.20\,\text{m})(1.08 \times 10^{-3}\,\text{N-s/m}^2)} = 3.2 \times 10^5$$

From Figure 10.6, at this Reynolds number, $K_0 \approx 0.625$. The solution is converged, so we conclude that $Q = 0.054\,\text{m}^3/\text{s}$.

EXAMPLE 10.4

Air flows at 20°C through a 6-cm pipe. A square-edged orifice plate with $\beta = 0.4$ is chosen to meter the flow rate. A pressure drop of 250 cm H_2O is measured at flange taps with an upstream pressure of 93.7 kPa abs. Find the flow rate.

KNOWN

$d_1 = 6\,\text{cm}$ $p_1 = 93.7\,\text{kPa abs}$

$\beta = 0.4$ $T_1 = 20°C = 293\,\text{K}$

$H = 250\,\text{cm}\,H_2O$

Properties (found in the Appendix)

   Air: $\nu = 1.0 \times 10^{-5}\,\text{m}^2/\text{s}$
   Water: $\rho_{H_2O} = 999\,\text{kg/m}^3$

ASSUMPTIONS   Steady air flow
      Ideal gas ($p = \rho RT$)

FIND Q

SOLUTION   Orifice flow rate is governed by equation (10.14), which will require information about E, C, and Y. From the given information, we estimate both the orifice area, $A_{d_0} = 4.52 \times 10^{-4}\,\text{m}^2$, and the velocity of approach factor, $E = 1.013$. The air density is found from the ideal gas equation of state:

$$\rho_1 = \frac{p_1}{RT_1} = \frac{93,700\,\text{N/m}^2}{(287\,\text{N} - \text{m/kg})(293\,\text{K})} = 1.114\,\text{kg/m}^3$$

The pressure drop is found from equation (9.3) to be $p_1 - p_2 = \rho_{H_2O}gH = 24,500\,\text{N/m}^2$.

The pressure ratio for this gas flow, $(p_1 - p_2)/p_1 = 0.26$. For pressure ratios greater than 0.1 the compressibility of the air should be considered. Figure 10.7 indicates that $Y = 0.92$, for $k = 1.4$ (air) and a pressure ratio of 0.26.

As in the previous example, the discharge coefficient cannot be found explicitly unless the flow rate is known, since $C = f(\text{Re}_{d_1}, \beta)$. So a trial-and-error approach is used. From Figure 10.6, guess a value for $K_0 = CE \approx 0.61$ (or $C = 0.60$). Then,

$$Q = CEYA\sqrt{\frac{2\Delta p}{\rho_1}} = (0.60)(1.013)(0.92)(4.52 \times 10^{-4}\,\text{m}^2)\sqrt{\frac{(2)(24,500\,\text{N/m}^2)}{1.114\,\text{kg/m}^3}}$$

$$= 0.053\,\text{m}^3/\text{s}$$

Check: For this flow rate, $\text{Re}_{d_1} = 4Q/\pi d_1 \upsilon = 7.4 \times 10^4$ and from Figure 10.6, $K_0 \approx 0.61$ as assumed. The flow rate through the orifice is taken to be 0.053 m³/s.

Sonic Nozzles

Sonic nozzles are used to meter and to control the flow rate of compressible gases [6]. They may take the form of any of the previously described obstruction meters. If the gas flow rate through an obstruction meter becomes sufficiently high, the sonic condition will be reached at the meter throat. At the sonic condition, the gas velocity equals the speed of sound of the gas. At that point the throat is considered to be choked, i.e., the mass flow rate through the throat is at a maximum for the given inlet conditions. Any further increase in pressure drop across the meter will not increase the mass flow rate. The theoretical basis for such a meter stems from the early work of Bernoulli, Venturi, and St. Venant (1797–1886). In 1866, Julius Weisbach (1806–1871) developed a direct relation between pressure drop and a maximum mass flow rate.

For a perfect gas undergoing an isentropic process, the pressure drop corresponding to the onset of the choked flow condition at the meter minimum area, the meter throat, is given by the *critical pressure ratio*:

$$\left(\frac{p_0}{p_1}\right)_{\text{critical}} = \left(\frac{2}{k+1}\right)^{k/(k-1)} \tag{10.16}$$

where p_0 is the throat pressure. If $(p_0/p_1) \leq (p_0/p_1)_{\text{critical}}$ the meter throat is choked and the gas flows at the sonic condition mass flow rate.

The steady-state energy equation written for a perfect gas is given by

$$c_p T_1 + \frac{\overline{U}_1^2}{2} = c_p T_0 + \frac{\overline{U}_0^2}{2} \tag{10.17}$$

where c_p is the constant pressure specific heat, which is assumed constant. Combining equations (10.6), (10.16), and (10.17) with the gas equation of state, $p = \rho RT$, yields the mass flow at and below the critical pressure ratio:

$$\dot{m}_{\text{max}} = \rho_1 A \sqrt{2RT_1} \sqrt{\frac{k}{k+1}\left(\frac{2}{k+1}\right)^{2/(k-1)}} \tag{10.18}$$

where k is the specific heat ratio of the gas. Equation (10.18) provides a measure of the ideal mass flow rate for a perfect gas. As with all obstruction meters, this ideal rate must be modified using a discharge coefficient to account for losses. However, the ideal and actual flow rates tend to differ by no more than 3%. When calibrations cannot be run, a $C = 0.99 \pm 2\%$ (95%) should be assumed [2].

The sonic nozzle provides a very convenient method to meter and to regulate a gas flow. The judicious selection of throat diameter can establish any desired fluid flow rate provided that the flow is sonic at the throat. This capability makes the sonic nozzle attractive as a local calibration standard for gases. Special orifice plate designs exist for metering at very low flow rates [2]. Both large pressure drops and system pressure losses must be tolerated with the technique, although low loss venturi designs exist that minimize losses [2].

EXAMPLE 10.5

A flow nozzle is to be used at choked conditions to regulate the flow of nitrogen N_2 at 1.3 kg/s through a 6-cm-i.d. pipe. The pipe is pressurized at 690 kPa abs and gas flows at 20°C. Determine the maximum β ratio nozzle that can be used. $R_{N_2} = 297\,\text{N–m/kg-K}$.

KNOWN

$$N_2(k = 1.4) \qquad p_1 = 690\,\text{kPa abs} \quad \dot{m} = 1.3\,\text{kg/s}$$
$$d_1 = 6\,\text{cm} \qquad T_1 = 293\,\text{K}$$

ASSUMPTIONS   Perfect gas (ideal gas with constant specific heats)
Steady, choked flow
$C = 0.99$ (with systematic uncertainty, $B = 2\%$)

FIND   Find $\beta_{\max} = d_{0_{\max}}/d_1$

SOLUTION   Equation (10.18) can be rewritten in terms of nozzle throat area, A:

$$A_{\max} = \frac{\dot{m}}{p_1\sqrt{2RT}\sqrt{[k/(k+1)][2/(k+1)]^{2/(k-1)}}}$$
$$= 8.41 \times 10^{-4}\,\text{m}^2$$

Since $A = \pi d_0^2/4$, this yields a maximum nozzle throat diameter, d_0:

$$d_{0_{\max}} = 3.27\,\text{cm}$$

Hence, $\beta = d_0/d_1 \leq 0.545$ is required to maintain the specified mass flow rate.

Obstruction Meter Selection

Selecting between obstruction meter types depends on a number of factors that require some engineering compromises. Primary considerations include meter placement, overall pressure loss, accuracy, overall costs.

It is important to provide sufficient upstream and downstream pipe lengths on each side of a flow meter for proper flow development. The flow development dissipates swirl, promotes a symmetric velocity distribution, and allows for proper pressure recovery downstream of the meter. For most situations, these installation effects can be minimized by placing the flow meter in adherence to the considerations outlined in Figure 10.13. However, the engineer needs to be wary that installation effects are difficult to predict, particularly downstream of elbows (out-of-plane double elbow turns are notoriously

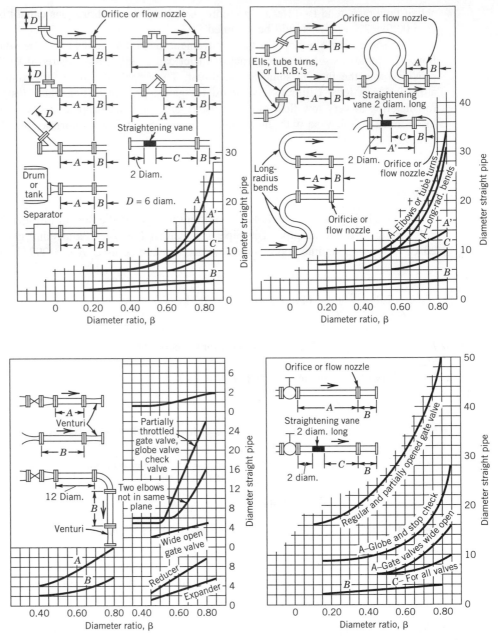

Figure 10.13 Recommended placements for flow meters in a pipeline. (Courtesy of American Society of Mechanical Engineers, New York; reprinted from [2].)

difficult), and Figure 10.13 is to be used as a guide only [11, 12]. Even with recommended lengths, an additional 0.5% systematic uncertainty should be added to the uncertainty in the discharge coefficient to account for swirl effects [12]. Unorthodox installations require meter calibration to determine flow coefficients or use of special correction procedures [11, 12]. The physical size of a meter may become an important consideration in itself, since in large-diameter pipelines, venturi and nozzle meters can take up a considerable length of pipe.

The unrecoverable overall pressure loss, Δp_{loss}, associated with a flow meter depends on β ratio and flow rate. These losses must be overcome by the system prime mover (e.g., a pump, fan, or compressor driven by a motor) in addition to any other pipe system losses and are to be considered during prime mover sizing. In established systems, flow meters must be chosen on the basis of the pressure loss that the prime mover can accommodate and still maintain a desired system flow rate. The power, $\dot{W}$, required to overcome any loss in a system for flow rate Q is given by

$$\dot{W} = Q\frac{\Delta p_{loss}}{\eta} \tag{10.19}$$

where η is the prime mover efficiency.

The actual cost of any meter depends on the initial capital cost of the meter, the costs of installing the meter (including system down time), calibration costs, and the added capital and operating costs associated with flow meter pressure losses to the system. Indirect costs may include product losses due to the meter flow rate uncertainty.

The ability of an obstruction flow meter to accurately estimate the volume flow rate in a pipeline depends on both the method used to calculate flow rate and factors inherent to the meter. If standard tabulated values for meter coefficients are used, factors that contribute to the overall uncertainty of the measurement enter through the following data-acquisition elemental errors: (1) Actual β ratio error and pipe eccentricity, (2) pressure tap position error, (3) temperature effects leading to relative thermal expansion of components, and (4) actual upstream flow profile. These errors are in addition to the errors inherent in the coefficients themselves and errors associated with the estimation of the upstream fluid density [2]. Direct in situ calibration of any flow meter installation can reduce these contributions to overall uncertainty. For example, the actual β ratio value, pressure tap locations, and other installation effects would be included in a system calibration and accommodated in the computed flow coefficient.

The selection of any meter should consider whether the system into which the meter is to be installed will be used at more than one flow rate. If so, considerations as to meter performance and its effect on system performance over the entire anticipated flow rate range should be included. The range over which any one meter can be used is known as the meter *turndown*.

EXAMPLE 10.6

For the orifice meter in Example 10.3 having a $\beta = 0.5$ and a pressure drop of 50 cm Hg, calculate the permanent pressure loss due to the meter that must be overcome by a pump.

KNOWN

$H = 50 \, \text{cm Hg}$

$\beta = 0.5$

ASSUMPTION   Steady flow

FIND

Δp_{loss}

SOLUTION   For a properly designed and installed orifice meter, Figure 10.8 indicates that the permanent pressure loss of the meter will be about 75% of the pressure drop across the meter. Hence,

$$\Delta p_{loss} = 0.75 \times \gamma_{Hg}H = 49.9 \, \text{kPa} \; (\text{or } H_{loss} = 38 \, \text{cm Hg})$$

COMMENT In comparison, for $Q = 0.053 \, \text{m}^3/\text{s}$ a typical venturi (say $15°$ outlet) with a $\beta = 0.5$ would provide a pressure drop equivalent to 19.6 cm Hg with a permanent loss of only 3.1 cm Hg. Since the pump needs to supply enough power to create the flow rate and to overcome losses, a smaller pump could be used with the venturi.

EXAMPLE 10.7

For the orifice of Example 10.6, estimate the operating costs required to overcome these permanent losses if electricity is available at $0.08/kW-h, the pump is used 6000 h/year, and the pump-motor efficiency is 60%.

KNOWN

$$\text{Water at } 16°\text{C} \qquad Q = 0.053 \, \text{m}^3/\text{s}$$
$$\Delta p_{\text{loss}} = 49.9 \, \text{kPa} \qquad \eta = 0.60$$

ASSUMPTIONS Steady flow

Meter installed according to standards [2]

FIND Annual cost due to Δp_{loss} alone

SOLUTION The pump power required to overcome the orifice meter permanent pressure loss is estimated from equation (10.19),

$$\dot{W} = Q \frac{\Delta p_{\text{loss}}}{\eta}$$

$$= \frac{(0.55 \, \text{m}^3/\text{s})(49,900 \, \text{Pa})(3600 \, \text{s/h})(1 \text{N}/\text{m}^2/\text{Pa})(1 \, \text{W}/\text{N} - \text{m/s})}{0.6} = 4530 \, \text{W} = 4.53 \, \text{kW}$$

The additional annual pump operating cost required to overcome this is

$$\text{Cost} = (4.53 \, \text{kW})(6000 \, \text{h/year})(\$0.08/\text{kW-h}) = \$2,175/\text{year}$$

COMMENT In comparison, the venturi discussed in the Comment to Example 10.6 would cost about $180/year to operate. If installation space allows, the venturi meter may be the better long-term solution.

EXAMPLE 10.8

An ASME long radius nozzle ($\beta = 0.5$) is to be installed into a horizontal section of 12-in. (schedule 40) pipe. A long, straight length of pipe exists just downstream of a fully open gate valve but upstream of an in-plane $90°$ elbow. Determine the minimum lengths of straight unobstructed piping that should exist upstream and downstream of the meter.

KNOWN

$$d_1 = 11.94 \, \text{in.} \; (30.33 \, \text{cm})$$
$$\beta = 0.5$$

Installation layout

ASSUMPTION Steady flow

FIND Meter placement

SOLUTION From Figure 10.13, the meter should be placed a minimum of eight pipe diameters ($A = 8$) downstream of the valve and a minimum of three pipe diameters ($B = 3$) upstream of any flow fitting. Note that since a typical flow nozzle has a length of about 1.5 pipe diameters, this installation will require a straight pipe section of at least 12.5 pipe diameters $(8 + 3 + 1.5)$.

Laminar Flow Elements

Laminar flow meters take advantage of the linear relationship between volume flow rate and the pressure drop over a length for laminar pipe flow. For a pipe of diameter d with pressure drop over a length L, the governing equations of motion lead to

$$Q = \frac{\pi d^2}{128\mu} \frac{p_1 - p_2}{L} \qquad \text{Re}_d < 2000 \qquad (10.20)$$

This relation was first demonstrated by Jean Poiseulle (1799–1869) who conducted meticulous tests documenting the resistance of flow through capillary tubes.

The simplest type of laminar flow meter consists of two pressure taps separated by a length of piping.[1] However, because the Reynolds number must remain low, this restricts either the size of pipe diameter that can be used or flow rate to which equation (10.20) can be applied for a given fluid. This limitation is overcome in commercial units by using laminar flow elements (Figure 10.14), which consist of a bundle of small-diameter tubes or passages, placed in parallel. The strategy of a laminar flow element is to divide up the flow by passing it through the tube bundle so as to reduce the flow rate per tube such that the Reynolds number in each tube remains below 2000. Pressure drop is measured between the entrance and the exit of the laminar flow element. Because of the additional entrance and exit losses associated with the laminar flow element, a flow coefficient is used to modify equation (10.20). While standard tables for these coefficients do not exist, commercial units come supplied with individual calibration charts. The coefficient is essentially constant over the useful meter range.

Because of the laminar flow restriction, any laminar flow element will have an upper limit on usable flow rate. Various meter sizes and designs are available to accommodate user needs. Turndowns up to 100:1 are available.

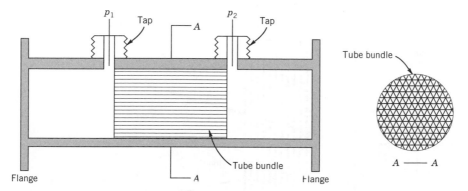

Figure 10.14 Laminar flow element flow meter.

---

[1] Alternately, if flow rate can be measured or is known, equation (10.20) provides the basis for a capillary tube viscometer.

Laminar flow elements offer some distinct advantages over other pressure differential meters. These include (1) a high sensitivity even at extremely low flow rates, (2) an ability to measure pipe system flows in either meter direction, (3) a wide usable flow range, and (4) the ability to indicate an average flow rate in pulsating flows. The instrument systematic uncertainty in flow rate determination is as low as ±0.25% of the flow rate. However, these meters are very susceptible to clogs, restricting their use to clean fluids. All of the measured pressure drop remains a system pressure loss.

10.6   INSERTION VOLUME FLOW METERS

Dozens of volume flow meter types based on a number of different principles have been proposed, developed, and sold commercially. A large group of meters is based on some phenomenon that is actually sensitive to the average velocity across a control surface of known area, i.e., $Q = f(\overline{U}, A) = \overline{U}A$. Several of these designs are included in the discussion below. Another common group, called positive displacement meters, actually measure parcels of a volume of fluid per unit time, i.e., $Q = f(\forall, t) = \forall/t$.

Electromagnetic Flow Meters

The operating principle of an *electromagnetic flow meter* [7] is based on the fundamental principle that an electromotive force (emf) of electric potential, E, is induced in a conductor of length, **L**, which moves with a velocity, **U**, through a magnetic field of magnetic flux, **B**. This physical behavior was first recorded by Michael Faraday (1791–1867). In principle, we write

$$E = \mathbf{U} \times \mathbf{B} \cdot \mathbf{L} \qquad (10.21)$$

A practical utilization of the principle is shown in Figure 10.15 for use with electrically conductive fluids. From equation (10.21) the magnitude of E is affected by the average

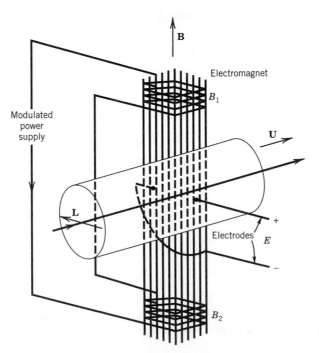

Figure 10.15 Electromagnetic principle as applied to a working flow meter.

velocity, $\overline{U}$, as

$$E = \overline{U}BL\sin\alpha = f(\overline{U})$$

where α is the angle between the mean velocity vector and the magnetic flux vector, usually at 90°. In general, electrodes are located either in or mounted on the pipe wall in a diametrical plane that is normal to the known magnetic field. As an electrically conductive fluid moves through the magnetic field, the induced electric potential is detected and measured by the electrodes, which are separated by the length, L. The average magnitude of the velocity, $\overline{U}$, across the pipe is thus inferred through the measured emf. The flow rate is found by

$$Q = \overline{U}\frac{\pi d_1^2}{4} = \frac{E}{BL}\frac{\pi d_1^2}{4} = K_1 E \tag{10.22}$$

The value of L is on the order of the pipe diameter, the exact value depending on the meter construction and magnetic flux lines. The static sensitivity K_1 is a meter constant found by calibration and supplied by a manufacturer. The relationship between flow rate and measured potential is linear.

The electromagnetic flow meter comes commercially as a packaged flow device, which is installed directly inline and connected to an external electronic output unit. Units are available using either permanent magnets, called dc units, or variable flux strength electromagnets, called ac units. The magnetic flux strength of an ac unit can be increased on site for a strong signal at low flow rates of low conductivity fluids such as water. Special designs include a flow sensor unit which actually can clamp over (not in-line with) a nonmagnetic pipe, a design, favored to monitor blood flow rate through major arteries during surgery or for diagnostic measurements on existing pipelines.

The electromagnetic flow meter has a very low pressure loss associated with its use due to its open tube, no obstruction design and is suitable for installations that can tolerate only a small pressure drop. This absence of internal parts is very attractive for metering corrosive and "dirty" fluids. The operating principle is independent of fluid density and viscosity, responding only to average velocity, and there is no difficulty with measurements in either laminar or turbulent flows, provided that the velocity profile is reasonably symmetrical. It can be used in either steady or pulsatile flows, providing either time-averaged or instantaneous data in the latter. Data-acquisition errors down to $\pm0.25\%$ of the measured flow rate can be attained, although values from ±1 to $\pm5\%$ are more common for these meters. The use of any meter is limited to fluids having a threshold value of electrical conductivity, the actual value of which depends on a particular meter's design, but fluids with values as low as 0.1 μsieman/cm have been metered. The addition of salts to a fluid will increase its conductivity.

Vortex Shedding Meters

Nearly everyone has observed an oscillating street sign or heard the "singing" of power lines on a windy day. These are examples of the effects induced by vortex shedding from bluff-shaped bodies, a natural phenomenon in which alternating vortices are shed in the wake of the body. The vortices formed on opposite sides of the body are carried downstream in the body's wake forming a "vortex street," each vortex having an opposite sign of rotation. This behavior is seen in Figure 10.16, a photograph that captures the vortex shedding downstream of a section of an aircraft wing. The aerodynamicist Theodore von Karman (1881–1963) first deduced the existence of a vortex street, although Leonardo da Vinci appears to have been the first to actually record the phenomenon [1].

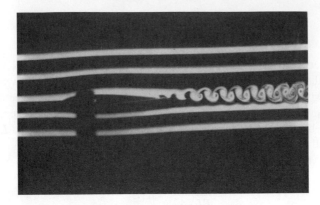

Figure 10.16 Smoke lines in this photograph reveal the vortex shedding behind a streamlined wing-shaped body in a moving flow.

A *vortex flow meter* operates on the principle that the frequency of vortex shedding depends on the average velocity of the flow past the body and the body shape. The basic relationship between shedding cyclical frequency, f, where $f = \omega/2\pi$, and average velocity, $\overline{U}$, for a given shape is given by the Strouhal number,

$$\text{St} = fd/\overline{U} \tag{10.23}$$

where d is a characteristic length for the body.

A typical design is shown in Figure 10.17. The shedder spans the pipe, so its length $l \approx d_1$, and $d/l \approx 0.3$, so as to provide for strong, stable vortex strength. In general, the Strouhal number is a function of Reynolds number but various geometrical shapes, known as shedders, can be used to produce a stable vortex flow that has a constant Strouhal number over a broad range of flow Reynolds number (for $\text{Re}_d > 10^4$). Examples are given in Table 10.1. The quality and strength of the shedding can be improved by manipulation of the design of the tail end of the body and by providing a slightly concave upstream body face that traps the stagnation streamline at a point giving way to a stable oscillation. Accordingly, there are a number of proprietary designs in existence. In general, abrupt edges on the shedder restrict the dependence on Reynolds number by fixing the flow separation points.

For a fixed body and constant Strouhal number, the flow rate for a pipe of inside diameter d_1 is

$$Q = \overline{U}A = (C\pi d_1^2/4\,\text{St})fd = K_1 f \tag{10.24}$$

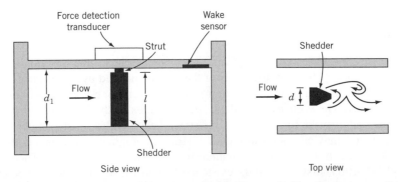

Figure 10.17 Vortex shedding flow meter. Different shedder shapes are available.

Table 10.1 Shedder Shape and Strouhal Number

Cross Section	Strouhal Number[a]
	0.16
	0.19
	0.16
	0.15
	0.12

[a]For Reynolds number $Re_d \geq 10^4$. Strouhal number $St = fd/\overline{U}$.

where the constant C accounts for shedder blockage effects that tend to increase the average velocity sensed. The value of K_1, known as the K-factor, is the meter static sensitivity and is a meter coefficient. K_1 remains essentially constant for $10^4 < Re_d < 10^7$. Shedding frequency can be measured in many ways. The shedder strut can be instrumented to detect the force oscillation by using strut-mounted strain gauges or capacitance sensor, for example, or a piezoelectric crystal wall sensor can be used to detect the pressure oscillations in the flow.

The lower flow rate limit on vortex meters appears to be at Reynolds numbers (based on d in Figure 10.17) near 10,000, below which the Strouhal number can vary nonlinearly with flow rate and shedding becomes unstable regardless of shedder design. This can be a problem in metering high-viscosity pipe flows ($\mu > 20$ cp). The upper flow bound is limited only by the onset of cavitation in liquids and by the onset of compressibility effects in gases at Mach numbers exceeding 0.2. Property variations affect meter performance only indirectly. Density variations affect the strength of the shed vortex, and this places a lower limit on fluid density, which is based on the sensitivity of the vortex shedding detection equipment. Viscosity affects the operating Reynolds number. Otherwise, within bounds, the meter is insensitive to property variations.

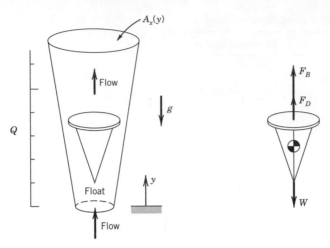

Figure 10.18 Rotameter.

The meter has no moving parts and relatively low pressure losses compared to obstruction meters. A single meter can operate over a flow range of up to 20:1 above its minimum with a linearity in K_1 of $\pm 0.5\%$. Because d can be replaced by a constant times d_1 in equation (10.24), we see that the strength of the shedding frequency sensed over the pipe area decreased as the pipe diameter cubed. This suggests that a meter's ability to resolve flow rate drops off as pipe diameter increases, placing an upper limit on the meter size.

Rotameters

The rotameter is a widely used insertion meter for flow rate indication. As depicted in Figure 10.18, the meter consists of a float within a vertical tube, tapered to an increasing cross-sectional area at its outlet. Flow entering through the bottom passes over the float, which is free to move. The equilibrium height of the float indicates the flow rate.

The operating principle of a rotatmeter is based on the balance between the drag force, F_D, and the weight, W, and buoyancy forces, F_B, acting on the float in the moving fluid. It is the drag force that varies with the average velocity over the float.

The force balance in the vertical direction y yields

$$\sum F_y = 0 = -F_D + W - F_B$$

with $F_D = C_D \rho \overline{U}^2 A_x / 2$, $W = \rho_b g \forall_b$, and $F_B = \rho g \forall_b$. The average velocity sensed by the float depends on its height in the tube and is given by

$$\overline{U} = \overline{U}(y) = \sqrt{2(\rho_b - \rho)g\forall_b / C_D \rho A_x} \qquad (10.25)$$

where

ρ_b = density of float (body)
ρ = density of fluid
C_D = drag coefficient of the float
A_x = tube cross-sectional area
$\overline{U}$ = average velocity past the float
$\forall_b$ = volume of float (body)

In operation, the float rises to an equilibrium position. The height of this position increases with flow velocity and, hence, flow rate. This flow rate is found by

$$Q = \overline{U}A_a(y) = K_1 A_a(y) \tag{10.26}$$

where $A_a(y)$ is the annular area between the float and the tube and K_1 is a meter constant. Both the average velocity and the annular area depend on the height of the float in the tube. So the float's vertical position is a direct measure of flow rate which can be read from a graduated scale, electronically sensed with an optical cell, or detected magnetically. Floats with sharp edges are less sensitive to fluid viscosity changes with temperature. A typical meter turndown is 10:1 with an instrument systematic uncertainty of $\sim\pm2\%$ of flow rate.

Turbine Meters

Turbine meters make use of angular momentum principles to meter flow rate. In a typical design (Figure 10.19), a rotor is encased within a bored housing through which the fluid to be metered is passed. Its housing contains flanges or threads for direct insertion into a pipeline. In principle, the exchange of momentum between the flow and the rotor turns the rotor at a rotational speed that is proportional to the flow rate. Rotor rotation can be measured in a number of ways. For example, a reluctance pickup coil can sense the passage of magnetic rotor blades producing a pulse train signal at a frequency that is directly related to rotational speed. This can be directly output as a TTL pulse train or the frequency can be converted to an analog voltage.

The rotor angular velocity, ω, will depend on the average flow velocity, $\overline{U}$, and the fluid kinematic viscosity, v, through the meter bore of diameter, d_1. Dimensionless analysis of these parameters [8] yields the functional relation $Q/\omega d_1^2 = f(\omega d_1^2/v)$ from which we get

$$Q = K_1 \omega \tag{10.27}$$

where K_1 is a meter constant that depends on fluid properties and is called the meter K-factor. In practice, there will be a region in which the rotor angular velocity will vary linearly with flow rate and this becomes the meter operating range.

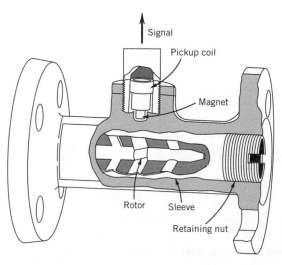

Figure 10.19 Cutaway view of a turbine flow meter. (Courtesy of Actaris Gas Division, Owenton, KY.)

Turbine meters offer a low-pressure drop and very good accuracy. A typical instrument systematic uncertainty in flow rate is ∼±0.25% with a turndown of 20 to 1. They are exceptionally repeatable making them good candidates for local flow rate standards. However, their use must be restricted to clean fluids because of possible fouling of their rotating parts. The turbine meter rotational speed is sensitive to temperature changes, which affect fluid viscosity and density. Some compensation for viscosity variations can be made electronically [9]. The turbine meter is very susceptible to installation errors caused by pipe flow swirl [11], and a careful selection of installation position is suggested.

Transit Time and Doppler (Ultrasonic) Flow Meters

Ultrasonic meters use sound waves to determine flow rate. *Transit time flow meters* use the travel time of ultrasonic waves to estimate average flow velocity. Referring to Figure 10.20, a pair of transducers, separated by some distance, are fixed to the outside of a pipe wall. A reflector may be applied to the opposite outside wall of the pipe to increase signal-to-noise ratio. Each transducer acts as a transmitter and a receiver for ultrasonic waves. An ultrasonic wave emitted by one transducer passes through the fluid, reflects off the pipe wall, and is received by the other transducer. The difference in transit time for a wave to travel from transducer 1 to transducer 2 and from transducer 2 to transducer 1 is directly related to the average velocity of flow in the pipe. For a fluid with speed of sound a, for flow with average velocity $\overline{U}$ based on a flow rate Q, and for a beam oriented at angle θ relative to the pipe flow axis,

$$t_1 = \frac{2L}{a + \overline{U}\cos\theta} \qquad t_2 = \frac{2L}{a - \overline{U}\cos\theta}$$

When $\overline{U} \ll a$, then with $L = d_1/\sin\theta$

$$Q = \overline{U}A = \frac{K_1 \pi d_1 a^2 (t_2 - t_1)}{16\cot\theta} \tag{10.28}$$

where K_1 is a meter constant.

Transit time meters are noninvasive and so offer no pressure drop. Portable models allow themselves to be strapped to the outside of the pipe, making them useful for field diagnostics. They can be set up to measure time-dependent velocity, including flow direction with time. Instrument relative systematic uncertainty ranges from ∼1 to 5% of the flow rate.

Doppler flowmeters use the Doppler effect to measure the average velocity of fluid particles (contaminants, particulate, or small bubbles) suspended in the pipe flow. In one

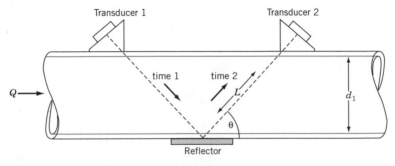

Figure 10.20 Principle of a transit time (ultrasonic) flow meter.

approach, a single ultrasonic transducer is mounted to the outside of a pipe wall. A wave of frequency, f, on the order of 100 kHz to 1 MHz is emitted. The emitted wave is bounced back by particles suspended in the moving fluid, and this scattered wave is detected by the transducer. The scattered waves travel at a slightly different frequency, the difference being the Doppler frequency. The flow rate is

$$Q = \overline{U}A = \frac{\pi d_1^2 a f_d}{8f \cos\theta} \tag{10.29}$$

These devices can measure time-dependent flow rates. Instrument systematic uncertainty is ~2% of the flow rate.

Positive Displacement Meters

Positive displacement meters contain mechanical elements that define a known volume filled with the fluid being metered. The free-moving elements are displaced or rotated by the action of the moving fluid. A geared counting mechanism counts the number of element displacements to provide a direct reading of volume of fluid passed through the meter, $\forall$. This metering method is common to water, gasoline, and to natural gas meters.

To be used as volume flow rate meters, volume per unit time can be used in conjunction with a timer, so that

$$Q = \forall/t \tag{10.30}$$

These units serve applications needing ruggedness and accuracy in steady flow metering.

Diaphragm meters contain two opposing flexible bellows coupled by linkages to exhaust valves that alternately displace a known volume of fluid through the meter. Through coupled linkage, the flow of fluid through one bellows drives the motion of the opposite bellows. So as entering gas causes one bellows to expand within its chamber sealed by a closed valve, the gas in the opposite bellows exhausts through an open valve to allow its bellows to collapse; this alternating motion cycles the gears of its counter register (output display). Such 'dry gas' meters are common to natural gas or propane lines for volume measurement. Thomas Glover invented the first diaphragm meter in 1843, and modern meters retain the major elements of his design.

Wobble meters contain a disk that is seated in a chamber. The flow of liquid through the chamber causes the disk to oscillate so that a known volume of fluid moves through the chamber with each oscillation. The disk is connected to a counter that records each oscillation as volume is displaced. Wobble meters are used for domestic water applications where they must have an uncertainty of no greater than 1% of actual delivery. Frequently found on oil trucks and at the gasoline pump, *rotating vane meters* use rotating cups or vanes that move about an annular opening displacing a known volume with each rotation.

With any of these meters, uncertainty can be as low as ±0.2% of actual delivery. Because of the good accuracy of these meters, they are often used as local working standards to calibrate other types of volume flow meters.

10.7   MASS FLOW METERS

There are many situations in which the mass flow rate is the quantity of interest. Mass flow rate demands measuring the momentum per unit volume, $\rho\overline{U}$, to yield $\dot{m} = \rho\overline{U}A$.

If the density of the metered fluid is known under the exact conditions of measurement, then a direct estimate can be made based on volume flow rate measurements. In such applications one can assume that both the errors in the tabulated values for fluid density and the fluid density variations within a volume meter are small enough to ignore. But not all fluids have readily established equations of state to provide a known value of density (e.g., petroleum products, polymers, and cocoa butter), and many processes are subject to significant changes in density. When volume flow methods are used, the uncertainty in the density increases the overall uncertainty in mass flow rate above the uncertainty in volume flow rate. As such, it is not uncommon to use density measuring devices just upstream of the volume flow measurement to improve the uncertainty in the mass flow determination in critical applications. The direct measurement of mass flow rate is desirable because it eliminates the uncertainties associated with estimating or measuring actual density.

The difference between a meter that is sensitive to Q as opposed to $\dot{m}$ is not trivial. Prior to the 1970s reliable commercial mass flow meters with sufficient accuracy to circumvent volume flow rate corrections were generally not available, even though the basic principles and implementation schemes for such meters had been understood in theory for several decades. United States patents dating back to the 1940s record schemes for using heat transfer, Coriolis forces, and momentum methods to infer mass flow rate directly.

Thermal Flow Meter

The rate at which energy, $\dot{E}$, must be added to a flowing fluid to raise its temperature between two control surfaces is directly related to the mass flow rate by

$$\dot{E} = \dot{m}c_p\Delta T \tag{10.31}$$

where c_p is the fluid specific heat. Methods to utilize this effect to directly measure mass flow rate incorporate an inline meter having some means to input energy to the fluid over the meter length. The passing of a current through an immersed filament is a common method. Fluid temperatures are measured at the upstream and the downstream locations of the meter. This type of meter is quite easy to use and appears reliable and is widely used for gas flow applications. In fact, in the 1980s, the technique was adapted for use in automobile fuel injection systems to provide an exact air–fuel mixture to the engine cylinders despite short-term altitude, barometric, and seasonal environmental temperature changes.

The operating principle of this meter assumes that c_p is known and remains constant over the length of the meter. For common gases, such as air, this assumption is quite valid at reasonable temperatures and pressures. Flow rate turndown of up to 100: 1 is possible with uncertainties down to ±0.5% of flow rate with very little pressure drop. But the assumptions become restrictive for liquids and for many gases for which c_p either may be a strong function of temperature or may not be well established.

A second type of thermal mass flow meter is actually a velocity-sensing meter and thermal sensor together in one direct insertion unit. The meter uses both hot-film anemometry methods to sense fluid velocity through a conduit of known diameter and an adjacent RTD sensor for temperature measurement. For sensor and fluid temperatures, T_s and T_f, respectively, mass flow rate is inferred from the correlation

$$\dot{E} = [C + B(\rho\overline{U})^{1/n}](T_s - T_f) \tag{10.32}$$

where C, B, and n, are constants which depend on fluid properties [10] and are determined through calibration. In a scheme to reduce the fluid property sensitivity of the meter, the RTD may be used as an adjacent resistor leg of the anemometer Wheatstone bridge circuit to provide a temperature-compensated velocity output over a wide range of fluid temperatures, with excellent repeatability (±0.25%). Gas velocities of up to 12,000 ft/min and flow rate turndown of 50 to 1 are possible with uncertainties down to ±2% of the flow rate and very little pressure drop. A series of these mass flow probes can be combined to span across large ducts or chimneys to provide better averaging information.

Coriolis Flow Meter

The term "Coriolis flow meter" refers to the family of insertion meters that meter mass flow rate by inducing a Coriolis acceleration on the flowing fluid and measuring the resulting developed force [14]. The developed force is directly related to the mass flow rate independent of the fluid properties. The Coriolis effect was proposed by Gaspard de Coriolis (1792–1843) following his studies of accelerations in rotating systems. Coriolis meters pass a fluid through a rotating or vibrating pipe system to develop the Coriolis force. A number of methods to utilize this effect have been proposed since the first U.S. patent for a Coriolis effect meter was issued in 1947. This family of meters has seen steady growth in market share since the mid-1980s.

The most common scheme for commercially available units are based on a scheme in which the pipe flow is diverted from the main pipe and divided between two bent, parallel, adjacent tubes of equal diameter, such as shown for the device in Figure 10.21. The tubes themselves are mechanically vibrated in a relative out-of-phase sinusoidal oscillation by an electromagnetic driver. In general, a fluid particle passing through the meter tube, which is rotating (due to the oscillating tube) relative to the fixed pipe, experiences an acceleration at any arbitrary position S. The total acceleration at S, $\ddot{\mathbf{r}}$, is composed of several components (see Figure 10.22),

$$\ddot{\mathbf{r}} = \ddot{\mathbf{R}}_{O'} + \dot{\boldsymbol{\omega}} \times \mathbf{r}_{S/O'} + \boldsymbol{\omega} \times \boldsymbol{\omega} \times \mathbf{r}_{S/O'} + \ddot{\mathbf{r}}_{S/O'} + 2\boldsymbol{\omega} \times \dot{\mathbf{r}}_{S/O'} \qquad (10.33)$$

where

$\ddot{\mathbf{R}}_{O'}$ = translation acceleration of rotating origin O' relative to fixed origin O

$\boldsymbol{\omega} \times \boldsymbol{\omega} \times \mathbf{r}_{S/O'}$ = centripetal acceleration of S relative to O'

$\dot{\boldsymbol{\omega}} \times \mathbf{r}_{S/O'}$ = tangential acceleration of S relative to O'

$\ddot{\mathbf{r}}_{S/O'}$ = translational acceleration of S relative to O'

$2\boldsymbol{\omega} \times \dot{\mathbf{r}}_{S/O'}$ = Coriolis acceleration at S relative to O'

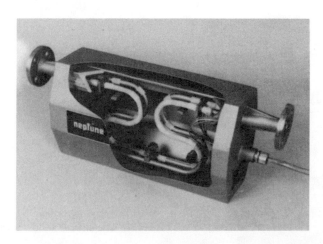

Figure 10.21 Cutaway view of a Coriolis mass flow meter. (Courtesy of Actaris Neptune Liquid Measurement Division, Greenwood, SC.)

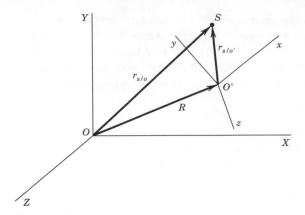

Figure 10.22 Fixed and rotating reference frames.

with ω the angular velocity of point S relative to O'. In these meters, the tubes are rotated but not translated, so that the translational accelerations are zero. A fluid particle will experience forces due to the remaining accelerations that cause equal and opposite reactions on the meter tube walls. The Coriolis acceleration distinguishes itself by acting in a plane perpendicular to the tube axes and develops a force gradient that creates a twisting motion or oscillating rotation about the tube plane.

The utilization of the Coriolis force depends on the shape of the meter. However, the basic principle is illustrated in Figure 10.23. Rather than rotating the tubes a complete 360° about the pipe axis, the meter tubes are vibrated continuously at a drive frequency, ω, with amplitude displacement, z, about the pipe axis. This eliminates rotational seal problems. The driving frequency is selected at the tube resonant frequency that places the driven tube into what is called a limit cycle, a continuous, single-frequency oscillation. This configuration is essentially a self-sustaining tuning fork because the meter will naturally respond to any disturbance at this frequency with a minimum in input energy.

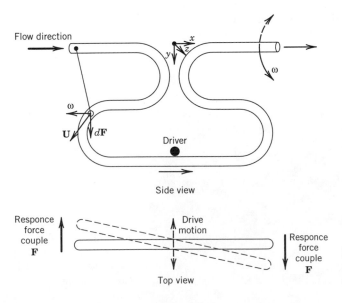

Figure 10.23 Concept of the operating principle of a Coriolis mass flow meter.

As a fluid particle of elemental mass, dm, flows through a section of the flow meter, it experiences the Coriolis acceleration, $2\omega \times \dot{\mathbf{r}}_{S/O'}$, and an inertial Coriolis force

$$dF = (2\omega \times \dot{\mathbf{r}}_{S/O'})\,dm \qquad (10.34)$$

acting in the z direction. For the flow meter design of Figure 10.21 and as depicted in Figure 10.23, as the particle travels along the meter the direction of the velocity vector changes. This results, using the right-hand rule, in a change in direction of vector force, $d\mathbf{F}$. The resultant forces experienced by the tube are of equal magnitude but opposite sign of those experienced by the particle. Each tube segment senses a corresponding differential torque, $d\mathbf{T}$, a rotation about the y axis at a frequency ω_c,

$$d\mathbf{T} = x' \times d\mathbf{F} = x' \times (2\omega \times \dot{\mathbf{r}}_{S/O'})\,dm \qquad (10.35)$$

where x' refers to the x distance between the elemental mass and the y axis, $\mathbf{x}' = x'\hat{e}_x$. The total torque experienced by each tube is found by integration along the total path length of the tube, L,

$$\mathbf{T} = \int_0^L d\mathbf{T} \qquad (10.36)$$

A differential element of fluid will have the mass, $dm = \rho A\,dl$, for an elemental cross section of fluid, A, of differential length, dl, and of density, ρ. If this mass moves with an average velocity, $\overline{U}$, then the differential mass can be written as

$$dm = \rho A\,dl = \dot{m}(dl/\overline{U}) \qquad (10.37)$$

The Coriolis cross-product can be expressed as

$$2\,\omega \times \dot{\mathbf{r}}_{S/O'} = (2\omega \times \dot{\mathbf{r}}_{S/O'}\sin\theta)\hat{e}_z = (2\,\omega_c\overline{U}\sin\theta)\hat{e}_z \qquad (10.38)$$

where θ is the angle between the Coriolis rotation and the velocity vector. Then,

$$\dot{m} = \frac{T}{2\int_0^L (\rho x'\omega_c \sin\theta)dl}\,\hat{e}_y \qquad (10.39)$$

Since the velocity direction changes by 180°, the Coriolis developed torque will act in opposite directions on each side of the tube.[2] The meter tubes will twist about the y axis of the tube at an angle δ. For small angles of rotation the twist angle is related to torque by

$$\delta = k_s T = \text{constant} \times \dot{m} \qquad (10.40)$$

where k_s is related to the stiffness of the tube. The objective becomes to measure the twist angle, which can be accomplished in many ways. For example, by driving the two meter tubes 180° out of phase, the relative phase at any time between the two tubes will be directly related to the mass flow rate. The exact relationship is linear over a wide flow range and determined by calibration with any fluid [15].

The tangential and the centripetal accelerations remain of minor consequence due to the tube stiffness in the directions in which they act. However, at high mass flow rates they can excite modes of vibration in addition to the driving mode, ω, and response mode, ω_c. In doing so, they affect the meter linearity and zero error (drift), which affects mass flow uncertainty and meter turndown. Essentially, the magnitude of these undesirable

---

[2]It is not difficult to envision a design in which the velocity does not reverse direction along the flow path but ω does.

effects are inherent to the particular meter shape and are controlled, if necessary, through tube-stiffening members.

Another interesting problem occurs mostly in meters of small tube diameter where the tube mass may approach the mass of the fluid in the tube. At flow rates that correspond to flow transition from a laminar to turbulent regime, the driving frequency can excite the flow instabilities responsible for the flow transition. The fluid and tube can go out of phase reducing the response amplitude and its corresponding torque. This affects the meter's calibration linearity, but a good design can contain this effect to within 0.5% of the meter reading.

The meter principle is unaffected by changing fluid properties but temperature changes will affect the overall meter stiffness, an effect that can be compensated for electronically. A very desirable feature is an apparent insensitivity to installation position. Commercially available Coriolis flow meters can measure flow rate with an instrument systematic uncertainty to $\pm0.25\%$ of mass flow rate, but $\pm0.10\%$ is achievable. Turndown is about 20 to 1. The meter is also used as an effective densitometer.

10.8   FLOW METER CALIBRATION AND STANDARDS

While a fundamental primary standard for flow rate does not exist, there are a number of calibration test code procedures in place. The general procedure for the calibration of in-line flow meters requires establishing a steady flow in a calibration flow loop and then determining the volume or mass of flowing fluid passing through the flow meter in an accurately determined time interval. Such flow loop calibration systems are known as *provers*. Several methods to establish the flow rate are discussed.

In liquids, variations of a "catch-and-weigh" technique are often employed in flow provers. One variation of the technique consists of a calibration loop with catch tank as depicted in Figure 10.24. Tank A is a large tank from which fluid is pumped back to a constant head reservoir which supplies the loop with a steady flow. Tank B is the catch-and-weigh tank into which liquid can be diverted for an accurately determined period of time. The liquid volume is measured, either directly using a positive displacement meter, or indirectly through its weight, and flow rate deduced through time. The ability to determine the volume and the uncertainty in the initial and final time of the event are fundamental limitations to the accuracy of this technique. Neglecting installation effects,

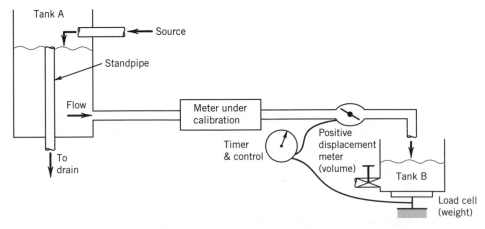

Figure 10.24 Flow diagram of a flow meter prover for liquids.

the ultimate limits of uncertainty (at 95%) in flow rate of liquids are on the order of $\pm 0.03\%$, a number based on $u_{\text{meter}} \sim 0.02\%$, $u_{\forall} \sim 0.02\%$, and $u_t \sim 0.01\%$.

Flow meter calibration by determining the pipe velocity profile is particularly effective for in situ calibration in both liquids and gases, provided that the gas velocity does not exceed about 70% of the sonic velocity. Velocity traverses at any cross-sectional location some 20–40 pipe diameters downstream of any pipe fitting in a long section of straight pipe are preferred.

Comparison calibration against a local standard flow meter is another common means of establishing the flow rate through a prover. Flow meters are installed in tandem with the standard and directly calibrated against it. Turbine and vortex meters and Coriolis mass flow meters have consistent, highly accurate calibration curves and are often used as local standards. Other provers use an accurate positive displacement meter to determine flow volume over time. Of course, these standards must be periodically recalibrated. In the United States, NIST maintains flow meter calibration facilities for this purpose. However, installation effects in the end-user facility will not be accounted for in an NIST calibration. This last point accounts for much of the uncertainty in a flow meter calibration. Lastly, a sonic nozzle can also be used as a standard to establish the flow rate of a gas in a comparison calibration. In any of these methods, the calibration uncertainty is limited by the standard used, installation effects, and the inherent limitations of the flow meter calibrated.

10.9   ESTIMATING STANDARD FLOW RATE

When the range of pressures and temperatures for flow processes vary in use, the measured "actual" flow rates can be adjusted to a standard temperature and pressure for comparison. This adjusted flow rate is called the *standard flow rate*. Standard flow rates are often reported in units such as SCMM (standard cubic meters per minute) or standard cubic feet per minute (SCFM). These are found by converting from the actual conditions at the measurement point, with the actual flow rate noted with corresponding units, such as ACMM or ACFM. The mass flow rate remains the same regardless of conditions, so using subscript s for standard conditions and "a" for the actual measured conditions, we have $\dot{m}_s = \dot{m}_a$, so that

$$Q_s = Q_a \frac{\rho_a}{\rho_s} \tag{10.41}$$

For example, the actual flow rates of common gases are "standardized" to 760 mm Hg absolute and 20°C. Assuming ideal gas behavior, $\rho = p/RT$, the standard flow rate is found from equation (10.41) as

$$Q_s = Q_a \frac{T_s\, p_a}{T_a\, p_s} = Q_a \left(\frac{293}{273 + T_a}\right)\left(\frac{760 + p_a}{760}\right) \tag{10.42}$$

10.10   SUMMARY

Flow quantification has been an important engineering task for well over two millennia. In this chapter, methods to determine the volume rate of flow and the mass rate of flow were presented. The engineering decision involving the selection of a particular meter was found to depend on a number of constraining factors. In general, flow rate can be determined to within about $\pm 0.25\%$ of actual flow rate with the best of present technology. However, new methods may push these limits even further, provided that calibration standards can be developed to document method uncertainties and the effects of installation.

REFERENCES

1. da Vinci, L., *Del Moto e Misura Dell'Aqua*, E. Carusi and A. Favaro, eds., Bologna, 1924.
2. *Flow Meters, Theory and Applications*, 6th ed., American Society of Mechanical Engineers, New York, 1971.
3. *Flow of Fluids through Valves and Fittings*, Technical Paper No. 410, Spiral Edition, Crane Co., Chicago, 1998.
4. *ASHRAE Fundamentals*, American Society of Heating, Refrigeration and Air Conditioning Engineers, Rev. ed., 1997.
5. Rouse, H., and S. Ince, *History of Hydraulics*, Dover, New York, 1957.
6. Arnberg, B. T., A review of critical flowmeters for gas flow measurements, *Transactions of the ASME*, 84:447, 1962.
7. Shercliff, J. A., *Theory of Electromagnetic Flow Measurement*, Cambridge University Press, New York, 1962.
8. Hochreiter, H. M., Dimensionless correlation of coefficients of turbine-type flowmeters, *Transactions of the ASME*, October 1958.
9. Lee, W. F., and H. Karlby, A study of viscosity effects and its compensation on turbine flow meters, *Journal of Basic Engineering*, September 1960.
10. Hinze, J. O., *Turbulence*, McGraw-Hill, New York, 1953.
11. Mattingly, G., *Fluid measurements: Standards, calibrations and traceabilities*, Proc. ASME/AIChE National Heat Transfer Conference, Philadelphia, PA, 1989.
12. *Measurement of Fluid Flow in Pipes Using Orifice*, Nozzle and Venturi, ASME Standard MFC-3M-1985, American Society of Mechanical Engineers, New York, 1985.
13. *Fan Engineering*, 8th edition, Buffalo Forge Co., Buffalo, New York, 1983.
14. *Measurement of Fluid Flow by Means of Coriolis Mass Flow Meters*, MFC-11M, American Society of Mechanical Engineers, 2003.
15. Corwon, M., and R. Oliver, *Omega-shaped, Coriolis-type Mass Flow Meter System*, U.S. Patent 4,852,410, 1989.

NOMENCLATURE

c_p, c_v	specific heats $[l^2/t^{2\circ}]$	$\dot{m}$	mass flow rate $[mt^{-1}]$
d	diameter $[l]$	p	pressure $[ml^{-1}t^{-2}]$
d_0	flow meter throat or minimum diameter $[l]$	$p_1 - p_2$, Δp	pressure drop $[ml^{-1}t^{-2}]$
d_1	pipe diameter $[l]$	Δp_{loss}	permanent pressure loss $[ml^{-1}t^{-2}]$
d_2	vena contracta diameter (see Figure 10.4) $[l]$	r	radial coordinate $[l]$
		r_1	pipe radius $[l]$
f	cyclical frequency $(\omega/2\pi)$ $[t^{-1}]$	r_H	hydraulic radius $[l]$
g	gravitational acceleration constant $[lt^{-2}]$	A	area $[l^2]$
		A_0	area based on d_0 $[l^2]$
$h_{L_{1-2}}$	energy head loss between points 1 and 2 $[l]$	A_1	area based on pipe diameter d_1 $[l^2]$
		A_2	vena contracta area $[l^2]$
k	ratio of specific heats, c_p/c_v	**B**	magnetic field flux vector

C	discharge coefficient
C_D	drag coefficient
E	velocity of approach factor; voltage
Gr	Grashof number
H	manometer deflection $[l]$
K_0	flow coefficient ($= CE$)
K_1	flow meter constant; static sensitivity
L	length $[l]$
Q	volume flow rate $[l^3 t^{-1}]$
R	gas constant $[l^2 t^{-2} \text{-} \o]$
Re_{d_1}	Reynolds number (based on d_1)

S	specific gravity
T	temperature $[°]$; torque $[ml^2 t^{-2}]$
U	velocity $[lt^{-1}]$
$\forall$	volume $[l^3]$
Y	expansion factor
β	diameter ratio
ρ	density $[ml^{-3}]$
δ	twist angle
ω	frequency $[t^{-1}]$; angular velocity $[t^{-1}]$
μ	absolute viscosity $[ml^{-1} t^{-1}]$
ν	kinematic viscosity ($= \mu/\rho$) $[l^2 t^{-1}]$

PROBLEMS

10.1 Determine the average mass flow rate of 5°C air at 1 bar abs. through a 5-cm-i.d. pipe whose velocity profile is found to be symmetric and described by $U(r) = 25[1 - (r/r_1)^2]$ cm/s

10.2 A 10-cm-i.d. pipe of flowing 10°C air is traversed along three radial lines with measurements taken at five equidistant stations along each radial. Determine the best estimate of the pipe flow rate. Systematic uncertainties are $\pm 2\%$ of the mean flow rate.

Radial		U(r) [cm/s]		
Location	r (cm)	Line 1	Line 2	Line 3
1	1.0	25.31	24.75	25.10
2	3.0	22.48	22.20	22.68
3	5.0	21.66	21.53	21.79
4	7.0	15.24	13.20	14.28
5	9.0	5.12	6.72	5.35

10.3 A manometer is used in a 5.1-cm water line to measure the pressure drop across a flow meter. Mercury is used as the manometric fluid ($S = 13.57$). If the manometer deflection is 10.16 cm Hg, determine the pressure drop in N/m^2.

10.4 A bourdon tube pressure gauge indicates a 69 kPa drop across an orifice meter located in a 25.4-cm. pipe containing flowing air at 32°C. Find the equivalent pressure head in cm H$_2$O.

10.5 For a square-edged orifice plate using flange taps to meter the flow of O$_2$ at 60°F through a 3-in.-i.d. pipe, the upstream pressure is 100 lb/in.^{2} and the downstream pressure is 76 lb/in.^{2}. If the orifice diameter is 1.5 in., can this flow be treated as incompressible for the purposes of estimating the flow rate? Demonstrate your reasoning. $R_{O_2} = 48.3$ ft-lb/lbm-°R

10.6 Find the discharge coefficient of a 5-cm-diameter, square-edged orifice plate using flange taps and located in a 15-cm pipe if the Reynolds number is 250,000.

10.7 At what flow rate of 20°C water through a 10-cm-i.d. pipe would the discharge coefficient of a square-edged orifice plate ($\beta = 0.4$) become essentially independent of Reynolds number?

10.8 Estimate the flange tap pressure drop across a square-edged orifice plate ($\beta = 0.5$) within a 12-cm-i.d. pipe if 25°C water flows at 50 L/s. Compute the permanent pressure loss associated with the plate.

10.9   Determine the flow rate of 38°C air through a 6-cm pipe if an ASME flow nozzle with a 3-cm throat and throat taps are used, the pressure drop measured is 75 cm H_2O, and the upstream pressure is 94.4 kPa abs.

10.10   A square-edged orifice ($\beta = 0.5$) is used to meter N_2 at 520°R through a 4-in. pipe. If the upstream pressure is 20 psia and the downstream pressure is 15 psia, determine the flow rate through the pipe. Flange taps are used. $R_{N_2} = 55.13$ ft-lb/lbm-°R

10.11   Size a suitable orifice plate to meter the steady flow of water at 20°C flowing through a 38-cm pipe if the nominal flow rate expected is 200 kg/s and the pressure drop must not exceed 15 cm Hg.

10.12   A cast venturi meter is to be used to meter the flow of 15°C water through a 10-cm pipe. For a maximum differential pressure of 76 cm H_2O and a nominal 0.5-m^3/min flow rate, select a suitable throat size.

10.13   For 120 ft^3/m of 60°F water flowing through a 6-in. pipe, size a suitable orifice, venturi, and nozzle flow meter if the maximum pressure drop cannot exceed 20 in. Hg. Estimate the permanent losses associated with each meter. Compare the annual operating cost associated with each if power cost is \$0.10/kW-h, a 60% efficient pump motor is used, and the meter is operated 6000 hr/year.

10.14   Estimate the flow rate of water through a 15-cm-i.d. pipe that contains an ASME long radius nozzle ($\beta = 0.6$) if the pressure drop across the nozzle is 25.4 cm Hg. Water temperature is 27°C.

10.15   Select a location for a 9-cm square-edged orifice along a 4-m straight run of 18-cm pipe if the orifice must be situated downstream but in-plane of a 90° elbow. The orifice is to be operated uncalibrated under ASME codes.

10.16   An orifice ($\beta = 0.4$) is used as a sonic nozzle to meter a 40°C air flow in a 5-cm pipe. If upstream pressure is 695 mm Hg abs and downstream pressure is 330 mm Hg abs, estimate the mass flow rate through the meter. $R = 287$ N-m/kg-K, $k = 1.4$.

10.17   Compute the flow rate of 20°C air through a 0.5-m square-edged orifice plate that is situated in a 1.0-m-i.d. pipe. The pressure drop across the plate is 90 mm H_2O, and upstream pressure is 2 atm.

10.18   An ASME long radius nozzle ($\beta = 0.5$) is to be used to meter the flow of 20°C water through a 8-cm-i.d. pipe. If flow rates will range from 0.6 to 1.6 L/s, select the range required of a single pressure transducer if that transducer will be used to measure pressure drop over the flow rate range. If the transducer to be used has a typical error of 0.25% of full scale, estimate the uncertainty in flow rate at the design stage.

10.19   A square-edged orifice plate is selected to meter the flow of water through a 2.3 in.-i.d. pipe over the range of 10 to 50 gal/min at 60°F. It is desired to operate the orifice in the range where C is independent of the Reynolds number for all flow rates. A pressure transducer having an accuracy to within 0.5% of reading is to be used. Select a suitable orifice plate and estimate the design-stage uncertainty in measured flow rate at 10, 25, and 50 gal/min. Assume flange pressure taps and reasonable values for the systematic uncertainty on pertinent parameters.

10.20   Estimate the error contribution to the uncertainty in flow rate due to the effect of the relative humidity of air on air density. Consider the case of using an orifice plate to meter air flow if the relative humidity of the 70°F air can vary from 10 to 80% but a density equivalent to 45% relative humidity is used in the computations.

10.21   For Problem 10.20, suppose the air flow rate is 17 m^3/hr at 20°C through a 6-cm-i.d. pipe, a square-edged orifice ($\beta = 0.4$) is used with flange taps, and the pressure drop can be measured to within ±0.5% of reading for all pressures above 5 cm H_2O using a manometer. If basic dimensions are maintained to within 0.1 mm, estimate a design-stage uncertainty in the flow rate. Use $p_1 = 96.5$ kPa abs., $R = 287$ N-m/kg-K.

10.22   It is desired to regulate the flow of air through an air sampling device using a sonic nozzle. A flow rate of 45 ft³/min relative to 70°F is to be maintained. For an ASME long radius nozzle, determine the downstream pressure required to choke the nozzle if air is supplied at 14.1 psia. Select a suitable nozzle throat diameter. $k = 1.4$, $R = 53.3$ (ft-lb)/ (lbm-°R).

10.23   An orifice meter is to be installed between flanges in a 10-cm diameter pipe. It is desired that the meter develop a 100 mm Hg pressure head at 2 m³/min of 20°C water using flange taps. Specify an appropriate size (d_o) for the orifice meter. $S_{Hg} = 13.6$

10.24   An ASME long radius nozzle is used to meter the flow of 20°C water through a 20-cm diameter pipe. The operating flow rate expected is between 5000 cm³/s and 50,000 cm³/s. For $\beta = 0.5$, specify the input range required of a pressure transducer used to measure the expected pressure drop. Estimate the permanent pressure loss associated with this nozzle. If a catalog is available, select an appropriate pressure transducer from its listings based on input range.

10.25   Select a pressure transducer from a catalog to meet the needs of Problem 10.24 based on input and output range. If its output is to be measured using a data-acquisition system which uses a ± 5 V, 12-bit A/D converter, estimate the percent relative quantization error at the low and high flow rates. Do these numbers seem appropriate? If not, how can they be improved?

10.26   A vortex flow meter uses a shedder having a Strouhal number of 0.20. Estimate the mean duct velocity if the shedding frequency indicated is 77 Hz and the shedder characteristic length is 1.27 cm.

10.27   A thermal mass flow meter is used to meter 30°C air flow through a 2-cm-i.d. pipe. If 25 W of power are required to maintain a 1°C temperature rise across the meter, estimate the mass flow rate through the meter. Clearly state any assumptions. $c_p = 1.006$ kJ/kg-K.

10.28   When used with air (or any perfect gas), a sonic nozzle can be used to regulate volume flow rate. However, uncertainty arises due to variations in density brought on by local atmospheric pressure and temperature changes. For a range of 101.3 ± 7 kPa abs and $10 \pm 5°C$, estimate the error in regulating a 1.4-m³/min flow rate at critical pressure ratio due to these variations alone.

10.29   Estimate an uncertainty in the determined flow rate in Example 10.4 assuming that dimensions are known to within 0.025 mm, pressure is known to within 0.25 cm H_2O, and the pressure drop shows a standard deviation of 0.5 cm H_2O in 20 readings. Upstream pressure is constant. All assumptions should be justified and reasonable.

10.30   A thermal mass flow meter is used to meter air in a 1-cm-i.d. tube. The meter adds 10 W of energy to the air passing through the meter from which the meter senses a 3°C temperature gain. What is the mass flow rate? $c_p = 1.006$ kJ/kg-K.

10.31   A vortex meter is to use a shedder having a profile of a forward facing equilateral triangle (Table 10.1) with a characteristic length of 10 mm. Estimate the shedding frequency developed for 20°C air at 30 m/s in a 10-cm-i.d. pipe. Estimate the meter constant and measured flow rate.

10.32   The flow of air is measured to be 30 m³/min at 50 mm Hg and 15°C. What is the flow rate in SCMM?

10.33   A 6 in. × 4 in. i.d. cast venturi is used to measure the flow rate of water. Estimate the flow rate and its uncertainty at 95%. The following values are known from a large sample:

x	Value	b_x	$S_{\bar{x}}$
C	0.984	0.00375	0
d_o	3.995 in.	0.0005	0
d_1	6.011 in.	0.001	0
ρ	62.369 lb_m/ft^3	0.002	0.002
H	100 in H_2O @ 68°F	0.15	0.4

10.34   A 60-mm-i.d. pipe will be used to transport alcohol ($\rho = 790$ kg/m^3; $\mu = 1.2 \times 10^{-3}$ N-s/m^2) from storage to its point of use in a manufacturing process. A flow nozzle with flange taps is selected to meter the flow. If a pressure drop of 4kPa is desired at 0.003 m^3/s, specify the nozzle diameter, d_o.

Chapter 11

Strain Measurement

11.1 INTRODUCTION

The design of load-carrying components for machines and structures requires information about the distribution of forces within the particular component. Proper design of devices such as shafts, pressure vessels, and support structures must consider load-carrying capacity and allowable deflections. Mechanics of materials provides a basis for predicting these essential characteristics of a mechanical design and provides the fundamental understanding of the behavior of load-carrying parts. However, theoretical analysis is often not sufficient and experimental measurements are required to achieve a final design.

Our interest in this chapter is the measurement of physical displacements in engineering components. Engineering designs are based on a safe level of stress within a material. In an object that is subject to loads, forces within the object act to balance the external loads.

As a simple example, consider a slender rod that is placed in uniaxial tension, as shown in Figure 11.1. If the rod is sectioned at B—B, a force within the material at B—B is necessary to maintain static equilibrium for the sectioned rod. Such a force within the rod, per unit area, is called *stress*. Design criteria are based on stress levels within a part. In most cases stress cannot be measured directly. But the length of the rod in Figure 11.1 will change when the load is applied, and such changes in length or shape of a material can be measured. The stress is calculated from these measured deflections. Before we can proceed to develop techniques for these measurements, we will briefly review the relationship between deflections and stress.

11.2 STRESS AND STRAIN

The experimental analysis of stress is accomplished by measuring the deformation of a part under load, and inferring the existing state of stress from the measured deflections. Again, consider the rod in Figure 11.1. If the rod has a cross-sectional area of A_c, and the load is applied only along the axis of the rod, the normal stress is defined as

$$\sigma_a = F_N/A_c \qquad (11.1)$$

where A_c is the cross-sectional area and F_N is the tension force applied to the rod, normal to the area A_c. The ratio of the change in length of the rod (which results from applying the load) to the original length is the *axial strain*, defined as

$$\varepsilon_a = \delta L/L \qquad (11.2)$$

where ε_a is the average strain over the length L, δL is the change in length, and L is the original unloaded length. For most engineering materials, strain is a small quantity; strain is usually reported in units of 10^{-6} m/m or 10^{-6} in./in. These units are equivalent to a dimensionless unit called a microstrain (μs).

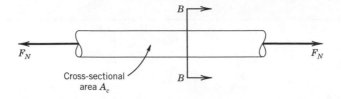

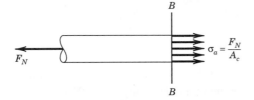

Figure 11.1 Free-body diagram illustrating internal forces for a rod in uniaxial tension.

Stress-strain diagrams are very important in understanding the behavior of a material under load. Figure 11.2 is such a diagram for mild steel (a ductile material). For loads less than that required to permanently deform the material, most engineering materials display a linear relationship between stress and strain. The range of stress over which this linear relationship holds is called the *elastic region*. The relationship between uniaxial stress and strain for this elastic behavior is expressed as

$$\sigma_a = E_m \varepsilon_a \tag{11.3}$$

where E_m is the modulus of elasticity, or Young's modulus, and the relationship is called Hooke's law. Hooke's law applies only over the range of applied stress where the relationship between stress and strain is linear. Different materials respond in a variety of ways to loads beyond the linear range, largely depending on whether the material is ductile or brittle. For almost all engineering components, stress levels are designed to remain well below the elastic limit of the material; thus, a direct linear relationship may

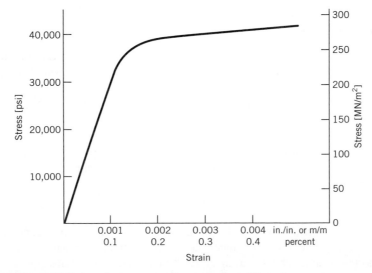

Figure 11.2 A typical stress-strain curve for mild steel.

be established between stress and strain. Under this assumption, Hooke's law forms the basis for experimental stress analysis, through the measurement of strain.

Lateral Strains

Consider the elongation of the rod shown in Figure 11.1, which occurs as a result of the load F_N. As the rod is stretched in the axial direction, the cross-sectional area must decrease, since the total mass (or volume for constant density) must be conserved. Similarly, if the rod were compressed in the axial direction, the cross-sectional area would increase. This change in cross-sectional area is most conveniently expressed in terms of a lateral (transverse) strain. For a circular rod, the lateral strain is defined as the change in the diameter divided by the original diameter. In the elastic range, there is a constant rate of change in the lateral strain as the axial strain increases. In the same sense that the modulus of elasticity is a property of a given material, the ratio of lateral strain to axial strain is also a material property. This property is called *Poisson's ratio*, defined as

$$v_P = \frac{|\text{lateral strain}|}{|\text{axial strain}|} = \frac{\varepsilon_L}{\varepsilon_a} \tag{11.4}$$

Engineering components are seldom subject to one-dimensional axial loading. The relationship between stress and strain must be generalized to a multidimensional case. Consider a two-dimensional geometry, as shown in Figure 11.3, subject to tensile loads in both the x and y directions, resulting in normal stresses σ_x and σ_y. In this case, for a biaxial state of stress the stresses and strains are

$$\varepsilon_y = \frac{\sigma_y}{E_m} - v_P \frac{\sigma_x}{E_m} \qquad \varepsilon_x = \frac{\sigma_x}{E_m} - v_P \frac{\sigma_y}{E_m}$$

$$\sigma_x = \frac{E_m(\varepsilon_x + v_P \varepsilon_y)}{1 - v_p^2} \qquad \sigma_y = \frac{E_m(\varepsilon_y + v_P \varepsilon_x)}{1 - v_p^2} \tag{11.5}$$

$$\tau_{xy} = G\gamma_{xy}$$

In this case, all of the stress and strain components lie in the same plane. The state of stress in the elastic condition for a material is similarly related to the strains in a complete three-dimensional situation [1, 2]. *Since stress and strain are related, it is possible to determine stress from measured strains under appropriate conditions.* However, strain measurements are made at the surface of an engineering component. The measurement

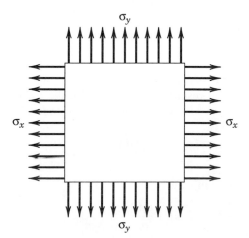

Figure 11.3 Biaxial state of stress.

yields information about the state of stress on the surface of the part. The analysis of measured strains requires application of the relationship between stress and strain at a surface. Such analysis of strain data is described in [3]. Our emphasis in this chapter will be on techniques for the measurement of strain.

11.3 RESISTANCE STRAIN GAUGES

The measurement of the small displacements that occur in a material or object that is subject to a mechanical load determines the strain. Strain can be measured by methods as simple as observing the change in the distance between two scribe marks on the surface of a load carrying member, or as advanced as optical holography. In any case, the ideal sensor for the measurement of strain would (1) have good spatial resolution, implying that the sensor would measure strain at a point; (2) be unaffected by changes in ambient conditions; and (3) have a high-frequency response for dynamic (time-resolved) strain measurements. A sensor that closely meets these characteristics is the *bonded resistance strain gauge*.

The resistance of a strain gauge changes when it is deformed, and this is easily related to the local strain. Both metallic and semiconductor materials experience a change in electrical resistance when they are subjected to a strain. The amount that the resistance changes depends on how the gauge is deformed, the material it is made of, and the design of the gauge. Gauges can be made quite small for good resolution and with a low mass to provide a high frequency response. With some ingenuity, ambient effects can be minimized or eliminated.

In an 1856 publication in the *Proceedings of the Royal Society* in England, Lord Kelvin (William Thomson) [4] laid the foundations for understanding the changes in electrical resistance that metals undergo when subjected to loads, which eventually led to the strain gauge concept. Two individuals, Edward Simmons at the California Institute of Technology and Arthur Ruge at the Massachusetts Institute of Technology, began the modern development of strain measurement in the late 1930s. Their development of the bonded metallic wire strain gauge led to commercially available strain gauges. The resistance strain gauge also forms the basis for a variety of other transducers, such as load cells, pressure transducers, and torque meters.

Metallic Gauges

To understand how metallic strain gauges work, consider a conductor having a uniform cross-sectional area, A_c, and a length, L, made of a material having an electrical resistivity, ρ_e. For this electrical conductor, the resistance, R, is given by

$$R = \rho_e L / A_c \qquad (11.6)$$

If the conductor is subjected to a normal stress along the axis of the wire, the cross-sectional area and the length will change, resulting in a change in the total electrical resistance, R. The total change in R is due to several effects, as illustrated in the total differential:

$$dR = \frac{A_c(\rho_e dL + L d\rho_e) - \rho_e L dA_c}{A_c^2} \qquad (11.7)$$

which may be expressed in terms of Poisson's ratio as

$$\frac{dR}{R} = \frac{dL}{L}\left(1 + 2v_p\right) + \frac{d\rho_e}{\rho_e} \qquad (11.8)$$

Hence, the changes in resistance are caused by two basic effects: the change in geometry as the length and cross-sectional area change, and the change in the value of the resistivity, ρ_e. The dependence of resistivity on mechanical strain is called piezoresistance and may be expressed in terms of a *piezoresistance coefficient*, π_1 defined by

$$\pi_1 = \frac{1}{E_m}\frac{d\rho_e/\rho_e}{dL/L} \tag{11.9}$$

With this definition, the change in resistance may be expressed

$$dR/R = (dL/L)\left(1 + 2v_p + \pi_1 E_m\right) \tag{11.10}$$

EXAMPLE 11.1

Determine the total resistance of a copper wire having a diameter of 1 mm and a length of 5 cm. The resistivity of copper is $1.7 \times 10^{-8}\,\Omega\,\text{m}$.

KNOWN

$$D = 1\,\text{mm}$$
$$L = 5\,\text{cm}$$
$$\rho_e = 1.7 \times 10^{-8}\,\Omega\,\text{m}$$

FIND The total electrical resistance

SOLUTION The resistance may be calculated from equation (11.6) as

$$R = \rho_e L/A_c$$

where

$$A_c = \frac{\pi}{4}D^2 = \frac{\pi}{4}\left(1 \times 10^{-3}\right)^2 = 7.85 \times 10^{-7}\,\text{m}^2$$

The resistance is then

$$R = \frac{(1.7 \times 10^{-8}\,\Omega\,\text{m})(5 \times 10^{-2}\,\text{m})}{7.85 \times 10^{-7}\,\text{m}^2} = 1.08 \times 10^{-3}\,\Omega$$

COMMENT If the material were nickel instead of copper, for the same diameter and length of wire, what would be the resistance? The resistivity of nickel is $7.8 \times 10^{-8}\,\Omega\,\text{m}$, which results in a resistance of $5 \times 10^{-3}\,\Omega$.

EXAMPLE 11.2

A very common material for the construction of strain gauges is the alloy constantan (55% copper with 45% nickel), having a resistivity of $49 \times 10^{-8}\,\Omega\,\text{m}$. A typical strain gauge might have a resistance of 120 Ω. What length of constantan wire of diameter 0.025 mm would yield a resistance of 120 Ω?

KNOWN The resistivity of constantan is $49 \times 10^{-8}\,\Omega\,\text{m}$.

FIND The length of constantan wire needed to produce a total resistance of 120 Ω

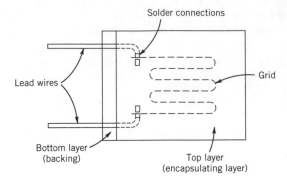

Figure 11.4 Detail of a basic strain gauge construction. (Courtesy of Micro-Measurements Division, Measurements Group, Inc. Raleigh, NC 27611.)

SOLUTION   From equation 11.6, we may solve for the length, which yields in this case

$$L = \frac{RA_c}{\rho_e} = \frac{(120\,\Omega)(4.91 \times 10^{-10}\,\text{m}^2)}{49 \times 10^{-8}\,\Omega\,\text{m}} = 0.12\,\text{m}$$

The wire would then be 12 cm in length to achieve a resistance of 120 Ω.

As shown by Example 11.2, a single straight conductor is normally not practical for strain measurement with meaningful resolution. Instead, a simple solution is to bend the wire conductor so that several lengths of wire are oriented along the axis of the strain gauge, as shown in Figure 11.4.

Figure 11.5 illustrates the construction of a typical metallic-foil bonded strain gauge. Such a strain gauge consists of a metallic foil pattern that is formed in a manner similar to the process used to produce printed circuits. This photo-etched metal foil pattern is mounted on a plastic backing material. The *gauge length*, as illustrated in Figure 11.5, is an important specification for a particular application. Because strain is usually measured at the location on a component where the stress is a maximum and the stress gradients are high, the fact that the strain gauge averages the measured strain over the gauge length becomes important. Because the maximum strain is the quantity of interest, errors can result from improper choice of a gauge length [5].

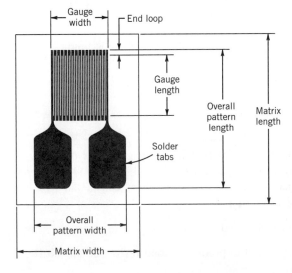

Figure 11.5 Construction of a typical metallic foil strain gauge. (Courtesy of Micro-Measurements Division, Measurements Group, Inc. Raleigh, NC 27611.)

Strain Gauge Construction and Bonding

The variety of conditions encountered in particular applications require special construction and mounting techniques, including design variations in the backing material, the grid configuration, bonding techniques, and total gauge electrical resistance. The adhesives used in the bonding process and the mounting techniques for a particular gauge and manufacturer will vary according to the specific application. However, there are fundamental aspects that are common to all bonded resistance gauges.

The strain gauge backing serves several important functions. It electrically isolates the metallic gauge from the test specimen and transmits the applied strain to the sensor. A bonded resistance strain gauge must be appropriately mounted to the specimen for which the strain is to be measured. The backing provides the surface used for bonding with an appropriate adhesive. Backing materials are available that are useful over temperatures that range from -270 to $290°C$.

The adhesive bond serves as a mechanical and thermal coupling between the metallic gauge and the test specimen. As such, the strength of the adhesive should be sufficient to accurately transmit the strain experienced by the test specimen, and should have thermal conduction and expansion characteristics suitable for the application. If the adhesive shrinks or expands during the curing process, apparent strain can be created in the gauge. A wide array of adhesives are available for bonding strain gauges to a test specimen. Among these are epoxies, cellulose nitrate cement, and ceramic-based cements.

If a strain gauge is placed in a state of stress where there are both normal and lateral (transverse) components of strain, the total change in resistance of the gauge will be a result of both the axial and lateral components of strain. Strictly speaking, the relationship for change in resistance expressed in equation (11.8) is true only for a single conductor in a state of uniaxial tension or compression. The goal of the design of electrical resistance strain gauges is to create a sensor that is sensitive only to strain along one axis, and therefore has zero sensitivity to lateral and shearing strains. Actual strain gauges are sensitive to lateral strains to some degree; for most gauges the sensitivity to shearing strains may be neglected.

Gauge Factor

The change in resistance of a strain gauge is normally expressed in terms of an empirically determined parameter called the *gauge factor*, GF. For a particular strain gauge, the gauge factor is supplied by the manufacturer. The gauge factor is defined as

$$GF \equiv \frac{\delta R/R}{\delta L/L} = \frac{\delta R/R}{\varepsilon_a} \tag{11.11}$$

Relating this definition to equation (11.10), we see that the gauge factor is dependent on the Poisson ratio for the gauge material and its piezoresistivity. For metallic strain gauges, the Poisson ratio is approximately 0.3 and the resulting gauge factor is ~ 2.

The gauge factor represents the total change in resistance for a strain gauge, under a calibration loading condition. The calibration loading condition generally creates a biaxial strain field, and the lateral sensitivity of the gauge influences the measured result. Strictly speaking then, the sensitivity to normal strain of the material used in the gauge and the gauge factor are not the same. Generally gauge factors are measured in a biaxial strain field that results from the deflection of a beam having a value of Poisson's ratio of 0.285. Thus, for any other strain field there is an error in strain indication due to the transverse sensitivity of the strain gauge. The following equation expresses the percentage error due to transverse sensitivity for a strain gauge mounted on any material, at any orientation in the strain field:

$$e_L = \frac{K_t\left(\varepsilon_L/\varepsilon_a + v_{p0}\right)}{1 - v_{p0}K_t} \times 100 \tag{11.12}$$

where

$\varepsilon_a, \varepsilon_L = $ axial and lateral strains, respectively (with respect the axis of the gauge)

$v_{p0} = $ Poisson's ratio of the material on which the manufacturer measured GF (usually 0.285 for steel)

$e_L = $ error as a percentage of axial strain (with respect to the axis of the gauge)

$K_t = $ lateral (transverse) sensitivity of the strain gauge

Typical values of the transverse sensitivity for commercial strain gauges range from 0.05 to −0.19. Figure 11.6 shows a plot of the percentage error for a strain gauge as a function of the ratio of lateral loading to axial loading and the lateral sensitivity. It is possible to correct for the lateral sensitivity effects [6].

Semiconductor Strain Gauges

When subjected to a load, a semiconductor material exhibits a change in resistance, and therefore can be used for the measurement of strain. Silicon crystals are the basic material

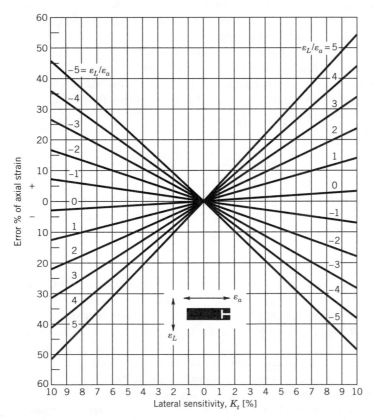

Figure 11.6 Strain measurement error due to strain gauge (transverse) sensitivity. (Courtesy of Measurements Group, Inc. Raleigh, NC 27611.)

for semiconductor strain gauges; the crystals are sliced into very thin sections to form strain gauges. Mounting such gauges in a transducer, such as a pressure transducer, or on a test specimen requires backing and adhesive techniques similar to those used for metallic gauges. Because of the large piezoresistance coefficient, the semiconductor gauge exhibits a very large gauge factor, as large as 200 for some gauges. These gauges also exhibit higher resistance, longer fatigue life, and lower hysteresis under some conditions than metallic gauges. However, the output of the semiconductor strain gauge is nonlinear with strain, and the strain sensitivity or gauge factor may be markedly dependent on temperature.

For a semiconductor, the effect of applied stress is to change the number and the mobility of the charge carriers within the material, thus causing large changes in the resistivity. Resistivity is a direct measure of the charge carrier density. The semiconductor crystals from which strain gauges are constructed contain a certain amount of impurities. These impurities control the number and mobility of the charge carriers within the gauge, and thus allow control of the gauge characteristics during the manufacturing process. Semiconductor materials for strain gauge applications have resistivities ranging from 0.001 to 1.0 Ω cm. Semiconductor strain gauges may have a relatively high or low density of charge carriers [3, 7]. Those made of materials having a relatively high density of charge carriers ($\sim 10^{20}$ carriers/cm^3) exhibit little variation of their gauge factor with strain or temperature. On the other hand, for the case where the crystal contains a low number of charge carriers ($< 10^{17}$ carriers/cm^3), the gauge factor may be approximated as

$$GF = \frac{T_0}{T} GF_0 + C_1 \left(\frac{T_0}{T}\right)^2 \varepsilon \tag{11.13}$$

where GF_0 is the gauge factor at the reference temperature T_0, under conditions of zero strain [8], and C_1 is a constant for a particular gauge. The behavior of a high-resistivity P-type semiconductor is shown in Figure 11.7.

Because of the capability for producing small gauge lengths, silicon semiconductor strain gauge technology provides for the production of very small transducers. For

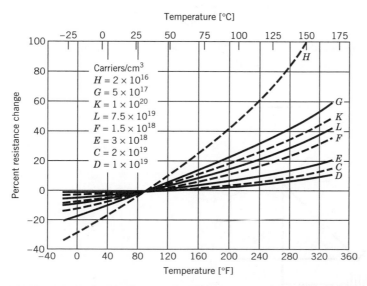

Figure 11.7 Temperature effect on resistance for various impurity concentrations for P-type semiconductors (reference resistance at 81°F). (Courtesy of Kulite Semiconductor Products, Inc.)

example, flush-mount pressure transducers having diameters of less than 8 mm provide pressure measurements up to 15,000 psi, with excellent frequency response characteristics. However, silicone diaphragm pressure transducers require special procedures for measuring in liquid environments. Semiconductor strain gauges are somewhat limited in the maximum strain that they can measure, approximately 5000 $\mu\epsilon$ for tension, but larger in compression. Because of the possibility of an inherent sensitivity to temperature, careful consideration must be made for each application to provide appropriate temperature compensation or correction. Temperature effects can result, for a particular measurement, in zero drift during the duration of a measurement.

11.4   STRAIN GAUGE ELECTRICAL CIRCUITS

A Wheatstone bridge is generally used to detect the small changes in resistance that are the output of a strain gauge measurement circuit. A typical strain gauge measuring installation on a steel specimen will have a sensitivity of 10^{-6} $\Omega/(kN\ m^2)$. As such, a high-sensitivity device such as a Wheatstone bridge is desirable for measuring resistance changes for strain gauges. The fundamental relationships for the analysis of such bridge circuits are discussed in Chapter 6. Equipment is commercially available that can measure changes in gauge resistance of less than 0.0005 Ω (0.000001 $\mu\epsilon$).

A simple strain gauge Wheatstone bridge circuit is shown in Figure 11.8. The bridge output under these conditions is given by equation (6.15):

$$E_0 + \delta E_0 = E_i \frac{(R_1 + \delta R)R_4 - R_3 R_2}{(R_1 + \delta R + R_2)(R_3 + R_4)} \tag{6.15}$$

where E_0 is the bridge output at initial conditions, δE_0 is the bridge deflection, and δR is the change in the strain gauge resistance. Consider the case where all the fixed resistors and the strain gauge resistance are initially equal, and the bridge is balanced such that $E_0 = 0$. If the strain gauge is then subjected to a state of strain, the change in the output voltage, δE_0, from equation (6.15) reduces to

$$\frac{\delta E_o}{E_i} = \frac{\delta R/R}{4 + 2(\delta R/R)} \approx \frac{\delta R/R}{4} \tag{11.14}$$

under the assumption that $\delta R/R \ll 1$. The simplest form of equation (11.14) is suitable for all but those measurements that demand the highest accuracy, and remains valid for

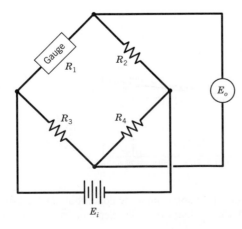

Figure 11.8 Basic strain gauge Wheatstone bridge circuit.

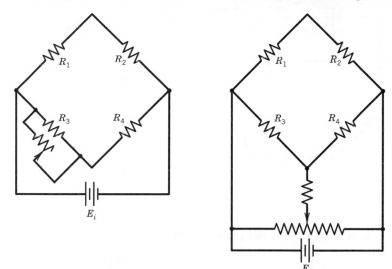

Figure 11.9 Balancing schemes for bridge circuits.

values of $\delta R/R \ll 1$. Using the relationship from equation (11.11) that $\delta R/R = GF\varepsilon$,

$$\frac{\delta E_o}{E_i} = \frac{GF\varepsilon}{4 + 2GF\varepsilon} \approx \frac{GF\varepsilon}{4} \qquad (11.15)$$

Equations (11.14) and (11.15) yield two practical equations for strain gauge measurements using a single active gauge in a Wheatstone bridge.

The Wheatstone bridge has several distinct advantages for use with electrical resistance strain gauges. The bridge may be balanced by changing the resistance of one arm of the bridge. Therefore, once the gauge is mounted in place on the test specimen under a condition of zero loading, the output from the bridge may be zeroed. Two schemes for circuits to accomplish this balancing are shown in Figure 11.9. Shunt balancing provides the best arrangement for strain gauge applications, since the changes in resistance for a strain gauge are small. Also, the strategic placement of multiple gauges in a Wheatstone bridge can both increase the bridge output and cancel out certain ambient effects and unwanted components of strain, as discussed in the next two sections.

EXAMPLE 11.3

A strain gauge, having a gauge factor of 2, is mounted on a rectangular steel bar $(E_m = 200 \times 10^6 \text{ kN/m}^2)$, as shown in Figure 11.10. The bar is 3 cm wide and 1 cm high, and is subjected to a tensile force of 30 kN. Determine the resistance change of the strain gauge, if the resistance of the gauge is 120 Ω in the absence of the axial load.

KNOWN

$$GF = 2 \qquad E_m = 200 \times 10^6 \text{ kN/m}^2 \qquad F_N = 30 \text{ kN}$$
$$R = 120\,\Omega \qquad A_c = 0.03\,\text{m} \times 0.01\,\text{m}$$

FIND   The resistance change of the strain gauge for a tensile force of 30 kN

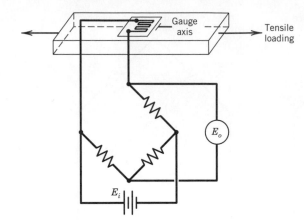

Figure 11.10 Strain gauge circuit subject to uniaxial tension.

SOLUTION The stress in the bar under this loading condition is

$$\sigma_a = \frac{F_N}{A_c} = \frac{30\,\text{kN}}{(0.03\,\text{m})(0.01\,\text{m})} = 1 \times 10^5 \,\text{kN/m}^2$$

and the resulting strain is

$$\varepsilon_a = \frac{\sigma_a}{E_m} = \frac{1 \times 10^5 \,\text{kN/m}^2}{200 \times 10^6 \,\text{kN/m}^2} = 5 \times 10^{-4} \,\text{m/m} \tag{11.16}$$

For strain along the axis of the strain gauge, the change in resistance from equation (11.11) is

$$\delta R/R = \varepsilon GF$$

or

$$\delta R = R\varepsilon GF = (120\,\Omega)(5 \times 10^{-4})(2) = 0.12\,\Omega$$

EXAMPLE 11.4

Suppose the strain gauge described in Example 11.3 is to be connected to a measurement device capable of determining a change in resistance to an uncertainty of $\pm 0.005\ \Omega$ (95%). This uncertainty includes a resolution of $0.001\ \Omega$. What uncertainty in stress would result when using this resistance measurement device?

KNOWN A stress is to be inferred from a strain measurement using a strain gauge having a gauge factor of 2 and a zero load resistance of 120 Ω. The measurement of resistance has an uncertainty of $\pm 0.005\ \Omega$ (95%).

FIND The design-stage uncertainty in stress

SOLUTION The design-stage uncertainty in stress, $(u_d)_\sigma$, is given by

$$(u_d)_\sigma = \frac{\partial \sigma}{\partial(\delta R)} (u_d)_{\delta R}$$

with

$$\sigma = \varepsilon E_m = \frac{\delta R/R}{GF} E_m$$

Then with

$$\frac{\partial \sigma}{\partial (\delta R)} = \frac{E_m}{R(GF)}$$

we can express the uncertainty as

$$(u_d)_\sigma = \frac{E_m}{R(GF)} (u_d)_{\delta R} = \frac{200 \times 10^6 \, \text{kN/m}^2}{120 \, \Omega (2)} (0.005 \, \Omega)$$

This results in a design-stage uncertainty in stress of $(u_d)_\sigma = 4.17 \times 10^3 \, \text{kN/m}^2$ (95%) or $\sim 2.4\%$ of the expected stress.

11.5   PRACTICAL CONSIDERATIONS FOR STRAIN MEASUREMENT

This section describes some characteristics of strain gauge applications that allow practical implementation of strain measurement.

The Multiple Gauge Bridge

The output from a bridge circuit can be increased by the appropriate use of more than one active strain gauge. This increase can be related to a *bridge constant* as illustrated in the following discussion. In addition, multiple gauges can be used to compensate for unwanted effects, such as temperature or specific strain components. Consider the case when all four resistances in the bridge circuit of Figure 11.8 represent active strain gauges. In general, the bridge output is given by

$$E_0 = E_i \left[\frac{R_1}{R_1 + R_2} - \frac{R_3}{R_3 + R_4} \right] \tag{11.17}$$

The strain gauges R_1, R_2, R_3, and R_4 are assumed initially to be in a state of zero strain. If these gauges are now subjected to strains such that the resistances change by dR_i, where $i = 1, 2, 3$, and 4, then the change in the bridge output voltage can be expressed as

$$dE_0 = \sum_{i=1}^{4} \frac{\partial E_0}{\partial R_i} dR_i \tag{11.18}$$

Evaluating the appropriate partial derivatives from equation (11.17) yields

$$dE_0 = E_i \left[\frac{R_2 dR_1 - R_1 dR_2}{(R_1 + R_2)^2} + \frac{R_3 dR_4 - R_4 dR_3}{(R_3 + R_4)^2} \right] \tag{11.19}$$

Then from equations (11.2) and (11.11), $dR_i = R_i \varepsilon_i GF_i$, and the value of dE_0 can be determined. Assuming that $dR_i \ll R_i$, the resulting change in the output voltage, δE_0, may now be expressed as

$$\delta E_0 = E_i \left[\frac{R_1 R_2}{(R_1 + R_2)^2} (\varepsilon_1 GF_1 - \varepsilon_2 GF_2) + \frac{R_3 R_4}{(R_3 + R_4)^2} (\varepsilon_4 GF_4 - \varepsilon_3 GF_3) \right] \tag{11.20}$$

If $R_1 = R_2 = R_3 = R_4$, then

$$\frac{\delta E_0}{E_i} = \frac{1}{4} (\varepsilon_1 GF_1 - \varepsilon_2 GF_2 + \varepsilon_4 GF_4 - \varepsilon_3 GF_3) \tag{11.21}$$

It is possible and desirable to purchase matched sets of strain gauges for a particular application, so that $GF_1 = GF_2 = GF_3 = GF_4$, and

$$\frac{\delta E_0}{E_i} = \frac{GF}{4}(\varepsilon_1 - \varepsilon_2 + \varepsilon_4 - \varepsilon_3) \qquad (11.22)$$

Equation (11.22) is important and forms the basic working equation for a strain gauge bridge circuit using multiple gauges (compare this equation with 11.15).

Equation (11.22) shows that for a bridge containing one or more active strain gauges, equal strains on opposite bridge arms sum, whereas equal strains on adjacent arms of the bridge cancel. These characteristics can be used to increase the output of the bridge, to provide temperature compensation, or to cancel unwanted components of strain. Practical means of achieving these desirable characteristics will be explored further, after the concept of the bridge constant is developed.

Bridge Constant

Commonly used strain gauge bridge arrangements may be characterized by a *bridge constant*, κ, defined as the ratio of the actual bridge output to the output of a single gauge sensing the maximum strain, ε_{max} (assuming the remaining bridge resistances remain fixed). The output for a single gauge experiencing the maximum strain may be expressed

$$\frac{\delta R}{R} = \varepsilon_{max} GF \qquad (11.23)$$

So that, again for a single gauge,

$$\frac{\delta E_0}{E_i} \cong \frac{\varepsilon_{max} GF}{4} \qquad (11.24)$$

The bridge constant, κ, is found from the ratio of the actual bridge output given by equation (11.22) to the output for a single gauge given by equation (11.24). When more than one gauge is used in the bridge circuit, equation (11.15) becomes

$$\frac{\delta E_0}{E_i} = \frac{\kappa \delta R/R}{4 + 2\delta R/R} = \frac{\kappa GF\varepsilon}{4 + 2GF\varepsilon} \approx \frac{\kappa GF\varepsilon}{4} \qquad (11.25)$$

The simplest form of equation (11.25) is suitable for all but those measurements demanding the greatest accuracy possible, and remains valid for values of $\delta R/R \ll 1$. The bridge constant concept is illustrated in Example 11.5.

EXAMPLE 11.5

Determine the bridge constant for two strain gauges mounted on a structural member, as shown in Figure 11.11. The member is subject to uniaxial tension, which produces an axial strain ε_a and a lateral strain $\varepsilon_L = -\upsilon_p \varepsilon_a$. Assume that all the resistances in Figure 11.11 are initially equal, and therefore the bridge is initially balanced. Let $(GF)_1 = (GF)_2$.

KNOWN   Strain gauge installation shown in Figure 11.11

FIND   The bridge constant for this installation

ASSUMPTION   The change in the strain gauge resistances are small compared to the initial resistance (see explanation as follows).

SOLUTION   If the gauges are mounted so that gauge 1 is aligned with the axial tension and gauge 2 is mounted transversely on the member, the output of the bridge will be greater than for

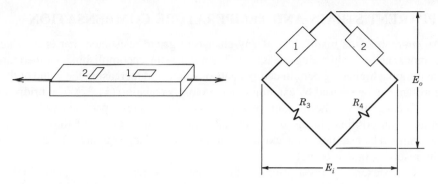

Figure 11.11 Bridge circuit with two active arms; strain gauge installation for increased sensitivity.

gauge 1 alone. Since strain gauges generally experience small resistance changes, it is often convenient to develop approximate relationships for specific bridge circuit arrangements and strain gauge installations under this assumption.

The changes in resistance for the gauges may be expressed as

$$\frac{\delta R_1}{R_1} = \varepsilon_a (GF)_1 \tag{11.26}$$

and

$$\frac{\delta R_2}{R_2} = -v_p \varepsilon_a (GF)_2 = -v_p \frac{\delta R_1}{R_1} \tag{11.27}$$

With only one gauge active, equation (11.14) is applicable. If $\delta R_1 / R_1 \ll 1$

$$\frac{\delta E_0}{E_i} = \frac{\delta R_1 / R_1}{4 + 2(\delta R_1 / R_1)} \approx \frac{\delta R_1 / R_1}{4} \tag{11.14}$$

But with both gauges installed and active, as shown in Figure 11.11, the output of the bridge is determined from an analysis of the bridge response, which results in

$$\frac{\delta E_0}{E_i} = \frac{(\delta R_1 / R_1)(1 + v_p)}{4 + 2(\delta R_1 / R_1)(1 + v_p)}$$

In practical applications, it is most often the case that changes in resistance are small in comparison to the resistance values; thus,

$$\frac{\delta E_0}{E_i} = \frac{(\delta R_1 / R_1)(1 + v_p)}{4} \tag{11.28}$$

Therefore the bridge constant is the ratio of equations (11.28) to (11.14)

$$\frac{(\delta R_1 / R_1)(1 + v_p)/4}{(\delta R_1 / R_1)/4} \tag{11.29}$$

and the bridge constant is

$$\kappa = 1 + v_p$$

Comparing equation (11.28) to (11.14) shows that the use of two gauges oriented as described has increased the output of the bridge by a factor of $1 + v_p$ over that of using a single gauge.

11.6 APPARENT STRAIN AND TEMPERATURE COMPENSATION

Apparent strain is manifested as any change in gauge resistance that is not due to the component of strain being measured. Techniques for accomplishing temperature compensation, eliminating certain components of strain, and increasing the value of the bridge constant can be devised by examining more closely equation (11.22). The bridge constant is influenced by (1) the location of strain gauges on the test specimen and (2) the gauge connection positions in the bridge circuit. The combined effect of these two factors is determined by examining the existing strain field and using equation (11.22) to determine the resulting bridge output.

Let us examine how a component of strain can be removed (compensation) from the measured signal. Consider a beam having a rectangular cross section and subject to the loading condition shown in Figure 11.12, where the beam is subject to an axial load F_N and a bending moment, M. The stress distribution in this cross section is given by

$$\sigma_x = \frac{-12My}{bh^3} + \frac{F_N}{bh} \tag{11.30}$$

To remove the effects of bending strain, identical strain gauges are mounted to the top and bottom of the beam as shown in Figure 11.12, and they are connected to bridge locations 1 and 4 (or any two opposite bridge arms). The gauges experience equal but opposite bending strains [see equation (11.30)], and both strain gauges are subject to the same axial strain caused by F_N. The bridge output under these conditions is

$$\frac{\delta E_0}{E_i} = \frac{GF}{4}(\varepsilon_1 + \varepsilon_4) \tag{11.31}$$

where $\varepsilon_1 = \varepsilon_{a1} + \varepsilon_{b1}$ and $\varepsilon_4 = \varepsilon_{a4} - \varepsilon_{b4}$, with subscripts a and b referring to axial and bending strain, respectively. Hence, the bending strains cancel but the axial strains will sum, giving

$$\frac{\delta E_0}{E_i} = \frac{GF}{2}\varepsilon_a \tag{11.32}$$

For a single gauge experiencing the maximum strain,

$$\frac{\delta E_0}{E_i} = \frac{GF}{4}\varepsilon_a \tag{11.33}$$

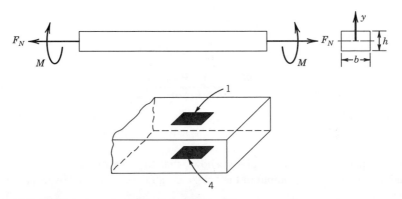

Figure 11.12 Bridge arrangement for bending compensation.

Table 11.1   Common Gauge Mountings

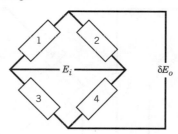

	Arrangement	Compensation Provided	Bridge Constant κ
	Single gauge in uniaxial stress	None	$\kappa = 1$
	Two gauges sensing equal and opposite strain—typical bending arrangement	Temperature	$\kappa = 2$
	Two gauges in uniaxial stress	Bending only	$\kappa = 2$
	Four gauges with pairs sensing equal and opposite strains	Temperature and bending	$\kappa = 4$
	One axial gauge and one poisson gauge		$\kappa = 1 + \nu$
	Four gauges with pairs sensing equal and opposite strains— sensitive to torsion only. Typical shaft arrangement.	Temperature and axial	$\kappa = 4$

The ratio of the output of represented in (11.32) to that represented in (11.33) has a value of 2, which is the bridge constant ($\kappa = 2$) for the strain gauge arrangement in Figure 11.12. So this arrangement compensates for the bending strain and doubles the output value.

A guide for some practical bridge-gauge configurations is provided in Table 11.1.

Temperature Compensation

Differential thermal expansion between the gauge and the specimen on which it is mounted creates an apparent strain in the strain gauge. Thus, temperature sensitivity of strain gauges is a result of both the changes in resistance caused by temperature changes in the gauge itself, and the strain experienced by the gauge as a result of differential thermal expansion between the gauge and the material on which it is mounted. Using gauges of identical alloy composition as the specimen would minimize this latter effect.

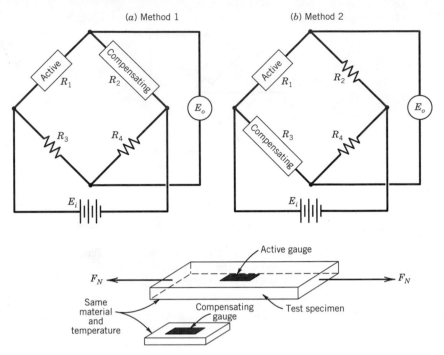

Figure 11.13 Strain gauge installation for temperature compensation.

However, even keeping the specimen at a constant temperature may not be enough to eliminate the effect of gauge thermal expansion. Heating of the strain gauge as a result of current flow from the measuring device may be a source of significant error because the gauge is also a temperature-sensitive element. The temperature sensitivity of a strain gauge is an obstacle to accurate mechanical strain measurement that must be considered. Fortunately, there are effective ways to deal with it.

Figure 11.13 shows two circuit arrangements that provide temperature compensation for a strain measurement. The strain gauge mounted on the test specimen experiences changes in resistance caused by temperature changes and by applied strain, whereas the compensating gauge experiences resistance changes caused only by temperature changes. As long as the compensating gauge (see Figure 11.13) experiences an identical thermal environment as the measuring gauge, temperature effects will be eliminated from the circuit. To show this, consider the case in which all the bridge resistances are initially equal, and the bridge is balanced. If the temperature of the strain gauges now changes, their resistance will change as a result of thermal expansion, creating an apparent thermal strain. Under an applied axial load, the output of the bridge is derived from equation (11.22),

$$\delta E_0 = E_i \frac{GF(\varepsilon_1 - \varepsilon_2)}{4} = E_i \frac{GF\varepsilon_a}{4} \tag{11.34}$$

where $\varepsilon_1 = \varepsilon_a + \varepsilon_T$ and $\varepsilon_2 = \varepsilon_T$, where ε_T refers to the apparent strain. Under this condition, the output value will not be affected by changes in temperature. The result will be the same if the compensating gauge is mounted in arm 3 instead, as shown in Figure 11.13(b). Further, any two active gauges mounted on adjacent bridge arms will compensate for temperature changes.

Because this relationship holds in general, we can state that as long as two gauges, which are mounted to a specimen or to similar specimens, remain at the same temperature and are connected to adjacent arms of a Wheatstone bridge, they will provide temperature compensation for each other.

Looking back, the arrangement in Figure 11.12 will not provide temperature compensation, as gauges 1 and 4 are on opposite bridge arms. However, temperature compensation could be provided for this installation by having two additional strain gauges that are at the same temperature as gauges 1 and 4, and occupy arms 2 and 3 of the bridge.

Bridge Static Sensitivity

The static sensitivity of the bridge arrangement in Figure 11.13(a) (method 1) is

$$K_B = \frac{E_0}{\varepsilon} = E_i \frac{R_1 R_2}{(R_1 + R_2)^2} GF \tag{11.35}$$

and with $E_i = (2R_g)I_g$ and $R_g = R_1 = R_2$, the static sensitivity may be expressed in the terms of the current flowing through the gauge, (I_g), as

$$K_B = \frac{1}{2} GF \sqrt{(I_g^2 R_1)R_1} \tag{11.36}$$

Note that $(I_g)^2 R_g$ is the power dissipated in the strain gauge as a result of the bridge current. Excessive power dissipation in the gauge would cause temperature changes and introduce uncertainty into a strain measurement. These effects can be minimized by good thermal coupling between the strain gauge and the object to which it is bonded, to allow effective dissipation of thermal energy.

Consider the static sensitivity of the bridge arrangement in Figure 11.13(b). With identical gauges at positions R_1 and R_3, and equal resistance changes for the two gauges, no change in bridge output would occur. However, the static sensitivity for this arrangement is not the same as for method 1, but is given by

$$K_B = \frac{R_1/R_2}{1 + R_1/R_2} GF \sqrt{(I_g^2 R_1)R_1} \tag{11.37}$$

Here the sensitivity is the same as for a bridge having a single active gauge and without temperature compensation. However, the sensitivity depends on the choice of the fixed resistor R_2. If $R_1 = R_2$, the resulting sensitivity will be the same as for method 1. However, resistor R_2 can be chosen to provide the desired static sensitivity for the circuit, within the limitations of measurement capability and allowable bridge current.

Construction and Installation

Figure 11.14 shows a variety of strain gauge configurations. Standard gauge resistances are 120 and 350 Ω. Because of the accuracy of the process that produces the metal foil, a variety of patterns for the metal film are possible.

The operating assumption that leads to the definition of the gauge factor is that the change in resistance of the gauge is linear with applied strain, for a particular gauge. However, a strain gauge installed in a measurement environment will exhibit some nonlinearity. Also, in cycling between a loaded and unloaded condition, there will be some degree of hysteresis and a shift in the resistance for a state of zero strain. A typical

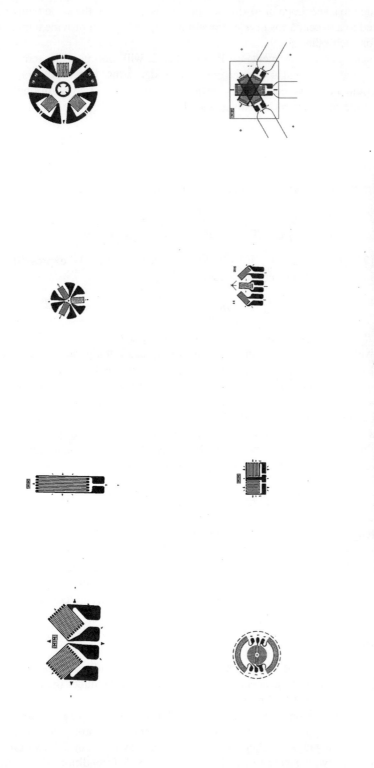

Figure 11.14 Strain gauge configurations. (Courtesy of Micro-Measurements Division, Measurements Group, Inc. Raleigh, NC 27611.)

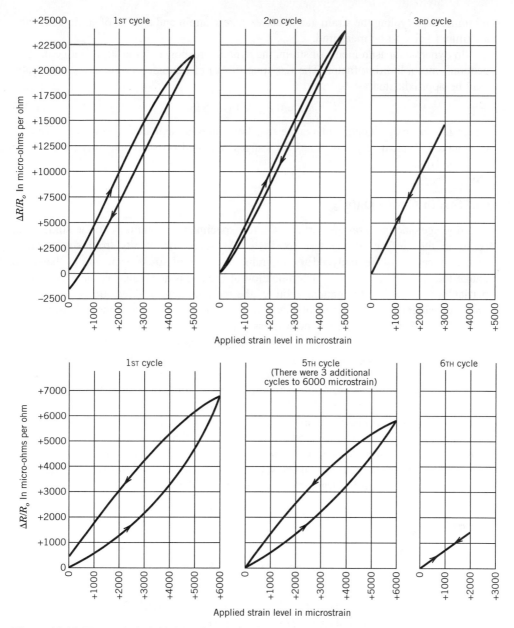

Figure 11.15 Hysteresis in initial loading cycles for two strain gauge materials. (Courtesy of Micro-Measurements Division, Measurements Group, Inc. Raleigh, NC 27611.)

cycle of loading and unloading is shown in Figure 11.15. The strain gauge will typically indicate lower values of strain during unloading than are measured as the load is increased. The extent of these behaviors is determined not only by the strain gauge characteristics, but also by the characteristics of the adhesive, and by the previous strains that the gauge has experienced. For properly installed gauges, the deviation from linearity should be on the order of 0.1% [3]. On the other hand, first-cycle hysteresis and zero shift are difficult to predict. The effects of first-cycle hysteresis and zero shift can be minimized by cycling the strain gauge between zero strain and a value of strain above the maximum value to be measured.

In dynamic measurements of strain, the dynamic response of the strain gauge itself is generally not a limiting factor. The rise time (90%) of a bonded resistance strain gauge may be approximated as [9]

$$t_{90} \approx 0.8(L/a) + 0.5 \,\mu\text{s} \tag{11.38}$$

where L is the gauge length and a is the speed of sound in the material on which the gauge is mounted. Typical response times for gauges mounted on steel specimens are on the order of 1 μs.

Analysis of Strain Gauge Data

Strain gauges mounted on the surface of a test specimen respond only to the strains that occur at the surface of the test specimen. As such, the results from strain gauge measurements must be analyzed to determine the state of stress occurring at the strain gauge locations. The complete determination of the stress at a point on the surface of a particular test specimen in general requires the measurement of three strains at the point under consideration. The result of these measurements yields the principal strains and allow determination of the maximum stress [3].

If more information is available concerning the expected state of stress, less than three strain gauges may be employed. When the directions of the principal axes are known in advance, only two strain measurements are necessary to calculate the maximum stress at the measured point. The multiple-element strain gauges used to measure more than one strain at a point are called *strain gauge rosettes*. An example of a two-element rosette is shown in Figure 11.16. For general measurements of strain and stress, commercially available strain rosettes can be chosen that have a pattern of multiple-direction gauges that is compatible with the specific nature of the particular application [5]. In practice, the applicability of the single-axis strain gauge is extremely limited, and improper use can result in large errors in the measured stress.

Signal Conditioning

The most common form of signal conditioning in strain gauge bridge circuits is to amplify the signal by using a low noise amplifier. For an amplifier of gain G_A, equation (11.25) becomes

$$\frac{\delta E_0}{E_i} = \frac{G_A \kappa (\delta R/R)}{4 + 2(\delta R/R)} = \frac{G_A \kappa GF \varepsilon}{4 + 2(\delta R/R)} \approx \frac{G_A \kappa GF \varepsilon}{4} \tag{11.39}$$

The simplest form of equation (11.39) is suitable for all but those measurements that demand the highest accuracy, and remains valid for values of $\delta R/R \ll 1$.

A common means of recording strain gauge bridge circuit signals is the automated data-acquisition system. A schematic diagram of such a setup is shown in Figure 11.17. Also, a typical interface card was presented in Figure 7.28.

Uncertainties in Multichannel Measurements

Uncertainties caused by an automated multichannel strain measurement system can be estimated in a system calibration. Example 11.6 describes an initial test of such a system, often called a shakedown test, and discusses the information that results from this test.

(a)   Single-plane type.

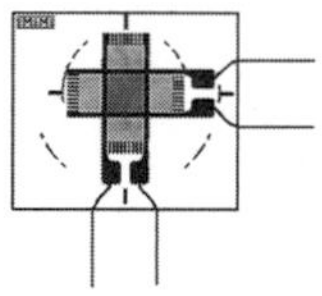

Figure 11.16 Biaxial strain gauge rosettes. (Courtesy of Micro-Measurements Division, Measurements Group, Inc. Raleigh, NC 27611.) (a) Single-plane type. (b) Stacked type.

(b)   Stacked type.

EXAMPLE 11.6

The strain developed at M different locations about a large test specimen is to be measured at each location using a setup similar to that shown in Figure 11.17. At each location similar strain gauges are appropriately mounted and connected to a Wheatstone bridge that is powered by an external supply. The bridge deflection voltage is to be amplified and measured. A similar setup is used at all M locations. The output from each amplifier is input through an M-channel multiplexer to an automated data-acquisition system. If N (say 30) readings for each of the M setups are taken while the test specimen is maintained in a condition of uniform zero strain, what information is obtained?

KNOWN   Strain setup of Figure 11.17:

$$M(j = 1, 2, \ldots, M) \text{ setups}$$
$$N(i = 1, 2, \ldots, N) \text{ readings per setup}$$

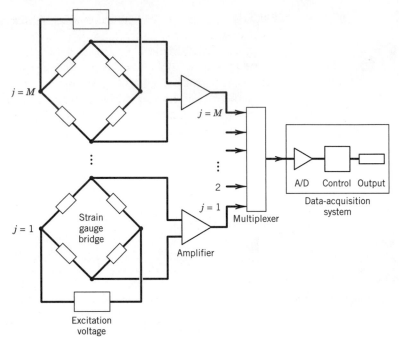

Figure 11.17 Data acquisition and reduction system for Example 11.6.

ASSUMPTIONS   All setups are to be operated in a similar manner. Gauges are operated in their linear regimes.

SOLUTION   Consider the data that will become available. First, each setup is exposed to a similar strain. Hence, each setup should indicate the same strain. We can calculate the pooled mean value of the $M \times N$ readings to obtain

$$\langle \bar{\varepsilon} \rangle = \frac{1}{MN} \sum_{j=1}^{M} \sum_{i=1}^{N} \varepsilon_{ij} = \frac{1}{M} \sum_{j=1}^{M} \bar{\varepsilon}_j$$

The difference between the pooled mean strain and the applied strain (zero here) is an estimate of the systemation uncertainty that can be expected from any channel during data acquisition.

Random errors will manifest themselves through scatter in the data set. The pooled standard deviation of the mean

$$S_{\bar{\varepsilon}} = \frac{\langle S_\varepsilon \rangle}{\sqrt{MN}} \qquad \text{where} \qquad \langle S_\varepsilon \rangle = \sqrt{\frac{\sum_{j=1}^{M} \sum_{i=1}^{N} \left(\varepsilon_{ij} - \bar{\varepsilon}_j \right)^2}{M(N-1)}}$$

will provide a representative estimate of the random uncertainty to be expected from any channel that is due to the data acquisition and reduction instrumentation.

COMMENT   This test yields an estimate of the systematic and random uncertainties due to propagation of the elemental errors amid M setups due to:

- Excitation voltage errors (differences in settings; variations)
- Amplifier error (differences in gain; noise)
- A/D converter, multiplexer, and conversion errors
- Computer errors (noise and roundoff)

- Bridge null errors
- Apparent strain error during the test
- Variations in gauge factors and gauge heating

The test data do not include temporal variation of the measurands or procedural variations under loading, instrument calibration errors, temperature variation effects and electrical noise induced by operation of the loading test of the specimen, dynamic effects on the gauges, including differences in creep and fatigue, or reduction curve fit errors.

11.7   OPTICAL STRAIN MEASURING TECHNIQUES

Optical methods for experimental stress analysis can provide fundamental information concerning directions and magnitudes of the stresses in parts under design loading conditions. Optical techniques have been developed for the measurement of stress and strain fields, either in models made of materials having appropriate optical properties or through coating techniques for existing specimens. Photoelasticity takes advantage of the changes in optical properties of certain materials that occur when these materials are strained. For example, some plastics display a change in optical properties when strained that causes an incident beam of polarized light to be split into two polarized beams that travel with different speeds and that vibrate along the principal axes of stress. As the two light beams are out of phase, they can be made to interfere; measuring the resulting light intensity yields information concerning applied stress. To implement this method, a model is constructed of an appropriate material, or a coating is applied to an existing part.

A second optical method of stress analysis is based on the development of a moiré pattern, which is an optical effect resulting from the transmission or reflection of light from two overlaid grid patterns. The fringes that result from relative displacement of the two grid patterns can be used to measure strain; each fringe corresponds to the locus of points of equal displacement.

Recent developments in strain measurement include the use of lasers and holography to very accurately determine whole field displacements for complex geometries.

Basic Characteristics of Light

To utilize optical strain measurement techniques, we must first examine some basic characteristics of light. Electromagnetic radiation, such as light, may be thought of as a transverse wave with sinusoidally oscillating electric and magnetic field vectors that are at right angles to the direction of propagation. In general, a light source emits a series of waves containing vibrations in all perpendicular planes, as illustrated in Figure 11.18. A light wave is said to be plane-polarized if the transverse oscillations of the electric field are parallel to each other at all points along the direction of propagation of the wave.

Figure 11.18 illustrates the effect of a polarizing filter on an incident light wave; the transmitted light will be plane polarized, with a known direction of polarization. Complete extinction of the light beam could be achieved by introduction of a second polarizing filter, with the axis of polarization at 90° to the first filter (labeled an Analyzer in Figure 11.18). These behaviors of light are employed to measure direction and magnitude of strain in photoelastic materials.

Photoelastic Measurement

Photoelastic methods of stress analysis take advantage of the anisotropic optical characteristics of some materials, notably plastics, when subject to an applied load to

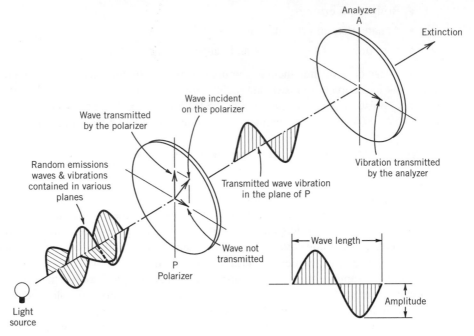

Figure 11.18 Polarization of light. (Courtesy of Measurements Group, Inc. Raleigh, NC 27611.)

determine the strain field. Stress analysis may be accomplished either by constructing a model of the part to be analyzed from a material selected for its optical properties or by coating the actual part or prototype with a photoelastic coating. If a model is constructed from a suitable plastic, the required loads for the model are significantly less than the service loads of the actual part, which reduces effort and expense in testing.

The changes in optical properties, known as artificial birefringence, which occur in certain materials subject to a load or loads was first observed by Sir David Brewster [10] in 1815. He observed that when light passes through glass that is subject to uniaxial tension, such that the stress is perpendicular to the direction of propagation of the light, the glass becomes doubly refracting, with the axes of polarization in the glass aligned with and perpendicular to the stress. Maxwell [11] and Neumann [12] first put forward the mathematical observation that the relationship between artificial birefringence and applied stress or strain is linear. These relations are known as the stress-optic law.

The anisotropy that occurs in photoelastic materials results in two refracted beams of light and one reflected beam, produced for a single incident beam of appropriately polarized light. The two refracted beams propagate at different velocities through the material because of an anisotropy in the index of refraction. In an appropriately designed photoelastic (two-dimensional) model, these two refracted components of the incident light travel in the same direction and can be examined in a polariscope. The degree to which the two light waves are out of phase is related to the stress by the stress-optic law:

$$\delta \propto n_x - n_y \qquad\qquad (11.40)$$

where

δ = relative retardation between the two light beams

n_x, n_y = indices of refraction in the directions of the principal strains

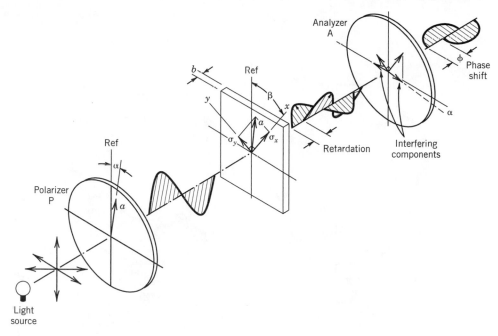

Figure 11.19 Construction of a plane polariscope. (Courtesy of Measurements Group, Inc. Raleigh, NC 27611.)

The index of refraction changes in direct proportion to the amount the material is strained, such that

$$n_x - n_y = K\left(\varepsilon_x - \varepsilon_y\right) \tag{11.41}$$

The strain optical coefficient, K, is generally assumed to be a material property that is independent of the wavelength of the incident light. However, if the photoelastic material is strained beyond the elastic limit, this constant may become wavelength dependent, a phenomenon known as *photoelastic dispersion*.

Figure 11.19 shows the use of a plane polariscope to examine the strain in a photoelastic model. Plane polarized light enters the specimen and emerges with two planes of polarization along the principal strain axes. This light beam is then passed through a polarizing filter, called the analyzer, which transmits only the component of each of the light waves that is parallel to the plane of polarization. The transmitted waves will interfere, since they are out of phase, and the resulting light intensity will be a function of the angle between the analyzer and the principal strain direction and the phase shift between the beams. The variations in strain in the specimen produce a pattern of fringes, which can be related to the strain field through the strain-optic relation.

When a photoelastic model is observed in a plane polariscope, a series of fringes is observed. The complete extinction of light occurs at locations where the principal strain directions coincide with the axes of the analyzer or where either the strain is zero or $\varepsilon_x - \varepsilon_y = 0$. These fringes are termed *isoclinics*, and are used to determine the principal strain directions at all points in the photoelastic model. Figure 11.20 shows the isoclinics in a ring subject to a compression load (as shown in the figure). A reference direction is selected along the horizontal compression load and labeled 0°. For each measurement angle, one of the principal strains at a point on an isoclinic is parallel to the specified angle and the other is perpendicular. For the 0° isoclinics, the principal strains are oriented at 0 and 90°.

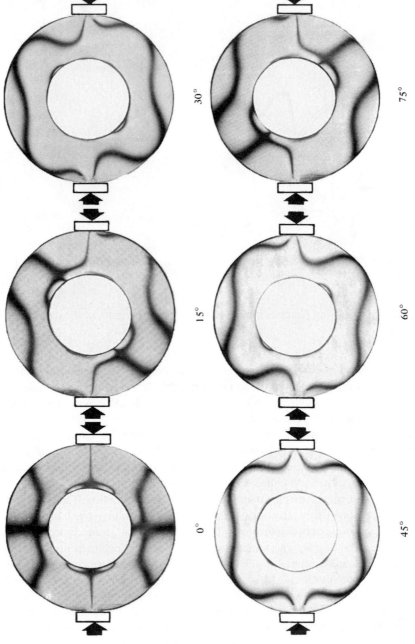

30° 75° 15° 60° 0° 45°

Figure 11.20 Isoclinic fringes in a ring loaded in compression. (Courtesy of Measurements Group, Inc., Raleigh, NC.)

Using the fact that the direction of the principal axes is known at a free surface and the fact that the shear stress is zero on a free surface, the magnitude of the stress on the boundary can be determined. The primary applications for photoelasticity, especially in a historical sense, have been in the study of stress concentrations around holes or reentrant corners. In these cases, the maximum stress is at the boundary and corresponds to one of the principal stresses. This maximum stress can be obtained directly by the optical method, since the shear stress is zero on the boundary.

Optical methods provide information about the strain and stress at every point in the object being examined, in contrast to a strain gauge that supplies information about the strain at a single location on the object. The optical methods provide the possibility of identifying stress concentration locations and may allow for an improvement in design or guide detailed measurements with strain gauges.

Moiré Methods

A moiré pattern results from two overlaid, relatively dense patterns that are displaced relative to each other. This observable optical effect occurs, for example, in color printing where patterns of dots form an image. If the printing is slightly out of register, a moiré pattern will result. Another common example is the striking shimmering effect that occurs with some patterned clothing on television. This effect results when the size of the pattern in the fabric is essentially the same as the resolution of the television image.

In experimental mechanics, moiré patterns are used to measure surface displacements, typically in a model constructed specifically for this purpose. The technique uses two gratings, or patterns of parallel lines spaced equally apart. Figure 11.21 shows two line gratings. There are two important properties of line gratings for moiré techniques. The pitch is defined as the distance between the centers of adjacent lines in the grating, and for typical gratings has a value of from 1 to 40 lines/mm. The second characteristic of gratings is the ratio of the open, transparent area of the grating to the total area, or, for a line grating the ratio of the distance between adjacent lines to the center-to-center distance, as illustrated in Figure 11.21. Clearly, a greater density of lines per unit width allows a greater sensitivity of strain measurement; however, as line densities increase, coherent light is required for practical measurement.

To determine strain using the moiré technique, a grating is fixed directly to the surface to be studied. This can be accomplished through photoengraving, cementing film copies of a grating to the surface, or interferometric techniques. The master or reference grating is next placed in contact with the surface, forming a reference for determining the relative displacements under loaded conditions. A series of fringes result when the gratings are displaced relative to each other; the bright fringes are the loci of points where

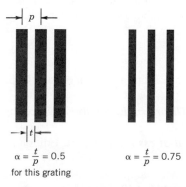

$$\alpha = \frac{t}{p} = 0.5$$
for this grating

$$\alpha = \frac{t}{p} = 0.75$$

Figure 11.21 Moiré gratings.

the projected displacements of the surface are integer multiples of the pitch. The technique then is two-dimensional, providing information concerning the projection of the displacements into the plane of the master grating. Once a fringe pattern is recorded, data-reduction techniques are employed to determine the stress and strain field. Graphical techniques exist that allow the strain components in two orthogonal directions to be determined. Further information on moiré techniques, and additional references may be found in the review article by Sciammarella [13].

Recently, techniques such as moiré-fringe multiplication have greatly increased the sensitivity of moiré techniques, with possible grating density of 1200 lines/mm. Moiré interferometry is an extension of moiré-fringe multiplication that uses coherent light and has sensitivities on the order of 0.5 mm/fringe [14]. A reflective grating is applied to the specimen, which experiences deformation under load conditions. The technique provides whole field readings of in-plane strain, with a four-beam optical arrangement currently in use [15].

11.8 SUMMARY

Experimental stress analysis can be accomplished through several practical techniques, including electrical resistance, photoelastic, and moiré strain measurement techniques. Each of these methods yields information concerning the surface strains for a test specimen. The design and selection of an appropriate strain measurement system begin with the choice of a measurement technique.

Optical methods are useful in the initial determination of a stress field for complex geometries and the determination of whole-field information in model studies. Such whole-field methods provide the basis for design and establish information necessary to make detailed local strain measurements.

The bonded electrical resistance strain gauge provides a versatile means of measuring strain at a specific location on a test specimen. Strain gauge selection involves the specification of strain gauge material, the backing or carrier material, and the adhesive used to bond the strain gauge to the test specimen, as well as the total electrical resistance of the gauge. Other considerations include the orientation and pattern for a strain gauge rosette, and the temperature limit and maximum allowable elongation. In addition, for electrical resistance strain gauges, appropriate arrangement of the gauges in a bridge circuit can provide temperature compensation and elimination of specific components of strain.

The techniques for strain measurement described in this chapter provide the basis for determining surface strains for a test specimen. While the focus here has been the measurement techniques, the proper placement of strain gauges and the interpretation of the measured results require further analysis. The inference of load-carrying capability and safety for a particular component is a result of the overall experimental program.

REFERENCES

1. Hibbeler, R. C., *Mechanics of Materials*, 5th Edition, Prentice-Hall, Upper Saddle River, NJ, 2002.
2. Timoshenko, S. P., and J. M. Goodier, *Theory of Elasticity*, Engineering Society Monographs, McGraw-Hill, New York, 1970.
3. Dally, J. W., and W. F. Riley, *Experimental Stress Analysis*, 3rd Edition., McGraw-Hill, New York, 1991.
4. Thomson, W. (Lord Kelvin), On the electrodynamic qualities of metals, *Philosophical Transactions of the Royal Society* (London) 146: 649–751, 1856.

5. Micro-Measurements Division, Measurements Group, Inc., *Strain Gauge Selection: Criteria, Procedures, Recommendations*, Technical Note 505-1, Raleigh, NC, 1989.

6. Micro-Measurements Division, Measurements Group, Inc., *Strain Gauge Technical Data, Catalog 500, Part B, and TN-509: Errors Due to Transverse Sensitivity in Strain Gauges*, Raleigh, NC, 1982.

7. Kulite Semiconductor Products, Inc., Bulletin KSG-5E: *Semiconductor Strain Gauges*, Leonia, NJ.

8. Weymouth, L. J., J. E. Starr, and J. Dorsey, Bonded resistance strain gauges, *Experimental Mechanics* 6(4):19A, 1966.

9. Oi, K., Transient response of bonded strain gauges, *Experimental Mechanics* 6(9):463, 1966.

10. Brewster, D., On the effects of simple pressure in producing that species of crystallization which forms two oppositely polarized images and exhibits the complementary colours by polarized light, *Philosophical Transactions A (GB)* 105:60, 1815.

11. Maxwell, J. C., On the equilibrium of elastic solids, *Transactions of the Royal Society* 20(Part I):87, 1853.

12. Neumann, F. E., Uber Gesetze der Doppelbrechung des Lichtes in comprimierten oder ungleichformig Erwamten unkrystallischen Korpern, *Abh. Akad. Wiss.* Berlin, Part II: 1, 1841 (in German).

13. Sciammarella, C. A., The moiré, method-A review, *Experimental Mechanics* 22(11): 418, 1982.

14. Post, D., Moiré interferometry at VPI & SU, *Experimental Mechanics* 23(2):203, 1983.

15. Post, D., Moiré interferometry for deformation and strain studies, *Optical Engineering* 24(4):663, 1985.

NOMENCLATURE

b	width $[l]$		δR	resistance change $[\Omega]$
c	speed of sound $[l\ t]$		M	bending moment $[m\ l^2\ t^{-2}]$
h	height $[l]$		R	electrical resistance $[\Omega]$
n_i	index of refraction in the direction of the principal strain in the i direction		T	temperature $[°]$
			T_0	reference temperature $[°]$
$(u_d)_x$	design-stage uncertainty in the variable x		α	Moiré grating width to spacing parameter
A_c	cross-sectional area $[l^2]$		γ_{xy}	shear strain in the xy plane $[m\ l^{-1}\ t^{-2}]$
D	diameter $[l]$		δ	relative retardation between two light beams in a photoelastic material
E_m	modulus of elasticity $[ml^{-1}t^{-2}]$			
E_i	input voltage $[V]$		ε_a	axial strain
E_o	output voltage $[V]$		ε_i	strain in the i coordinate direction (i.e., x direction)
e_l	strain gauge lateral sensitivity error as a percentage of axial strain		ε_l	lateral or transverse strain
F_N	force normal to $A_c[m\ l\ t^{-2}]$		κ	bridge constant
G	shear modulus $[m\ t^{-2}]$		v_p	Poisson ratio
GF	gauge factor		v_{po}	Poisson ratio for gauge factor calibration test specimen
K	strain optical coefficient			
K_B	bridge sensitivity $[V]$		π_1	piezoresistance coefficient $[t^2\ l\ m]$
K_l	strain gauge lateral sensitivity		ρ_e	electrical resistivity $[\Omega\ l]$
L	length $[l]$		σ	stress $[m\ l^{-1}\ t^{-2}]$
δE_o	voltage change $[V]$		σ_a	axial stress $[m\ l^{-1}\ t^{-2}]$
δL	change in length $[l]$		τ_{xy}	shear stress in the xy plane $[m\ l^{-1}\ t^{-2}]$

PROBLEMS

11.1   Calculate the change in length of a steel rod $\left(E_m = 30 \times 10^6 \, \text{psi}\right)$ having a circular cross section, a length of 10 in., and a diameter of $1/4$ in. The rod supports a mass of 40 lb_m in such a way that a state of uniaxial tension is created in the rod.

11.2   Calculate the change in length of a steel rod $(E_m = 20 \times 10^{10} \, \text{Pa})$ that has a length of 0.3 m and a diameter of 5 mm. The rod supports a mass of 50 kg in a standard gravitation field in such a way that a state of uniaxial tension is created in the rod.

11.3   An electrical coil is made by winding copper wire around a core. What is the resistance of 20,000 turns of 16-gauge wire (0.051-in. diameter) at an average radius of 2.0 in.?

11.4   Compare the resistance of a volume of $\pi \times 10^{-5} \, \text{m}^3$ of aluminum wire having a diameter of 2 mm, with the same volume of aluminum formed into 1-mm-diameter wire. (The resistivity of aluminum is $2.66 \times 10^{-8} \, \Omega\text{m}$.)

11.5   A conductor made of nickel $(\rho_e = 6.8 \times 10^{-8} \, \Omega\,\text{m})$ has a rectangular cross section $5 \times 2 \, \text{mm}$ and is 5 m long. Determine the total resistance of this conductor. Calculate the diameter of a 5-m-long copper wire having a circular cross section that yields the same total resistance.

11.6   Consider a Wheatstone bridge circuit having all resistances equal to 100 Ω. The resistance R_1 is a strain gauge that cannot sustain a power dissipation of more than 0.25 W. What is the maximum applied voltage that can be used for this bridge circuit? At this level of bridge excitation, what is the bridge sensitivity?

11.7   A resistance strain gauge with $R = 120 \, \Omega$ and a gauge factor of 2 is placed in an equal-arm Wheatstone bridge in which all the resistances are equal to 120 Ω. If the maximum gauge current is to be 0.05 A, what is the maximum allowable bridge excitation voltage?

11.8   A strain gauge having a nominal resistance of 350 Ω and a gauge factor of 1.8 is mounted in an equal-arm bridge, which is balanced at a zero applied strain condition. The gauge is mounted on a 1-cm^2 aluminum rod, having $E_m = 70 \, \text{GPa}$. The gauge senses axial strain. The bridge output is 1 mV for a bridge input of 5 V. What is the applied load, assuming the rod is in uniaxial tension?

11.9   Consider a structural member subject to loads that produce both axial and bending stresses, as shown in Figure 11.12. Two strain gauges are to be mounted on the member and connected in a Wheatstone bridge in such a way that the bridge output indicates the axial component of strain only (the installation is bending compensated). Show that the installation of the gauges shown in Figure 11.12 will not be sensitive to bending.

11.10   A steel beam member $\left(v_p = 0.3\right)$ is subjected to simple axial tensile loading. One strain gauge aligned with the axial load is mounted on the top and center of the beam. A second gauge is similarly mounted on the bottom of the beam. If the gauges are connected as arms 1 and 4 in a Wheatstone bridge (Figure 11.11), determine the bridge constant for this installation. Is the measurement system temperature compensated? (Clearly explain why or why not). If $\delta E_0 = 10 \mu\text{V}$ and $E_i = 10 \, \text{V}$, determine the axial and transverse strains. The gauge factor for each gauge is 2, and all resistances are initially equal to 120 Ω.

11.11   An axial strain gauge and a transverse strain gauge are mounted to the top surface of a steel beam that experiences a uniaxial stress of 2222 psi. The gauges are connected to arms 1 and 2 (Figure 11.11) of a Wheatstone bridge. With a purely axial load applied, determine the bridge constant for the measurement system. If $\delta E_0 = 250 \, \mu\text{V}$ and $E_i = 10 \, \text{V}$, estimate the average gauge factor of the strain gauges. For this material, Poisson's ratio is 0.3 and the modulus of elasticity is $29.4 \times 10^6 \, \text{psi}$.

11.12   A strain gauge is mounted on a steel cantilever beam of rectangular cross section. The gauge is connected in a Wheatstone bridge; initially $R_{\text{gauge}} = R_2 = R_3 = R_4 = 120 \, \Omega$. A gauge resistance change of 0.1 Ω is measured for the loading condition and gauge

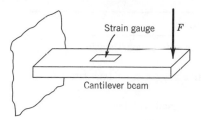

Figure 11.22 Loading for Problem 11.12.

orientation shown in Figure 11.22. If the gauge factor is $2.05 \pm 1\%$ (95%), estimate the strain. Suppose the uncertainty in each resistor value is 1% (95%). Estimate an uncertainty in the measured strain due to the uncertainties in the bridge resistances and gauge factor. Assume that the bridge operates in a null mode, which is detected by a galvanometer. Also assume reasonable values for other necessary uncertainties and parameters, such as input voltage or galvanometer sensitivity.

11.13 Two strain gauges are mounted so that they sense axial strain on a steel member in uniaxial tension. The 120-Ω gauges form two legs of a Wheatstone bridge and are mounted on opposite arms. For a bridge excitation voltage of 4 V and a bridge output voltage of 120 μV under load, estimate the strain in the member. What is the resistance change experienced by each gauge? The gauge factor for each of the strain gauges is 2, and E_m for this steel is 29×10^6 psi.

11.14 A rectangular bar is instrumented with strain gauges and subjected to a state of uniaxial tension. The bar has a cross-sectional area of 2 in.^{2}, and the bar is 12 in. long. The two strain gauges are mounted such that one senses the axial strain, while the other senses the lateral strain. For an axial load of 1500 lb, the axial strain is measured as 1500 $\mu\varepsilon$ (μin./in.), and the lateral gauge indicates a strain of $-465\,\mu\varepsilon$. Determine the modulus of elasticity and Poisson's ratio for this material.

11.15 A round member having a cross-sectional area of 3 cm^2 experiences an axial load of 10 kN. Two strain gauges are mounted on the member, one measuring an axial strain of 600 $\mu\varepsilon$ (μin./in.), and the other measuring a lateral strain of $-163\,\mu\varepsilon$. Determine the modulus of elasticity and Poisson's ratio for this material.

11.16 Show that the use of a dummy gauge together with a single active gauge compensates for temperature but not for bending. Consider the case in which the active gauge is subjected to axial loading with minimal bending and both gauges experience the same temperature.

11.17 Show that four strain gauges mounted to a shaft such that the gauge pairs measure equal and opposite strain can be used to measure torsional twist (as suggested in Table 11.1). Show that this method compensates for axial and bending strains and for temperature.

11.18 For each bridge configuration shown in Figure 11.23 determine the bridge constant. Assume that all of the active gauges are identical, and that all of the fixed resistances are equal.

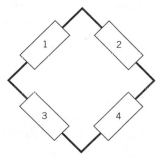

Figure 11.23 Bridge arrangement for Problem 11.19.

Bridge Arrangement	Description
(a) 1: Active gauge 2–4: Fixed resistors	Single gauge in uniaxial tension.
(b) 1: Active gauge 2: Poisson gauge	Two active gauges in uniaxial stress field.
3, 4: Fixed resistors	Gauge 1 aligned with maximum axial stress: gauge 2 lateral.

(c) 1: Active gauge
    3: Active gauge
    2, 4: Fixed resistors
	Equal and opposite strains applied to the
	    active gauges (bending compensation).

(d) Four active gauges	Gauges 1 and 4 aligned with uniaxial stress; gauges 2 and 3 transverse.

(e) Four active gauges	Gauges 1 and 2 subject to equal and
	    opposite strains; gauges 3 and 4 subject
	    to the same equal but opposite strains.

11.19 A bathroom scale uses four internal strain gauges to measure the displacement of its diaphragm as a means for determining load. Four active gauges are used in a bridge circuit. The gauge factor is 2, and gauge resistance is 120 Ω for each gauge. If an applied load to the diaphragm causes a compression strain on R_1 and R_4 of 20 $\mu\varepsilon$, while gauges R_2 and R_3 experience a tensile strain of 20 $\mu\varepsilon$, estimate the bridge deflection voltage. The supply voltage is 9 V. Refer to Figure 11.23 for gauge position.

11.20 Suppose in Problem 11.19 that the lead wires of two gauges (R_2 and R_4) are accidentally interchanged on the assembly line such that a compression strain is now sensed on R_1 and R_2 of 20 $\mu\varepsilon$, while gauges R_4 and R_3 experience a tensile strain of 20 $\mu\varepsilon$. Should this really matter? Estimate the bridge deflection voltage. Refer to Figure 11.23 for gauge position.

11.21 A single-strain gauge is mounted on the surface of a thin-walled pressure vessel, which has a diameter of 1 m. For a strain gauge oriented along the tangential stress direction, determine the percentage error in tangential strain measurement caused by lateral sensitivity as a function of pressure and vessel wall thickness. The lateral sensitivity is 0.03.

11.22 Consider a Wheatstone bridge circuit having all fixed resistances equal to 100 Ω and with a strain gauge located at the R_1 position, which has a value of 100 Ω under conditions of zero strain. The strain gauge is mounted so as to sense longitudinal strain for a thin-walled pressure vessel made of steel having a wall thickness of 2 cm and a diameter of 2 m. The strain gauge has a gauge factor of 2 and cannot sustain a power dissipation of more than 0.25 W. What is the maximum static sensitivity that can be achieved with this proposed measurement system? Is this static sensitivity constant with input pressure? If not, under what conditions would it be reasonable to assume a constant static sensitivity? The static sensitivity should be expressed in units of V/kPa.

11.23 Design a Wheatstone bridge measurement system to measure the tangential strain in the wall of the pressure vessel described above, and develop a reasonable estimate of the resulting uncertainty. You may assume that the strain does not vary with time, and that the bridge will be operated in a balanced mode. Your selection of specified values for the input voltage, the fixed resistances in the bridge and the galvanometer sensitivity, and their associated uncertainties will allow completion of this design.

11.24 A steel cantilever beam is fixed at one end and free to move at the other. A load F of 980 N is applied to the free end. Four axially aligned strain gauges (GF = 2) are mounted to the beam a distance L from the applied load, two on the upper surface, R_1 and R_4, and two on the lower surface, R_2 and R_3. The bridge deflection output is passed through an amplifier (Gain, $G_A = 1000$) and measured. For a cantilever, the relationship between applied load and strain is

$$F = \frac{2E_m I \varepsilon}{Lt}$$

where I is the beam moment of inertia $(= bt^3/12)$, t is the beam thickness, and b is the beam width. Estimate the measured output for the applied load if $L = 0.1$ m, $b = 0.03$ m, $t = 0.01$ m, and the bridge excitation voltage is 5 V. $E_m = 200$ GPa.

11.25 A cantilever beam is to be used as a scale. The beam, made of 2024-T4 aluminum, is 21 cm long, 0.4 cm thick, and 2 cm wide. The scale load of between 0 and 200 g is to be concentrated at a point along the beam centerline 20 cm from its fixed end. Strain gauges are to be used to measure beam deflection and mounted to a Wheatstone bridge to provide an electrical signal that is proportional to load. Either 1/4-, 1/2-, or full-bridge gauge arrangements can be used. Design the sensor arrangement and its location on the beam. Specify appropriate signal conditioning to excite the bridge and to measure its output on a data-acquisition system using a ± 5 V, 12-bit A/D converter. Will the system achieve a 4% uncertainty at the design stage? Specifications and system choices below are stated at 95% confidence levels.

Sensors: one, two, or four axial gauges at $120\,\Omega \pm 0.2\%$ (selectable)

Gauge factor of $2.0 \pm 1\%$

Bridge excitation: 1, 3, 5 V $\pm 0.5\%$ (selectable)

Bridge null: within 5 mV

Signal conditioning:

Amplifier gain: 1X, 10X, 100X, 1000X $\pm 1\%$ (selectable)

Low-pass filter: f_c at 0.5, 5, 50, 500 Hz @ -12dB/octave (selectable)

Data-acquisition system:

Conversion errors: $<0.1\%$ reading

Sample rate: 1, 10, 100, 1000 Hz (selectable)

Chapter 12

Mechatronics[1]: Sensors, Actuators, and Controls

12.1 INTRODUCTION

Rapid advances in microprocessors have lead to a dramatic increase in electronically controlled devices and systems. All of these systems require sensors and actuators to be interfaced with the electronics. Understanding the operating principles and limitations of sensors is essential to selecting and interfacing sensors for linear motion, rotary motion, and engineering variables such as force torque and power. *Actuators* are required to produce motion, whether it is for the motion of a electric car seat, for a fly-by-wire throttle for automotive applications, or for positioning a precision laser welder. We will treat sensors and actuators, and provide a limited introduction to linear control theory.

12.2 SENSORS

Throughout the previous chapters, a wide variety of measurement sensors have been described, and the fundamentals of their operation described. Thermocouples, strain gauges, flow meters, and pressure sensors represent the means to measure very important engineering process variables of temperature, strain, flow rate, and pressure. In this chapter we add to this base by introducing methods and sensors for the measurement of linear and rotary displacement, acceleration and vibration, velocity measurement, force or load, torque, and mechanical power.

Displacement Sensors

Methods to measure position or displacement are often important elements in mechatronic systems. Often potentiometric or LVDT transducers are employed in these applications.

Potentiometers

A potentiometer is a device employed to measure linear or rotary displacement. The principle of operation relies on an increase in electrical resistance with displacement.

A wire-wound potentiometer[2] or variable electrical resistance transducer is depicted in Figure 12.1. This transducer is composed of a sliding contact and a winding. The

---

[1]The term "mechatronics" is derived from *mechanical* and *electronic*, and refers to the integration of mechanical and electronic devices.

[2]The potentiometer-transducer should not be confused with the potentiometer-instrument. Although both are based on voltage-divider principles, the latter measures low-level voltages (10^{-6} to 10^{-3} V) as described in Chapter 6.

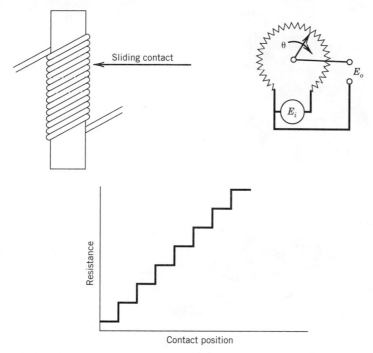

Figure 12.1 Potentiometer construction.

winding is made of many turns of wire, wrapped around a nonconducting substrate. Output signals from such a device can be realized by imposing a known voltage across the total resistance of the winding, and measuring the output voltage, which is proportional to the fraction of the distance the contact point has moved along the winding. Potentiometers can also be configured in a rotary form, with numerous total revolutions of the contact possible in a helical arrangement. The output from the sliding contact as it moves along the winding is actually discrete, as illustrated in Figure 12.1; the resolution is limited by the number of turns per unit distance. The loading errors associated with voltage-dividing circuits, discussed in Chapter 6, should be considered in choosing a measuring device for the output voltage.

Wire-wound potentiometers display a stepwise output as the wiper contacts successive turns of the wire winding. Conductive plastic potentiometers were developed to eliminate this stepwise output, and are now widely employed in mechatronic systems. Two such potentiometers are shown in Figure 12.2. The key components are the structural support and mechanical interface, the conductive plastic resistor, and the wiper where electrical contact occurs. A linear output with displacement is most often the design objective for these sensors. Typical linearity errors have uncertainties ranging from 0.2 to 0.02%.

Linear Variable Differential Transformers

The linear variable differential transformer (LVDT), as shown in Figure 12.3, produces an ac output with an amplitude that is proportional to the displacement of a movable core. The waveform of the output from the LVDT is sinusoidal; the amplitude of the sine wave is proportional to the displacement of the core for a limited range of core motion. The movement of the core causes a mutual inductance in the secondary coils for an ac voltage

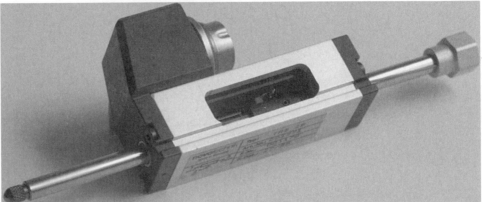

Figure 12.2 Conductive plastic potentiometers. (Courtesy of Novotechnic U.S., Inc.).

applied to the primary coil. In 1831, the English physicist Michael Faraday (1791–1867) demonstrated that a current could be induced in a conductor by a changing magnetic field. An interesting account of the development of the transformer may be found in [2].

Recall that for two coils in close proximity, a change in the current in one coil will induce an emf in the second coil according to Faraday's law. The application of this inductance principle to the measurement of distance begins by applying an ac voltage to the primary coil of the LVDT. The two secondary coils are connected in a series circuit, such that when the iron core is centered between the two secondary coils the output voltage amplitude is zero (see Figure 12.3). Motion of the magnetic core changes the mutual inductance of the coils, which causes a different emf to be induced in each of the two secondary coils. Over the range of operation, the output amplitude is essentially linear with core displacement, as first noted in a U.S. patent by G. B. Hoadley in 1940 [1].

The output of a differential transformer is illustrated in Figure 12.4. Over a specific range of core motion the output is essentially linear. Beyond this linear range, the output amplitude will rise in a nonlinear manner to a maximum, and eventually fall to zero. The output voltages on either side of the zero displacement position are 180° out of phase. With appropriate circuitry it is possible to determine positive or negative displacement of the core. However, note that due to harmonic distortion in the supply voltage and the fact that the two secondary coils are not identical, the output voltage with the coil centered is not zero but instead reaches a minimum. Resolution of an LVDT depends strongly on the resolution of the measurement system used to determine its output. Resolutions in the microinch $(10^{-8}\,\text{m})$ range can be accomplished.

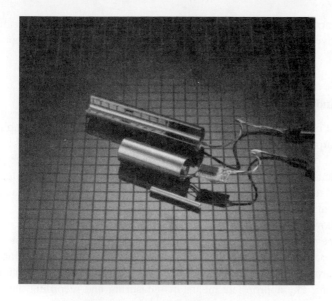

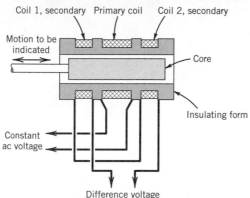

Figure 12.3 Construction of an LVDT. (Courtesy of Schaevitz Engineering; from [1].)

The differential voltage output of an LVDT, as shown in Figure 12.4, may be analyzed by assuming that the magnetic field strengths are uniform along the axis of the coils, neglecting end effects, and limiting the analysis to the case where the core does not move beyond the ends of the coils [1]. Under these conditions the differential voltage may be expressed in terms of the core displacement. The sensitivity of the LVDT in the linear range is a function of the number of turns in the primary and secondary coils, the root-mean-square (rms) current in the primary coil and the physical size of the LVDT. The

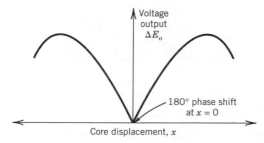

Figure 12.4 LVDT output as a function of core position.

output of an LVDT is not linear for all positions of the core, but in practical applications a range of core motion over which the nonlinearity is negligible is specified as the operating range.

Excitation Voltage and Frequency

The dynamic response of an LVDT is directly related to the frequency of the applied ac voltage because the output voltage of the secondary coil is induced by the variation of the magnetic field induced by the primary coil. For this reason the excitation voltage should have a frequency at least 10 times the maximum frequency in the measured input. An LVDT can be designed to operate with input frequencies ranging from 60 Hz up to 25 kHz (for specialized applications, frequencies in the megahertz range can be used).

The maximum allowable applied voltage for an LVDT is determined by the current-carrying capacity of the primary coil, typically in the 1- to 10-V range. A constant current source is preferable for an LVDT, to limit temperature effects. For other than a sine wave input voltage form, harmonics in the input signal will increase the voltage output at the null position of the core. The appropriate means of measuring and recording the output

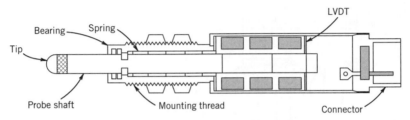

(a) Cross-section of typical LVDT gauge head.

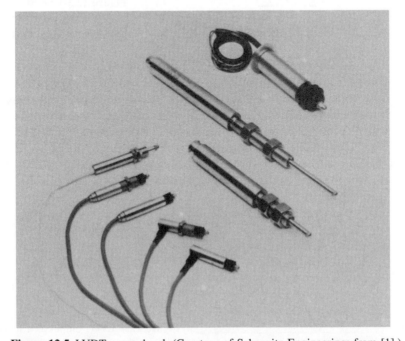

Figure 12.5 LVDT gauge head. (Courtesy of Schaevitz Engineering; from [1].)

signal from an LVDT and the ac frequency applied to the primary coil should be chosen based on the highest frequencies present in the input signal to the LVDT. For example, for static measurements and signals having frequency contents much lower than the excitation frequency of the primary coil, an ac voltmeter may be an appropriate choice for measuring the output signal. In this case, it is likely that the frequency response of the measuring system would be limited by the averaging effects of the ac voltmeter. For higher frequency signals, it is possible to create a dc voltage output that follows the input motion to the LVDT through demodulation and amplification of the resulting signal, using a dedicated electronic circuit. Alternatively, the output signal can be sampled at a sufficiently high frequency using a computer data-acquisition system to allow signal processing for a variety of purposes including quality assurance or control.

The measurement of distance using an LVDT is accomplished using an assembly known as an LVDT gauge head. Such devices are widely used in machine tools and various types of gauging equipment. Control applications will have similar transducer designs. The basic construction is shown in Figure 12.5, which can yield instrument errors as low as 0.05% and repeatability of 0.0001 mm.

Angular displacement can also be measured using inductance techniques employing a rotary variable differential transformer (RVDT). The output curve of an RVDT and a typical construction are shown in Figure 12.6, where the linear output range is approximately $\pm 40°$.

Measurement of Acceleration and Vibration

The measurement of acceleration is required for a variety of purposes, ranging from machine design to guidance systems. Because of the range of applications for acceleration and vibration measurements, there exists a wide variety of transducers and measurement techniques, each associated with a particular application. In this section we address some fundamental aspects of these measurements along with some common applications.

Displacement, velocity, or acceleration measurements are also referred to as shock or vibration measurements depending on the waveform of the forcing function that causes the acceleration. A forcing function that is periodic in nature generally results in accelerations that are analyzed as vibrations. On the other hand, a force input having a short duration and a large amplitude would be classified a shock load.

The fundamental aspects of acceleration, velocity, and displacement measurements can be discerned through an examination of the most basic device for measuring acceleration and velocity, a seismic transducer.

Seismic Transducer

A seismic transducer consists of three basic elements, as shown in Figure 12.7: a spring-mass-damper system, a protective housing, and an appropriate output transducer. Through the appropriate design of the characteristics of this spring-mass-damper system, the output is a direct indication of either displacement or acceleration. To accomplish a specific measurement, this basic seismic transducer is rigidly attached to the object experiencing the motion that is to be measured.

Consider the case where the output transducer senses the position of the seismic mass; a variety of transducers could serve this function. Under some conditions, the displacement of the seismic mass serves as a direct measure of the acceleration of

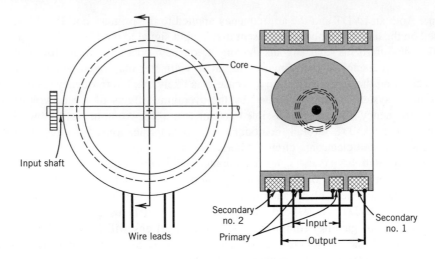

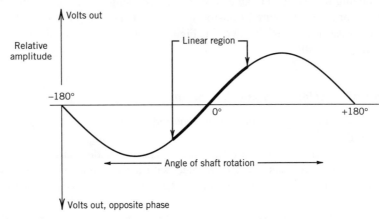

Figure 12.6 Rotary variable differential transformer. (Courtesy of Schaevitz Engineering; from [6].)

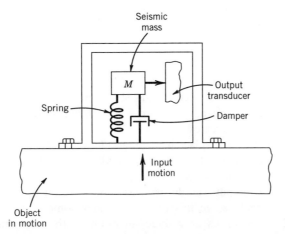

Figure 12.7 Seismic transducer.

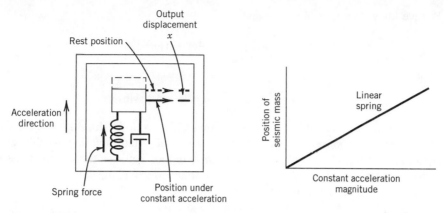

Figure 12.8 Response of a seismic transducer to a constant acceleration.

the housing, and the object to which it is attached. To illustrate the relation between the relative displacement of the seismic mass and acceleration, consider the case in which the input to the seismic instrument is a constant acceleration. The response of the instrument is illustrated in Figure 12.8. At steady-state conditions, under this constant acceleration, the mass will be at rest with respect to the housing. The spring deflects an amount proportional to the force required to accelerate the seismic mass, and since the mass is known, Newton's second law yields the corresponding acceleration. The relationship between a constant acceleration and the displacement of the seismic mass is linear for a linear spring (where $F = kx$).

We wish to measure not only constant accelerations, but also complex acceleration waveforms. Recall from Chapter 2 that a complex waveform can be represented as a series of sine or cosine functions, and that by analyzing a measuring system response to a periodic waveform input we can discern the allowable range of frequency inputs.

Consider an input to our seismic instrument such that the displacement of the housing is a sine wave, $y_h = A \sin \omega t$, and the absolute value of the resulting acceleration of the housing is $A\omega^2 \sin \omega t$. If we consider a free-body diagram of the seismic mass, the spring force and the damping force must balance the inertial force for the mass. (Notice that gravitational effects could play an important role in the analysis of this instrument, for instance, if the instrument were installed in an aircraft.) The spring force and damping force are proportional to the relative displacement and velocity between the housing and the mass, whereas the inertial force is dependent only on the absolute acceleration of the seismic mass.

Since

$$y_m = y_h + y_r \tag{12.1}$$

Newton's second law may be expressed as

$$m\frac{d^2 y_m}{dt^2} + c\frac{dy_r}{dt} + ky_r = 0 \tag{12.2}$$

Substituting equation (12.1) in equation (12.2) yields

$$m\left(\frac{d^2 y_h}{dt^2} + \frac{d^2 y_r}{dt^2}\right) + c\frac{dy_r}{dt} + ky_r = 0 \tag{12.3}$$

But we know that $y_h = A \sin \omega t$ and

$$\frac{d^2 y_h}{dt^2} = -A\omega^2 \sin \omega t \tag{12.4}$$

Thus,

$$m\frac{d^2 y_r}{dt^2} + c\frac{dy_r}{dt} + ky_r = mA\omega^2 \sin \omega t \tag{12.5}$$

This equation is identical in form to equation (3.12). As in the development for a second-order system response, we will examine the steady-state solution to this governing equation. The transducer will sense the relative motion between the seismic mass and the instrument housing. Thus, for the instrument to be effective, the value of y_r must provide indication of the desired output.

For the input function $y_h = A \sin \omega t$, the steady-state solution for y_r is

$$(y_r)_{\text{steady}} = \frac{(1/\omega_n^2)A\omega^2 \cos(\omega t - \phi)}{\left\{ [1 - (\omega/\omega_n)^2]^2 + [2\zeta(\omega/\omega_n)]^2 \right\}^{1/2}} \tag{12.6}$$

where

$$\omega = \sqrt{\frac{k}{m}} \quad \zeta = \frac{c}{2\sqrt{km}} \quad \phi = \tan^{-1}\frac{2\zeta(\omega/\omega_n)}{1 - (\omega/\omega_n)^2} \tag{12.7}$$

The characteristics of this seismic instrument can now be discerned by examining equations (12.6) and (12.7). The natural frequency and damping will be fixed for a particular design. We wish to examine the motion of the seismic mass and the resulting output for a range of input frequencies.

Vibrometer

For vibration measurements, it is desireable to measure the amplitude of the displacements associated with the vibrations; thus, the desired behavior of the seismic instrument would be to have an output that gave a direct indication of y_h. For this to occur, the seismic mass should remain essentially stationary in an absolute frame of reference, and

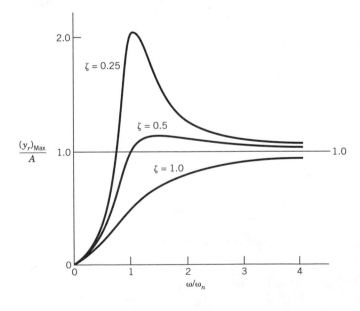

Figure 12.9 Displacement amplitude at steady state as a function of input frequency for a seismic transducer.

the housing and output transducer should move with the vibrating object. To determine the conditions under which this behavior would occur, the amplitude of y_r at steady state is examined. The ratio of the maximum amplitude of the output divided by the equivalent static output in the present case is $(y_r)_{max}/A$. For vibration measurements, this ratio should have a value of 1. Figure 12.9 shows $(y_r)_{max}/A$ as a function of the ratio of the input frequency to the natural frequency. Clearly, as the input frequency increases, the output amplitude, y_r, approaches the input amplitude, A, as desired. Thus, a seismic instrument that is used to measure vibration displacements should have a natural frequency smaller than the expected input frequency. Damping ratios near 0.7 are common for such an instrument. The seismic instrument designed for this application is called a *vibrometer*.

EXAMPLE 12.1

A seismic instrument like the one shown in Figure 12.7 is to be used to measure a periodic vibration having an amplitude of 0.05 in. and a frequency of 15 Hz.

1. Specify an appropriate combination of natural frequency and damping ratio such that the amplitude error in the output is less than 5%.

2. What spring constant and damping coefficient would yield these values of natural frequency and damping ratio?

3. Determine the phase lag for the output signal. Would the phase lag change if the input frequency were changed?

KNOWN Input function $y_h = 0.05 \sin 30\pi t$.

FIND

Values of $\omega_n, \zeta, k, m,$ and c to yield a measurement with less than 5% magnitude error
Examine the phase response of the system

SOLUTION Numerous combinations of the mass, spring constant, and damping coefficient would yield a workable design. Let's choose $m = 0.05$ lb$_m$ and $\zeta = 0.7$. We know that for a spring-mass-damper system

$$\omega = \sqrt{\frac{k}{m}} \quad c_c = 2\sqrt{km}$$

where c_c is the critical damping coefficient. With $\zeta = c/c_c = 0.7$, the damping coefficient, c, is found as

$$c = 2(0.7)\sqrt{km}$$

We can now examine the values of $(y_r)_{max}/A$ for $\zeta = 0.7$. The results are shown in tabular form:

$\dfrac{\omega}{\omega_n}$	$\dfrac{(y_r)_{max}}{A}$
10	1.000
8	1.000
6	1.000
4	0.999
3	0.996
2	0.975
1.7	0.951

Acceptable behavior is achieved for values of $\omega/\omega_n \geq 1.7$ for $\zeta = 0.7$, and an acceptable maximum value of the natural frequency is

$$\omega_n = \omega/1.7 = 30\pi/1.7 = 55.4 \, \text{rad/s}$$

Since[3]

$$\omega_n = \sqrt{k/m}$$

we find that with $m = 0.05 \, \text{lb}_m$, the value of k is 4.8 lb/ft. Then the value of the damping coefficient is found as

$$c = 2(0.7)\sqrt{k/m} = 0.12$$

The phase behavior for this system is given by equation (12.7), and results in the phase response tabulated as

$\dfrac{\omega}{\omega_n}$	$\phi[°]$
10	172
8	169.9
6	166.5
4	159.5
3	152.3
2	137.0
1.7	128.5

COMMENT The design of such an instrument would have other constraints that need to be considered in the choice of design parameters. The design would be influenced by such factors as size, cost, and operating environment. The phase behavior determined for this combination of design parameters does not result in a linear relationship between phase shift and frequency. As such, the possibility of distortion of complex waveform inputs exists. These and other issues can be explored using the program file *seismic_transducer.vi* on the accompanying software.

Accelerometer

If it is desired to measure acceleration, the behavior of the seismic mass must be quite different. The amplitude of the acceleration input signal is $A\omega^2$. To have the output value y_r represent the acceleration, it is clear from equation (12.6) that the value of $y_r/A(\omega/\omega_n)^2$ must be a constant over the design range of input frequencies. If this is true, the output will be proportional to the acceleration. The amplitude of $y_r/A(\omega/\omega_n)^2$ can be expressed as

$$\frac{(y_r)_{\text{steady}}}{A(\omega/\omega_n)^2} = \frac{\cos(\omega t - \phi)}{\left\{[1 - (\omega/\omega_n)^2]^2 + [2\zeta(\omega/\omega_n)]^2\right\}^{1/2}} \tag{12.8}$$

and

$$M(\omega) = \frac{1}{\left\{[1 - (\omega/\omega_n)^2]^2 + [2\zeta(\omega/\omega_n)]^2\right\}^{1/2}} \tag{12.9}$$

---

[3]Note that if k has units of lb/ft, and mass is in units of lb_m, g_c is required in the expression for ω_n.

where $M(\omega)$ is the magnitude ratio as defined by equation (3.21). The magnitude ratio is plotted as a function of input frequency and damping ratio in Figure 3.16.

For the desired behavior to be achieved, it is clear that the magnitude ratio should be unity. Over a range of input frequency ratios from 0 to 0.4, as determined from Figure 3.16, the magnitude ratio is approximately 1. Typically, in an accelerometer the damping ratio is designed to be near 0.7, so that the phase shift is linear with frequency and distortion is minimized.

In summary, the seismic instrument can be designed so that the output can be interpreted in terms of either the input displacement or the input acceleration. Acceleration measurements may be integrated to yield velocity information; the differentiation of displacement data to determine velocity or acceleration introduces significantly more difficulties than the integration process.

Transducers for Shock and Vibration Measurement

In general, the destructive forces generated by vibration and shock are best quantified through the measurement of acceleration. Although a variety of accelerometers are available, strain gauge and piezoelectric transducers are widely employed for the measurement of shock and vibration [3].

A piezoelectric accelerometer employs the principles of a seismic transducer through the use of a piezoelectric element to provide a portion of the spring force. Figure 12.10 illustrates one basic construction of a piezoelectric accelerometer. A preload is applied to the piezoelectric element simply by tightening the nut that holds the mass and piezoelectric element in place. Upward or downward motion of the housing will change the compressive forces in the piezoelectric element, resulting in an appropriate output signal. Instruments are available with a range of frequency response from 0.03 to 10,000 Hz. Depending on the piezoelectric material used in the transducer construction, the static sensitivity can range from 1 to 100 mV/g. Steady accelerations cannot be effectively measured with such a piezoelectric transducer.

Strain gauge accelerometers are generally constructed using a mass supported by a flexure member, with the strain gauge sensing the deflection that results from an acceleration of the mass. The frequency response and range of acceleration of these instruments are related, such that instruments designed for higher accelerations have a wider bandwidth, but significantly lower static sensitivity. Table 12.1 provides typical performance characteristics for two strain gauge accelerometers that employ semiconductor strain gauges.

Many other accelerometer designs are available, including potentiometric, reluctive, or accelerometers that use closed-loop servo systems to provide a high output level.

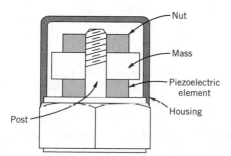

Figure 12.10 Basic piezoelectric accelerometer.

Table 12.1   Representative Performance Characteristics for
Piezoresistive Accelerometers[a]

Characteristic	25-g Range	2500-g Range
Sensitivity [mV/g]	50	0.1
Resonance frequency [Hz]	2700	30,000
Damping ratio	0.4–0.7	0.03
Resistance [Ω]	1500	500

[a]Adapted from reference [7].

Piezoelectric transducers have the highest frequency response and range of acceleration, but have relatively lower sensitivity. Amplification can overcome this drawback to some degree. Semiconductor strain gauge transducers have a lower frequency response limit than piezoelectric, but they can be used to measure steady accelerations. Other transducers generally have lower frequency response behaviors compared to piezoelectric and strain gauge units.

Velocity Measurements

Linear and angular velocity measurements utilize a variety of approaches ranging from radar and laser systems for speed measurement to mechanical counters to provide indication of a shaft rotational speed. For many applications, the sensors employed provide a scalar output of speed. However, sensors and methods exist that can provide indication of both speed and direction, when properly employed. Here we will consider techniques for the measurement of linear and angular speed.

Displacement, velocity, and acceleration measurements are made with respect to some frame of reference. Consider the case of a game of billiards in a moving railway car. Observers on the ground and on the train would assign different velocity vectors to the balls during play. The velocity vectors would differ by the relative velocity of the two observers. Simply differentiating the vector velocity equation, however, shows that the accelerations of the balls are the same in all reference frames moving relative to one another with constant relative velocity.

Linear Velocity Measurements

As previously discussed, measurements of velocity require a frame of reference. For example, the velocity of a conveyor belt might be measured relative to the floor of the building where it is housed. On the other hand, advantages may be realized in a control system if the velocity of a robotic arm, which "picks" parts from this same conveyor belt, is measured relative to the moving conveyor. In the present discussion, it is assumed that velocity is measured relative to a ground state, which is generally defined by the mounting point of a transducer.

Consider the measurement of velocity relative to a fixed frame of reference. Typically, if the measurement of linear velocity is to be made on a continuous basis, an equivalent angular rotational speed is measured, and the data are analyzed in such a way so as to produce a measured linear velocity. For example, a speedometer on an automobile provides a continuous record of the speed of the car, but the output is derived from measuring the rotational speed of the drive shaft or transmission.

Velocity from Displacement or Acceleration

Velocity can, in general, be directly measured by mechanical means only over very short times or small displacements, due to limitations in transducers. However, if the displacement of a rigid body is measured at identifiable time intervals, the velocity can be determined through differentiation of the time-dependent displacement. Alternatively, if acceleration is measured the velocity may be determined from integration of the acceleration signal. The following example demonstrates the effect of integration and differentiation on the uncertainty of velocities computed from acceleration or displacement.

EXAMPLE 12.2

Our goal is to assess the merits of measuring velocity through the integration of an acceleration signal as compared to differentiating a displacement signal. The following conditions are assumed to apply.

For both $y(t)$ and $y''(t)$ the data-acquisition system and transducers may be assumed to have the following characteristics: 8-bit A/D resolution and 1% accuracy for the measured variable (acceleration or displacement). The sensor outputs and uncertainties are described in Table 12.2. Note that the uncertainties are derived directly from the A/D resolution error and accuracy statement. Assume that the signals for both acceleration and displacement are sampled at 10 Hz and that numerical techniques are used to differentiate or integrate the resulting signals.

SOLUTION   Consider first determining the velocity through differentiation of the displacement signal. Displacement is measured digitally by the data acquisition, with a digitized value of displacement recorded at time intervals δt. The velocity at any time $n\delta t$ can be approximated as

$$v(t) = y'(t) = \frac{y_{n+1} - y_n}{\delta t} \tag{12.10}$$

Table 12.2   Specification and Uncertainty Analyses for Displacement and Acceleration

Measured Variable	Functional Form	Full-Scale Output Range
Displacement (cm)	$y(t) = 20 \sin 2t$	0–10 V
Acceleration (cm/s^2)	$y''(t) = -\dfrac{20}{4} \sin 2t$	−5 to 5 V

Uncertainty Values for Displacement and Acceleration	
Measured Variable	Uncertainty
Displacement	Accuracy: 1% full scale = ±0.2 cm
	A/D 8 bit (0.04 V) = ± 0.08 cm
	Total uncertainty = ±0.22 cm
Acceleration	Accuracy: 1% full scale = ± 0.05 cm/s^2
	A/D 8 bit (0.04 V) = ±0.04 cm/s^2
	Total uncertainty = ±0.064 cm/s^2

where

y_{n+1} = the $(n + 1)$ measurement of displacement, at time $(n + 1)\delta t$

y_n = the nth measurement of displacement, at time $n\delta t$

$v(t)$ = velocity at time t

If the signal is sampled at 10 Hz, δt is 0.1 s. For the present, we will assume that the uncertainty in time is negligible, so that the uncertainty in v, u_v can be expressed as

$$u_v = \left\{ \left[\frac{\partial v}{\partial y_{n+1}} u_{y_{n+1}} \right]^2 + \left[\frac{\partial v}{\partial y_n} u_{y_n} \right]^2 \right\}^{1/2} \tag{12.11}$$

The uncertainties in the measured displacements, y_n and y_{n+1}, will be equal and are listed in Table 12.2. Substituting these values for uncertainty in equation (12.11) yields an uncertainty of ± 3 cm/sec in the velocity measurement. Notice that this uncertainty magnitude is not a function of time or the measured velocity. This corresponds to a minimum uncertainty of 30% in the velocity measurement.

To determine velocity from acceleration, the measured values of acceleration must be integrated. Since we have a digital signal, the integration can be accomplished numerically as

$$v(t) = y'(t) = \sum_i y_i'' \delta t \tag{12.12}$$

Assuming the uncertainty in time is negligible, the uncertainty in velocity at any time t is simply

$$u_v = u_{y''} t \tag{12.13}$$

Clearly, the integration process tends to accumulate error as the calculation of velocity proceeds in time, as illustrated in Figure 12.11.

COMMENT   This example can be used to illustrate several useful principles for data analysis. Consider the effects of adding noise, or a degree of random error, to the measured data. In general, errors of this type tend to be minimized through a process of integration and

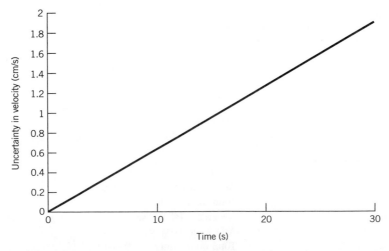

Figure 12.11 Uncertainty in velocity as a function of time.

amplified through differentiation. Differentiation tends to be extremely sensitive to low-amplitude, high-frequency noise, which can create very large errors in derivatives, especially at high sampling frequencies. Appropriate filtering or smoothing techniques are generally effective at reducing errors associated with noise in derivatives in cases where the noise has low amplitude.

Integration tends to eliminate low-amplitude, high-frequency noise from signals. However, unless a means is devised to prevent accumulation of error during integration, this advantage of integrating may be outweighed by error accumulation. In the present problem, for example, since the velocity is periodic, the integration could possibly be "reset" at each zero of the velocity, and the error accumulation eliminated.

Moving Coil Transducers

Moving coil transducers take advantage of the voltage generated when a conductor experiences a displacement in a magnetic field. This is the same phenomenon used to generate electric power in generators and alternators. An illustration of a moving coil velocity pickup is provided in Figure 12.12. Recall that a current carrying conductor experiences a force in a magnetic field, and that a force is required to move a conductor in a magnetic field. For the latter case, an emf is induced in the conductor. Consider the case in which the magnetic field strength is at right angles to the conductor. The induced emf is given by

$$\text{emf} = \pi B D_c l N \frac{dy}{dt} \qquad (12.14)$$

where

> $B =$ magnetic field strength
>
> $D_c =$ coil diameter
>
> emf $=$ induced electromotive force
>
> $N =$ number of turns in coil
>
> $dy/dt =$ velocity of coil linear motion
>
> $l =$ coil length

A moving coil transducer is appropriate for vibration applications in which the velocities of small amplitude motions are measured. The output voltage is proportional to the coil velocity, and the output polarity indicates the velocity direction. Static sensitivities on the order of 2 V s/m are typical. Moving coil transducers find application in seismic measurements as well as in vibration applications.

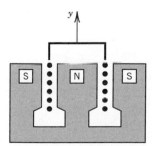

Figure 12.12 Moving coil transducer.

Angular Velocity Measurements

The measurement of angular velocity finds a wide range of applications, including such familiar examples as speedometers on automobiles. We will consider a variety of applications and measurement techniques.

Mechanical Measurement Techniques

Mechanical means of measuring angular velocity or rotational speed were developed primarily to provide feedback for control of engines and steam turbines. Mechanical governors and centrifugal tachometers [4] operate on the principle illustrated in Figure 12.13. Here the centripetal acceleration of the flyball masses result in a steady-state displacement of the spring, which provides a control signal or is a direct indication of rotational speed. For this arrangement, the spring force is proportional to the square of the angular velocity.

Stroboscopic Angular Velocity Measurements

A stroboscopic light source provides high-intensity flashes of light, which can be caused to occur at a precise frequency. A stroboscope is illustrated in Figure 12.14. Stroboscopes permit the intermittent observation of a periodic motion in a manner that appears to stop or slow the motion. Advances in electronics in the mid-1930s allowed the development of a stroboscope with a very well defined flashing rate, and led to the use of stroboscopic tachometers. Figure 12.15 illustrates the use of a strobe to measure rotational speed. A timing mark on the rotating object is illuminated with the strobe, and the strobe frequency is adjusted such that the mark appears to remain motionless, as shown in Figure 12.15(*a*). Thus, the highest synchronous speed is the actual rotational speed. At this speed, the output value is available from a calibration of the stroboscopic lamp flash frequency, with accuracies to less than 0.1%. Clearly, at rotational speeds higher than can be tracked by the human eye, the mark would appear motionless for integer multiples of the actual rotational speed and for integral submultiples, as illustrated in Figures 12.15(*b–e*).

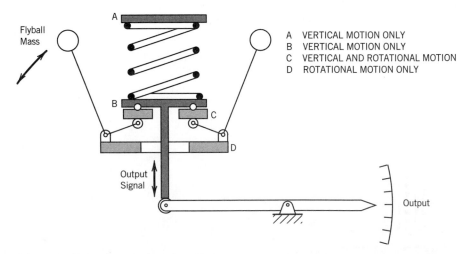

Figure 12.13 Mechanical angular velocity sensor.

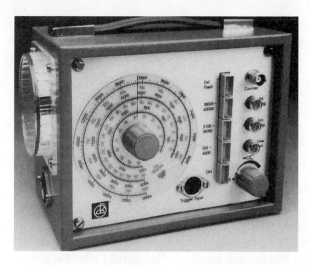

Figure 12.14 Stroboscope.
(Courtesy of Mill Devices Co.,
a division of A. B. Carter Inc.)

The synchronization of images at flashing rates other than the actual speed requires some practical approaches to ensuring the accurate determination of speed. Spurious images can easily result for symmetrical objects, and some asymmetric marking is necessary to prevent misinterpretation of stroboscopic data. To distinguish the actual speed from a submultiple, the flashing rate can be decreased until another single synchronous image appears. If this flashing rate corresponds to one-half the original rate, then the original rate is the actual speed. If it does not occur at one-half the original value, then the original value is a submultiple.

The upper limit of the flash rate of the strobe does not limit the ability of the stroboscope to measure rotational speed. For high speeds, synchronization can be

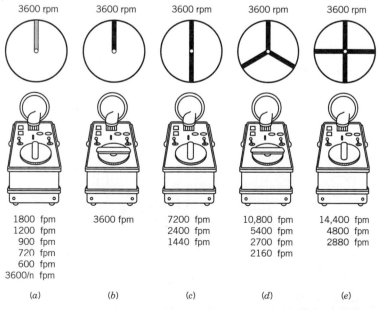

Figure 12.15 Images resulting from harmonic and subharmonic flashing rates (fpm) for stroboscopic angular speed measurement. (Courtesy of Mill Devices Co., a division of A. B. Carter Inc.)

achieved N times, with ω_1 representing the maximum achievable synchronization speed, and $\omega_2, \omega_3, \ldots$ representing successively lower synchronization speeds. The measured rotational speed is then calculated as

$$\omega = \frac{\omega_1 \omega_N (N-1)}{\omega_1 - \omega_N} \tag{12.15}$$

The program file *stroboscope.vi* with the companion software illustrates the basic principles behind interesting stroboscopic effects.

Electromagnetic Techniques

Several measurement techniques for rotational velocity utilize transducers that generate electrical signals, which are indicative of angular velocity. One of the most basic is illustrated in Figure 12.16. This transducer consists of a toothed wheel and a magnetic pickup, which consists of a magnet and a coil. As the toothed wheel rotates, an emf is induced in the coil as a result of changes in the magnetic field. As each ferromagnetic tooth passes the pickup, the reluctance of the magnetic circuit changes in time, yielding a voltage in the coil given by

$$E = C_B N_t \omega \sin N_t \omega t \tag{12.16}$$

where

$E =$ output voltage

$C_B =$ proportionality constant

$N_t =$ number of teeth

$\omega =$ angular velocity of the wheel

The angular velocity can be found either from the amplitude or the frequency of the output signal. The voltage amplitude signal is susceptible to noise and loading errors. Thus, less error is introduced if the frequency is used to determine the angular velocity;

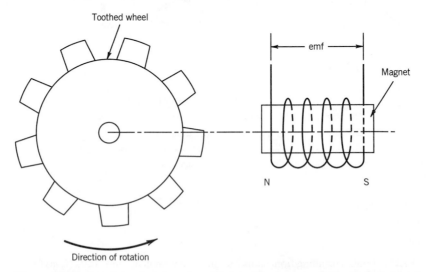

Figure 12.16 Angular velocity measurement employing a toothed wheel and magnetic pickup.

typically, some means of counting the pulses electronically is employed. This frequency information can be transmitted digitally for recording, which eliminates the noise and loading error problems associated with voltage signals.

Force Measurement

The measurement of force is most familiar as the process of weighing, ranging from weighing micrograms of a medicine to weighing trucks on the highway. Force is a quantity derived from the fundamental dimensions mass, length, and time. Standards and units of measure for these quantities are defined in Chapter 1. Some common techniques for force measurement are described in this section.

Load Cells

Load cell is a term used to describe a transducer that generates a voltage signal as a result of an applied force, usually along a particular direction. Such force transducers often consist of an elastic member and a deflection sensor. A technology overview for such devices is provided in [5]. These deflection sensors may employ changes in capacitance, resistance, or the piezoelectric effect to sense deflection. Consider first load cells, which are designed using a linearly elastic member instrumented with strain gauges.

Strain Gauge Load Cells   Strain gauge load cells are most often constructed of a metal, and have a shape such that the range of forces to be measured results in a measurable output voltage over the desired operating range. The shape of the linearly elastic member is designed to meet the following goals: (1) provide an appropriate range of force measuring capability with necessary accuracy and (2) provide sensitivity to forces in a particular direction, and have low sensitivity to force components in other directions.

A variety of designs of linearly elastic load cells are shown in Figure 12.17. In general, load cells may be characterized as beam-type load cells, proving rings, or columnar-type designs. Beam-type load cells may be characterized as bending beam load cells or shear beam load cells.

A bending beam load cell, as shown in Figure 12.18, is configured such that the sensing element of the load cell functions as a cantilever beam. Strain gauges are mounted on the top and bottom of the beam to measure normal or bending stresses. Figure 12.18 provides qualitative indication of the shear and normal stress distributions in a cantilever beam. In the linear elastic range of the load cell, the bending stresses are linearly related to the applied load.

In a shear beam load cell the beam cross section is that of an I-beam. The resulting shear stress in the web is nearly constant, allowing placement of a strain gauge essentially anywhere on the web with reasonable accuracy. Such a load cell is illustrated schematically in Figure 12.19, along with the shear stress distribution in the beam. In general, bending beam load cells are less costly due to their construction; however, the shear beam load cells have several advantages, including lower creep and faster response times. Typical load cells for industrial applications are illustrated in Figure 12.20.

Piezoelectric Load Cells   Piezoelectric materials are characterized by their ability to develop a charge when subject to a mechanical strain. The most common piezoelectric material is single-crystal quartz. The basic principle of transduction that occurs in a piezoelectric element may best be thought of as a charge generator and a capacitor. The

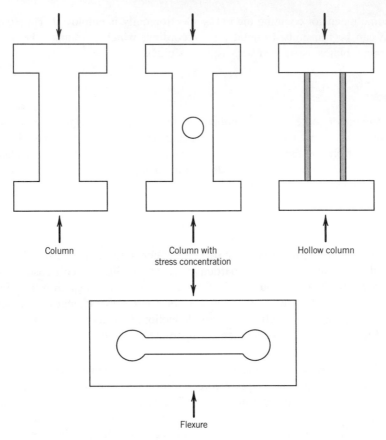

Column

Column with
stress concentration

Hollow column

Flexure

Figure 12.17 Elastic load cell designs.

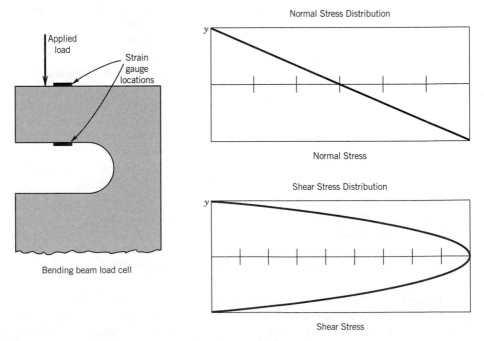

Applied
load

Strain
gauge
locations

Bending beam load cell

Normal Stress Distribution

Normal Stress

Shear Stress Distribution

Shear Stress

Figure 12.18 Bending beam load cell and stress distributions.

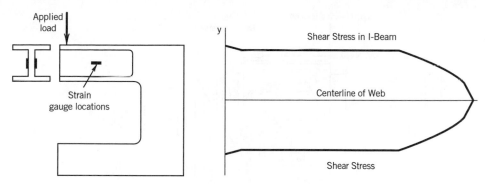

Figure 12.19 Shear beam load cell and shear stress distribution.

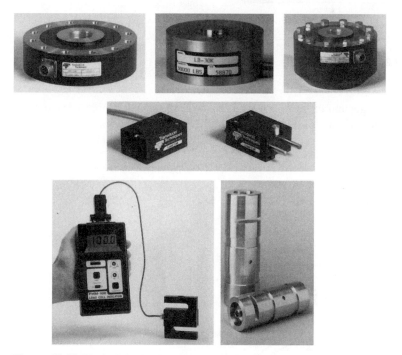

Figure 12.20 Typical load cells. (Courtesy of Transducer Techniques, Inc.)

frequency response of piezoelectric transducers is very high, because the frequency response is determined primarily by the size and material properties of the quartz crystal. The modulus of elasticity of quartz is approximately 85 GPa, yielding load cells with typical static sensitivities ranging from 0.05 to 10 mV/N, and frequency response up to 15,000 Hz. A typical piezoelectric load cell construction is shown in Figure 12.21.

Proving Ring   A ring-type load cell can be employed as a local force standard. Such a ring-type load cell, as shown in Figure 12.22, is often employed in the calibration of materials testing machines because of the high degree of precision and accuracy possible

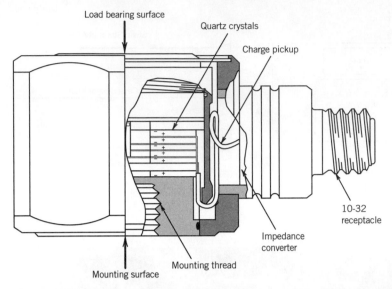

Figure 12.21 Piezoelectric load cell design. (Courtesy of The Kistler Instrument Co.)

with this arrangement of transducer and sensor. If the sensor is approximated as a circular right cylinder, the relationship between applied force and deflection is given by

$$\delta y = \left(\frac{\pi}{2} - \frac{4}{\pi}\right) \frac{F_n D^3}{16EI} \tag{12.17}$$

δy = deflection along the applied force
F_n = applied force

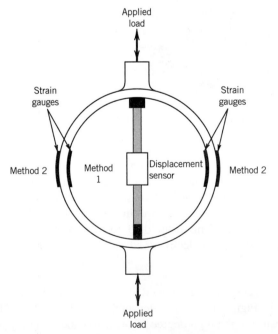

Figure 12.22 Ring-type load cell, or proving ring.

$D =$ diameter

$E =$ modulus of elasticity

$I =$ moment of inertia

The application of the proving ring involves measuring the deflection of the proving ring in the direction of the applied force. Typical methods for this displacement measurement include (1) displacement transducers, which measure overall displacement, and (2) strain gauges. These methods are illustrated in Figure 12.22.

Torque Measurements

Torque and mechanical power measurements are often associated with the energy conversion processes that serve to provide mechanical and electrical power to our industrial world. Such energy conversion processes are largely characterized by the mechanical transmission of power produced by prime movers such as internal combustion engines. From automobiles to turbine-generator sets, mechanical power transmission occurs through torque acting through a rotating shaft.

The measurement of torque is important in a variety of applications, including sizing of load-carrying shafts. This measurement is also a crucial aspect of the measurement of shaft power, such as in an engine dynamometer. Strain-gauge-based torque cells are constructed in a manner similar to load cells, in which a torsional strain in an elastic element is sensed by strain gauges appropriately placed on the elastic element. Figure 12.23 shows a circular shaft instrumented with strain gauges for the purpose of measuring torque.

Consider the stresses created in a circular shaft subject to a torque, T. The maximum shearing stress in a circular shaft occurs on the surface and may be calculated from the torsion formula[4]

$$\tau_{\max} = TR_0/J \tag{12.18}$$

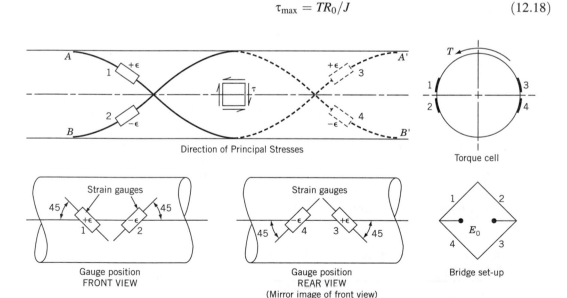

Figure 12.23 Direction of principal stresses for a shaft in pure torsion and corresponding shaft instrumented for torque measurement.

---

[4]Coulomb developed the torsion formula in 1775 in connection with electrical instruments.

where

τ_{max} = maximum shearing stress

T = applied torque

J = polar moment of inertia ($\pi R_o^4/2$ for a solid circular shaft)

For a circular shaft in pure torsion, there are no normal stresses, σ_x, σ_y, or σ_z. The principal stresses lie along a line, which makes a 45° angle with the axis of the shaft, as illustrated in Figure 12.23, and have a value equal to τ_{max}. Strains that occur along the helix-shaped curve labeled A–A' are opposite in sign from those that occur along B–B'. These locations allow placement of four active strain gauges in a Wheatstone bridge arrangement, and the direct measurement of torque in terms of bridge output voltage. A practical implementation is shown.

Mechanical Power Measurements

Almost universally, prime movers such as IC engines and gas turbines convert chemical energy in a fuel to thermodynamic work transmitted by a shaft to the end use. In automotive applications, the pistons create a torque on the crankshaft, which is ultimately transmitted to the driving wheels. In each case, the power is transmitted through a mechanical coupling. This section is concerned with the measurement of such mechanical power transmission.

Rotational Speed, Torque, and Shaft Power

Shaft power is related to rotational speed and torque as

$$\vec{P}_s = \vec{\omega} \times \vec{T} \tag{12.19}$$

where P_s is the shaft power, ω is the rotational velocity vector, and T the torque, vector. In general, the orientations of the torque and rotational velocity vectors are such that the equation may be written in scalar form as

$$P_s = \omega T \tag{12.20}$$

Table 12.3 provides a summary of useful equations related to shaft power, torque and speed as employed in mechanical measurements. Historically, a device called a Prony brake was used to measure shaft power. A typical Prony brake arrangement is shown in Figure 12.24. Consider using the Prony brake to measure power output for an IC engine. The Prony brake serves to provide a well-defined load for the engine, with the power output of the engine dissipated as thermal energy in the braking material. By adjusting the load, the power output over a range of speeds and throttle settings can be realized. The power is measured by recording the torque acting on the torque arm, and the rotational

Table 12.3   Shaft Power, Torque, and Speed Relationships

	SI	U.S. Customary
Shaft power, P	$P = \omega T$	$P = \frac{2\pi n T}{550}$
Power	$P(\text{W})$	$P(\text{hp})$
Rotational speed	ω (rad/s)	n(rev/sec)
Torque	T (N m)	T (ft lb)

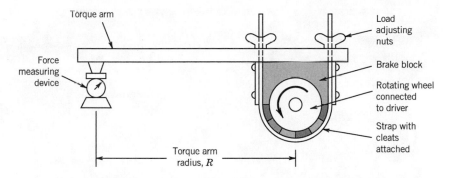

Figure 12.24 Prony brake. (Courtesy of the American Society of Mechanical Engineers, New York. Reprinted from PTC 19.7-1980 [10].)

speed of the engine. Clearly, this device is limited in speed and power, but does serve to demonstrate the operating principles of power measurement, and is historically significant as the first technique for measuring power.

Cradled Dynamometers

A Prony brake is an example of an absorbing dynamometer. The term *dynamometer* refers to a device that absorbs and measures the power output of a prime mover. Prime movers are large mechanical power-producing devices such as gasoline or diesel engines or gas turbines. Several methods of energy dissipation are utilized in various ranges of power, but the measurement techniques are governed by the same underlying principles. We will consider first the measurement of power, and then discuss means for dissipating the sometimes large amounts of power generated by prime movers.

The cradled dynamometer measures mechanical power by measuring the rotational speed of the shaft, which transmits the power, and the reaction torque required to prevent movement of the stationary part of the prime mover. This reaction torque is impressively illustrated as "wheelies" by motorcycle riders. A cradled dynamometer is supported in bearings, which are called trunion bearings, such that the reaction torque is transmitted to a torque or force-measuring device. The state-of-the-art dynamometer shown in Figure 12.25 is designed for emissions testing, with power absorption ratings above 200 hp and a top speed of 120 mph.

In principle, the operation of the dynamometer involves the steady-state measurement of the load F_r created by the reaction torque and the measurement of shaft speed. From equation (12.19) the transmitted shaft power can be calculated directly.

The ASME Performance Test Code [6] provides guidelines for the measurement of shaft power. According to [6], overall uncertainty in the measurement of shaft power by a cradled dynamometer results from: (a) trunnion bearing friction; (b) force measurement uncertainty (F_r); (c) moment arm length uncertainty (L_r); (d) static unbalance of dynamometer; and (e) uncertainty in rotational speed measurement.

A means of supplying a controllable load to the prime mover, and dissipating the energy absorbed in the dynamometer, are an integral part of the design of any dynamometer. Several techniques are described for providing an appropriate load.

Eddy Current Dynamometers   A direct current field coil and a rotor allow shaft power to be dissipated by eddy currents in the stator winding. The resulting conversion to

Figure 12.25 Dynamometer (Courtesy of Burke E. Porter Machinery Co., Grand Rapids, MI.)

thermal energy by Joulian heating of the eddy currents necessitates some cooling be supplied, typically using cooling water.

Alternating Currrent and DC Generators   Cradled ac and dc machines are employed as power absorbing elements in dynamometers. The ac applications require variable frequency capabilities to allow a wide range of power and speed measurements. The power produced in such dynamometers may be dissipated as thermal energy using resistive loads.

Waterbrake Dynamometers   A waterbrake dynamometer employs fluid friction and momentum transport to create a means of energy dissipation. Two representative designs are provided in Figure 12.26. The viscous shear type brake is useful for high rotational speeds, and the agitator type unit is used over a range of speeds and loads. Waterbrakes may be employed for applications up to 10,000 HP (7450 kW). The load absorbed by waterbrakes can be adjusted using water level and flow rates in the brake.

12.3   ACTUATORS

Linear Actuators

The task of a linear actuator is to provide motion in a straight line. We discuss three ways to achieve linear motion:

1. Conversion of rotary motion into linear motion. This can be accomplished using a linkage, as in the slider-crank mechanism, or using screw threads coupled to a rotary motion source.
2. Use of a fluid pressure to move a piston in a cylinder. When air or another gas is used as the working fluid, the system is called a *pneumatic system*. When a fluid such as oil is used as the working fluid, the system is termed *hydraulic*.
3. Electromagnetic

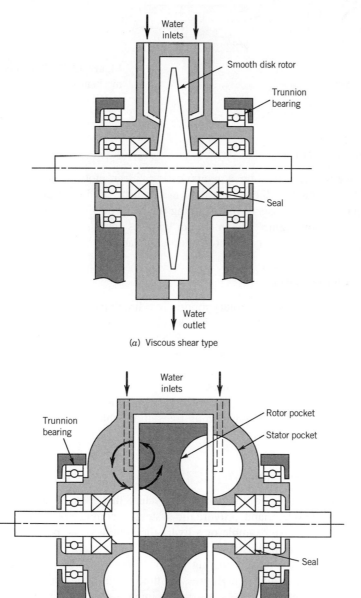

(a) Viscous shear type

(b) Momentum exchange type

Figure 12.26 Waterbrake dynamometers. (Courtesy of the American Society of Mechanical Engineers, New York. Reprinted from PTC 19.7-1980 [10].)

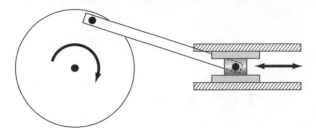

Figure 12.27 Slider-crank mechanism.

Slider-Crank Mechanism

A common means of generating a reciprocating linear motion, or converting linear motion to rotary motion, is the slider-crank mechanism, as illustrated in Figure 12.27. Such a mechanism is the basis of transforming the reciprocating motion of the piston in an internal combustion engine. This or similar linkages could also be applied in pick and place operations, or in a variety of automation applications.

Screw-drive Linear Motion

A common means for translating rotary motion into linear motion is a lead screw. A lead screw has helical threads that are designed for minimum backlash to allow precise positioning. Numerous designs exist for such actuating threads. The basic principle is illustrated in Figure 12.28. The rotary motion of the lead screw is translated into linear motion of the nut, with the torque required to drive the lead screw directly related to the thrust the particular application requires.

Common applications that employ a lead screw include the work table for a mill, and a variety of other precision positioning translation tables, such as the one shown in Figure 12.29.

Pneumatic and Hydraulic Actuators

The term *pneumatic* implies a component or system that uses compressed air as the energy source. On the other hand, a hydraulic system or component uses incompressible oil as the working fluid. An example of a hydraulic system is the power steering on an automobile; such a system is illustrated in Figure 12.30. Hydraulic fluid is supplied at an elevated pressure from the power steering pump. When a steering input is made from the driver, the rotary valve allows high-pressure fluid to enter the appropriate side of the piston and aid in turning the wheels. By maintaining a direct connection between the steering column and the rack and pinion, the car can be steered even if the hydraulic system fails.

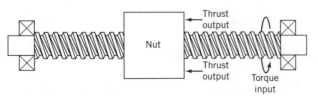

Figure 12.28 Linear actuation using a lead screw.

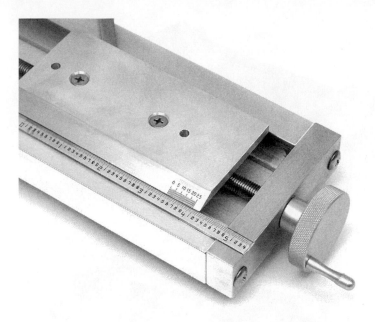

Figure 12.29 Precision translation table. (UniSlide® from Velmex, Inc).

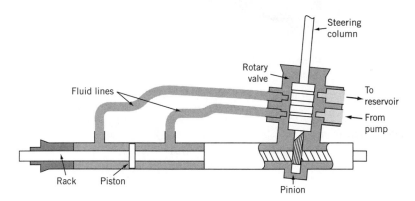

Figure 12.30 Schematic diagram of a power steering system.

Pneumatic Actuators

When compressed air is the energy source of choice, a pneumatic cylinder can create linear motion. In general, the purpose of a pneumatic cylinder is to provide linear motion between two fixed locations. Figure 12.31 shows a pneumatic cylinder and a cutaway of such a cylinder. By applying high-pressure compressed air to either side of the piston, linear actuation between two defined positions can easily be accomplished.

Solenoids

Solenoid is a term used to describe an electromagnetic that is employed to create linear motion of a plunger, as shown in Figure 12.32. The initial force available from a solenoid

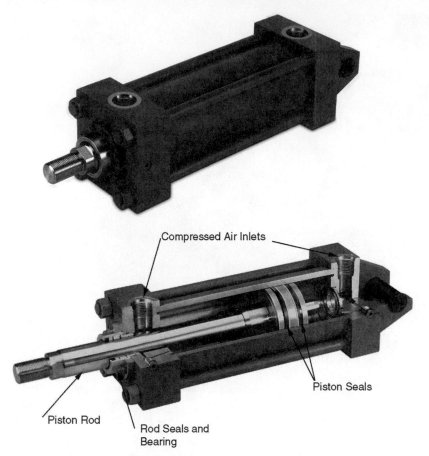

Figure 12.31 Construction of a pneumatic cylinder.

can be determined from

$$F = \frac{1}{2}(NI)^2 \frac{\mu A}{\delta^2}$$ (12.21)

where

F = force on plunger

N = number of turns of wire in the electromagnetic

I = current

μ = magnetic permeability of air $(4\pi \times 10^{-7}\,\mathrm{H/m})$

δ = size of the air gap

A = plunger cross-sectional area

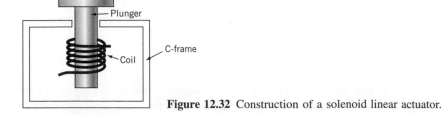

Figure 12.32 Construction of a solenoid linear actuator.

When the electromagnet is actuated, the resulting magnetic force pulls the plunger into the C-frame. Because the air gap is largest when the electromagnet is actuated, the minimum force occurs at actuation, and the force increases as the air gap decreases. Equation (12.21) is clearly not valid in the limit as $\delta \to 0$ and should only be applied to determine the initial force.

Rotary Actuators

There is a class of electric motors that has the primary purpose of providing power to a process. An example is the electric motor that drives an elevator, an escalator, or a centrifugal blower. In these applications the electric motor serves as a prime mover, with clear and specific requirements for rotational speed, torque, and power. However, some applications have stringent requirements for positioning.

Rotary positioning presents a significant engineering challenge, but one that is so ubiquitous that it has been addressed through a variety of design strategies. One design strategy uses a free-rotating dc motor to supply the motive power, and impose precise control on the resulting motion through gearing and some control scheme. DC motors that are subject to feedback control are generally described as servo-motors. While this may be appropriate and necessary for some applications, the stepper motor has found wide-ranging applications in precision rotary motion control, and is a better choice for many applications.

Stepper Motors

Stepper or stepping motors, as their name implies, are capable of moving a fraction of a rotation with a great degree of precision. This is accomplished by the design of a rotor that aligns with the magnetic field generated by energized coils. The step size can range from $90°$ to as little as $0.5°$. Two common types of stepper motors are variable reluctance and unipolar designs. The design of a variable reluctance stepping motor is illustrated in Figure 12.33. Let us consider the operation of this motor. There are three sets of windings, labeled 1, 2, and 3 in the figure, and there are two sets of teeth on the rotor, labeled X and Y. With the windings labeled 1 energized, the rotor snaps to a position where one set of the teeth is aligned with the windings. This motion is a result of the magnetic field generated by the windings. Suppose that winding 1 is de-energized and winding 2 is energized. The rotor will turn until the teeth marked Y are aligned with winding 2. This produces a $30°$ step.

A useful characteristic of stepper motors is holding torque. As long as one of the windings is energized, the rotor will resist motion, until the torque produced by the winding rotor interaction is overcome.

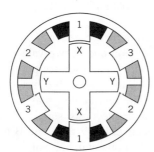

Figure 12.33 Variable reluctance stepper motor design.

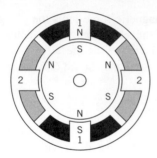

Figure 12.34 Variable reluctance stepper motor design having six poles and two windings.

The motor shown in Figure 12.34 is a variable reluctance design. Unipolar motors incorporate permanent magnets as the rotor. Figure 12.34 shows a rotor having six magnetic poles and two sets of windings. The motor will move in 30° increments as the windings are alternately energized.

Flow-Control Valves

Valves are mechanical devices intended to allow, restrict, throttle, or meter fluid flow through pipes or conduits. Flow-control valves are used to regulate either the flow or the pressure of a fluid by their electronic actuation. They generally function by allowing flow while in their open position, restricting flow when closed, and metering flow with a position that is somewhere between these settings. These valves contain a valve positioning element that is driven by an actuator, such as a solenoid. Any valve type can be controlled. The common control valve design offers either a single chamber body containing a poppet with valve seat or a multichamber body containing a sliding spool with multiple poppets. In either case, the position element motion is controlled by a solenoid.

Flow-control valves are used to transfer gases, liquids, and hydraulic fluids. The application ratings are as follows: general service, for working with common liquids and gases; cryogenic service, for fluids such as liquid oxygen; vacuum service, for low-pressure applications; and oxygen service, for a contamination-free flow of oxygen.

The control valve can respond to signals from any type of process variable transducer. The signal determines the position of the actuating solenoid. Solenoids are electromechanical devices that consist of a wire coil and a movable plunger, which seats against the coil. When current is applied to the coil, an actuating magnetic field is created. Electrical current is supplied to the solenoid coil, and the resulting magnetic field acts on the plunger, whose resulting motion actuates the valve positioning element.

A specific characteristic of any control valve refers to whether its nonenergized operating state is open or closed. This is referred to as its "fail position." The fail position of a control valve is determined by the nonenergized solenoid plunger position. This position is an important consideration for process safety.

These valves come in various configurations reflecting their number of ports. A two-way valve has two ports. Two-way position control takes on one of two values: open or closed. A two-way valve has two connections: supply port (P) and service port (A). Most common household valves fall into this category. A three-way valve has three port connections: supply (P), exhaust (T), and service (A). The service port may be switched between the supply and the exhaust. A four-way has four connections: supply (P), exhaust (T), and two service ports (A and B). The valve connects either P to A and B to T or P to B and A to T. In general, an N-way valve has N ports with N number of flow

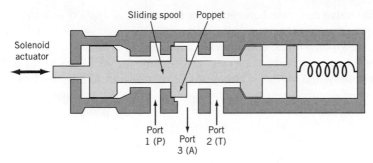

Figure 12.35 Three-way flow control valve (deactivated position shown).

directions available. An example of a three-port, sliding spool control valve is depicted in Figure 12.35. The solenoid drives the spool, which contains two valve seats. In the fully activated position, port P is open to service port A. When the solenoid is deactivated, port T is open to service port A. For example, in one application this valve can be used to pressurize a system (open the system to port P) for a period of time and then adjust the system pressure to another value (open the system to port A) for a period of time.

All valve ports offer some level of flow resistance, and this is specified through a flow coefficient, C_v. Flow resistance can be adjusted in design by varying the internal dimensions of the valve chamber and can be set operationally by varying the element position within the chamber. The flow coefficient is easily found based on the formulation detailed in Chapter 10 and found as

$$Q = C_v \sqrt{\Delta p} \qquad (12.22)$$

where Q is the steady flow rate through the valve and Δp is the corresponding pressure drop. This loss is also expressed in terms of a K factor based on the average velocity through the ports:

$$\Delta p = K \rho \overline{U}^2 / 2 \qquad (12.23)$$

Flow-control valves are classified in a number of ways: the type of control, the number of ports in the valve housing, the specific function of the valve, and the type of valve element used in the construction of the valve. Directional-control valves allow or prevent the flow of fluid through designated ports. Flow can move in either direction. Check valves are a special class of directional valve that allow flow in only one direction. Proportional valves can be infinitely positioned to control the amount, pressure, and direction of fluid flow. In a proportional valve, the valve is opened by an amount proportional to the applied current. The valves are termed proportional because their output flow is not exactly linear in relation to the input signal. These valves provide a way to control pressure or flow rate with a high response rate.

In the simplest application, a solenoid is used to turn a valve either on or off. In a more demanding application, the solenoid is expected to cycle rapidly to open and close the valve. The time between each signal cycle coupled with the internal flow loss character of the valve determines the average flow. Valve time response can be defined in several ways but all are consistent with the methods used in Chapter 3. The 90% response time, t_{90}, is the time required to either fill or exhaust a target device chamber through a valve port, in effect a step function response. There will be a separate time response for filling or exhausting. Either way,

$$t_{90} = m + F \forall \qquad (12.24)$$

where m is the valve lag time between when the signal is applied and steady flow is established at the designated port, F is the reciprocal of the average flow rate through the port, and $\forall$ is the volume of the target device chamber. For example, a valve having an F of 0.54 ms/cm^3 and a lag time of 20 ms will require $t_{90} = 155$ ms to fill a 250-cm^3 chamber. Alternately, the valve frequency response can be found by cycling the valve with a sine wave electrical signal and measuring the flow rate through the valve. The valve frequency bandwidth is thus established by its -3-dB point.

12.4 CONTROLS

Control of a process or system can be exerted in a wide variety of ways. Suppose our goal is to create a healthy lawn by appropriate watering. Each day we could monitor the weather forecast, take into account the probability of precipitation, and choose whether to water and for how long. We could choose to water all of our lawn or just those parts most subject to stress from heat and lack of moisture. If we choose to water, we could place the sprinklers and turn on the faucet (hopefully remembering to shut off the flow at an appropriate later time!).

All of the functions described above for lawn care are completely reasonable for a person to accomplish, and they represent the functioning of an intelligent controller. Suppose we wish to introduce some automation into the process.

At the simplest level, a timer-based control system could be implemented, as shown in Figure 12.36. The functioning of this system would be to open and close the faucet at predetermined times of the day. At the simplest level, this could be a mechanical timer that watered the lawn once each 24-hr period for a predetermined length of time. This type of control is called **open-loop control**. For this control system there are no sensors to monitor the amount of water applied to the lawn; in fact, all the control system is accomplishing is to open the faucet.

More advanced automatic control systems implement **closed-loop control**. For the present example, it might be desired to apply 100 gallons of water to the lawn. A flow meter that sensed the total water flow that had occurred could be used to provide feedback to the control system to allow the faucet to be closed when the flow totaled 100 gallons. Such flow meters are common and serve as your water meter. The term *closed-loop* or *feedback* control simply implies that the variable that is to be controlled is being

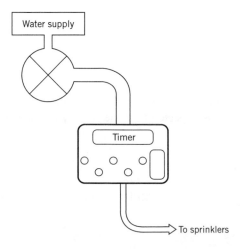

Figure 12.36 Open-loop control of a sprinkler system.

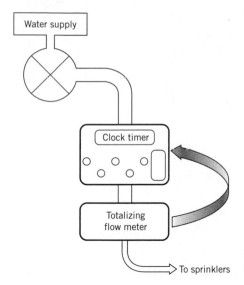

Figure 12.37 Feedback control of a sprinkler system.

measured, and that the control system in some way uses this measurement to exert the control. Another familiar example of a closed-loop control scheme is a motion-sensing outdoor light. The actuation of the light is based on the presence or absence motion in the range of the sensor and the light will turn off after a pre-set interval if motion is no longer sensed.

Figure 12.37 illustrates a control system described for applying 100 gallons of water to our lawn. There are two inputs to the controller, the time of day, and the output of the totalizing flow meter. At the appropriate time of day, the controller opens the valve. The totalizing flow meter output is used by the controller to close the valve after the total flow reaches 100 gallons. This type of control scheme is termed **on–off control**. The valve controlling the water flow is either fully open or fully closed.

Probably the most familiar form of an on–off control system is the thermostat for a home furnace or air conditioner. Figure 12.38 shows the status of a home furnace and a time trace of the inside temperature during a winter day. A schematic representation of this control system is shown in Figure 12.39. A key element here is that there is the possibility of a **disturbance** that would influence the rate of change of the inside temperature. Suppose a delivery arrives, and the door remains open for a period of time. The thermostat must then respond to this disturbance and attempt to maintain the inside temperature at the set point.

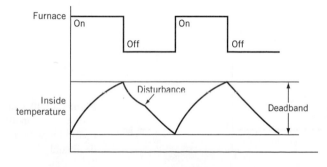

Figure 12.38 Operation of on–off controller with a deadband.

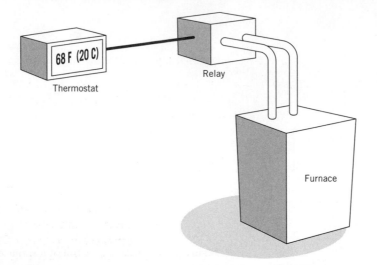

Figure 12.39 Components of a thermostatic control for a home furnace.

Essentially all practical implementations of on–off control systems require a dead-band that creates a hysteresis loop in the control action. This is illustrated in Figure 12.40. The deadband is centered around zero error, and the action of the controller depends on the magnitude of the error. Here the error is defined as

$$e = T_{\text{set point}} - T_{\text{room}}$$

Recall that we are considering a furnace thermostat under winter conditions. As the room temperature falls relative to the set point, the error becomes a larger positive number. When the error reaches the value corresponding to "Furnace ON" in Figure 12.40, the furnace begins to add heat to the conditioned space. Room temperature begins to rise, and the error decreases toward zero. The furnace remains on until the temperature reaches a predetermined value that is greater than the set point. Here the error is negative. At this temperature, the furnace is turned off, and room temperature begins once again to decrease. Because of the deadband in the controller, no further control action occurs until the error reaches the "Furnace ON" error magnitude.

Many control systems are designed to maintain a specified set point, without a deadband. Suppose that due to varying water pressure our lawn was being watered nonuniformly. We might choose to control both the water flow rate and the total water

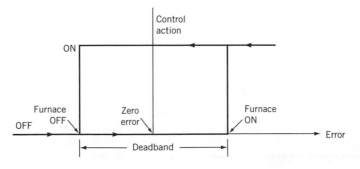

Figure 12.40 Control action of an on–off controller with a deadband.

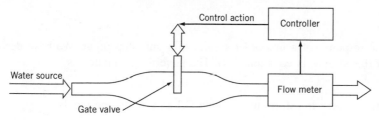

Figure 12.41 Flow rate control system.

flow applied to the lawn. A scheme for accomplishing this is shown in Figure 12.41. The key components of the control system are a flow meter, the controller, and an actuator that can control the position of the gate valve. Here a desired flow rate is set, say 2 gal/min. The task of the controller is to vary the position of the gate valve in order to maintain the flow rate at the set point.

Dynamic Response

As another example, consider the cruise control system on an automobile. Once a desired speed is set, the system varies the throttle position to ensure that the set point is maintained. A central issue in the analysis and design of control systems lies in the dynamic response of the physical process, the controller, and the actuators. This is another example of closed-loop control. But an automobile does not respond instantaneously to changes in the throttle position, just as a thermal sensor needs time to respond to changes in ambient temperature, and time is required for a stepper motor to change the position of a valve. The complexities of the combined responses of the physical and control systems, especially in the presence of disturbances, are the subject of the remainder of this section.

Laplace Transforms

In Chapter 2, the Fourier transform allowed us to view time-dependent data in frequency space and provided insight into data that was simply impossible to quantify by examining a time trace of the signal. In a similar manner, we will find that the Laplace transform has distinct advantages for designing and analyzing control systems. In Chapter 3 we used Laplace transforms to model the response of simple systems. Now let us apply the discussion of Laplace transforms from Chapter 3 to systems control. The fundamental basis of the application of Laplace transforms to control systems is the solution of a mathematically well-posed initial value problem.

Consider an initial-value problem that is described by an ordinary differential equation having time as the independent variable. *If we apply the Laplace transform to such a differential equation, we convert it to an algebraic equation.* For partial differential equations in time and one spatial variable, the Laplace transform converts the partial differential equation into an ordinary differential equation in the space variable. Appendix C reviews the application of Laplace transforms and provides a table of Laplace transform pairs.

We illustrate the development of the Laplace transform by its application to first-order and second-order systems.

EXAMPLE 12.3

Consider the response of a first-order system to a unit step input. We have derived the response of this system in equation (3.4). The governing equation is

$$\tau \dot{y} + y = U_s(t) \tag{12.25}$$

Applying the Laplace transform using Table C.1 as a reference yields

$$\tau s Y(s) - y(0) + Y(s) = \frac{1}{s} \tag{12.26}$$

where we have employed the property of the Laplace transform that

$$\mathcal{L}\left[\tau \frac{dy(t)}{dt}\right] = \tau \mathcal{L}\left[\frac{dy(t)}{dt}\right]$$

The process of solving the differential equation involves:

1. Applying the Laplace transform to the governing differential equation (12.25)
2. Solving the resulting equation for $Y(s)$, equation (12.26)
3. Employing the table of Laplace transform pairs, Table C.1, to determine $y(t)$

We will assume that $y(0) = 0$. Solving equation (12.26) for $Y(s)$ yields

$$Y(s) = \frac{1}{s(\tau s + 1)} \tag{12.27}$$

An important tool for the inversion of Laplace transforms is partial fractions. To accomplish the inversion of equation (12.27) we start from the assumption

$$\frac{1}{s(\tau s + 1)} = \frac{A}{s} + \frac{B}{(\tau s + 1)} \tag{12.28}$$

Next we find a common denominator for the right-hand side of equation (12.28):

$$\frac{1}{s(\tau s + 1)} = \frac{A(\tau s + 1)}{s(\tau s + 1)} + \frac{Bs}{s(\tau s + 1)}$$

In this form we see that

$$1 = A(\tau s + 1) + Bs$$

Equating powers of s yields

$$1 = A$$
$$0 = A\tau + B \Rightarrow B = -\tau$$

We can now express $Y(s)$ as

$$Y(s) = \frac{1}{s} + \frac{-\tau}{(\tau s + 1)} = \frac{1}{s} + \frac{-1}{(s + 1/\tau)} \tag{12.29}$$

From transform pairs 1 and 5 in Table C.1 we find that

$$y(t) = 1 - e^{-t/\tau} \tag{12.30}$$

This is equivalent to equation (3.6)

$$\Gamma(t) = \frac{y(t) - y_\infty}{y_0 - y_\infty} = e^{-t/\tau}$$

where τ is the time constant for a first-order system.

EXAMPLE 12.4

Employing Laplace transforms, determine the solution of the second-order, ordinary linear differential equation,

$$\ddot{y} + 4\dot{y} - 5y = 0 \tag{12.31}$$

with the initial conditions

$$y(0) = -1$$
$$y'(0) = 4$$

SOLUTION   We recall that, (see equation C.5)

$$\mathcal{L}\left[\frac{d^n y(t)}{dt}\right] = s^n y(s) - s^{n-1}y(0) - s^{n-2}\frac{dy(0)}{dt} - \cdots - \frac{d^{n-1}y(0)}{dt^{n-1}}$$

If we apply this relationship to equation (12.31), we find that the Laplace transform is

$$s^2 Y(s) - sy(0) - \dot{y}(0) + 4[sY(s) - y(0)] - 5Y(s) = 0$$

Substituting the initial conditions yields

$$s^2 Y(s) - s - 4 + 4[sY(s) + 1] - 5Y(s) = 0$$

Solving for $Y(s)$ yields

$$Y(s) = \frac{-s}{s^2 + 4s - 5} \tag{12.32}$$

Now we are faced with the task of determining the inverse Laplace transform of this expression. Factoring the denominator yields

$$Y(s) = \frac{-s}{(s+5)(s-1)} \tag{12.33}$$

Applying partial fractions allows equation (12.33) to be expressed:

$$\frac{-s}{(s+5)(s-1)} = \frac{-5/6}{s+5} + \frac{-1/6}{s-1} \tag{12.34}$$

Finding the inverse Laplace transform from Table C.1 provides the solution for equation (12.34) as

$$y(t) = -\frac{5}{6}e^{-5t} - \frac{1}{6}e^{t} \tag{12.35}$$

Block Diagrams

A very useful representation of feedback control systems is accomplished using block diagrams. We'll first describe the basic elements of a block diagram.

Operational Blocks

Consider a public address system consisting of a microphone, an amplifier, and speakers. Figure 12.42 provides a single-input, single-output block representing the amplifier. An ideal amplifier would follow exactly the waveform of the input signal from the microphone, and simply multiply the voltage signal by a constant value, κ. This constant value is the **gain** of the amplifier.

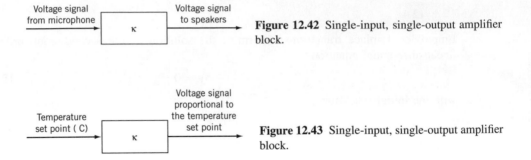

Figure 12.42 Single-input, single-output amplifier block.

Figure 12.43 Single-input, single-output amplifier block.

In Figure 12.43 a gain block is shown that supplies a reference voltage signal based on a temperature set point. In most practical applications, signals are transmitted in control systems as voltage or current. The gain represented in Figure 12.42 would have units of V/°C. **We note that a pure linear gain is equivalent to the static sensitivity of a zero-order system.**

To construct a block diagram that represents a control system, we need to introduce a second important type of block. This block is a comparator. Figure 12.44 illustrates the operation of a comparator. Two voltage signals are either added or subtracted by the comparator. In Figure 12.44 the signals represent the desired temperature set point and the measured temperature, in terms of voltages. The difference in these two values represents the error in the temperature value, the difference between the measured and set-point values.

We can now construct a block diagram of the thermostat for the home furnace described in Figures 12.38 through 12.40. Figure 12.45 shows the block diagram of the

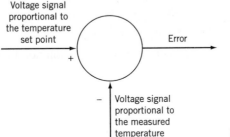

Figure 12.44 Comparator.

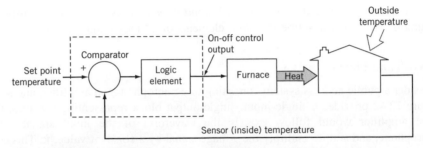

Figure 12.45 Block diagram representation of a furnace thermostat.

controller and the furnace and the house. Together the furnace and the house are usually referred to as the plant or the process.

The detailed design of a control system requires that we consider the time-dependent behavior of both the controller and the process. We will propose a process and derive the governing equation for the process. Then a controller will be implemented and the dynamic response of the system to a step change in the set point will be derived. A key point in the modeling of control systems is the concept of a stationary operating point.

Model for Oven Control

Plant Model

The system we wish to control is illustrated in Figure 12.46. An oven is maintained at a temperature above the ambient temperature by power input from an electric heater. The controller will be implemented to maintain a desired oven temperature.

A first law analysis of the oven at steady-state conditions yields

$$\dot{P} = \dot{Q}_{\text{loss}}$$

The energy loss from the oven may be expressed in terms of the oven temperature, T, the ambient temperature, T_∞, the surface area, A_s, and an overall loss coefficient, U, as

$$\dot{Q}_{\text{loss}} = UA_s(T - T_\infty) \tag{12.36}$$

Consider a steady-state operating condition for the oven, designated by the subscript "o." At steady state the power supplied by the heater is exactly balanced by the energy lost to the ambient,

$$\dot{P}_o = UA_s(T_o - T_\infty) \tag{12.37}$$

To aid in the mathematical analysis, we define a new temperature variable as

$$\theta = T - T_\infty \tag{12.38}$$

so that equation (12.37) can be expressed as

$$\dot{P}_o = UA_s\theta_o \tag{12.39}$$

The first law for the oven for a transient condition, one where the power input is changed, is

$$\dot{P} - UA_s\theta = mc\frac{d\theta}{dt} \tag{12.40}$$

where m is the mass of the oven and c is the average specific heat. This product represents the total heat capacity of the oven. Equation (12.40) now provides the governing equation for the plant in this example.

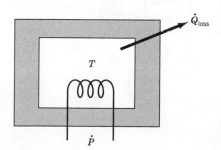

Figure 12.46 Energy flows for an oven.

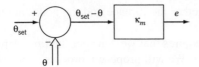

Figure 12.47 Proportional control.

Controller Model

We now wish to implement a controller. The first controller we will implement is termed a *proportional* controller and is illustrated in Figure 12.47. The error between the set-point temperature and the actual temperature of the oven is multiplied by a constant value to determine the power input to the heater. A block diagram of the controller and plant combined system is shown in Figure 12.48; two proportional gains are shown in this figure to emphasize that the error signal will most likely be a voltage.

The governing equation for the temperature of the oven may be expressed as

$$\frac{mc}{[UA_s + \kappa_p\kappa_m]}\frac{d\theta}{dt} + \theta = \frac{\kappa_p\kappa_m}{[UA_s + \kappa_p\kappa_m]}\theta_{set} \tag{12.41}$$

By comparison with equation (3.4),

$$\tau\dot{y} + y = KF(t)$$

this represents the response of a first-order system having a time constant of

$$\tau = \frac{mc}{[UA_s + \kappa_p\kappa_m]} \tag{12.42}$$

The analysis for this step change is best accomplished by using Laplace transforms.

Laplace Transform Analysis

Taking the Laplace transform of equation (12.42) yields

$$\frac{mc}{[UA_s + \kappa_p\kappa_m]}s\Theta(s) + \Theta(s) = \frac{\kappa_p\kappa_m\Theta_{set}(s)}{[UA_s + \kappa_p\kappa_m]} \tag{12.43}$$

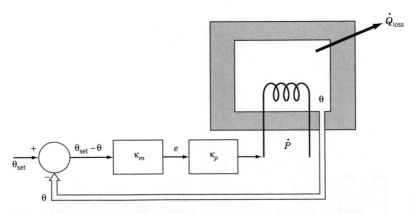

Figure 12.48 Proportional control of an oven.

which is of the form, as in Chapter 3,

$$(\tau s + 1)\Theta(s) = KF(s)$$

Solving equation (12.43) for the ratio $\Theta(s)/\Theta_{set}(s)$ yields

$$\frac{\Theta(s)}{\Theta_{set}(s)} = \frac{\dfrac{\kappa_p \kappa_m}{\left[UA_s + \kappa_p \kappa_m\right]}}{\left[\dfrac{mc}{UA_s + \kappa_p \kappa_m} s + 1\right]} = KG(s) \tag{12.44}$$

For convenience, define

$$C_1 = \frac{mc}{UA_s + \kappa_p \kappa_m}$$
$$C_2 = \frac{\kappa_p \kappa_m}{UA_s + \kappa_p \kappa_m} \tag{12.45}$$

With these definitions, equation (12.44) becomes

$$\frac{\Theta(s)}{\Theta_{set}(s)} = \frac{C_2}{[C_1 s + 1]} = G(s) \tag{12.46}$$

Equation (12.46) represents the transfer function, $G(s)$, for the system consisting of the oven and the PI controller. For the linear system consisting of the oven and controller, the transfer function represents the ratio of the Laplace transforms of the output to the input assuming a zero initial condition.

Step Response

Suppose we consider the start-up of the furnace, with an initial temperature equal to the ambient, or $\theta = 0$. From this condition we impose a value of θ_{set} that is larger than the ambient temperature, say θ_1; this represents a step-change input that is expressed in Laplace transform space as

$$\Theta_{set}(s) = \frac{\theta_1}{s} \tag{12.47}$$

The transform of the governing differential equation becomes

$$\Theta(s) = \frac{\theta_1 C_2}{s[C_1 s + 1]} \tag{12.48}$$

Using partial fractions yields

$$\Theta(s) = \theta_1 C_2 \left[\frac{1}{s} - \frac{1}{s + 1/C_1}\right] \tag{12.49}$$

Employing Table C.1 we can determine the time-domain solution corresponding to this Laplace transform as

$$\theta(t) = \theta_1 C_2 \left[1 - e^{-t/C_1}\right] \tag{12.50}$$

Recall that we have assumed $\theta(0) = 0$. In the limit as $t \to \infty$,

$$\theta(t) = \theta_1 C_2 = \frac{\kappa_p \kappa_m \theta_1}{UA_s + \kappa_p \kappa_m}$$

For a step-change input, we have found that $\theta(t) \neq \theta_1$ unless $[\kappa_p \kappa_m/(UA_s + \kappa_p \kappa_m)] = 1$. This will not in general be the case. *Thus we find that a proportional controller, in general, is characterized by a nonzero steady-state error.*

EXAMPLE 12.5

Consider the proportional control of an oven that has a total mass of 20 kg, with an average specific heat of 800 J/kg-K. The oven is initially at room temperature. A step input is supplied to the controller changing the set point temperature to 100°C above the ambient. Plot the temperature of the oven and the power supplied to the oven as a function of time, and determine the steady-state error if the values of the gains in the system are $\kappa_m = 20$ and $\kappa_p = 12$.

SOLUTION   Equation (12.50) provides the solution for oven temperature as a function of time for the case where there is a step change in the setpoint temperature. By examining Figure 12.48, we see that our controller provides a power to the oven given by

$$\kappa_p \kappa_m (\theta_1 - \theta)$$

The temperature response and the supplied power are provided in Figure 12.49.

COMMENT   The values of the gains in this problem are directly tied to the power required for the oven. The physical arrangement of heaters and the required electrical service will be a strong consideration in a practical implementation of this conceptual design. The effect of the controller gains and the system parameters can be further explored using the program *oven-1.vi*.

Proportional-Integral (PI) Control

The steady-state error that exists for a proportional controller may not be acceptable in many applications. Adding a control signal that is proportional to the time integral of the error improves the performance of the proportional controller. Let us first examine the behavior of a pure integral controller.

Integral Control

Figure 12.50 provides a block diagram in the Laplace transform domain for an integral controller applied to control the temperature of our oven. In the time domain, this

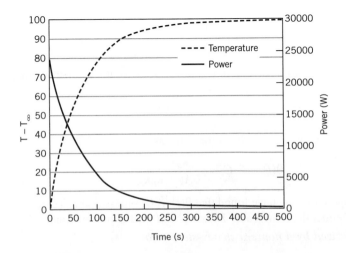

Figure 12.49 Response of oven-controller system.

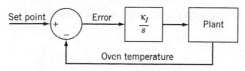

Figure 12.50 Block diagram for an integral controller.

results in

$$\dot{P}(t) = \kappa_I \int_0^t e(t)dt + \dot{P}(0) \tag{12.51}$$

We immediately see that as long as the error remains finite and positive, the power will continue to increase. Clearly, a pure integral controller would have very limited application.

Proportional-Integral (PI) Control

Suppose the actions of the proportional and integral controllers are combined, as shown in Figure 12.51. In Laplace transform space this may be expressed

$$P(s) = \kappa_p \kappa_m E(s) + \frac{\kappa_I \kappa_m}{s} E(s) = \left[\kappa_p \kappa_m + \frac{\kappa_I \kappa_m}{s} \right] E(s) \tag{12.52}$$

Expressed in terms of the set point, the closed-loop transfer function is

$$\frac{\Theta(s)}{\Theta_{\text{set}}(s)} = \frac{\kappa_m \left(\kappa_p + \frac{\kappa_I}{s} \right) \left(\frac{C_2}{C_1 s + 1} \right)}{1 + \kappa_m \left(\kappa_p + \frac{\kappa_I}{s} \right) \left(\frac{C_2}{C_1 s + 1} \right)} \tag{12.53}$$

Clearing fractions yields a form of the transfer function that allows the inverse transform to be determined:

$$\frac{\Theta(s)}{\Theta_{\text{set}}(s)} = \frac{\kappa_m \left(\kappa_p s + \kappa_I \right) C_2}{C_1 s^2 + s \left(1 + C_2 \kappa_m \kappa_p \right) + C_2 \kappa_m \kappa_I} \tag{12.54}$$

Once again we impose a step change in the set point temperature so that $\Theta_{\text{set}}(s) = \theta_1 / s$, and

$$\Theta(s) = \frac{\theta_1 \left[\kappa_m \left(\kappa_p s + \kappa_I \right) C_2 \right]}{s \left[C_1 s^2 + s \left(1 + C_2 \kappa_m \kappa_p \right) + C_2 \kappa_m \kappa_I \right]} \tag{12.55}$$

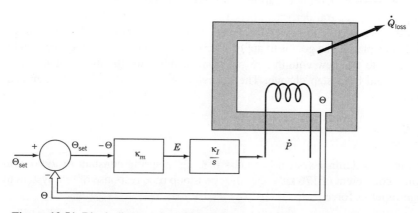

Figure 12.51 Block diagram for PI control of an oven.

Employing the final value theorem, we multiply by s and take the limit as $s \to 0$:

$$\lim_{s \to 0} s\Theta(s) = \lim_{s \to 0} \frac{s\left[\kappa_m(\kappa_p s + \kappa_I)C_2\right]\theta_1}{s\left[C_1 s^2 + s(1 + C_2\kappa_m\kappa_p) + C_2\kappa_m\kappa_I\right]} = \frac{C_2\kappa_m\kappa_I}{C_2\kappa_m\kappa_I}\theta_1 = \theta_1$$

Clearly, the steady-state error for this application of a PI controller is zero. This is a general result for PI control.

Time Response

The time response of this system to a step-change input could be determined by finding the inverse Laplace transform of equation (12.56):

$$\frac{\Theta(s)}{\Theta_{\text{set}}(s)} = \frac{\kappa_m(\kappa_p s + \kappa_I)C_2}{C_1 s^2 + s(1 + C_2\kappa_m\kappa_p) + C_2\kappa_m\kappa_I} \qquad (12.56)$$

However, by comparison with the Laplace transform of the governing differential equation for a second-order system, as provided in equation (3.13), Palm [7] shows that

$$\zeta = \frac{1 + C_2\kappa_m\kappa_p}{2\sqrt{C_1 C_2\kappa_m\kappa_I}} \qquad (12.57)$$

and that for $\zeta < 1$ the equivalent time constant is

$$\tau = \frac{2C_1}{1 + C_2\kappa_m\kappa_p} \qquad (12.58)$$

The behavior of the oven-controller system can be further explored using the program file *oven_2.vi*.

PID Control of a Second-Order System

Consider a spring-mass-damper system such as the one described in Figure 3.3. The governing equation describing the position of the mass as a function of time for a given forcing function $f(t)$ is

$$\frac{1}{\omega_n^2}\frac{d^2 y}{dt^2} + \frac{2\zeta}{\omega_n}\frac{dy}{dt} + y = f(t) \qquad (12.59)$$

This system and governing equation will serve as a model to demonstrate the properties of proportional-integral-derivative (PID) control when applied to a second-order system. Our first objective is to review the response of the second order system without a controller in place. The goal is to apply a step change in input to the system, and have the mass move to the new equilibrium position; in other words, the ideal response of the system would be a step change. The behavior of the system is now examined in two examples.

EXAMPLE 12.6

A spring-mass-damper system has mass of 2 kg, a spring constant of 7900 N/m, and a damping coefficient of 176 kg/s. Plot the open-loop time response of this system to a step-change input in force.

ASSUMPTIONS   The initial conditions for velocity, dy/dt and position y are zero.

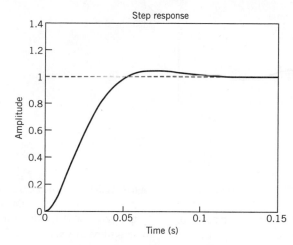

Figure 12.52 Second-order system response to a step-change input with a damping ratio of 0.7. Other parameters are described in Example 12.6.

SOLUTION   Although the solution to this example could be determined from the results in Chapter 3, we will determine this response using Laplace transforms. Taking the Laplace transform of equation (12.59) yields

$$\frac{1}{\omega_n^2}s^2Y(s) + \frac{2\zeta}{\omega_n}sY(s) + Y(s) = kF(s) \tag{12.60}$$

For the mass, spring constant, and damping ratio given in the problem statement, we compute the natural frequency and damping ratio as

$$\omega_n = \sqrt{\frac{k}{m}} = \sqrt{\frac{7900}{2}} = 62.85\,\text{rad/sec} \quad \text{and} \quad \zeta = \frac{c}{2\sqrt{km}} = \frac{176}{2\sqrt{7900 \times 2}} = 0.7$$

The open-loop transfer function defines the dynamic response of the system and is defined as

$$\frac{Y(s)}{kF(s)} = \left[\frac{1}{\frac{1}{\omega_n^2}s^2 + \frac{2\zeta}{\omega_n}s + 1}\right] = \left[\frac{1}{\frac{1}{3950}s^2 + 0.02235s + 1}\right] \tag{12.61}$$

where k is one for the present case. Several options exist for calculating and plotting the open-loop response of this system. Either the program file *open_loop.vi* or the function *step* in Matlab can be employed.

Figure 12.52 shows the response of this system to a step input; the output has been normalized so that the displacement, as characterized by the amplitude, is from 0 to 1.

EXAMPLE 12.7

Consider the second-order system described in Example 12.6. Change the damping ratio to 0.3 and to 1.3. Plot the open-loop response of the system to a step change in input.

SOLUTION   Figures 12.53 and 12.54 show the response of the system to a step change input for values of the damping coefficient of 0.3 and 1.3, respectively. By comparing Figures 12.52, 12.53, and 12.54, we can learn some important characteristics of second-order systems. Suppose our goal is to have the system respond quickly to the input, but without oscillation, and without exceeding the equilibrium value excessively. Let us characterize the response of the system having a damping ratio of 1.3 as being comparatively slow.

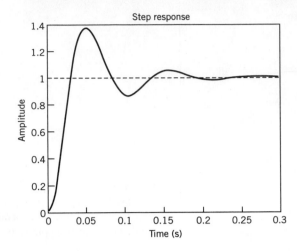

Figure 12.53 Second-order system response to a step-change input with a damping ratio of 0.3. Other parameters are described in Example 12.7.

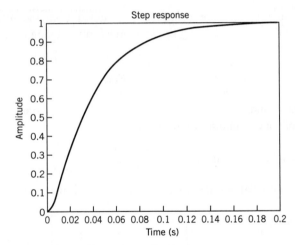

Figure 12.54 Second-order system response to a step-change input with a damping ratio of 1.3. Other parameters are described in Example 12.7.

This system requires 0.2 s to approach the equilibrium value of 1. The system having a damping coefficient of 0.3 first reaches 1 in a time less than 0.05 s, but there is both overshoot and oscillations. The system having a damping ratio of 0.7 has no oscillations, and minimal overshoot.

In the previous example, the response for a damping ratio of 0.7 appears to be the most likely candidate to meet our needs. But can we improve the situation by implementing a controller?

Proportional Control

Consider the application of a proportional controller to the second-order system described in Example 12.7 for the case where the damping coefficient is 0.3. The block diagram for the control system is shown in Figure 12.55.

The transfer function for this system when the damping ratio is 0.3 is

$$\frac{Y(s)}{F(s)} = \left[\frac{\kappa_p}{\dfrac{1}{\omega_n^2} s^2 + \dfrac{2\zeta}{\omega_n} s + 1} \right] = \left[\frac{\kappa_p}{\dfrac{1}{3950} s^2 + 0.02235 s + 1} \right] \tag{12.62}$$

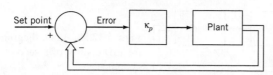

Figure 12.55 Block diagram for proportional control of a second-order plant.

We implement a proportional controller having a proportional gain of 10 and examine the step response of the system. Figure 12.56 shows the resulting response of the system. Clearly, we have not eliminated the overshoot and oscillations in this system. On the other hand, if we implement the same controller for the system having a damping ratio of 1.3, the response is shown in Figure 12.57. Here the proportional controller is helpful for the system's performance. Let us explore further whether by implementing a more sophisticated control scheme, we can better tune the response of the system having a damping ratio of 0.3.

Proportional-Integral-Differential Control (PID)

A block diagram representing PID control for our spring-mass-damper system is shown in Figure 12.58. The transfer function for this system is

$$\frac{Y(s)}{F(s)} = \left[\frac{\kappa_D s^2 + \kappa_p s + \kappa_I}{\frac{1}{\omega_n^2} s^3 + \left(\frac{2\zeta}{\omega_n} + \kappa_D \right) s^2 + \kappa_p s + \kappa_I} \right] \tag{12.63}$$

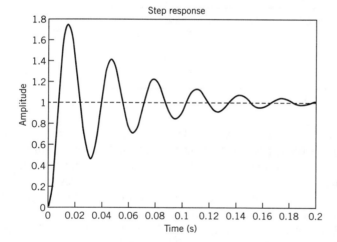

Figure 12.56 Response of a second-order system with a proportional controller having a gain of 10. The system has a damping ratio of 0.3 and the other parameters as described in Example 12.6.

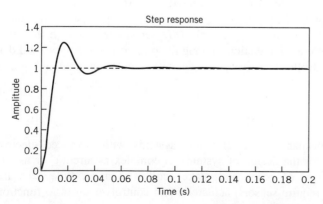

Figure 12.57 Proportional control of second-order system with a controller gain of 10, with damping 1.3.

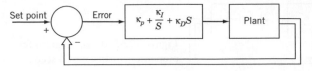

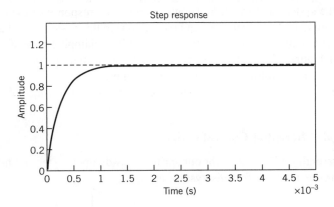

Figure 12.58 Block diagram for PID control of a second-order system.

Figure 12.59 Response of a second-order system having a damping ratio of 0.3 when subject to PID control.

Table 12.4 Effect of Increasing Gain on Rise Time, Overshoot, and Steady-State Error for PID Controller

Control Action	Rise Time	Overshoot	Steady-State Error
Proportional	⇩	⇧	⇩
Integral	⇩	⇧	ELIMINATE
Derivative		⇩	

Recall that taking a derivative in time space is multiplying by s in Laplace transform space, and integrating in time space is dividing by s in Laplace transform space. The choice of gains for the control actions, κ_p, κ_I, and κ_D, is a challenging task, and one for which much theory has been established [7]. However, for our purposes we wish only to demonstrate the possible improvement in the system performance. We select gains for the controllers of unity, and just test the performance. For our second-order system having a damping ratio of 0.3, the result is shown in Figure 12.59. We compare this figure with Figures 12.52 and 12.54, and clearly see that this control scheme has improved the performance of the system. The system behavior with a variety of parameters and control gains can be explored using the program file *PID_secondorder.vi* with the companion software. Although there are interactions among the proportional, integral, and derivative actions in a PID controller, the basic trends with increasing gain are provided in Table 12.4

12.5 SUMMARY

The integration of mechanical and electrical systems with advanced electronics is increasingly required for the design of systems as complex as aircraft, or as simple as a coffee maker that grinds beans and makes coffee at a set time each morning. Mechatronic systems require sensors, actuators, and control schemes to function as an integrated system. The present chapter presents the operating principles and designs of those sensors used most often for linear and rotary displacement in mechatronic systems. "Actuators" is a term that represents a broad range of mechanical, electrical, pneumatic, or hydraulic devices designed to create motion. Selection of an actuator should be done through a system-integration approach, so that the sensors and control scheme are appropriate for the chosen actuator.

We have provided an introduction to block diagram representation of dynamic systems based on Laplace transforms. We have also provided a review of the application of Laplace transforms to the solution of ordinary, linear differential equations. A discussion of open-and closed-loop control methods provided examples of proportional, proportional-integral, and proportional-integral-derivative control.

REFERENCES

1. Herceg, E. E., *Schaevitz Handbook of Measurement and Control*, Schaevitz Engineering, Pennsauken, NJ, 1976.
2. Coltman, J. W., The Transformer, *Scientific American*, 86, January 1988.
3. Bredin, H., Measuring Shock and Vibration, *Mechanical Engineering*, 30 February 1983.
4. Measurement of Rotary Speed, ASME Performance Test Codes, ANSI/ASME PTC 19.13–1961, The American Society of Mechanical Engineers, New York, 1961.
5. Gindy, S. S., Force and torque measurement, a technology overview, *Experimental Techniques*, 9:28, 1985.
6. Measurement of Shaft Power, ASME Performance Test Codes, ASME PTC 19.7–1980, (reaffirmed date: 1988), The American Society of Mechanical Engineers, New York, 1980.
7. Palm, William J., *Modeling Analysis and Control of Dynamic Systems*, 2nd ed., John Wiley and Sons, Inc., New York, 1999.

NOMENCLATURE

a	acceleration $[l\ t^2]$		y_r	relative displacement between seismic mass and housing in a seismic instrument $[l]$
c	damping coefficient $[m\ t]$			
c_c	critical damping coefficient $[m\ t]$		δy	deflection $[l]$
g	acceleration of gravity $[l\ t^{-2}]$		A	amplitude
k	spring constant $[m\ t^{-2}]$		B	magnetic field strength $[m\ C^{-1}\ t^{-1}]$
m	mass $[m]$		C_v	discharge coefficient
p	pressure $[m\ l^{-1}\ t^{-2}]$		D_c	coil diameter for magnetic pickup $[l]$
$\mathbf{r}$	radius, vector $[l]$		E_i	input voltage $[V]$
t	time $[t]$		E_o	output voltage $[V]$
δt	time interval for data sampling $[t]$		F	force, vector $[m\ l\ t^{-2}]$
u	uncertainty		J	polar moment of inertia $[l^4]$
v	linear velocity $[l\ t^{-1}]$		L	length $[l]$
y	displacement $[l]$		M	magnitude ratio
y_h	displacement of housing, seismic instrument $[l]$		M	moment, vector $[m\ l^2\ t^{-2}]$
			P	power $[m\ l^2\ t]$
y_m	displacement of seismic mass $[l]$		Q	volumetric flow rate $[l^3\ t^{-1}]$

R_o	outer radius [l]		τ_{max}	maximum shearing stress [$m\ t^2\ l^{-1}$]
T	temperature [$°$]		ϕ	phase angle
T	torque [$m\ l^2\ t^{-2}$]		κ	controller gain
$\forall$	volume [l^3]		ω	rotational speed [t^{-1}]
ζ	damping ratio		ω_n	natural frequency [t^{-1}]
τ	time constant [t]			

PROBLEMS

12.1 Consider a linear potentiometer as shown in Figure 12.1. The potentiometer consists of 0.1-mm copper wire ($\rho_e = 1.7 \times 10^{-8}$ Ω-m) wrapped around a core to form a total resistance of 1 kΩ. The sliding contact surface area is very small.

   a. Estimate the range of displacement that could be measured with this potentiometer for a 1.5-cm core.

   b. The circuit shown in Figure 12.60 is used to record position. On a single plot, show the loading error in an indicated displacement as a function of displacement, over the range found in (a), for values of the meter resistance, R_m, of 1, 10, and 100 kW. For practical meters, would the loading error be significant?

12.2 Mechatronic applications for linear displacement sensors are numerous. Select a particular application and develop specifications that would be required for a linear displacement sensor. Possible applications include automotive seat position, pick-and-place operations, and throttle position sensors.

12.3 Compare and contrast wire-wound and conductive plastic potentiometers. Are there advantages and disadvantages for either linear or rotary applications?

12.4 Compare and contrast the use of an LVDT and a resistance-based potentiometer for position measurement.

12.5 A seismic instrument, as shown in Figure 12.7, is used to measure a vibration given by

$$y(t) = 0.2 \cos 10t + 0.3 \cos 20t$$

where

   $y =$ displacement in inches

   $t =$ time in seconds

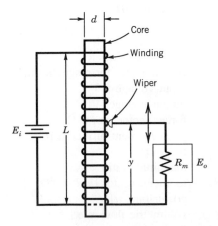

Figure 12.60 Circuit for Problem 12.1.

The seismic instrument is to have a damping ratio of 0.7 and a spring constant of 1.2 lb/ft.

a. Select a combination of seismic mass and damping coefficient to yield less than a 10% amplitude error in measuring the input signal. Under what conditions would the output signal experience significant distortion?

b. Describe the phase response of the system, either in a plot or tabular form.

12.6 A seismic instrument has a natural frequency of 20 Hz and a damping ratio of 0.65. Determine the maximum input frequency for a vibration such that the amplitude error in the indicated displacement is less than 5%.

12.7 A seismic instrument consists of a housing and a seismic mass. To measure vibration, the seismic mass should remain stationary in an absolute frame of reference. To measure acceleration, the magnitude ratio should be unity. Explain in detail the requirements for mass, spring constant, and damping ratio to satisfy these requirements.

12.8 A seismic instrument consists of a housing and a seismic mass. To measure vibration, the seismic mass should remain stationary in an absolute frame of reference. To measure acceleration, the magnitude ratio should be unity. Using Labview, develop a model of the seismic instrument for both vibration and acceleration measurement.

12.9 Determine the bandwidth for a seismic instrument employed as an accelerometer having a seismic mass of 0.2 g and a spring constant of 20,000 N/m, with very low damping. Discuss the advantages of a high natural frequency and a low damping ratio. Piezo-electric sensors are well suited for the construction of accelerometers because they possess these characteristics.

12.10 In Example 12.2, integration is identified as a method for reducing the effects of noise in a signal. Discuss how a moving average can be used to reduce the effects of noise in a velocity measurement, through the integration of an acceleration signal. Discuss the effects of averaging time on the elimination of noise. Assume that the noise has significant amplitude, but higher frequency content than the velocity being measured.

12.11 Consider a moving coil transducer having a coil diameter D_c of 0.8 cm and a coil length of 2 cm. The nominal range of velocities to be measured is from -1 to 10 cm/sec. The resulting emf will be measured by a PC-based data-acquisition system with an 8-bit A/D converter and a range of -1 to 1 V. The accuracy is 0.1% full scale. Plot the number of turns as a function of magnetic field strength to provide an accuracy of 1% in the resulting velocity measurement.

12.12 Consider measuring a rotational speed using a stroboscope. The rotational speed is higher than the flash rate of the stroboscope. The stroboscope is observed to synchronize at 10,000, 18,000, and 22,000 flashes per second. Determine the rotational speed.

12.13 Research the state-of-the art specifications for load cells designed to measure the smallest possible forces. What applications exist for such precise measurements of small forces?

12.14 Design a proving ring load cell appropriate to serve as a laboratory calibration standard in the range of 250–1000 N. The proving ring material is steel.

12.15 Power transmitted through the drive shaft of a car results in a rotational speed of 1800 rpm with a power transmission of 40 hp. Determine the torque that the driveshaft must support.

12.16 Discuss the importance of a dynamometer for automotive emissions testing.

12.17 Research applications for linear actuators. For each application that you identify suggest the most appropriate linear actuator. Was that the actuator chosen for the particular application? For many applications, a choice is made to convert rotary motion to linear motion.

12.18   Pneumatic actuators span a very large range of size and force. Research the range of commercially available pneumatic cylinders.

12.19   Estimate the flow coefficient for a flow control valve if the rating corresponds to 32 SCFM (0.906 SCMM) with a $\Delta p = 10$ psi (0.69 bar). The test is conducted using air with a line pressure of 100 psi (6.7 bar), $T = 68°F$ (20°C) and a relative humidity of 36% at the supply port. What is the pressure at the exhaust port? Base your answer on standard conditions.

12.20   A low-profile, three-port control valve with 6-mm-port diameters has an F value of 0.82 ms/cm^3 and a lag time of 8 ms to pressurize through ports P to A. The exhaust path through ports T to A has an F value of 0.70 ms/cm^3 with an 8-ms lag time. The valve is connected to a 500-mL vessel. Supply pressure is 1 bar; exhaust pressure is 0 bar. Estimate the time needed to pressurize this vessel to 0.9 bar. Estimate the time to exhaust this vessel to 0.1 bar.

12.21   Research the design of a totalizing flow meter, with particular emphasis on the meters used to generate water bills.

12.22   The companion software contains a program *stroboscope.vi* that models the behavior of a stroboscope.

   a. Set the bar rotation frequency to 40 rps, and then set the strobe frequency to 20 and 40 Hz. Explain your results.

   b. Set the bar rotation frequency to 100 rpa. Using $\omega_1 = 50$ Hz and $\omega_N = 10$ Hz apply equation (12.15) to show that we can measure rotation frequencies larger than the maximum frequency of the stroboscope. Be careful to find all N of the synchronous frequencies.

12.23   Describe the operating principle of a thermostat for residential applications, and the design of a bimetallic sensor for measuring the temperature. How is a deadband created in such an instrument?

12.24   Show that a proportional controller has a steady-state error. How would you quantify the steady-state error for a controller and a first-order system?

12.25   Using Labview, develop a model of the oven described in Example 12.5 coupled with a PI controller. Vary the proportional and integral gains and discuss the resulting behavior of this first-order system.

12.26   Using Labview, develop a model for PID control of the oven described in Example 12.5. Vary the gains and discuss the resulting behavior of the system. Quantify the power input required to achieve a given response.

Appendix A

A Guide for Technical Writing

A technical paper is a common way to communicate and to archive test results. Depending on the intended audience, papers usually take on one of three broad formats, with wording and purpose directed toward that audience: (1) Executive summary, (2) Laboratory report, and (3) Technical report.

The executive summary is intended typically for upper level management who will plan decisions around the input from the report. These consist of a one- or two-page (under 500-word) brief stating the reasons for the tests conducted, the important results, and the conclusions. The document composition is subdivided into *Objectives, Results*, and *Conclusions* regardless of whether or not the headings are used explicitly. A summary may include figures, charts and references, as needed. Because of its brevity and readership, a clarity in purpose, i.e., the strategic use of words and clearly directed and meaningful conclusions, is crucial.

Laboratory reports are internal reports usually accessed by members of your engineering group. These reports provide progress information or document useful information in a brief format that stresses results in detail. These reports should include the reasons for the tests, the pretest planning, the methods and test conditions used, with an emphasis on the discussion of the results and conclusions. The document is usually explicitly subdivided into *Objectives, Approach, Results and Discussion, Conclusion*. Related and prior works are clearly referenced. Data in the form of tables and graphs are cited and discussed within the report. A laboratory report may contain an *Abstract* as a brief snapshot of the report contents. Appendices containing supporting information, such as raw data and data analysis programs, may be attached but must be independent from the main body of the report.

Technical reports are intended for a technical audience outside of the immediate work group. They are written at a level and style that are appropriate for the receiving audience. Accordingly, these reports should include and/or cite sufficient background information so that the reader can understand the purpose of and the history behind the tests, the manner in which the tests were analyzed, and the results of the tests and their implications. A technical report must present enough information for the reader to follow the logic of the tests and to interpret the test results. Data should be presented only in the form of well-prepared tables and graphs cited from within the report. The report ends with a concise conclusion.

Each company, agency, or test laboratory has its own format for these reports. However, we provide suggested guidelines for preparing the following technical report. The usual subdivisions consist of *Abstract, Introduction/Background, Approach, Results and Discussion, Conclusion*. Related work is clearly referenced. Regardless of format or length, a report must communicate to the intended audience. Further information concerning technical reporting may be found in numerous guides for technical writing [1,2]

A.1   A GUIDE FOR TECHNICAL WRITING[3]

Competent engineers are able to communicate their ideas in both oral and written formats to both technical and nontechnical audiences. This primer is intended as a brief guide to constructing a technical report intended for a technical audience. The format of this section is itself representative of most technical reporting formats. The major headings, such as Abstract and Introduction, are typical, but are not meant to be exclusive. We provide this guide to show you the organization and approach to a technical report.

Abstract

Effective technical communication abilities are important in a technologically advancing society. The purposes of this document are to serve as a format example of a technical report and to provide specific procedures and ideas for generating sound technical reports. Guidelines for preparing each section of a report and detailed ideas for presentation of results in plots and tables are provided. The reader can draw from this outline to prepare a technical report on a chosen subject.

Introduction

In a 1980 study, the U.S. Department of Education and the National Science Foundation concluded that the majority of Americans are moving toward "virtual scientific and technological illiteracy" [4]. The technical person's ability to communicate effectively with both technical and nontechnical segments of society is essential for addressing myriad technological problems and decisions facing our society. In fact, career advancement in any profession is largely based on how well a person communicates. But it is especially important for engineers to develop effective technical writing skills that enable them to convey the results and significance of their work to a variety of audiences. The primary form of writing for practicing engineers and engineering managers (as well as researchers) is the technical report or one of its variations, the executive summary, or the laboratory report. In general, the technical report is directed toward a technically knowledgeable audience. The purpose of this guide is to provide reasonable ideas and suggestions for producing clear and concise technical manuscripts.

Specific Writing Steps

The following steps are recommended for effective technical writing:

1. Analyze your data, create figures and tables, and organize your thoughts. The scope of the material to be presented and the particular audience for whom you are writing should be kept in mind while deciding what to present.

2. Construct a thesis that adequately covers your material so as to create a detailed outline (note: A thesis tests a hypothesis or idea against the supporting data so as to lead to a conclusion). Then expand your outline to at least the level where topics and some sentences have been formulated. The use of this outline will help you avoid making a chronological or "stream of consciousness" presentation; The order in which you did the work and the order in which you report it will rarely match.

3. Begin by using the format or style guide for the specific publication. Work from those headings.

4. Write a first draft in which you concentrate on organization and documenting your ideas for the particular audience for which you are writing. Write this first draft as rapidly as possible to take advantage of your continuity of thought.

5. Read the draft, then rewrite it for presentation, clarity, and accuracy. Reread and edit as needed. Proofread the manuscript. Remember: Your audience cannot read your mind, just your report—put what you want to say in writing.

6. Use a typography that is easy to read. If not specified, try a 12-point serif font, such as Times. Use 12- or 14-point bold for titles and major headings, 12 point for subheadings, and use 10 point for footnotes. As for layout, use a format (headings and composition) as described below. Maintain margins of at least 1.25 in. (32 mm). Number all pages (exception: page 1 may be left unnumbered).

Approach to Writing Technical Papers

The key to effectively communicating the results of your work is to have a clear understanding of your current knowledge. Forget all previous misconceptions, mistakes, bad data, and so on. Do not report them. Instead, take a fresh look at the results and organize your thoughts to present the work in the best way. There are several supporting blocks for constructing a technical report: written text, figures and tables, and, if needed, appendices. Each figure and table must be described and explicitly referred to in the text. An appendix needs to be cited within the text. The following guidelines for preparing each section of a technical report provide the basics necessary for sound reporting procedures. *Just a note: These sections are not absolutes—use good judgment relative to your work and do what you must to communicate.*

The Abstract—A Summary of the Entire Report

An Abstract is a complete, concise distillation of the full report. It is always written last. It provides a brief (one sentence) introduction to the subject, a statement of the problem, highlights of the results (quantitative, if possible), and the major conclusion. It must stand alone without citing figures, tables, or references. A concise, clear approach is essential. Most abstracts are short and rarely exceed 150 words.

The Introduction—Why Did You Do What You Did?

An Introduction generally identifies the subject of the report, provides the necessary background information, including appropriate literature review, and provides the reader with a clear rationale for the work described. This is where you develop and state the hypothesis or concept tested. The Introduction does not contain results, and generally does not contain equations.

Analysis—Is There A Model?

An Analysis section develops a descriptive model related to the hypothesis. Sufficient detail (mathematical or otherwise) should be provided for the reader to clearly understand the physical assumptions associated with a theory or model.

Experimental Program—What Did You Measure and How?

The Experimental Program section is intended to describe how the experimental model was developed to support the analysis and test the hypothesis, and detail how the results were obtained. Provide an overview of the approach, test facilities, validations, and range of measurements. As a rule of thumb, provide only sufficient detail to allow the experiment to be conducted by someone else. Do not give instructions, this is not a recipe; rather report what was done. If a list of equipment is included in the report, it should be a table in the body of the report, or should be placed in an appendix. Uncertainty analysis information can be described either here or in the Results section, or both. In cases where both an analysis and experiment are described, these two sections of the report should complement and support each other. The relationship of the analysis to the experiment should be clearly stated.

Numerical Model—What Did You Simulate and How?

If a numerical simulation was performed, it should be described under a separate heading using the same guidelines as presented for the Experimental Program.

Results and Discussion—So What Did You Find?

Here you present and discuss your test results and tie them back to your original objectives or hypothesis. Data must be interpreted to be useful, and this is where you transform raw data into useful results. When presenting your results remember that even though you are usually writing to an experienced technical audience, what may be clear to you may not be obvious to the reader. Guide the reader toward your interpretation. Tell the story.

Often the most important vehicles for the clear presentation of results are figures and tables. Since you have spent significant time in preparing the plots and tables and you are intimately familiar with their trends and implications, the reader needs your insight to understand the results as well as you. As a good rule, spend at least one paragraph discussing each. Remember to number and provide a caption for each. Column heads in tables should accurately describe the data that appear in that column. Each table and figure must be explicitly and individually cited and described in the text.

Figure A.1 shows an example of an appropriately prepared plot. Note that this plot has a figure number and a descriptive caption, and clearly labeled axes. As an example of what to write, your report might start with: "*This figure shows the measured temperature as a function of elevation (x/L) within the heated cylindrical packed particle bed. Two curves are shown, one taken along the centerline ($R/R_0 = 0.0$) and the other along the bed wall ($R/R_0 = 1$). The bed temperature is essentially uniform from x/L $\sim$ 0.5 to the top of the bed (x/L = 1.0) but shows spatial gradients elsewhere.*" Then the extended discussion of this figure might focus on the differences in the curves, why they are different, how they relate to a model or prediction, and assess whether their shape and magnitudes make sense.

The visual impact of a plot conveys considerable information about the relationship between the plotted dependent and independent variables. Suggestions for creating effective plots are:

1. The independent parameter is always plotted on the x axis; the dependent parameter is always plotted on the y axis.

2. Try to use at least four tics or increments for each coordinate, but fewer than ten. Multiples of 1, 2, or 5 are good increments because they make it easy to

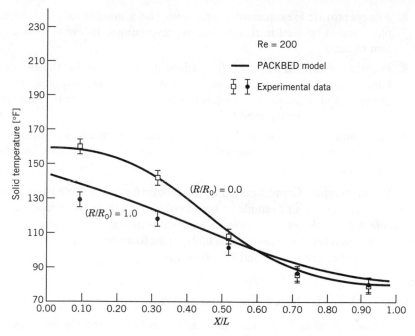

Figure A.1 Comparison of experimentally measured and numerically predicted (PACKBED) results for axial temperature distributions in a thermal energy storage bed.

interpolate. Watch out for automatic scaling features from software: It is hard to justify strange increments such as 0, 7, 14.

3. In drawing smooth curves through experimental data points, try to follow these rules:

   (a) Always show the data points as symbols, such as open or filled squares, circles, or triangles.

   (b) Do not extend a curve beyond the ends of the data points. If you need to extrapolate, then use a dashed line outside the known range. If appropriate, indicate the curve-fit equations.

   (c) If you are certain that $y = 0$ when $x = 0$, put the curve through the origin. But if you are uncertain, don't use the origin; stop the line at the lowest point (see 3b).

4. Always include a symbol legend. Consistently use the same symbols for the same variables between graphs. Pick a symbol font that will be legible.

5. To compare experimentally determined data with model predictions, show the experimental points as symbols and the model as a smooth curve. Similarly, show numerical predictions as a smooth curve. Clear labeling is essential for this kind of plot.

6. Whenever possible, label curves individually, even if it means using arrows to match the lines with the labels.

7. The figure should contain enough information to stand alone, especially if there's a chance it will later be used by itself (e.g., as a slide). However, each figure must still be discussed within the text.

8. Always provide axes quantities with units. Use a suitable font size. Figures and plots tend to be used in reports and in presentations, so use at least a 12 point font or larger.

9. Position the data on the graph (i.e., adjust the scales) so that the curves are not bunched near the top, bottom, or one of the sides. The only time white space serves a valid function in a graph is when you are comparing it to another graph that makes appropriate use of the same scale.

10. Uncertainty limits should be indicated for a number of measured points on a given plot using interval bands about the mean value, as shown in Figure A.1.

Table A.1 is an example of results presented in a tabular form. The table has a number and a descriptive caption. As an example of how to describe a table, your report could start with, *"Table A.1 lists the measured thermistor voltage output as a function of an applied velocity."* The extended discussion of this table might focus on its range, curve fit, how it relates to a model or prediction, and how it is used.

Conclusions—What Do I Now Know and So What?

The Conclusions section is where you should concisely restate your answer to the question: "What do I know now?" It must support or refute your hypothesis and provide a useful closure to the report. In a short summary, restate why the work was done, how it was done, and why these results are significant. This is not the place to offer new facts or discussion. A conclusion will normally have a quantified outcome. An example might be: *"The temperature measuring system was calibrated against a laboratory standard RTD. The system was found to indicate the correct temperature over the range of 0 to 150°C with no more than a ±0.5°C uncertainty at 95% confidence. This is acceptable for our intended use."* It would not normally be appropriate to conclude simply, "The temperature measuring system was tested and worked well." Conclusions should be clear and concise statements of the important findings of a particular study; most conclusions require some quantitative statement to be useful.

Table A.1   Characteristics of a Thermistor Anemometer in a Uniform Flow Field

Velocity [m/s]	[V]
0.467	3.137
0.950	3.240
2.13	3.617
3.20	3.811
3.33	3.876
4.25	3.985
5.00	4.141
6.67	4.299
8.33	4.484
10.0	4.635
12.0	4.780

Appendices

Appendices provide noncritical, supplemental, or supporting information. They should stand on their own. Examples of supporting information might include raw data, example calculations, or programs.

Uncertain as to whether or not some information should be in an Appendix? Ask yourself: If I place this material in the Appendix and the reader did not read the Appendix, would the main body of the report be sufficient? It should be!

References

The references cited in a formal list should be available to the reader and described in sufficient detail for the reader to obtain the source with a reasonable effort. Pick a format and be consistent. References for this document provide an acceptable format guide.

Writing Tips

1. Communicate! You are telling a story, albeit a highly technical one. Clarity and consistency are important. Define all nonstandard terms the first time they are used and stick to those terms and definitions throughout all writing on that subject. Carefully read it, then rewrite it! Have someone else read it, then rewrite it!

2. Avoid the use of contractions and possessives and jargon.

3. Don't overdo significant figures. See the discussion in Chapter 1.

On Writing Style

Engineers often write reports in the third person in deference to impartiality and to focus attention on the subject matter at hand. The idea is to disassociate the writer from the action and make the equipment/model/test the "doer of the action." [5] This is a notable goal but one that takes time and practice to do effectively. But achieving this goal does not require extensive use of the passive voice, despite traditional beliefs. Do use the active voice where possible to liven up your report. Even the occasional impersonal construction in the first person is fine if it improves readability. Frankly, we find it difficult to write extensively in the third person, passive voice without the writing coming off as clumsy . . . and long. But choose the style that suits your writing best. It is important to communicate effectively!

Conclusions

Technical authors must be cognizant of coupling the intended audience with the test goals when presenting the results of their writing. This report has outlined the essential features germane to technical report writing, and it serves as an example of both style and format. We describe the purpose and content for each section of a report. We also pass along some useful advice drawn from our experience in helping young engineers learn to write technically. We conclude that an effective, polished, and professional product can be produced only through careful and persevering revision of manuscript, by incorporating effective figures and tables, and by targeting the intended audience.

REFERENCES

1. Tichy, H. J., and S. Foudrinier, "Effective Writing for Engineers, Managers, and Scientists," 2nd ed., Wiley Interscience, New York, 1988.
2. Donnell, J., and S. Jeter, *Writing Style and Standards in Undergraduate Reports*, College Pub., 2003.
3. Henry, M. H., and H. K. Lonsdale, "The researcher's writing guide," *Journal of Membrane Science* 13:101–107, 1983.
4. Naisbitt, J., *Megatrends*, Warner Books, New York, 1984. (See also, U.S. report fears most Americans will become scientific illiterates, *New York Times*, Oct. 23, 1984.)
5. Daniell, B., R. Figliola, A. Young, and D. Moline, "Learning to Write: Experiences with Technical Writing Pedagogy within the Mechanical Engineering Laboratories, *Proc. ASEE*, Paper 1141, 2003.

Appendix B

Property Data and Conversion Factors

Table B.1 Properties of Pure Metals and Selected Alloys

	Density [kg/m^3]	Modulus of Elasticity [GPa]	Coefficient of Thermal Expansion [10^{-6}m/m-K]	Thermal Conductivity [W/m-K]	Electrical Resistivity [10^{-6} Ω-cm]
Pure Metals					
Aluminum	2 698.9	62	23.6	247	2.655
Beryllium	1 848	275	11.6	190	4.0
Chromium	7 190	248	6.2	67	13.0
Copper	8 930	125	16.5	398	1.673
Gold	19 302	78	14.2	317.9	2.01
Iron	7 870	208.2	15.0	80	9.7
Lead	11 350	12.4	26.5	33.6	20.6
Magnesium	1 738	40	25.2	418	4.45
Molybdenum	10 220	312	5.0	142	8.0
Nickel	8 902	207	13.3	82.9	6.84
Palladium	12 020	—	11.76	70	10.8
Platinum	21 450	130.2	9.1	71.1	10.6
Rhodium	12 410	293	8.3	150.0	4.51
Silicon	2 330	112.7	5.0	83.68	1×10^5
Silver	10 490	71	19.0	428	1.47
Tin	5 765	41.6	20.0	60	11.0
Titanium	4 507	99.2	8.41	11.4	42.0
Zinc	7 133	74.4	15.0	113	5.9
Alloys					
Aluminum (2024, T6)	2 770	72.4	22.9	151	4.5
Brass (C36000)	8 500	97	20.5	115	6.6
Brass (C86500)	8 300	105	21.6	87	8.3
Bronze (C90700)	8 770	105	18	71	1.5
Constantan annealed (55% Cu 45% Ni)	8 920	—	—	19	44.1
Steel (AISI 1010)	7 832	200	12.6	60.2	20
Stainless Steel (Type 316)	8 238	190	—	14.7	—

Source: Complied From *Metals Handbook*, 9th ed., American Society for Metals, Metals Park, OH, 1978, and other sources.

Table B.2 Thermophysical Properties of Selected Metallic Solids

Composition	Melting point [K]	Properties at 300 K				Properties at various temperatures [K]							
		ρ [kg/m³]	c_p [J/kg·K]	k [W/m·K]	$\alpha \times 10^4$ [m²/s]	k[W/m·K]				c_p[J/kg·K]			
						100	200	400	600	100	200	400	600
Aluminum													
Pure	933	2702	903	237	97.1	302	237	240	231	482	796	949	1033
Alloy 2024-T6 (4.5% Cu, 1.5% Mg, 0.6% Mn)	755	2770	875	177	73.0	65	163	186	186	473	787	925	1042
Alloy 195, cast (4.5% Cu)	—	2790	883	168	68.2	—	—	174	185	—	—	—	—
Chromium	2118	7160	449	93.7	29.1	159	111	90.9	80.7	192	384	484	542
Copper													
Pure	1358	8933	385	401	117	482	413	393	379	252	356	397	417
Commercial bronze (90% Cu, 10% Al)	1293	8800	420	52	14	—	42	52	59	—	785	460	545
Phosphor gear bronze (89% Cu, 11% Sn)	1104	8780	355	54	17	—	41	65	74	—	—	—	—
Cartridge brass (70% Cu, 30% Zn)	1188	8530	380	110	33.9	75	95	137	149	—	360	395	425
Constantan (55% Cu, 45% Ni)	1493	8920	384	23	6.71	17	19	—	—	237	362	—	—
Iron													
Pure	1810	7870	447	80.2	23.1	134	94.0	69.5	54.7	216	384	490	574
Armco (99.75%)	—	7870	447	72.7	20.7	95.6	80.6	65.7	53.1	215	384	490	574
Carbon steels													
Plain carbon (Mn ≤ 1%, Si ≤ 0.1%)	—	7854	434	60.5	17.7	—	—	56.7	48.0	—	—	487	559
AISI 1010	—	7832	434	63.9	18.8	—	—	58.7	48.8	—	—	487	559

Composition	Melting Point (K)	ρ (kg/m³)	c_p (J/kg·K)	k (W/m·K)	$\alpha \cdot 10^6$ (m²/s)	k (100 K)	k (200 K)	k (400 K)	k (600 K)	c_p (100 K)	c_p (200 K)	c_p (400 K)	c_p (600 K)
Carbon-silicon (Mn ≤ 1%, 0.1% < Si ≤ 0.6%)	—	7817	446	51.9	14.9	—	—	49.8	44.0	—	—	501	582
Carbon—manganese—silicon (1% < Mn ≤ 1.65%, 0.1% < Si ≤ 0.5%)	—	8131	434	41.0	11.6	—	—	42.2	39.7	—	—	487	559
Chromium (low) steels 1/2 Cr-1/4 Mo-Si (0.18% C, 0.65% Cr, 0.23% Mo, 0.6% Si)	—	7822	444	37.7	10.9	—	—	38.2	36.7	—	—	492	575
1 Cr—1/2 Mo (0.16% C, 1% Cr, 0.54% Mo, 0.39% Si)	—	7858	442	42.3	12.2	—	—	42.0	39.1	—	—	492	575
1 Cr—V (0.2% C, 1.02% Cr, 0.15% V)	—	7836	443	48.9	14.1	—	—	46.8	42.1	—	—	492	575
Stainless steels													
AISI 302	—	8055	480	15.1	3.91	—	—	17.3	20.0	—	—	512	559
AISI 304	1670	7900	477	14.9	3.95	9.2	12.6	16.6	19.8	272	402	515	559
AISI 316	—	8238	468	13.4	3.48	—	—	15.2	18.3	—	—	504	550
AISI 347	—	7979	480	14.2	3.71	—	—	15.8	18.9	—	—	513	559
Lead	601	11 340	129	35.3	24.1	39.7	36.7	34.0	31.4	118	125	132	142
Magnesium	923	1740	1024	156	87.6	169	159	153	149	649	934	1074	1170
Molybdenum	2894	10 240	251	138	53.7	179	143	134	126	141	224	261	275
Nickel													
Pure	1728	8900	444	90.7	23.0	164	107	80.2	65.6	232	383	485	592
Nichrome (80% Ni, 20% Cr)	1672	8400	420	12	3.4	—	—	14	16	—	—	480	525
Inconel X—750 (73% Ni, 15% Cr, 6.7% Fe)	1665	8510	439	11.7	3.1	8.7	10.3	13.5	17.0	372	—	473	510

Table B.3   Thermophysical Properties of Saturated Water (Liquid)

T [K]	ρ [kg/m^3]	c_p [kJ/kg·K]	$\mu \times 10^6$ [N·s/m^2]	k [W/m·K]	Pr	$\beta \times 10^6$ [K^{-1}]
273.15	1000	4.217	1750	0.569	12.97	−68.05
275.0	1000	4.211	1652	0.574	12.12	−32.74
280	1000	4.198	1422	0.582	10.26	46.04
285	1000	4.189	1225	0.590	8.70	114.1
290	999	4.184	1080	0.598	7.56	174.0
295	998	4.181	959	0.606	6.62	227.5
300	997	4.179	855	0.613	5.83	276.1
305	995	4.178	769	0.620	5.18	320.6
310	993	4.178	695	0.628	4.62	361.9
315	991	4.179	631	0.634	4.16	400.4
320	989	4.180	577	0.640	3.77	436.7
325	987	4.182	528	0.645	3.42	471.2
330	984	4.184	489	0.650	3.15	504.0
335	982	4.186	453	0.656	2.89	535.5
340	979	4.188	420	0.660	2.66	566.0
345	977	4.191	389	0.664	2.46	595.4
350	974	4.195	365	0.668	2.29	624.2
355	971	4.199	343	0.671	2.15	652.3
360	967	4.203	324	0.674	2.02	679.9
365	963	4.209	306	0.677	1.90	707.1
370	961	4.214	289	0.679	1.79	728.7
373.15	958	4.217	279	0.680	1.73	750.1
400	937	4.256	217	0.688	1.34	896
450	890	4.40	152	0.678	0.99	
500	831	4.66	118	0.642	0.86	
550	756	5.24	97	0.580	0.88	
600	649	7.00	81	0.497	1.14	
647.3	315	00	45	0.238	00	

Formulas for interpolation (T = absolute temperature)
$$f(T) = A + BT + CT^2 + DT^3$$

$f(T)$	A	B	C	D	Standard deviation, σ
		273.15 < T < 373.15 K			
ρ	766.17	1.80396	-3.4589×10^{-3}		0.5868
c_p	5.6158	-9.0277×10^{-3}	14.177×10^{-6}		4.142×10^{-3}
k	−0.4806	5.84704×10^{-3}	-0.733188×10^{-5}		0.481×10^{-3}
		273.15 < T < 320 K			
$\mu \times 10^6$	0.239179×10^6	-2.23748×10^3	7.03318	-7.40993×10^{-3}	4.0534×10^{-6}
$\beta \times 10^6$	-57.2544×10^3	530.421	−1.64882	1.73329×10^{-3}	1.1498×10^{-6}
		320 < T < 373.15 K			
$\mu \times 10^6$	35.6602×10^3	−272.757	0.707777	-0.618833×10^{-3}	1.0194×10^{-6}
$\beta \times 10^6$	-11.1377×10^3	84.0903	−0.208544	0.183714×10^{-3}	1.2651×10^{-6}

Source: From F. P. Incropera and D. P. DeWiott. *Fundamentals of Heat and Mass Transfer*, Wiley, New York, 1985.

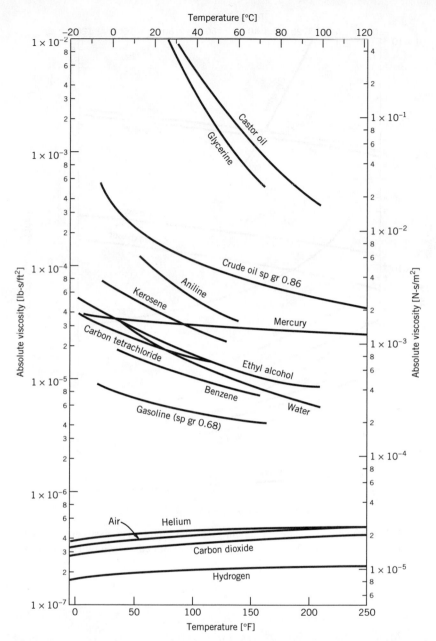

Figure B.1 Absolute viscosities of certain gases and liquids. (From V. L. Streeter and E. B. Wylie, *Fluid Mechanics*, 8th ed., McGraw-Hill, New York, 1985.)

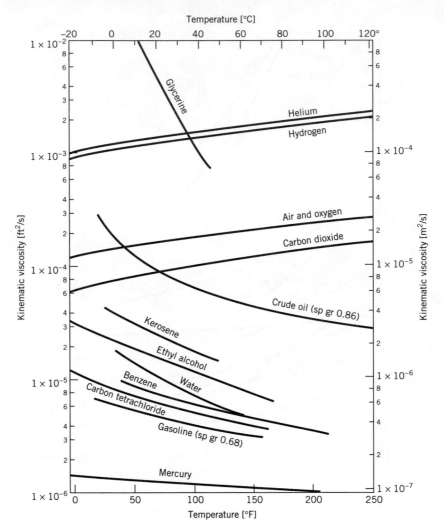

Figure B.2 Kinematic viscosities of certain gases and liquids. The gases are at standard pressure. (From V. L. Steeter and E. B. Wylie, *Fluid Mechanics* 8th ed., McGraw-Hill, New York, 1985.)

Appendix C

Laplace Transform Basics

We offer the following primer of Laplace transforms for review purposes to meet the needs of this text. Extensive treatment of this subject is available in [1]. The Laplace transform of a function $y(t)$ is defined as

$$Y(s) = \mathcal{L}[Y(t)] = \int_0^\infty y(t)e^{-st}dt \tag{C.1}$$

There are many functions that are useful in modeling linear systems that have analytical forms for their Laplace transforms. Consider the unit step function, defined by

$$U_s(t) = \left\{ \begin{matrix} 0 & t < 0 \\ 1 & t > 0 \end{matrix} \right\} \tag{C.2}$$

Applying the definition of the Laplace transform yields

$$\mathcal{L}[U_s(t)] = \int_0^\infty e^{-st}dt = \frac{1}{s} \tag{C.3}$$

The unit step function and its Laplace transform form a Laplace transform pair. Table C.1 provides a list of common Laplace transform pairs.

One of the most important properties of the Laplace transform results from the application of the Laplace transform to a derivative, as

$$\mathcal{L}\left[\frac{dy(t)}{dt}\right] = sY(s) - y(0) \tag{C.4}$$

Differentiation in the time domain is the same as multiplication in the Laplace transform domain. This relationship is derived through application of integration by parts of the definition of the Laplace tansform, can be generalized as

$$\mathcal{L}\left[\frac{d^n y(t)}{dt^n}\right] = s^n Y(s) - s^{n-1}y(0) - s^{n-2}\frac{dy(0)}{dt} - \cdots - \frac{d^{n-1}y(0)}{dt^{n-1}} \tag{C.5}$$

Final Value Theorem

A useful property of the Laplace transform allows us to determine the value of the function $y(t)$ knowing only the Laplace transform of the function. The final value theorem states

$$\lim_{t \to \infty} y(t) = \lim_{s \to \infty} [sY(s)] \tag{C.6}$$

Let's apply this to the unit step function

$$\lim_{s \to 0} sY(s) = \lim_{s \to 0} s\frac{1}{s} = 1 \tag{C.7}$$

Clearly this is the correct limit!

Table C.1   Laplace Transform Pairs

	$f(t)$	$\mathcal{L}(f)$
1	1	$\dfrac{1}{s}$
2	t	$\dfrac{1}{s^2}$
3	t^2	$\dfrac{2!}{s^3}$
4	t^n	$\dfrac{n!}{s^{n+1}}$
	where n is integer	
5	e^{-at}	$\dfrac{1}{s+a}$
6	te^{-at}	$\dfrac{1}{(s+a)^2}$
6	$\dfrac{e^{-at}-e^{-bt}}{b-a}$	$\dfrac{1}{(s+a)(s+b)}$
7	$\sin at$	$\dfrac{a}{s^2+a^2}$
8	$\cos at$	$\dfrac{s}{s^2+a^2}$

Laplace Transform Pairs

Table C.1 provides Laplace transform pairs for a variety of important engineering functions. More extensive lists and methods for more complex transformations may be found in most advanced engineering mathematics texts.

REFERENCE

1. Schiff, Joel L., *The Laplace Transform: Theory and Applications*, Springer, New York, 1999.

Glossary of New Terms

A/D converter   A device that converts an analog voltage into a digital number.

absolute temperature scale   A temperature scale referenced to the absolute zero of temperature; equivalent to thermodynamic temperature.

accelerometer   An instrument for measuring acceleration.

accuracy   The closeness with which a measuring system indicates the actual value.

advanced-stage uncertainty   The uncertainty due to measurement system, measurement procedure, and measurand variation effects.

alias frequency   A frequency appearing in a discrete data set that is not in the original signal.

apparent strain   A false output from a strain gauge as a result of input from extraneous variables, such as temperature.

astable multivibrator   Switching circuit that toggles on or off continuously in response to an applied voltage command.

bandpass filter   A device that allows a range of frequencies in a signal to pass unaltered, while frequencies outside this range, both higher and lower, are attenuated.

Bessel filter   Filter design having a linear phase shift over its frequency passband.

bias   An offset value.

bias error   See systematic error.

bimetallic thermometer   A temperature-measuring device that utilizes differential thermal expansion of two materials.

bistable multivibrator   See flip-flop.

bit   The smallest unit in binary representation.

bit number   The number of bits used to represent a word in a digital device.

bonded resistance strain gauge   A strain sensor that exhibits an electrical resistance change indicative of strain. The sensor is bonded to the member in which strain is being measured in such a way that the gauge experiences the same strain conditions.

bridge constant   The ratio of the actual bridge voltage to the output of a single strain gauge sensing the maximum strain.

buffer   Memory set aside to be accessed by a specific I/O device, such as to temporarily store data.

bus   A major computer line used to transmit information.

Butterworth filter   Filter design having a flat magnitude ratio over its frequency passband.

byte   A collection of 8 bits.

calibration   The act of applying a known input to a system to observe the system output.

calibration curve   Plot of the input versus the output signals.

central tendency   Tendency of data scatter to group about a central (mean or most likely) value.

charge amplifier   Amplifier used to convert a high impedance charge into a voltage.

circular frequency   Frequency specified in radians per second; angular frequency.

combined standard uncertainty   Uncertainty estimate that includes both systematic and random standard uncertainties at a confidence level of 68%; normally used with Type A and Type B error methods.

common-mode voltage   Voltage difference between two different ground points. See ground loop.

concomitant method   An alternate method to assess the value of a variable.

confidence interval   Interval evaluated from the combined effects of random and systematic errors that defines the limits about the mean value within which the true value is expected to fall.

confidence level   Probability that the true value falls within the stated interval.

control   A means to set and maintain a value during a test.

conversion error   The errors associated with converting an analog value into a digital number.

conversion time   The duration associated with converting an analog value into a digital number.

coupled system   The combined effect of interconnected instruments used to form a measurement system.

d'Arsonval movement   The basis for most analog electrical meters, which senses the flow of current through torque on a current carrying loop.

D/A converter   A device that converts a digital number into an analog voltage.

damping ratio A measure of system damping—a measure of a system's ability to absorb or dissipate energy.

data-acquisition board A plug-in board for a PC that contains a minimum of devices required for data acquisition.

data-acquisition system A system that quantifies and stores data.

deflection mode Measuring method that uses signal energy to cause a deflection in the output indicator. See loading error.

dependent variable A variable whose value itself depends on the value of one or more other variables.

design-stage uncertainty The uncertainty due to instrument error and instrument resolution; an uncertainty value based on a minimum amount of information, typical at the early stage of test design.

deterministic signal A signal that is predictable in time or in space, such as a sine wave or ramp function.

differential-ended connection Dual-wire connection scheme that measures the difference between two signals with no regard to ground.

direct memory access Transfer protocol that allows a direct transmission path between a device and computer memory.

dispersion The scattering of values of a variable around a mean or reported value.

dynamic pressure Difference between the total and static pressures at any point; kinetic energy per unit volume.

dynamic response Response of a system to a well-defined sine wave input.

dynamometer Device for measuring shaft power.

earth ground A ground path whose voltage level is zero.

end standard Length standard that describes the distance between two flat square ends of a bar.

error Difference between a measured value, as indicated by a measurement system, and its true (actual) value.

error fraction A measure of the time-dependent error in the time response of a first-order system.

expanded uncertainty Uncertainty due to the combined effects of random and systematic uncertainties at a stated confidence level (typically 95%).

extraneous variable A variable that is not or cannot be controlled during a measurement, but one that affects the measured value.

firewire A serial method for data transmission between a computer and peripherals using a thin cable and based on the IEEE 1394 standard.

first-order system A system characterized as having time-dependent storage or dissipative ability but having no inertia.

fixed point temperature Temperatures specified through a reproducible physical behavior, such as a melting or boiling point.

flip-flop A switching circuit that toggles either on or off on command. A bistable multivibrator.

frequency bandwidth The range of frequencies in which the magnitude ratio remains within 3 db of unity.

frequency response The output signal amplitude versus input frequency relation characteristic to a measurement system.

full-scale output (FSO) The arithmetic difference between the end points (range) of a calibration expressed in units of output variable.

galvanometer Device used to sense the flow of current from a zerostate.

gauge blocks Working length standard for machining and calibration.

gauge factor Property of a strain gauge, which relates changes in electrical resistance to strain.

ground Signal return path to earth.

ground loop A signal circuit formed by grounding a signal at more than one point with the ground points having a different voltage potential.

Hall effect Voltage induced across a current-carrying conductor situated within a magnetic field. The magnetic field exerts a transverse force pushing the charged electrons to one side of the conductor.

handshake An interface procedure to control data flow between different devices.

high-pass filter A device that allows frequencies in a signal to pass unaltered, while frequencies lower than a certain threshold are attenuated.

Hooke's law The fundamental linear relation between stress and strain. The proportionality constant is the modulus of elasticity.

hysteresis The difference in the indicated value for any particular input when that input is approached in an increasing input direction versus when approached in a decreasing input direction.

immersion errors Temperature measurement errors, which result from differences between sensor temperature and the temperature being measured.

impedance matching   Ensures that loading errors are kept to an acceptable level; requires assessing signal levels and instrument characteristics.

independent variable   A variable whose value can be changed directly as opposed to a dependent variable.

indicated value   The value displayed or indicated by the measurement system.

input   Process information sensed by the measurement system.

interference   Extraneous effect that imposes a deterministic trend on the measured signal.

interferometer   Device for measuring length, which uses interference patterns of light sources; provides very high resolution.

interrupt   A signal to initiate a different procedure. Analogous to raising one's hand.

junction   For thermocouples, an electrical connection, which measures temperature.

least squares regression   Analysis tool used in curve fitting that minimizes the sum of the squares of the deviations between the data set and the predicted curve fit.

line standard   Standard for length made by scribing two marks on a dimensionally stable material.

linearity   The closeness of a calibration curve to a straight line.

load cell   Sensor for measuring force or load; a variety of principles may be employed.

loading error   Error that arises in measurements of all types due to energy being extracted from the system being measured by the act of measurement.

low-pass filter   Device that allows low frequencies in a signal to pass unaltered, while frequencies higher than a certain threshold are attenuated.

linear variable differential transformer (LVDT)   A sensor/transducer that provides an emf output as a function of core displacement.

Mach number   Ratio of local speed to the speed of sound of the media.

magnitude ratio   The ratio of output amplitude to the input amplitude of a dynamic signal.

measurand   See measured variable.

measured value   The value measured.

measured variable   A variable whose value is measured; also known as the measurand.

mechatronics   Subject area focused on the interaction between electrical and mechanical

components of devices and the control of such devices.

metrology   The science of weights and measures.

moiré patterns   An optical effect resulting from overlaying two dense grids, which are displaced slightly in space from one another.

monostable   One-shot device that toggles on and then off on a single signal voltage command.

moving coil transducer   A sensor that uses a displacement of a conductor in a magnetic field to sense velocity.

multiplexer   A input/output (I/O) switch with multiple input lines but only one output line.

natural frequency   The frequency of the free oscillations of an undamped system. A system property.

noise   An extraneous effect that imposes random variations on the measured signal.

null mode   Measuring method that balances a known amount against an unknown amount to determine the measured value.

Nyquist frequency   One-half of the sample rate (sample frequency).

ohmmeter   An instrument for measuring resistance.

operating conditions   Conditions of the test, including all equipment settings, sensor locations, and environmental conditions.

optical pyrometer   Nonintrusive temperature measurement device. Measurement is made through an optical comparison.

oscilloscope   An instrument for measuring a time-varying voltage and displaying a trace; very high frequency signals can be examined.

output   The value indicated by the measurement system.

overall error   The square root of the sum of the squares (RSS) of all known errors that could affect a measuring system.

parallel communication   Technique that transmits information in groups of bits simultaneously.

parameter   The value defined by some functional relationship between variables related to the measurement.

passband   Range of frequencies over which the signal amplitude is attenuated by no more than 3 dB.

Peltier effect   Describes the reversible conversion of energy from electrical to thermal at a junction of dissimilar materials through which a current flows.

phase shift A shift or offset between the angle of the input signal and the corresponding measured signal.

photoelastic Describes a change in optical properties with strain.

Poisson's ratio Ratio of lateral strain to axial strain.

pooled Statistical values determined from separate data sets, such as replications, that are grouped together.

population The set of all possible values for a variable.

potentiometer Refers to a variable resistor, or to an instrument for measuring small voltages with high accuracy.

potentiometer transducer Variable electrical resistance transducer for measurement of length or rotation angle.

precision The amount of dispersion expected with repeated measurements; the random error in a measurement; term often used to suggest a high resolution with low random error.

precision (instrument specification) Ultimate performance claim of instrument repeatability (random error) based on multiple tests conducted at multiple labs and on multiple units.

precision error See random error.

precision indicator Statistical measure that quantifies the dispersion in a reported value.

precision interval Interval evaluated from data scatter that defines the probable range of a statistical value and ignoring the effects of systematic errors.

primary standard The defining value of a unit.

Prony brake A historically significant dynamometer design, which uses mechanical friction to dissipate and measure power.

proving ring An elastic load cell, which may be used as a local calibration standard.

pyranometer Optical instrument used to measure irradiation on a plane.

quantization Process of converting an analog value into a digital number.

quantization error An error brought on by the resolution of an A/D converter.

radiation shield A device which reduces radiative heat transfer to a temperature sensor, to reduce insertion errors associated with radiation.

random error The portion of error that varies randomly on repeated measurements of the same variable.

random standard uncertainty The random uncertainty estimated at 68% probability.

random uncertainty Uncertainty assigned to a random error; contribution to combined uncertainty due to random errors.

randomization methods Test approaches used to break up the interference effects of extraneous variables.

range The lower to upper limits of an instrument or test.

recovery errors Temperature measurement error in a flowing gas stream related to the conversion of kinetic energy into temperature rise; most important in high-speed flows.

repeatability Random error estimate based on multiple tests within a given lab on a single unit.

repetition Repeating a measurement (or repeated measurements) during the same test.

replication The duplication of a test under similar operating conditions.

reproducibility Random error estimate based on multiple tests performed in different labs on a single unit.

resistivity Material property describing the basic electrical resistance of a material.

resolution The smallest detectable change in measured value as indicated by the measuring system.

resonance frequency The frequency at which the magnitude ratio reaches its maximum value greater than unity.

result The value calculated from a functional relation.

resultant The variable calculated from the values of other variables.

ringing frequency The frequency of the free oscillations of a damped system; a function of the natural frequency and damping ratio.

rise time The time required for a first-order system to respond to 90% of a step change in signal.

rms value Root-mean-square value; derived from power dissipation by an ac current.

roll-off slope Decay rate associated with a filter.

RTD Resistance temperature detector; senses temperature through changes in electrical resistance of a conductor.

root-sum-square value (RSS) The net value obtained by taking the square root of the sum of a series of squared values. Correct method used to estimate the combined error from different sources.

sample   A data point or single value of a population.

sample rate   The rate or frequency at which data points are acquired. Reciprocal of sample time increment.

sample statistics   Statistics obtained from a set of data points that does not include the entire population of that variable.

sample time increment   The time interval between two successive data measurements. Reciprocal of sample rate.

saturation error   Error due to an input value exceeding the device maximum.

second-order system   A system whose behavior includes time-dependent inertia.

Seebeck effect   Source of open circuit emf in thermocouple circuits.

seismic transducer   Instrument for measuring displacement in time, velocity, or acceleration; based on a spring-mass-damper system.

sensitivity   The rate of change of a variable (y) relative to a change in another variable (x), for example, $\partial y/\partial x$.

sensitivity index   The derivative between a resultant and a measured variable ($\partial R/\partial x$) that describes the rate of change in the result due to a change in the variable.

serial communication   Technique that transmits information 1 bit at a time.

sensor   The portion of the measurement system that senses or responds directly to the process variable being measured.

settling time   The time required for a second-order system to settle to within $\pm10\%$ of the final value of a step change in input value.

shielded twisted pair   Describes electrical wiring, which reduces mutual induction between conductors through twisting the conductors around each other, and which reduces noise through shielding.

single-ended connection   Two-wire connection scheme that measures the signal relative to ground.

signal-to-noise ratio   Ratio of signal power to noise power.

span   The difference between the maximum and minimum values of operating range of an instrument.

standard   The known value used as the basis of a calibration.

static pressure   Pressure sensed by a fluid particle as it moves with the flow.

static sensitivity   The rate of change of the output signal relative to a change in input for a static signal. The slope of the static calibration curve at a point. Also called the static gain. See sensitivity.

steady response   A portion of the time-dependent system response, which either remains constant or repeats its waveform with time.

step response   System response to a step function input.

stagnation pressure   Pressure at the point of which the flow is brought to rest. See total pressure.

strain   Elongation per unit length of a member subject to an applied force.

stress   Internal force per unit area. These internal forces maintain in equilibrium the applied external forces.

stroboscope   High-intensity source of light, which can be made to flash at a precise rate, for a precise duration. It is used to measure the frequency of rotation.

Student-*t* distribution   Sampling distribution of the mean value for small numbers of samples.

systematic error   The portion of error that remains constant in repeated measurements of the same variable. A constant offset between the indicated value and the actual value measured.

systematic uncertainty   Uncertainty assigned to a systematic error. The uncertainty due to all the systematic errors affecting the result.

systematic standard uncertainty   The systematic uncertainty at 68% probability; nominally this value is one-half the systematic uncertainty.

temperature scale   Establishes a universal means of assigning a quantitative value to temperatures.

thermistor   Temperature-sensitive semiconductor resistor.

thermocouple   Junction of two dissimilar conductors used to measure temperature.

thermoelectric   A type of device that indicates a thermally induced emf.

thermopile   A multiple junction thermocouple circuit, designed to measure a particular temperature or temperature difference.

Thomson effect   Describes the creation of an emf through a temperature difference in a homogeneous conductor.

time constant   System property defining the time required for a first-order system to respond to 63.2% of a step input.

time delay   The lag time between an applied input signal and the measured output signal.

time response   The complete time-dependent system response.

total pressure   Pressure sensed at a point at which the flow is brought to rest in an isentropic (no losses) manner. It is the sum of the static and dynamic pressures. Isentropic value of stagnation pressure.

total sample period   Duration of the measured signal represented by the data set.

transducer   The portion of measurement system, which transforms the sensed information into a different form. Also loosely refers to a device that houses the sensor, transducer, and often, signal conditioning stages of a measurement system.

transient response   Portion of the time-dependent system response that decays to zero with time.

true rms   Data-reduction technique that can correctly provide the rms value for a nonsinusoidal signal. A signal conditioning method that integrates the signal.

true value   The actual or exact value of the measured variable.

TTL (true-transistor-logic)   Switched signal that toggles between a high (e.g., 5 V) and a low (e.g., 0 V) state.

turndown   Ratio of the highest flow rate to the lowest flow rate that a particular flow meter can measure.

type A uncertainty   An uncertainty value estimated from the statistics of a variable.

type B uncertainty   An uncertainty value estimated by other than statistical methods.

uncertainty   An estimate of the probable error in a reported value.

uncertainty analysis   A process of identifying the errors in a measurement and quantifying their effects.

uncertainty interval   An interval about a variable or result that is expected to contain the true value.

USB   Universal serial bus. A serial method of data transmission between a computer and peripherals using a four-wire cable.

vernier calipers   Tool for measuring both inside and outside dimensions.

vernier scale   Allows increased resolution in reading a length scale.

VOM   Acronym for volt-ohm meter.

wheatstone bridge   Electrical circuit for measuring resistance with high precision; can be used to measure static or dynamic signals.

word   A collection of bits used to represent a number.

zero drift   A shift away from zero output under a zero input value condition.

zero-order system   A system whose behavior is independent of the time-dependent characteristics of storage or inertia.

zero-order uncertainty   Random uncertainty estimate based only on a measurement system's resolution errors.

Index

PREFIXES FOR QUANTITY STATEMENTS

10^{18}	exa	E	1,000,000,000,000,000,000
10^{15}	peta	P	1,000,000,000,000,000
10^{12}	tera	T	1,000,000,000,000
10^{9}	giga	G	1,000,000,000
10^{6}	mega	M	1,000,000
10^{3}	kilo	k	1,000
10^{2}	hecto	h	100
10^{1}	deca	da	10
1			1
10^{-1}	deci	d	0.1
10^{-2}	centi	c	0.01
10^{-3}	milli	m	0.001
10^{-6}	micro	μ	0.000001
10^{-9}	nano	n	0.000000001
10^{-12}	pico	p	0.000000000001
10^{-15}	femto	f	0.000000000000001
10^{-18}	atto	a	0.000000000000000001